Numerical Methods

Numerical methods are the cornerstone of most engineering and science programmes. However, their applications vary greatly for different streams. This book focuses on mathematical and process applications that are relevant specifically to chemical engineers, or to allied streams like biochemical, mechanical, cryogenic, and energy engineering.

Each chapter starts with the formulation and graphical representation of the numerical method. This is followed by the sequential steps and algorithms required to create computer-assisted solutions and simulations. Algorithms chosen apply to real-world examples or case studies so as to show how exactly they are used. Finally, the strengths and weaknesses of the numerical method under discussion are explained, thus helping the reader choose the best method for a specific problem at hand. The text incorporates extensive mathematical problems, illustrative examples, industrially relevant case studies, etc., so that readers gain insights into the ground realities.

Amiya K. Jana is an associate professor in the Department of Chemical Engineering, Indian Institute of Technology (IIT) Kharagpur. He has authored over 145 articles in several international journals of repute, and two other textbooks, *Process Simulation and Control using ASPEN™* (2012) and *Chemical Process Modeling and Computer Simulation* (2018). Jana has received numerous awards and recognition, including Fellow of the Royal Society of Chemistry (FRSC) (UK, 2023), Faculty Excellence Award (IIT Kharagpur, 2021), Fellow of the Alexander von Humboldt Foundation (Germany, 2017), World's Top 2% Scientist for four consecutive years (Stanford University, 2020–23), and editorial board member of *Scientific Reports* (Nature Group) and *Frontiers in Control Engineering* (Frontiers Publisher, Switzerland). His research focuses on renewable energy, clean fuel, desalination, process integration and control.

Numerical Methods

Theory and Engineering Applications

Amiya K. Jana

Shaftesbury Road, Cambridge CB2 8EA, United Kingdom

One Liberty Plaza, 20th Floor, New York, NY 10006, USA

477 Williamstown Road, Port Melbourne, VIC 3207, Australia

314–321, 3rd Floor, Plot 3, Splendor Forum, Jasola District Centre, New Delhi – 110025, India

103 Penang Road, #05–06/07, Visioncrest Commercial, Singapore 238467

Cambridge University Press is part of Cambridge University Press & Assessment, a department of the University of Cambridge.

We share the University's mission to contribute to society through the pursuit of education, learning and research at the highest international levels of excellence.

www.cambridge.org
Information on this title: www.cambridge.org/9781009211802

First published 2024

Printed in India by Avantika Printers Pvt. Ltd.

A catalogue record for this publication is available from the British Library

ISBN 978-1-009-21180-2 Paperback

To my
***Dad** Sovachand (Late)*
***Daughter** Lekhoni Rose*

Mathematics is the gate and key to science

– Roger Bacon

Mathematics is the language with which God has written the universe

– Galileo

Contents

Part III Systems of Differential Equations

Preface

"Mathematics is the language with which God has written the universe" – Galileo

Analytical methods are used to find exact solutions of various physical and engineering problems. But these methods have certain limitations in that they are generally able to solve linear (or linearized) systems or those having low dimensionality and simple geometry. It is a fact that the natural systems (i.e., practical problems) are inherently non-linear and complex. For them, thus numerical methods are inevitable for finding approximate solution. In the pre-computer era, the application of numerical methods was limited as they involved repeated arithmetic operations. Today, with the advent of fast and cheaper digital computers, the role of numerical techniques in solving scientific and engineering problems has exploded.

It is fairly correct that there are several excellent books available on *numerical methods* in the market. Among them, many of which are specific to only science stream, and a few are relevant to specific engineering discipline. Usually, the theoretical part is covered more in the science streams, whereas application part gets more importance in the engineering disciplines. At this juncture, we felt the necessity of a book for readers who are keen to investigate the wide applicability of various numerical approaches to interesting examples and case studies from both the science and engineering streams. Hence attempt is made here to bridge this gap to some extent.

This text book covers various elegant and advanced numerical techniques that are commonly employed to solve the algebraic and differential equations (ordinary and partial differential equations). To easily understand the theory of numerical methods, they are presented in a simple manner in that the Lemma, Corollary etc. are not included. Furthermore, to gain insights into these methods, all involved calculations in them are shown in details. The issues related to their usability, limitations, and conservativeness (if any) are concisely discussed with beautiful worked-out examples from both science and engineering fields. This apart, there are a couple of advanced and effective numerical techniques covered with their relevant applications.

An appealing part of this book is that a wide range of process units are used to illustrate the numerical methods. At first, the numerical solutions of some mathematically worked-out examples are discussed that are typically solved by the students of all disciplines at the early stage of their undergraduate (UG) course. Further, for the interest of senior-level (post graduate, PG) students, and practicing engineers and scientists, special care is taken in this book such that the interested readers can understand the advanced numerical methods, enjoy their applications in many industrially relevant case studies and gain physical insights with ground realities. A few such notable process examples include:

- Multiple reactors system
- Electrical network
- Traffic flow
- Phase equilibrium (boiling point and dew point temperature)
- Cooling fin
- Gravity pendulum
- Separator (e.g., adsorption and distillation column)
- Reactor (e.g., continuous stirred tank reactor, CSTR; plug flow reactor; batch reactor; bioreactor; and membrane reactor)
- Heat exchanger
- Flash drum
- Coupled reaction–diffusion in a catalyst pellet
- Unsteady state heat conduction
- Coupled heat and mass transfer

Most of the chapters of this text book are designed with the following methodology for several numerical techniques:

- At first, the formulation of the numerical method is shown followed by its graphical representation
- Computer-assisted solution/simulation algorithm is developed with showing sequential steps
- The algorithm so developed is applied to a mathematical example or case study or both, to illustrate the numerical approach
- For the worked-out problems, produced results are typically reported in terms of numerical values along with plots, wherever applicable, so that one can match their solutions in both ways.
- Finally, we end the topic with making important remarks based on our observation. The relative merits and demerits of the concerned numerical scheme are also highlighted over the other methods. It helps the readers to pick up a suitable method for a specific type of problem. Walking in this direction, we have made a general recommendation (guideline) for the readers at the end of the chapter.

Along with numerical solution, analytical solution for many problems, if easily available, are also sought mainly for comparison purpose. This gives a clear understanding about the ability and simplicity of the numerical approaches, and motivates to numerically solve the difficult problems (e.g., non-linear problems), where analytical solution is really impossible.

Apart from simulating/solving various case studies (total 38), there are several worked-out mathematical examples (total 107) solved to illustrate the numerical methods for both science and engineering streams. At the end of each chapter, along with numerous theoretical questions, a large number of mathematical problems (for both science and engineering streams) and case studies are included for further practice.

ACKNOWLEDGEMENTS

I started writing this book in February, 2020. Just a few weeks into it, we were shocked to face the consequences of a completely new set of constraints in our daily life. From the very beginning of this ongoing global pandemic of coronavirus disease 2019 (COVID-19) caused by severe acute respiratory syndrome coronavirus 2 (SARS-CoV-2), like many global universities, my institute was also closed for a reasonably long period of time in various phases. Despite this, I got the access for writing this text book at my department's desk and gathered mental strength to complete it within time. For this, first of all, I would like to thank my institute (IIT Kharagpur) and one of its beloved daughters, the Chemical Engineering Department.

I am really thankful to all my colleagues here, including S. Ray, A. N. Samanta, S. Chakraborty, S. Neogi, N. C. Pradhan, S. De, S. Dasgupta, and of course, B. C. Meikap.

Heartfelt thanks go to my PhD students, especially Niraj and Avinash, who helped by verifying a couple of my numerical solutions.

This book would not have been finished without the support, patience and forbearance of my wife (Mahua), mother (Minati) and sister (Lily).

I am greatly indebted to the staff at Cambridge University Press who worked hard to ensure the top quality publication of this book on time.

Part I

Introduction to Numerical Simulation

This is the introductory part of this text book and it consists of just one chapter, Chapter 1, titled *Introduction to Numerical Method and Process Simulation*. Here, our goal is to mainly cover some basic concepts related to:

- Numerical method
- Process simulation

Broadly speaking, these are the two main topics of this book, along with engineering applications of the numerical algorithms. They are further detailed in subsequent parts of the book, having a total of eight chapters.

1

Introduction to Numerical Method and Process Simulation

"Without mathematics, there's nothing you can do. Everything around you is mathematics. Everything around you is numbers." – Shakuntala Devi

Key Learning Objectives

- Knowing the difference between analytical and numerical solution
- Learning the basics of process modeling and simulation
- Necessity of numerical simulator in process engineering
- What is an error and what role it plays in numerical simulation

1.1 Introduction to Numerical Method

Numerical methods are widely used to solve mathematical problems that arise in natural sciences, social sciences, engineering, medicine, and business. These methods can provide approximate solutions of sufficient accuracy for engineering purposes. Numerical analysis of the engineering problems had started in the mid-1960s when advancement in high-speed digital computing was made. Since then, the numerical approaches have gained attention for solving practical problems, for which, closed form analytical solutions are either intractable or do not exist.

Why numerical method?
We typically follow two ways to solve mathematical equations. They include:

- Analytical method
- Numerical method

The former approach is usually preferred to solve the linear equations, whether they are in algebraic or in differential form. On the other hand, the numerical methods are very common in solving the non-linear equations, which constitute most of our natural systems.

Moreover, the numerical methods are also used for linear systems when analytical solution is complicated and time-consuming.

Let us compare the merits and demerits between the analytical and numerical methods (Table 1.1) to understand why and when the numerical method gets preference over the analytical one.

TABLE 1.1 Basic differences between the analytical and numerical method

Analytical method	Numerical method
1. Analytical method yields exact solution.	1. Numerical method yields approximate solution.
2. Analytical method is applied to a limited range of problems.	2. Numerical method is applied to a much wider range of problems.
3. Knowledge of higher mathematics is a pre-requisite for finding analytical solution, particularly of complex problems	3. Knowledge of higher mathematics or physics is not mandatory
4. If any error occurs in a direct analytical method, there is no remedy to recover it.	4. One can improve the accuracy level in numerical results by increasing the number of iterations.

1.2 Introduction to Process Simulation

1.2.1 Process Modeling

Prior to process simulation, let us first know how to define the mathematical model in a simple way.

> *Definition of 'mathematical model'*: The mathematical description of a real-life system or phenomenon (whether physical, sociological, or even economic) is referred to as mathematical model.

The way and the purpose of developing a mathematical model for a process (i.e., a *process model*) are described step-by-step in the following section.

Step 1: Make necessary assumptions or hypotheses

Step 2: Formulate the mathematical equations for a process to represent its various phenomena

Step 3: Solve the modeling equations (which is referred as *process simulation*)

Step 4: Verify the model predictions by comparing with known facts. If the model fails to predict the process behavior precisely, one can repeat from Step 1 with relaxing a few assumptions or making alternative (more relevant) assumptions. On the other hand, if the model works fine (with a reasonable accuracy), one can extend its applications to various conditions, even at which experimental investigations are not available so far.

Structure of mathematical model

Industrial processes are represented by mathematical models that typically consist of:

- Algebraic equation[1]
- Differential equation[2]

These differential algebraic equation (DAE) systems are solved (better to say, *simulated*) by using the numerical method. The structure of such systems is demonstrated below (Table 1.2) in accordance with their constituent equations solved with numerical methods.

Table 1.2 Structure of the DAE system

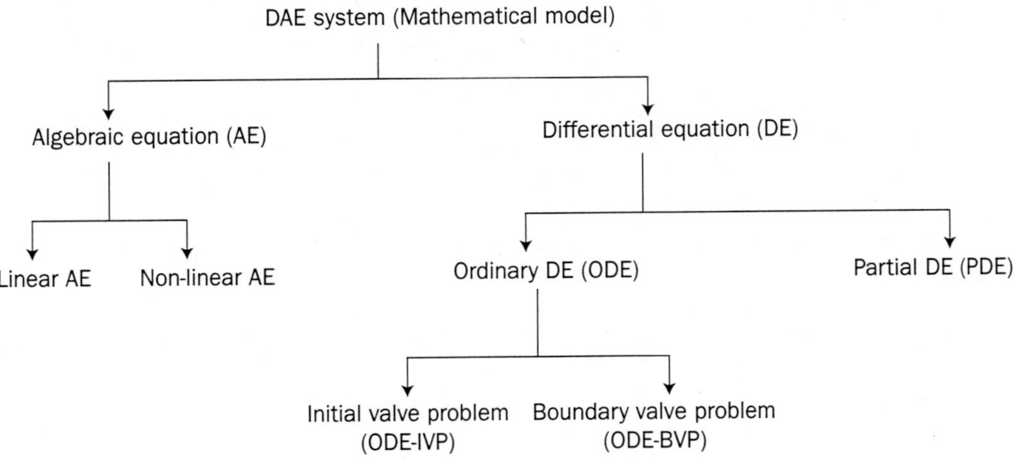

Let us now develop the mathematical model of a simple chemical reactor that operates in batch mode (i.e., no reactant input and product output).

Case Study 1.1 Developing the model of a batch reactor

The schematic of a batch reactor is demonstrated in Figure 1.1. As shown, there is no input and output stream involved. At the beginning, reactants are basically taken in and then the reactor is sealed before raising its operating condition to the desired level.

Let us consider the following second-order, irreversible reaction:

$$A \rightarrow P$$

where, A is the reactant and P is the product.

1 According to dictionary definition, it is an equation in the form of a polynomial containing a finite number of terms and equated to zero (https://www.dictionary.com/browse/algebraic-equation). Sometimes, the algebraic equations may include transcendental functions (e.g., trigonometric and exponential functions).

2 An equation which contains the derivatives of one or more dependent variables with respect to one or more independent variables is called *differential equation* (DE).

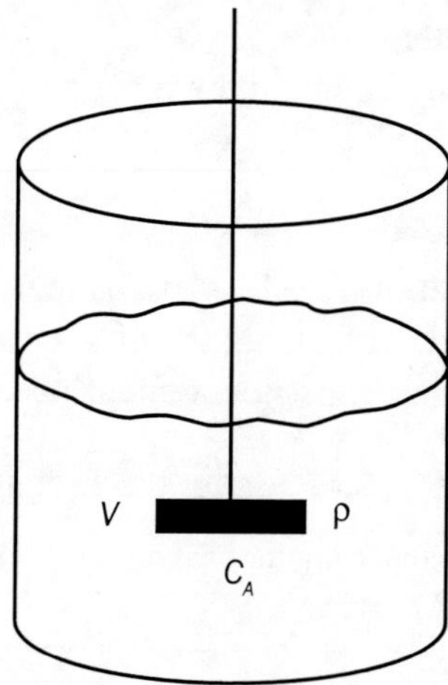

FIGURE 1.1 A batch reactor

Assumptions

Prior to developing the model of a process, one needs to make necessary assumptions and considerations. For the sample batch reactor, they include:

1. Perfect mixing in the reactor (i.e., no variation of concentration of species A, C_A, in spatial direction)
2. Isothermal reactor (no variation of temperature, T)
3. Constant density, ρ
4. Constant reactor volume, V

Deriving model

The conservation principle has the following representation:

$$\begin{pmatrix}\text{Rate of}\\ \text{accumulation}\end{pmatrix}=\begin{pmatrix}\text{Rate of}\\ \text{input}\end{pmatrix}-\begin{pmatrix}\text{Rate of}\\ \text{output}\end{pmatrix}+\begin{pmatrix}\text{Rate of}\\ \text{generation}\end{pmatrix}-\begin{pmatrix}\text{Rate of}\\ \text{depletion}\end{pmatrix} \tag{1.1}$$

The example batch reactor (input = output = 0) *does not* involve:

- Total mass balance equation (since volume and density are constant)
- Energy balance equation (since temperature is constant)

The component mass balance is only left and it is derived below. Here, the concentration C_A is the sole variable of interest. Accordingly, we have:

Rate of accumulation of species $A = \dfrac{d}{dt}(VC_A)$

Rate of input of species $A = 0$

Rate of output of species $A = 0$

Rate of generation of species $A = -(-r_A)\ V$

where, $(-r_A)$ is the rate of disappearance of species A, and it has the following form:

$$(-r_A) = kC_A^2$$

in which, k denotes the reaction rate constant.

Substituting in Equation (1.1),

$$\frac{d}{dt}(VC_A) = -(-r_A)V$$

Rearranging,

$$\frac{dC_A}{dt} = -k\, C_A^2 \tag{1.2}$$

This is the final form of the batch reactor model.

1.2.2 Numerical Simulation

To observe the behavior of a process at various situations (or conditions), one needs to solve (i.e., simulate) the developed model equations of that process. For example, reactive processes are typically characterized in terms of component composition, operating temperature and pressure, and reaction conversion with respect to time or space, or both.

The numerical simulation one can understand through the simple definition given below.

> *Definition of 'numerical simulation'*: Solving the mathematical model by the use of numerical technique (usually by employing a computer) is called numerical simulation.

With this, let us discuss the simulation of the model equation derived for a batch reactor (see Case Study 1.1).

Case Study 1.2 Simulating batch reactor model

Numerical simulation

The numerical simulation of an ODE (and coupled ODEs) is detailed in Chapters 4 and 5 of this book. At this stage, let us start with solving this simple problem represented by Equation (1.2) by applying our knowledge gained in mathematics.

Discretizing Equation (1.2),

$$\frac{\Delta C_A}{\Delta t} = -k\, C_A^2$$

We can write further,

$$\frac{C_{A,\,k+1}-C_{A,\,k}}{\Delta t}=-k\ C_{A,\,k}^2 \tag{1.3}$$

Here, the suffix k denotes the present time step and Δt the integration time interval.

Rearranging,

$$C_{A,\,k+1}=C_{A,\,k}-\Delta t\ k\ C_{A,\,k}^2 \tag{1.4}$$

This is the final numerical form of the batch reactor model ready for computer simulation. Now, one needs to develop a computer code and find the numerical solutions in terms of $C_{A,\,k+1}$ for $k = 0, 1, 2, 3, \ldots$. One can run the code for the specified operational time or until the stopping criteria (see Sub-section 1.3.3) is met.

Analytical solution

To validate the results of the numerical method used above, we would like to solve the reactor model equation analytically. From Equation (1.2),

$$\int \frac{dC_A}{C_A^2}=-k\int dt \tag{1.5}$$

Let us adopt the following conditions:

at $t = 0$, the reactor concentration is C_{A0}

at any time t, the reactor concentration is C_A

Accordingly, we can rewrite Equation (1.5) in the definite integral form:

$$\int_{C_{A0}}^{C_A} \frac{dC_A}{C_A^2}=-k\int_0^t dt \tag{1.6}$$

Integrating and rearranging,

$$C_A(t)=\frac{C_{A0}}{1+k\,C_{A0}t} \tag{1.7}$$

This is the final form of analytical solution of the batch reactor model.

Comparing analytical and numerical solutions

For the batch reactor, the following information are available:

$C_{A0} = 1$ kmol/m^3

$k = 2$ m^3/kmol.min

$\Delta t = 0.001$ min

Using these values, the numerical and analytical solutions are obtained based on Equations (1.4) and (1.7), respectively. They are compared for the first 0.005 min in Table 1.3.

TABLE 1.3 Comparing analytical and numerical results

Time	Analytical solution	Numerical solution
0.0	1.0	1.0
0.001	0.99800399	0.99800000
0.002	0.99601594	0.99600799
0.003	0.99403579	0.99402393
0.004	0.99206349	0.99204776
0.005	0.99009901	0.99007944

Recall that the analytical solution is exact or true solution and the numerical solution is approximate one.

1.2.3 NECESSITY OF NUMERICAL SIMULATOR

At this point of time, one question naturally comes in our mind:

> *When we have the experimental setup, what is the need of developing its model and simulating that model numerically?*

To get the answer, let us know the use of the numerical simulator. Among a wide variety of its multi-purpose usage, a few notable ones are included below.

- **To train operating personnel**: The trainee engineers typically spend from a couple of weeks to a few months to gain hands-on experience on the process units (design, operation, and troubleshooting) where they are supposed to be posted. During this training period, along with the normal run of the plant, the start-up and shut down operations are also visualized by running the process simulator as a virtual unit. This apart, several dangerous/abnormal situations can be generated in the simulator, which are most unlikely to be experienced in the real-time operations.
- **To investigate the process behavior**: For developing a simulator, we follow the two common steps: formulation of model and process simulator, and then its validation with experimental data. Verifying its predictability with a reasonable accuracy, the simulator is subsequently extended to use at different conditions (beyond the known operating condition) and more importantly, for scale-up of the process. By this way, it is possible to save money, and of course time, by avoiding the experimentation.

  Now-a-days, there are several *virtual labs* being developed by employing the dynamic process simulators. The beauty of such a lab is that if you have remote access, run it at your convenient time and place (e.g., during train journey) with a feeling of real-time experimentation.
- **To find optimal operating condition**: One of the principal objectives of a manufacturing unit is to make it more profitable. Process simulator is a crucial tool employed to find an optimal operating condition that should lead to maximizing the profit after satisfying several constraints.

- **To identify stability condition**: There are a few industrial processes, which are unstable beyond a certain operating region. To identify this region, a process simulator plays an active role and helps in devising the process technology.
- **To design advanced controller**: Because of an unbelievable advancement made in digital computing, the use of model-based controllers has increased greatly. As the name suggests, these advanced controllers are formulated by the use of process model. They are simulated to produce control actions, which are then physically implemented to maintain the process variables of interest (e.g., product purity and productivity).

  This apart, the process simulator plays a vital role in estimating the states typically required for controller simulation. It is fairly true that measuring all required states is not only expensive, but also adds dead-time to the controller output. This dead-time leads to degrade the control performance and thus, the necessity of state estimation arises.

  Moreover, the controller tuning requires the process model. Examples of such tuning methods include: Ziegler–Nichols, Cohen–Coon, and tuning based on the time-integral performance criterion.

1.3 Error

The term *error* is very well-known to us since our childhood, in different forms of it, namely mistake, inaccuracy, blunder, and so on. We would however be interested to know about it precisely for its use in numerical simulation.

> *Definition of 'error'*: A measure of the difference between a computed or measured value and a true or theoretically correct value.

It indicates that there is some inaccuracy involved in computation (i.e., numerical simulation) and of course in measurement. We will subsequently call this error as *true error*.

Let us consider an equation of the form:

$$f(x) = 0 \tag{1.8}$$

By solving it, one can find its root in terms of x. Now, we define:

True error = approximate value – true value

With notations,

$$e_k = x_k - x^* \tag{1.9}$$

Recall that suffix k represents the present or kth time step. Here, x^* is the true or exact solution of Equation (1.8) and accordingly, one can write:

$$f\left(x^*\right) = 0 \tag{1.10}$$

Note that in this book we will denote the *true error* by e.

On the other hand, we mean the *error*, denoted by E, as the difference of x between two consecutive steps as:

$$E_k = x_k - x_{k-1} \tag{1.11}$$

It is worth noticing that in computation, we estimate the absolute value of the error (i.e., $|e|$ and $|E|$).

A short note
As mentioned, the error is estimated for computed or measured value with reference to the true value. This error is typically characterized by the terms, *accuracy* and *precision*, and they both are formally defined below.

> *Definition of 'accuracy'*: It refers to the closeness of the computed (or measured) value with its true value.[3]

> *Definition of 'precision'*: It refers to how close the computed value is with its measured value.

It is worth emphasizing that we do not worry about each small error, but they can add up and eventually lead to large errors in the final result. Thus it is treated very seriously in numerical simulation.

Case Study 1.3 Batch reactor (revisited)

Here, we would like to perform the error analysis involved in the simulation of the batch reactor example discussed before (see Case Study 1.2). Calculation of both the error and true error in their absolute forms is done based on the solutions produced in Table 1.3. For this, one can follow Table 1.4. Recall that the analytical solution represents the true value (x^*).

Table 1.4 Performing error analysis

Time	Analytical solution	Numerical solution	Absolute true error	Absolute error
0.0	1.0	1.0	0.0	–
0.001	0.99800399	0.99800000	3.990E-6	2.00000E-3
0.002	0.99601594	0.99600799	7.950E-6	1.99201E-3
0.003	0.99403579	0.99402393	1.186E-5	1.98406E-3
0.004	0.99206349	0.99204776	1.573E-5	1.97617E-3
0.005	0.99009901	0.99007944	1.957E-5	1.96832E-3

---

3 When a measured variable shows 5% true error, then we can say that its accuracy level is 95%.

1.3.1 Error in Process Modeling and Simulation

There are typically two sources of error involved in process simulation; one is process modeling and the other one is simulation. They are, respectively, called the *modeling error* and the *simulation error*. Further classification of these two errors is made in the following schematic diagram.

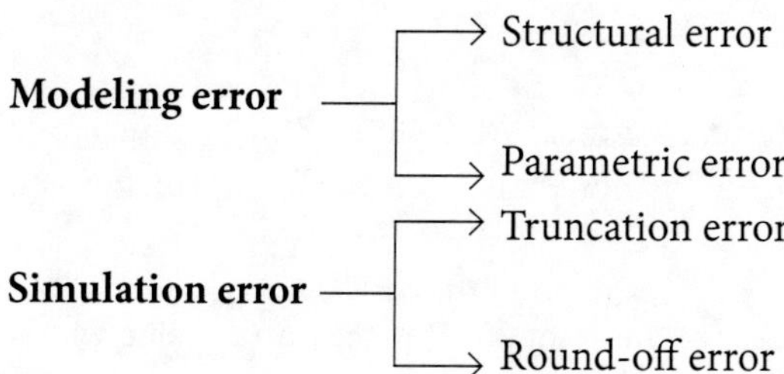

Now, we will know a little more about all these errors.

Structural error

Most of the physical phenomena in nature are inherently non-linear. To represent those phenomena precisely, we need non-linear model and correct parameter values. In reality, it is indeed tough to have these information with cent percent accuracy. In fact, we make several assumptions and simplifications to derive the model; sometimes we do not have proper understanding of a physical system and consequently, wrong concept comes into picture. By this way, the mathematical model of a plant leads to a plant/model mismatch. This mismatch is referred here as the *structural error* in process modeling. To understand this, let us take the following examples.

Example 1: It is not so uncommon to represent a non-linear system by its linearized model. This simplification makes a structural discrepancy.

Example 2: A simple compartmental distillation model (i.e., a lumped parameter model) is sometimes preferred to use as a predictor for a distillation plant. This typically makes a difference between the structure of the actual process and its approximate model, which leads to form the structural error.

Parametric error

Values of process parameters when differ from their actual values, the *parametric error* arises. It also includes the measurement error[4] since the measured values are usually involved in finding the process parameter.

Examples: There are several process parameters that are not exactly known to us and thus, they offer parametric error. A few of them include: overall heat transfer coefficient, mass transfer coefficient, pre-exponential factor, and so on.

4 It is sometimes called *measurement noise*.

Example 1.1 Structural and parametric error

Structural error

Let us assume that the following non-linear model truly represents a plant:

$$\textit{Actual plant}: \frac{dx}{dt} = \alpha\sqrt{x} \quad \text{(non-linear model)} \tag{1.12}$$

Notice that in Equation (1.12), $\alpha\sqrt{x}$ is the non-linear term. To linearize it, we use the Taylor series expansion about the operating point, say, $x_0 = 1$:

$$\alpha\sqrt{x} = \alpha\sqrt{x_0} + \left[\frac{d}{dx}(\alpha\sqrt{x})\right]_{x=x_0} \frac{(x-x_0)}{1\,!} + \left[\frac{d^2}{dx^2}(\alpha\sqrt{x})\right]_{x=x_0} \frac{(x-x_0)^2}{2\,!} + ..$$

Neglecting the second- and higher-order terms, and substituting $x_0 = 1$, we get:

$$\alpha\sqrt{x} \approx \alpha + \frac{\alpha}{2}(x-1)$$

This leads to the following linearized form:

$$\textit{Approximate model}: \frac{dx}{dt} = \alpha + \frac{\alpha}{2}(x-1) \quad \text{(linearized model)} \tag{1.13}$$

The true representation of the plant is made by Equation (1.12), whereas we have the approximate model [Equation (1.13)] for our use. Thus, there exists a difference in structure between the non-linear plant and its linearized form of the model, which leads to the structural error.

Parametric error

In the actual plant [Equation (1.12)], there is a parameter involved, namely, α. Suppose that its actual value is 2.0, whereas our estimated value is 1.8. This difference is the parametric error and it has direct effect on the simulation results.

Truncation error

In numerical analysis, truncation error (TE) is made by truncating an infinite series and approximating it by a finite number of terms.

Let us consider an infinite Taylor series as:

$$e^x = 1 + x + \frac{x^2}{2\,!} + \frac{x^3}{3\,!} + ... + \frac{x^n}{n\,!} + ... \tag{1.14}$$

Neglecting the third- and higher-order terms,

$$e^x = 1 + x + \frac{x^2}{2\,!} + \text{ truncation error}$$

Clearly, a finite number of terms are adopted here to describe an infinite series, and thus, the truncation error arises. In general, this error originates due to mathematical approximations and in this respect, one can consider it as a modeling error. However, today, there is no need to make any such approximation during model development. Because we can directly include exp(x) in our computer code developed to simulate the associated model and the truncation error can come out as the simulation error.

Note that incorporating an increasing number of higher-order terms may provide higher accuracy in numerical result by reducing the truncation error. However, using a large number of terms, for example, in the series expansion of derivatives, may lead to instability problem if one applies a difference equation of an order higher than a partial differential equation (PDE) being examined.

Further, the truncation error includes the *discretization error*, which arises from approximating continuous functions by sets of discrete data points. It is worth noting that this truncation error can be reduced by using the finer meshes (e.g., finer time increment, Δt) in numerical simulation.

Local Truncation Error versus Global Truncation Error (LTE vs GTE)

There are two types of truncation error, namely, local and global truncation errors, dealt in initial value problems. To understand them, let us consider an ordinary differential equation (ODE) system represented by:

$$\frac{dx}{dt} = f(t, x) \tag{1.15}$$

Using the Taylor expansion,

$$x_{k+1} = x_k + \left.\frac{dx}{dt}\right|_{t_k} (t_{k+1} - t_k) + R$$

This gives:

$$x_{k+1} = x_k + h f(t_k, x_k) + R$$

We can also write,

$$x_{k+1} = x_k + hf(t_k, x_k) + o(h^2) \tag{1.16}$$

Here, h is the step size $(= \Delta t = t_{k+1} - t_k)$ and R is the *local truncation error* (LTE). It is now obvious that there is a LTE of $o(h^2)$ in each time step.[5]

On the other hand, the *global truncation error* (GTE) is defined as:

$$\text{GTE} = x(t_N) - x_N$$

in which, $x(t_N)$ is the actual value of x at $t = t_N$ (t_N be the value of t where the integration terminates) and x_N is its estimated/approximate value.

5 This notation $o(h^2)$ [pronounced as "*oh* of h^2"] conveys that the error term is approximately a multiple of h^2.

As indicated, there are total N number of integration steps as:

$$N = \frac{t_N - t_0}{h}$$

This means the total number of steps is proportional to $\frac{1}{h}$. It is observed before that the magnitude of LTE in each time step is $o(h^2)$. The total error accumulating from each time step scales with $h^{-1}o(h^2)$ and thus, the magnitude of GTE is $o(h)$ $[=h^{-1}o(h^2)]$. It becomes obvious that the magnitude of GTE is reduced by 1 over that of LTE.

Geometric Interpretation (LTE vs GTE)

Let us draw Figure 1.2 to understand the LTE vs GTE in a better way. As mentioned, $x(t_{k+1})$ denotes the correct solution, whereas the approximate solution x_{k+1} we can compute from Equation (1.16) that typically represents a straight line as shown in Figure 1.2. Moreover, the upper curve represents the correct solution of Equation (1.15) and the lower curve represents the correct solution in terms of $z(t)$ when one starts from approximate x_k, instead of $x(t_k)$.

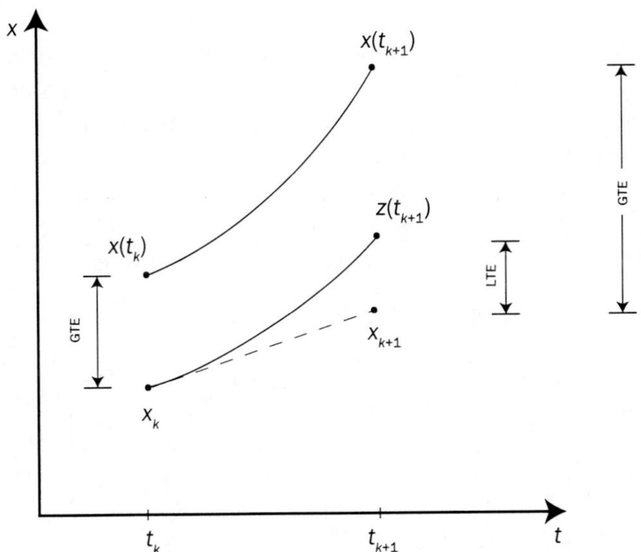

FIGURE **1.2** Illustrating the local and global truncation errors for a differential system

From Figure 1.2,

$$(\text{GTE})_{t_{k+1}} = x(t_{k+1}) - x_{k+1} = x(t_{k+1}) - z(t_{k+1}) + z(t_{k+1}) - x_{k+1}$$

It gives:

$$(\text{GTE})_{t_{k+1}} = \underbrace{x(t_{k+1}) - z(t_{k+1})}_{\text{amplified error}} + \underbrace{z(t_{k+1}) - x_{k+1}}_{(\text{LTE})_{t_{k+1}}}$$

As shown, the LTE is specific to a single step, whereas the GTE is the sum of the local error and the amplified error from previous steps. At this point, we should remember that the GTE is not the sum of local truncation errors from previous time steps.

Remarks

1. We will see later that the approximation made in Equation (1.16) corresponds to the explicit Euler method (detailed in Chapter 4).
2. It is shown above that this method (Euler) has a global truncation error of $o(h)$. It means if we reduce the step size h by a factor of 1/2, the GTE will accordingly decrease by a factor of 1/2. Let us take another method, which has the GTE of, say, $o(h^4)$. For this, the GTE will be reduced by a factor of 1/16 when h is reduced by a factor of 1/2.

Order of Approximation

Let us revisit Equation (1.16), in which, the notation $o(h^2)$ stands for the *order of approximation*. Accordingly, it is said that Equation (1.16) leads to second-order accuracy.[6]

Round-off error

Digital computers do arithmetic within a certain precision limit, which leads to some errors. Basically, this error is caused owing to the limited size of registers in the arithmetic unit of the computer. For a single calculation, as stated, this error might be small. But when we simulate a process, it typically involves thousands of calculations. As a result, those small errors get added, leading to a large error in the final result.

To understand this point (how the computers do arithmetic) better, let us touch upon a few concepts in computer science. The following floating point format is typically used in computers to write a number: $m \times b^e$ with $\frac{1}{b} \leq m \leq 1$. Here, m represents the *mantissa* or *significand*, b the *base* or *radix* (2 in binary, 8 in octal, 10 in decimal and 16 in hexadecimal) and e the *exponent*.

Let us consider a real number, $a = \frac{22}{7}$. Computer cannot store it as 22/7. First it is transformed to the following form,

$$a = \frac{22}{7} = 3.142857142857142857 14.....$$

and then the corresponding *normalized form* is stored. But remember, there is a number (i.e., 142857) involved infinite times in a and a computer can only store up to its maximum capacity. For example, for the mantissa limit of $m <= 6$, a will be stored in the normalized form as 0.314286×10^1.

6 In case of differential equation system, however Equation (1.16) is written as: $\frac{x_{k+1} - x_k}{h} = f(t_k, x_k) + o(h)$. Thus, it is considered first-order accurate and this will be followed in Chapters 4, 5 and 6.

Let us adopt the following six-digit representation: 0.314286×10^1. This value obviously deviates from the actual value and round-off error is generated in a single calculation. Notice that in *rounding*, the normalized floating point number is chosen in such a way that it is nearest to the number *a*.

On the other hand, due to *chopping*, the real number '*a*' has the following six-digit representation: 0.314285×10^1. Notice that in chopping, the number *a* is retained up to six-digits and the remaining digits are simply chopped off. Thus, for the number *a*, we denote the rounded and chopped floating point representation, respectively, as:

$$\text{fl}_{\text{round}}(a) = 0.314286\times10^1$$

$$\text{fl}_{\text{chop}}(a) = 0.314285\times10^1$$

In general, one can reduce the round-off error by the use of double-precision arithmetic.[7]

In our own hand calculation, similarly the round-off error is generated. For example:

$1.408 + 0.006 = 1.414 \approx 1.41$ (rounding off to the second decimal)

$1.408 + 0.009 = 1.417 \approx 1.42$ (rounding off to the second decimal)

The following rules we typically follow to round off a number having the format of $m\times b^e$:

i) All digits positioned right of m th digit should be discarded
ii) m th digit remains unchanged if the $m + 1$ th digit is less than $b/2$
iii) m th digit is increased by 1 if the $m + 1$ th digit is greater than $b/2$
iv) m th digit is increased by 1 if it is odd and the $m + 1$ th digit is equal to $b/2$. When the m th digit is even, it remains unchanged if the $m + 1$ th digit is equal to $b/2$ and we call this as symmetric rounding around even number. Similarly, one can consider symmetric rounding around odd number.

A note on floating point representation of numbers

As indicated, the following form represents an n-digits floating point number in the base b as:

$$a = \pm(0.\ d_1d_2\cdots d_n)\ b^e$$

in which, $(0.\ d_1d_2\cdots.d_n)$ is referred as mantissa. Now question is: how to count the number of digits in the mantissa? To answer this, let us first note that the countable digits are called *significant digits* or *significant figures* and the rest are *non-significant* ones. Actually, the significant digits are used to constitute the mantissa. With this, let us go through the following rules to know how the digits are counted:

i) All non-zero digits are considered significant digits. For example, 1.2345 includes five significant digits.
ii) All zeros present in between non-zero digits are also significant. Thus, in 1.20305, the number of significant digits is counted as six.

7 In double precision (float variable, 64 bits), C-programming allots 52 bits for mantissa (m), 11 bits for exponent (e) and 1 bit for sign (+ or –). In single-precision (float variable, 32 bits), $m = 23$, $e = 8$ and 1 bit for sign.

iii) Trailing zeros following a decimal point are considered significant digits. For example, 1.2030 and 1.2300 both include five significant digits.[8]
iv) Zeros between the decimal point and preceding a non-zero digit are non-significant. For example, 0.001234 and 0.0001234 have four significant digits each.

Eliminating the useless zeros (also called *spurious zeros*), one can write 0.001234 in the *normalized form* as $a = +(0.1234)10^{-2}$. Basically, to avoid the wastage of computer memory, computer typically stores all real numbers in their normalized forms.

1.3.2 Total Error

As mentioned before, reducing the mesh size leads to a decrease in the truncation error. But it happens at the expense of round-off error. This is because the mesh refinement leads to an increase in the number of arithmetic operations, thereby increasing the round-off error.

With this, the profile of total error (i.e., the sum of truncation and round-off error) is shown in Figure 1.3 with respect to mesh size. It is evident that there is a sweet point where the total error is minimum at a particular mesh size (indicated by point 'o') that one should adopt for the concerned numerical algorithm.

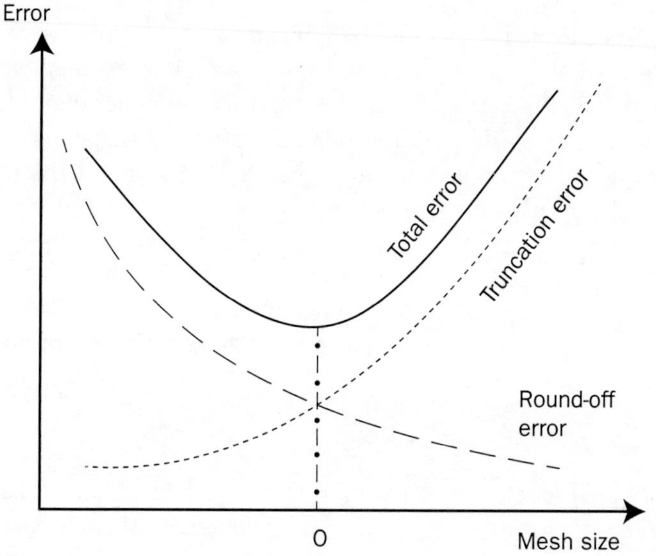

Figure 1.3 Error profile

Example 1.2 Total error

Using the Taylor series,

$$f(x_{k+1}) = f(x_k) + f'(x_k)h + \frac{f''(x_k)}{2!}h^2 + \ldots\ldots\ldots + \frac{f^{(n)}(x_k)}{n!}h^n + R$$

8 This is so because Problem 1.1 and Problem 1.10 in the exercise of Chapter 1 of a book are totally different.

One can use this formula in various forms, depending on how the right-hand terms are treated. For example, retaining only the first right-hand term, we get:

$$f(x_{k+1}) \approx f(x_k) + R$$

This is called *zero-order approximation*, which reveals that the value of f at the new point is the same as that at the old point (Figure 1.4). Notice that here the truncation error is of the order of h (i.e., $R \equiv o(h)$).

For a better prediction, one can adopt:

$$f(x_{k+1}) \approx f(x_k) + f'(x_k)h + o(h^2) \tag{1.17}$$

This is similarly called *first-order approximation* that includes an additional first-order term, consisting of a slope f' multiplied by h (Figure 1.4). Rearranging the terms in Equation (1.17), we get:

$$f'(x_k) \quad = \quad \frac{f(x_{k+1}) - f(x_k)}{h} \quad + \quad o(h)$$

True value	Finite difference approximation	Truncation error

Since we are using digital computers, the function values do include round-off error as:

$$f(x_k) = \bar{f}(x_k) + \varepsilon_k$$

$$f(x_{k+1}) = \bar{f}(x_{k+1}) + \varepsilon_{k+1}$$

Here $\bar{f}$ is the rounded function value and ε the associated round-off error. Combining the last three equations, we have:

$$f'(x_k) \quad = \quad \frac{\bar{f}(x_{k+1}) - \bar{f}(x_k)}{h} \quad + \quad \frac{\varepsilon_{k+1} - \varepsilon_k}{h} \quad + \quad o(h) \tag{1.18}$$

True value	Finite difference approximation	Round-off error	Truncation error

Remarks

1. Notice that although Equation (1.17) is second-order accurate, but its corresponding differential form [i.e., Equation (1.18)] is first-order accurate.
2. The total error of the finite difference approximation [see Equation (1.18)] includes the round-off error (that decreases with increasing step size) and truncation error (that increases with increasing step size).

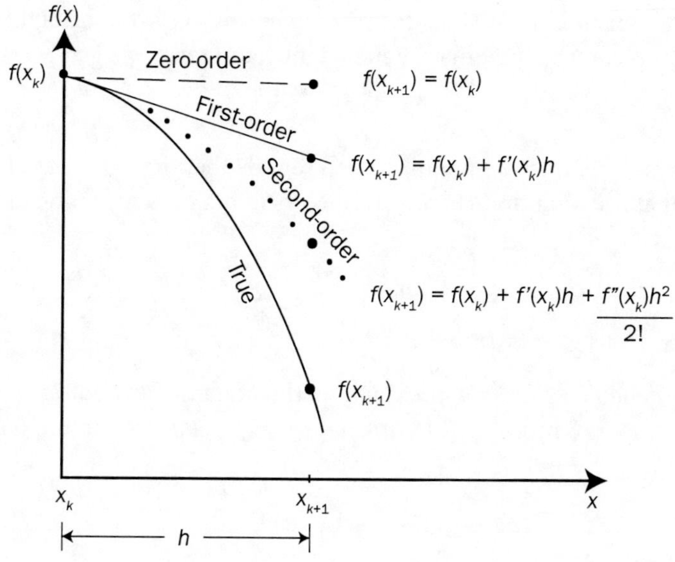

FIGURE 1.4 Comparing zero-, first- and second-order approximations of a true function

1.3.3 STOPPING CRITERIA

We would like to find the root of

$$f(x)=0 \tag{1.8}$$

by the use of, say, an *iterative convergence method* (i.e., a numerical approach). For this, one needs to continue the computation until $f(x)$ is close to zero. It is fairly true that we can very rarely satisfy Equation (1.8), getting exactly $f(x)$ equal to zero. Thus, we need to stop somewhere in our computation but question is: where should we stop?

For this, some stopping or termination criteria need to be specified for our use as a 'brake'. One such criterion is fixed based on the *absolute error* as:

$$\left|x_{k+1}-x_k\right| \leq \text{tol} \tag{1.19}$$

in which, x_k and x_{k+1} are the two consecutive iterates, and 'tol' denotes the *desired tolerance*.

Alternatively, the *relative absolute error*[9] may be constrained as:

$$\left|\frac{x_{k+1}-x_k}{x_{k+1}}\right| \leq \text{tol} \tag{1.20}$$

9 From the definition represented by Equation (1.9) [or (1.11)], we cannot find the error that is consistently meaningful for all cases. For example, an error in meter is significant when we measure the length of a lake but it has practically no meaning for the case of a river. The relative error is thus used that makes normalization as shown in Equation (1.20).

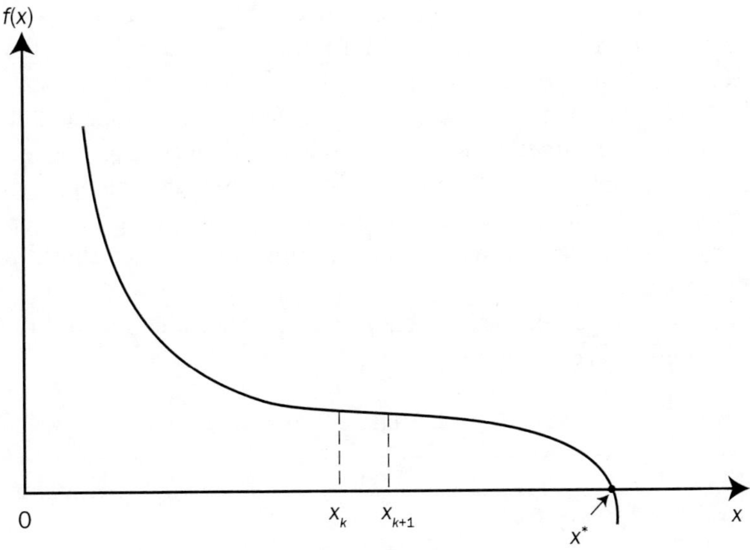

FIGURE 1.5 An example, for which Equation (1.19) or (1.22) is *not* recommended

Figure 1.5 describes an example, for which, Equation (1.19) may not be recommended (even though absolute error is < tol) since x_{k+1} is far away from the actual root x^*. In such a case, it is recommended to additionally use another criterion on the residual (i.e., the value of the function itself) as:

$$|f(x_k)| \leq \text{tol} \tag{1.21}$$

The reverse situation may also arise, which is that the absolute error is reasonably large but $|f(x_k)|$ satisfies the tolerance condition (see Example 3.3 solved by the *false position* method).

Remarks

1. To avoid any risk, it is recommended to check both the criteria [i.e., Equation (1.21) and either Equation (1.19) or (1.20)].
2. Similar to absolute error in Equation (1.19), there is a criterion as:

$$|f(x_{k+1}) - f(x_k)| \leq \text{tol} \tag{1.22}$$

This does not come out as an alternative to Equation (1.21) because of the same reason stated before. One can revisit Figure 1.5, which also indicates that Equation (1.22) may be satisfied but $f(x_{k+1})$ is far away from $f(x^*)$.

1.4 Summary and Concluding Remarks

This chapter covers the basics of process modeling and simulation using the numerical method. Describing the modeling protocol, the process simulation is briefed with the necessity of the simulator. For this, a simple example of a batch reactor is used. Defining the various forms of error, we have analyzed them for the same batch reactor. Finally, we learn the effect of total error and stopping criteria in computer simulation.

We should note down the point that computer is treated as a black box. It can basically do three limited jobs: arithmetic, storing of results, and retrieving the results from the memory. But many more one needs to do for computer simulation, apart from writing a set of instructions for a computer (called *programming*). Accordingly, a couple of relevant chapters are developed in the subsequent part of this book and you need to go through them with utmost care.

Before moving to answer exercise questions and problems, let us go through Table 1.5 that reports the prefixes used in a number.

Table 1.5 List of prefixes

Factor	Prefix	abbreviation
10^1	deka	da
10^2	hecto	h
10^3	kilo	k
10^6	mega	M
10^9	giga	G
10^{12}	tera	T
10^{15}	peta	P
10^{18}	exa	E
10^{21}	zetta	Z
10^{24}	yotta	Y
10^{-1}	deci	d
10^{-2}	centi	c
10^{-3}	milli	m
10^{-6}	micro	μ
10^{-9}	nano	n
10^{-12}	pico	p
10^{-15}	femto	f
10^{-18}	atto	a
10^{-21}	zepto	z
10^{-24}	yocto	y

Exercise

Review Questions

1.1 How to define the mathematical model?

1.2 What are the various applications of process simulator?

1.3 Why we need to compute error in numerical simulation?

1.4 Discuss the relative effect of the absolute error and absolute relative error.

1.5 How the total error can actually be calculated in the computer simulation?

1.6 A number 2.1 has two significant digits, whereas 2.10 has three. Do you think both the numbers are always same? Why?

Practice Problems

1.7 Simulate the batch reactor model developed in Case Study 1.1 for 10 min and then check whether Equation (1.8) is satisfied or not.

1.8 The true value of π in 6 decimal digits is 3.14159, while its approximate value is $\frac{22}{7}$. Determine the absolute error and absolute true error.

1.9 Using the Taylor series expansion:

$$e^x = 1 + x + \frac{x^2}{2!} + \frac{x^3}{3!} + \dots\dots\dots + \frac{x^n}{n!} + \dots\dots\dots$$

The true value of e is 2.71828. Tabulate the truncation errors for $n = 1, 2, 3, 4$, with $x = 1$.

1.10 Calculate the round-off and chopping error involved with $\frac{2}{\sqrt{\pi}}$ (consider six-digit representation).

1.11 Using 5 decimal digits floating-point round-off arithmetic, find the value of:

$$f(x) = \sin x - \cos x$$

at $x = 0.82346$.

1.12 Using 6 decimal digits floating-point round-off arithmetic, find the value of the hyperbolic cosine function:

$$\cosh(x) = \frac{e^x + e^{-x}}{2}$$

at $x = 1.0$.

1.13 Determine the linearized form of the following equations:

i) $y = x^2$ at $x_0 = 1$

ii) $\frac{dy}{dt} = -y^2 + \sqrt{x}$

1.14 The model of a liquid tank system has the following form:

$$A\frac{dh}{dt} + \alpha\sqrt{h} = F_i$$

where, A is the cross-sectional area of the tank, h the liquid height in the tank, F_i the inlet liquid flow rate and α a constant.

i) Identify the non-linear term and perform linearization of that term.

ii) Perform the qualitative and quantitative analysis of structural error.

iii) Is there any chance of having parametric error? How?

1.15 In Case Study 1.1, the model of an isothermal batch reactor is derived. For non-isothermal case, Equation (1.2) gets the following form:

$$\frac{dC_A}{dt} = -k_0 C_A^2 \exp\left[-\frac{E}{RT}\right]$$

where, the reaction rate constant,

$$k = k_0 \exp\left[-\frac{E}{RT}\right] \qquad \text{(Arrhenius law)}$$

in which,

k_0 = pre-exponential factor (constant)
E = activation energy (constant)
R = universal gas constant (constant)
T = reactor temperature (variable)

Transform the model into its linearized form.

Part II

Systems of Algebraic Equations

Algebraic equation is an equation in the form of a polynomial containing a finite number of terms and equated to zero (dictionary meaning). Sometimes, the algebraic equations may include transcendental functions (e.g., trigonometric, exponential and logarithmic functions).

We will cover two types of the systems of algebraic equations:

- Linear algebraic equation system
- Non-linear algebraic equation system

For solving the linear algebraic equations, various methods are to be discussed in Chapter 2, and in Chapter 3, the numerical methods are covered for solving the non-linear algebraic equations.

2

Linear Algebraic Equation

"The art of doing mathematics consists in finding that special case which contains all the germs of generality" – David Hilbert

Key Learning Objectives

- Revising basics of linear systems and their different types of solutions
- Learning various solution methods (direct and indirect)
- Applying these methods to a wide variety of systems examples
- Knowing how to choose a suitable solution method

2.1 Introduction

There are several systems whose physical operations are described solely by algebraic equations. The general form of such a system of n coupled *linear*[1] algebraic equations for n unknowns (i.e., x_i, $i = 1, 2, \ldots, n$) is:

$$\begin{aligned} a_{11}x_1 + a_{12}x_2 + \ldots\ldots\ldots + a_{1n}x_n &= b_1 \\ a_{21}x_1 + a_{22}x_2 + \ldots\ldots\ldots + a_{2n}x_n &= b_2 \\ &\vdots \\ a_{n1}x_1 + a_{n2}x_2 + \ldots\ldots\ldots + a_{n\,n}x_n &= b_n \end{aligned} \tag{2.1}$$

Here, a_{11}, $a_{12}, \ldots\ldots\ldots, a_{n\,n}$ are the coefficients of the system, and b_1, $b_2, \ldots\ldots\ldots, b_n$ are the constant terms.

One can represent Equation (2.1) in matrix form as:

$$[A]\ [X] = [B]$$

1 See the definition in Appendix A2.1.

or simply,

$$AX = B \tag{2.2}$$

in which, the coefficient matrix (i.e., an $n \times n$ square matrix),

$$A = \begin{bmatrix} a_{11} & a_{12} & \cdots\cdots & a_{1n} \\ a_{21} & a_{22} & \cdots\cdots & a_{2n} \\ \cdot & \cdot & & \cdot \\ \cdot & \cdot & & \cdot \\ \cdot & \cdot & & \cdot \\ a_{n1} & a_{n2} & \cdots\cdots & a_{nn} \end{bmatrix} \tag{2.3}$$

unknown vector (i.e., column vector),

$$X = \begin{bmatrix} x_1 \\ x_2 \\ \cdot \\ \cdot \\ \cdot \\ x_n \end{bmatrix} \tag{2.4}$$

and constant vector,

$$B = \begin{bmatrix} b_1 \\ b_2 \\ \cdot \\ \cdot \\ \cdot \\ b_n \end{bmatrix} \tag{2.5}$$

In Equation (2.3), a_{ij} denotes an element in the ith row and jth column of matrix A. Notice that here we use *capital/uppercase letters* to denote the matrices and *small/lowercase letters* for their entries/elements.

A linear system is said to be *homogeneous* when it is represented by a modified form of Equation (2.2) as:

$$AX = \mathbf{0} \tag{2.6}$$

Here, **0** denotes a *null* matrix (zero vector).

At this point we should note that aside from inherently linear systems of algebraic equations, these equations arise in the systems of differential equations (e.g., boundary value problems and partial differential equations) as well. Anyway, prior to reading the subsequent sections of this chapter, the reader is recommended to go through the Appendix 2A that covers the basics of vector and matrix algebra.

2.2 Types of Solution

For linear systems of algebraic equations, mainly three types of solutions exist:

i) a unique solution
ii) infinitely many solutions
iii) no solution

Let us briefly discuss them.

To solve a system represented by,

$$AX = B \tag{2.2}$$

instead of matrix A [Equation (2.3)], one can use the following *augmented matrix* (denoted by *aug A*) for the ease of computation:

$$aug\ A = \begin{bmatrix} a_{11} & a_{12} & \cdots\cdots & a_{1n} & b_1 \\ a_{21} & a_{22} & \cdots\cdots & a_{2n} & b_2 \\ \cdot & \cdot & & \cdot & \cdot \\ \cdot & \cdot & & \cdot & \cdot \\ \cdot & \cdot & & \cdot & \cdot \\ a_{n1} & a_{n2} & \cdots\cdots & a_{nn} & b_n \end{bmatrix} \tag{2.7}$$

Now to find the type of solution, let us consider that Equation (2.2) includes total m equations and there are n unknowns (i.e., $x_1, x_2, \ldots\ldots x_n$). This is obviously a general case, for which, $B = [b_1\ b_2, \ldots\ b_m]^T$ and A has the dimension of $m \times n$. To solve this problem, at first one needs to determine the *rank* of both the matrices A and *aug A*. The following conditions are subsequently checked for unique, infinite or no solution of the given system.

(a) A system of Equation (2.2) has solutions if the rank of A is same with that of *aug A*. If this necessary and sufficient condition is violated, the equations are incompatible, yielding *no solution*.
(b) If the rank of A and *aug A* is same ($= r$) (i.e., Condition (a) satisfied), one can assign arbitrary values to $n-r$ variables and find the values of the remaining r variables. It is further elaborated in Table 2.1.

Table 2.1 Different options under Condition (b)

$m = n$ ($r \leq n$)	$m < n$ ($r \leq m$)	$m > n$ ($r \leq n$)
$r = n$ (unique) $r < n$ (infinite)	(infinite)	$r = n$ (unique) $r < n$ (infinite)

(c) For a homogeneous system [Equation (2.6)], zero solution is always a solution (called *trivial* solution). One can obtain *non-trivial* solutions if and only if the rank of A is less than n.

In the following, a few cases are analyzed with examples and their geometric interpretations are given.

Example 2.1 Infinite solutions

Consider the following system:

$$\begin{aligned} x_1 + x_2 &= 3 \\ 2x_1 + 2x_2 &= 6 \end{aligned} \tag{2.8}$$

Solve it and find the type of solution existed.

Solution

For this system,

$$A = \begin{bmatrix} 1 & 1 \\ 2 & 2 \end{bmatrix}_{2\times 2}$$

$$Aug\,A = \begin{bmatrix} 1 & 1 & 3 \\ 2 & 2 & 6 \end{bmatrix}_{2\times 3}$$

It is easy to find the rank[2] of $A = 1$ and rank of $aug\,A = 1$. Obviously, it satisfies Condition (a). Next let us go through the options listed in Table 2.1 under Condition (b). In Equation set (2.8), the second equation is just double the first equation, indicating

total number of equations (m) = 1

total number of unknowns (n) = 2

Since $m < n$ and $r = m = 1$, it has *infinite solutions* (see Table 2.1).

It is true that we have a set of infinite solutions if more variables are there than the number of equations.

Geometric interpretation

Both the equations in Equation set (2.8) represent the *same line* and it is illustrated in Figure 2.1. Obviously, there are infinite intersections, which mean infinitely many solutions.

---

2 Consult any text book on *engineering mathematics*.

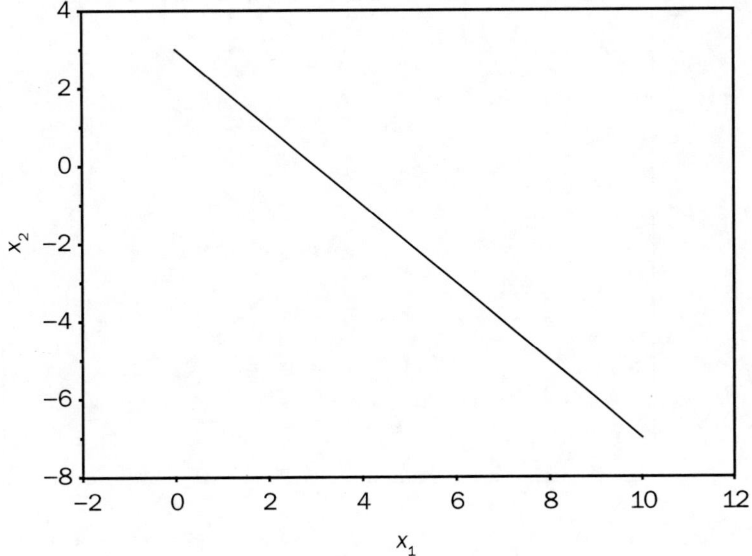

FIGURE 2.1 Illustrating a case of *infinite solutions*

Example 2.2 Unique solution

Consider the following system:

$$\begin{aligned} x_1 + x_2 &= 3 \\ x_1 + 2x_2 &= 6 \end{aligned} \tag{2.9}$$

Solve it and find the type of solution.

Solution

For the given system,

$$A = \begin{bmatrix} 1 & 1 \\ 1 & 2 \end{bmatrix}_{2\times 2}$$

$$Aug\,A = \begin{bmatrix} 1 & 1 & 3 \\ 1 & 2 & 6 \end{bmatrix}_{2\times 3}$$

It is easy to show that the rank of A = rank of $aug\ A$ = 2. Obviously, this case falls under Condition (b) with $m = n = 2$ and $r = n = 2$. Thus, it has a *unique solution* (Table 2.1).

Geometric interpretation

As shown in Figure 2.2, the two lines in Equation set (2.9) intersect at $x_1 = 0$ and $x_2 = 3$. This reveals that there exists a unique solution.

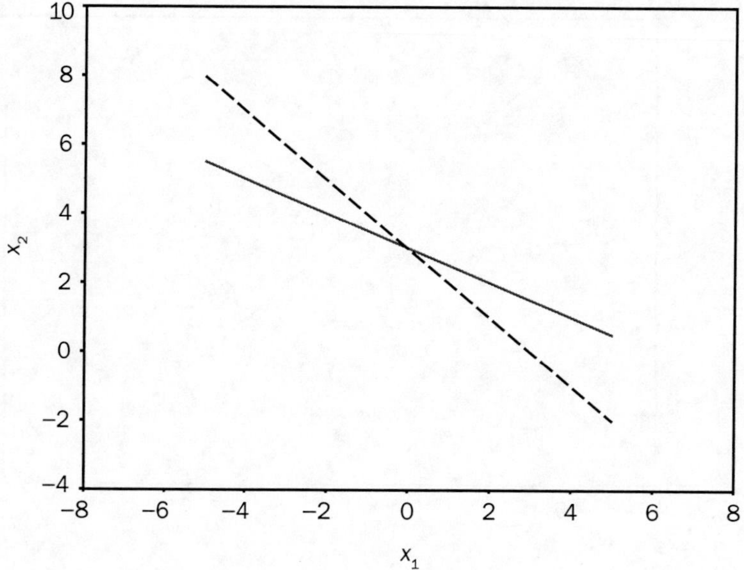

FIGURE 2.2 Illustrating a case of *unique solution*

Example 2.3 No solution

Consider the following system:

$$\begin{aligned} x_1 + x_2 &= 3 \\ x_1 + x_2 &= 5 \end{aligned} \tag{2.10}$$

Solve it and find the type of solution.

Solution

Here,

$$A = \begin{bmatrix} 1 & 1 \\ 1 & 1 \end{bmatrix}_{2\times 2}$$

and rank of $A = 1$.

$$Aug\,A = \begin{bmatrix} 1 & 1 & 3 \\ 1 & 1 & 5 \end{bmatrix}_{2\times 3}$$

Rank of $aug\,A = 2$.

Obviously, here Condition (a) is violated. Thus, Equations (2.10) are incompatible and the concerned system has *no solution*.

Geometric interpretation

As shown in Figure 2.3, the two lines in Equation set (2.10) are parallel and thus, they never intersect. Hence, there is no solution to the given system.

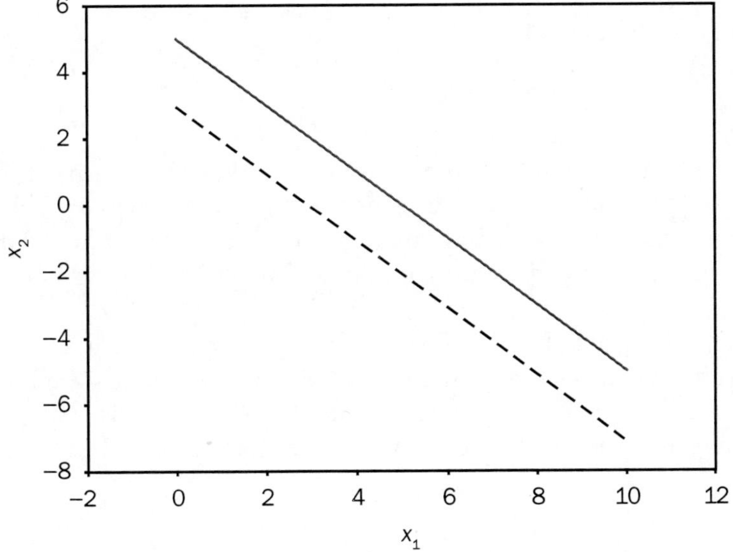

FIGURE 2.3 Illustrating a case of *no solution*

2.3 Methods of Solving Linear Algebraic Equations

Here we will learn a myriad of methods to solve the systems of linear algebraic equations. These linear methods are broadly classified into two categories, namely, direct methods and indirect methods.

Direct Methods

These methods produce the exact solution in a fixed number of steps, disregarding the round-off errors. Here, we would like to discuss the following direct methods:

- Matrix inversion method
- Cramer rule
- Gauss elimination
- Gauss elimination with pivoting
- LU decomposition
  - Crout method
  - Doolittle method
- Cholesky method
- Thomas algorithm
- Gauss–Jordan method

Indirect Methods

The indirect methods, which are also called *iterative methods*, provide a successive improvement of initial guesses and we converge to the solution when the number of steps tends to infinity. The following indirect methods are to be discussed here:

- Jacobi method
- Gauss–Seidel method
- Relaxation method

Computational Cost: A Short Note

The efficiency of a numerical algorithm is typically expressed in terms of *memory* that is quantified by the amount of space required in the computer to solve a problem. This space requirement is usually estimated for a numerical method by the *operational count*, which is defined as the number of divisions and multiplications involved in solving a system of equations.

Generally speaking, multiplication/division requires more computational time compared to addition/subtraction. The time requirement for a multiplication on a computer is almost equal to that for a division. Further note that the round-off error is mainly dependent on the total number of operational counts;[3] more counts lead to make the method more error-prone.

Direct Methods

In the following, we will discuss the direct methods one-by-one and the application of all those techniques to the example systems.

2.4 Matrix Inversion Method

Let us consider the linear system represented before by

$$AX = B \tag{2.2}$$

If matrix A is non-singular[4], inverse of A exists and the solution is simply obtained from:

$$X = A^{-1}B \tag{2.11}$$

where,

$$A^{-1} = \frac{\text{adj } A}{|A|} \tag{2.12}$$

In which,

adj A = adjoint of matrix A

$|A|$ = determinant of matrix A

This method is popularly called the *matrix inversion method.*

---

3 Sometimes called *floating-point operations* (or *flops*).

4 It means determinant of matrix A (denoted by $|A|$ or det A) is non-zero.

Remarks

1. This approach requires a large computational time, particularly to find A^{-1}.
2. Further, it suffers from the problem of *ill-conditioning* (detailed in Appendix A2.2.3).

To take care of these issues, several other techniques are developed for a large set of equations [Equation (2.2)] and they are discussed in the subsequent sections.

Example 2.4 Matrix inversion method

Solve the following system of equations:

$$-x_1 + 2x_2 - x_3 = 0$$
$$2x_1 + 3x_2 - 3x_3 = -1$$
$$3x_1 + x_2 - x_3 = 2$$

using the matrix inversion method.

Solution

Representing the given system in the form of Equation (2.2), we have,

$$A = \begin{bmatrix} -1 & 2 & -1 \\ 2 & 3 & -3 \\ 3 & 1 & -1 \end{bmatrix}$$

This matrix A is non-singular since

$$|A| = (-1)\times 0 - 2\times 7 - 1\times(-7) = -7$$

Again,

$$\text{adj } A = \begin{bmatrix} 0 & -7 & -7 \\ 1 & 4 & 7 \\ -3 & -5 & -7 \end{bmatrix}^{\mathrm{T}} = \begin{bmatrix} 0 & 1 & -3 \\ -7 & 4 & -5 \\ -7 & 7 & -7 \end{bmatrix}$$

From Equation (2.12),

$$A^{-1} = \frac{\text{adj } A}{|A|} = -\frac{1}{7}\begin{bmatrix} 0 & 1 & -3 \\ -7 & 4 & -5 \\ -7 & 7 & -7 \end{bmatrix} = \begin{bmatrix} 0 & -\frac{1}{7} & \frac{3}{7} \\ 1 & -\frac{4}{7} & \frac{5}{7} \\ 1 & -1 & 1 \end{bmatrix}$$

From Equation (2.11), finally we get:

$$X = A^{-1}B = \begin{bmatrix} 0 & -\frac{1}{7} & \frac{3}{7} \\ 1 & -\frac{4}{7} & \frac{5}{7} \\ 1 & -1 & 1 \end{bmatrix} \begin{bmatrix} 0 \\ -1 \\ 2 \end{bmatrix} = \begin{bmatrix} \frac{1}{7}+\frac{6}{7} \\ \frac{4}{7}+\frac{10}{7} \\ 1+2 \end{bmatrix} = \begin{bmatrix} 1 \\ 2 \\ 3 \end{bmatrix}$$

So the solution is obtained as:

$x_1 = 1$
$x_2 = 2$
$x_3 = 3$

Example 2.5 Matrix inversion method (infinitely many solutions)

Solve the following system of equations:

$$-x_1 + 2x_2 - x_3 = 0$$
$$2x_1 + 3x_2 - 3x_3 = -1$$
$$x_1 + 5x_2 - 4x_3 = -1$$

using the matrix inversion method.

Solution

In this problem,

$$A = \begin{bmatrix} -1 & 2 & -1 \\ 2 & 3 & -3 \\ 1 & 5 & -4 \end{bmatrix}$$

It gives:

$$|A| = -1(-12+15) - 2(-8+3) - 1(10-3) = -3+10-7 = 0$$

Since A is a singular matrix, the matrix inversion method is not applicable to this system.

Remark

Notice that the third equation is the addition of the first and second equations. Thus, the system is given with 2 independent equations for finding 3 variables. As a result, we have infinitely many solutions.

Example 2.6 Matrix inversion method (inconsistent system)

Let us make a little modification in the system of the last Example 2.5:

$$\begin{aligned} -x_1 + 2x_2 - x_3 &= 0 \\ 2x_1 + 3x_2 - 3x_3 &= -1 \\ x_1 + 5x_2 - 4x_3 &= 1 \end{aligned}$$

Solve it using the matrix inversion method.

Solution

For the same A matrix, we got $|A| = 0$ in the last example. Accordingly, as stated, matrix A is singular and the matrix inversion method is not applicable. Further, notice that adding the first two equations and then comparing with the third equation, we get: $-1 = 1$. It reveals that the system is inconsistent and it has no solution.

2.5 Cramer Rule

The Cramer rule[5] is used to solve the linear systems through determinants. For a system with Equation (2.2),

$$x_i = \frac{|A_i|}{|A|} \qquad i = 1, 2, \ldots\ldots, n \tag{2.13}$$

where,

$$A_i = \begin{bmatrix} a_{11} & \cdots & a_{1\ i-1} & b_1 & a_{1\ i+1} & \cdots & a_{1n} \\ a_{21} & \cdots & a_{2\ i-1} & b_2 & a_{2\ i+1} & \cdots & a_{2n} \\ \cdot & & \cdot & \cdot & \cdot & & \cdot \\ \cdot & & \cdot & \cdot & \cdot & & \cdot \\ a_{n1} & \cdots & a_{n\ i-1} & b_n & a_{n\ i+1} & \cdots & a_{nn} \end{bmatrix} \tag{2.14}$$

Notice that for determining A_i, the ith column of A needs to be replaced by the column vector B.

Further, one can get a unique solution of Equation (2.2) using the Cramer rule only if A is non-singular (i.e., $|A| \neq 0$) that is obvious in Equation (2.13). In other words, if A is singular, there is either no solution or infinitely many solutions and Equation (2.13) yields nonsense.

5 Named after the Genevan mathematician Gabriel Cramer (1704–1752).

Example 2.7 Cramer rule

Consider a simple 2 × 2 system,

$$\begin{aligned} a_{11}x_1 + a_{12}x_2 &= b_1 \\ a_{21}x_1 + a_{22}x_2 &= b_2 \end{aligned} \tag{2.15}$$

Solve this linear system, and find x_1 and x_2.

Solution

Here,

$$A = \begin{bmatrix} a_{11} & a_{12} \\ a_{21} & a_{22} \end{bmatrix}$$

and

$$|A| = a_{11}a_{22} - a_{12}a_{21} \tag{2.16}$$

Then

$$\begin{aligned} \det A_1 &= \det \begin{bmatrix} b_1 & a_{12} \\ b_2 & a_{22} \end{bmatrix} \\ &= b_1a_{22} - a_{12}b_2 \end{aligned} \tag{2.17}$$

Similarly,

$$\begin{aligned} \det A_2 &= \det \begin{bmatrix} a_{11} & b_1 \\ a_{21} & b_2 \end{bmatrix} \\ &= a_{11}b_2 - b_1a_{21} \end{aligned} \tag{2.18}$$

Using the Cramer rule [i.e., Equation (2.13)], we get the solution as:

$$x_1 = \frac{|A_1|}{|A|} = \frac{b_1a_{22} - a_{12}b_2}{a_{11}a_{22} - a_{12}a_{21}} \tag{2.19}$$

$$x_2 = \frac{|A_2|}{|A|} = \frac{a_{11}b_2 - b_1a_{21}}{a_{11}a_{22} - a_{12}a_{21}} \tag{2.20}$$

Remark

For homogeneous equations (i.e., $B = \mathbf{0}$), $|A_i| = 0$ [see Equations (2.17) and (2.18)]. It leads to make all the $x_i = 0$ (*trivial solution*).

Example 2.8 Cramer rule

Solve the following system of equations

$$
\begin{aligned}
3x_1 - 2x_2 + 3x_3 &= 8 \\
4x_1 + 2x_2 - 2x_3 &= 2 \\
x_1 + x_2 + 2x_3 &= 9
\end{aligned}
\tag{2.21}
$$

Solution

Here,

$$A = \begin{bmatrix} 3 & -2 & 3 \\ 4 & 2 & -2 \\ 1 & 1 & 2 \end{bmatrix}$$

$$B = \begin{bmatrix} 8 \\ 2 \\ 9 \end{bmatrix}$$

We find:

$$\det A = 44$$

$$\det A_1 = 44$$

$$\det A_2 = 88$$

$$\det A_3 = 132$$

With this, one obtains the solution as:

$$x_1 = \frac{\det A_1}{\det A} = \frac{44}{44} = 1$$

$$x_2 = \frac{\det A_2}{\det A} = \frac{88}{44} = 2$$

$$x_3 = \frac{\det A_3}{\det A} = \frac{132}{44} = 3$$

Remarks

1. For large problems with non-integer coefficients, the round-off error may become significant in case of Cramer rule.
2. The Cramer rule is perhaps the simplest technique to solve small systems. It is quite easy to show that this method requires approximately $n!$ multiplications and divisions. So, for a system of 100×100 matrix, the number of operations involved is approximately 10^{158} ($\approx 100!$). We are now understanding why the Cramer rule is not preferred to use in practice, despite its great simplicity.

In the following, we will learn a few other methods, starting with Gauss elimination, which actually works for large problems.

2.6 Gauss Elimination

The following system

$$AX = B \tag{2.2}$$

is to be solved by using the Gauss elimination[6] (also called *Naïve Gauss elimination*) method. This method is classified into two categories:

- Gauss elimination with backward substitution
- Gauss elimination with forward substitution

Note that there are two consecutive phases (elimination and substitution) involved and both the above techniques are discussed below.

2.6.1 Gauss Elimination with Back Substitution

The first target of this technique is to transform the original system $AX = B$ to $UX = B$, where U is the *upper triangular matrix*. Then the transformed system is solved through back substitution (starting from the last equation and working backwards) to find the final solution.

Accordingly, Equation (2.1) is first converted to:

$$\begin{aligned}
a_{11}x_1 + a_{12}x_2 + \; \cdots \; + a_{1\,n-1}x_{n-1} + a_{1n}x_n &= b_1 \\
a_{22}x_2 + \; \cdots \; + a_{2\,n-1}x_{n-1} + a_{2n}x_n &= b_2 \\
\vdots \\
a_{i\,i}x_i + \; \cdots \; + a_{i\,n-1}x_{n-1} + a_{i\,n}x_n &= b_i \\
\vdots \\
a_{n-1\,n-1}x_{n-1} + a_{n-1\,n}x_n &= b_{n-1} \\
a_{n\,n}x_n &= b_n
\end{aligned} \tag{2.22}$$

This set of equations one can write in matrix form as:

$$UX = B \tag{2.23}$$

where the upper triangular matrix,

$$U = \begin{bmatrix}
a_{11} & a_{12} & \cdots & \cdots & \cdots & a_{1\,n-1} & a_{1n} \\
0 & a_{22} & \cdots & \cdots & \cdots & a_{2\,n-1} & a_{2n} \\
. & . & . & . & . & . & . \\
. & . & . & . & . & . & . \\
0 & 0 & 0 & a_{i\,i} & \cdots & a_{i\,n-1} & a_{i\,n} \\
. & . & . & . & . & . & . \\
. & . & . & . & . & . & . \\
0 & 0 & 0 & 0 & \cdots & a_{n-1\,n-1} & a_{n-1\,n} \\
0 & 0 & 0 & 0 & \cdots & 0 & a_{nn}
\end{bmatrix} \tag{2.24}$$

6 Named after the German mathematician and physicist Carl Friedrich Gauss (1777–1855).

In the next step, solving the last equation followed by second-last, third-last and so on of Equation set (2.22), one can obtain the solution when all of the diagonal elements in U are non-zero as:

$$
\begin{aligned}
x_n &= \frac{b_n}{a_{n\,n}} \\
x_{n-1} &= \frac{b_{n-1} - a_{n-1\;n} x_n}{a_{n-1\;n-1}} \\
&\vdots \\
x_1 &= \frac{b_1 - a_{1n} x_n - a_{1\;n-1} x_{n-1} - \cdots - a_{12} x_2}{a_{11}} \\
&= \frac{b_1 - \sum_{j=2}^{n} a_{1j} x_j}{a_{11}}
\end{aligned}
\tag{2.25}
$$

In compact form, the solution can be expressed as:

$$
x_i = \frac{1}{a_{i\,i}} \left(b_i - \sum_{j=i+1}^{n} a_{i\,j} x_j \right) \tag{2.26}
$$

Since the unknowns are solved by back substitutions, this method is referred as the *back substitution* or the *backward sweep* approach.

Solution Algorithm

As stated, the first step of the Gauss elimination is to transform the original system into an upper triangular form. Then solve the transformed system of Equation (2.23) through back substitution. This two-stage algorithm (i.e., transformation followed by back substitution) is framed below.

Transformation

The following *elementary row operations* one can perform on the augmented matrix, *aug A*, without having any affect on the solution set:

Scalling: A row can be multiplied by a non-zero constant: $R_i \rightarrow aR_i \;\; (a \neq 0)$

Replacement: A row can be replaced by summing up that row with a multiple of any other row: $R_i \rightarrow R_i + aR_j$

Let us start with Equation (2.1):

$$
\begin{aligned}
&a_{11}x_1 + a_{12}x_2 + \dots\dots\dots + a_{1n}x_n = b_1 \\
&a_{21}x_1 + a_{22}x_2 + \dots\dots\dots + a_{2n}x_n = b_2 \\
&\vdots \\
&a_{n1}x_1 + a_{n2}x_2 + \dots\dots\dots + a_{n\,n}x_n = b_n
\end{aligned}
\tag{2.1}
$$

Now, eliminate the elements of kth column and here, i is the row number where elimination takes place. For this, the following steps can be carried out for computer implementation.

Step 1: For $k=1$, eliminate all coefficients in the first column below a_{11}
To perform the elimination in ith row, multiply kth row (here 1st row) by $(-a_{i1}/a_{11})$ and then add to ith row ($i = 2, 3, \ldots, n$). With this, we get:

$$\begin{aligned}
a_{11}^{(1)}x_1 + a_{12}^{(1)}x_2 + a_{13}^{(1)}x_3 + \quad \ldots \quad + a_{1\,n-1}^{(1)}x_{n-1} + a_{1n}^{(1)}x_n &= b_1^{(1)} \\
a_{22}^{(2)}x_2 + a_{23}^{(2)}x_3 + \quad \ldots \quad + a_{2\,n-1}^{(2)}x_{n-1} + a_{2n}^{(2)}x_n &= b_2^{(2)} \\
a_{32}^{(2)}x_2 + a_{33}^{(2)}x_3 + \quad \ldots \quad + a_{3\,n-1}^{(2)}x_{n-1} + a_{3n}^{(2)}x_n &= b_3^{(2)} \\
\vdots \\
a_{n2}^{(2)}x_2 + a_{n3}^{(2)}x_3 + \quad \ldots \quad + a_{n\,n-1}^{(2)}x_{n-1} + a_{n\,n}^{(2)}x_n &= b_n^{(2)}
\end{aligned} \tag{2.27}$$

Obviously, $a_{21}, a_{31}, a_{41}, \ldots, a_{n1}$ got eliminated. Here, superscript '1' refers to the original value of the coefficients, '2' refers to the coefficients updated once, '3' means updated twice, and so on. Accordingly,

For row 1: $a_{1j}^{(1)} = a_{1j}$ where, $j = 1, 2, 3, \ldots, n$

$$b_1^{(1)} = b_1$$

For row 2:

$$\begin{aligned}
a_{22}^{(2)} &= a_{22} - \frac{a_{21}}{a_{11}}a_{12} = a_{22}^{(1)} - \frac{a_{21}^{(1)}}{a_{11}^{(1)}}a_{12}^{(1)} \\
a_{23}^{(2)} &= a_{23} - \frac{a_{21}}{a_{11}}a_{13} = a_{23}^{(1)} - \frac{a_{21}^{(1)}}{a_{11}^{(1)}}a_{13}^{(1)} \\
&\vdots \\
a_{2n}^{(2)} &= a_{2n} - \frac{a_{21}}{a_{11}}a_{1n} = a_{2n}^{(1)} - \frac{a_{21}^{(1)}}{a_{11}^{(1)}}a_{1n}^{(1)} \\
b_2^{(2)} &= b_2 - \frac{a_{21}}{a_{11}}b_1 = b_2^{(1)} - \frac{a_{21}^{(1)}}{a_{11}^{(1)}}b_1^{(1)}
\end{aligned} \tag{2.28}$$

Likewise, one can continue the same operation for other rows. For this, one can verify that:

$$\begin{aligned}
a_{ij}^{(k+1)} &= a_{ij}^{(k)} - \frac{a_{ik}^{(k)}}{a_{kk}^{(k)}}a_{kj}^{(k)} \\
b_i^{(k+1)} &= b_i^{(k)} - \frac{a_{ik}^{(k)}}{a_{kk}^{(k)}}b_k^{(k)}
\end{aligned} \tag{2.29}$$

where,

$$i = 2, 3, \ldots, n$$
$$j = 2, 3, \ldots, n$$
$$k = 1$$

Step 2: For k =2, eliminate all coefficients in the second column below $a_{22}^{(2)}$

We will start this step with Equation (2.27). Multiply the current kth row (i.e., 2nd row) by $(-a_{i2}^{(2)}/a_{22}^{(2)})$ and then add to ith row (i = 3, 4, …, n). With this, we get:

$$\begin{aligned}
a_{11}^{(1)}x_1+a_{12}^{(1)}x_2+a_{13}^{(1)}x_3+ \quad \cdots \quad &+a_{1\,n-1}^{(1)}x_{n-1}+a_{1n}^{(1)}x_n=b_1^{(1)}\\
a_{22}^{(2)}x_2+a_{23}^{(2)}x_3+ \quad \cdots \quad &+a_{2\,n-1}^{(2)}x_{n-1}+a_{2n}^{(2)}x_n=b_2^{(2)}\\
a_{33}^{(3)}x_3+ \quad \cdots \quad &+a_{3\,n-1}^{(3)}x_{n-1}+a_{3n}^{(3)}x_n=b_3^{(3)}\\
\vdots\qquad\qquad &\\
a_{n3}^{(3)}x_3+ \quad \cdots \quad &+a_{n\,n-1}^{(3)}x_{n-1}+a_{n\,n}^{(3)}x_n=b_n^{(3)}
\end{aligned} \tag{2.30}$$

These operations can again be verified by Equation (2.29) with:

$i = 3, 4, \ldots, n$
$j = 3, 4, \ldots, n$
$k = 2$

Steps 3, 4, … n−1: Repeat the operations for k =3, 4, …, n−1 and finally obtain:

$$\begin{aligned}
a_{11}^{(1)}x_1+a_{12}^{(1)}x_2+a_{13}^{(1)}x_3+ \quad \cdots \quad &+a_{1\,n-1}^{(1)}x_{n-1}+a_{1n}^{(1)}x_n=b_1^{(1)}\\
a_{22}^{(2)}x_2+a_{23}^{(2)}x_3+ \quad \cdots \quad &+a_{2\,n-1}^{(2)}x_{n-1}+a_{2n}^{(2)}x_n=b_2^{(2)}\\
a_{33}^{(3)}x_3+ \quad \cdots \quad &+a_{3\,n-1}^{(3)}x_{n-1}+a_{3n}^{(3)}x_n=b_3^{(3)}\\
&\vdots\\
&a_{n-1\,n-1}^{(n-1)}x_{n-1}+a_{n-1\,n}^{(n-1)}x_n=b_{n-1}^{(n-1)}\\
&a_{n\,n}^{(n)}x_n=b_n^{(n)}
\end{aligned} \tag{2.31}$$

Further, one can verify by Equation (2.29) with:

$k = 3, 4, \ldots, n-1$
$i = k + 1, k + 2, \ldots\ldots, n$
$j = k + 1, k + 2, \ldots\ldots, n$

Back Substitution

It is now obvious that the set of transformed Equations (2.31) leads to a system having an upper triangular matrix (i.e., $UX = B$). One can obtain the solution, X by solving those equations sequentially in the reverse direction (backward sweep). This is followed since only x_n is present in the last equation, x_n and x_{n-1} in the second-last equation, etc. With this, we finally get the solution as:

$$x_n=\frac{b_n^{(n)}}{a_{n\,n}^{(n)}} \qquad (a_{n\,n}\neq 0) \tag{2.32}$$

$$x_i=\frac{1}{a_{i\,i}^{(i)}}\left(b_i^{(i)}-\sum_{j=i+1}^{n}a_{i\,j}^{(i)}x_j\right) \tag{2.33}$$

where, $a_{ii} \neq 0$ and $i = n-1, n-2, \ldots.., 1$.

In this back substitution method, the original system $AX = B$ is transformed into $UX = B$. Notice that the B matrix between the original system (i.e., $AX = B$) and the transformed system (i.e., $UX = B$) is different. To avoid any confusion, the transformed system is thus better to represent by $UX = C$ and it is followed in the subsequent discussion.

Remarks

1. The system with Equation (2.2) includes total n equations and there are n unknowns. However, *aug A* [Equation (2.7)] has the dimension of $n \times (n+1)$. One can follow the same algorithm developed above for back substitution to solve the system with *aug A* and in fact, it is followed in Example 2.9.
2. In Gauss elimination, we must avoid the division (of a row, for example) by zero. But this method (without pivoting) does not avoid this problem and thus, it is often called as '*naive*' *Gauss elimination.*

Example 2.9 Gauss elimination with backward sweep

Solve the following system of equations:

$$\begin{aligned} 3x + y &= 5 \\ x + 3y + z &= 10 \\ y + z &= 5 \end{aligned} \tag{2.34}$$

by using the Gauss elimination with backward sweep.

Solution

In this problem,

$$A = \begin{bmatrix} 3 & 1 & 0 \\ 1 & 3 & 1 \\ 0 & 1 & 1 \end{bmatrix}$$

$$aug\ A = \begin{bmatrix} 3 & 1 & 0 & 5 \\ 1 & 3 & 1 & 10 \\ 0 & 1 & 1 & 5 \end{bmatrix}$$

We would like to solve this problem with the use of *aug A* by applying the solution algorithm developed above.

Step 1: Eliminate all coefficients in the first column, except $a_{11}^{(1)}$ (= 3). Notice that here no operation needs to be conducted on third row since $a_{31}^{(1)}$ is already zero. With this, simply multiply first row by $-\frac{1}{3}$, add to second row and get,

$$\begin{bmatrix} 3 & 1 & 0 & 5 \\ 0 & \frac{8}{3} & 1 & \frac{25}{3} \\ 0 & 1 & 1 & 5 \end{bmatrix} \qquad R_2 \to R_2 - \frac{1}{3} R_1$$

Step 2: Eliminate the coefficients in the second column below $a_{22}^{(2)} \left(= \frac{8}{3}\right)$. For this, multiply second row by $-\frac{3}{8}$, add to third row and get,

$$\begin{bmatrix} 3 & 1 & 0 & 5 \\ 0 & \frac{8}{3} & 1 & \frac{25}{3} \\ 0 & 0 & \frac{5}{8} & \frac{15}{8} \end{bmatrix} \qquad R_3 \to R_3 - \frac{3}{8} R_2$$

Now we have the following form:

$$UX = \begin{bmatrix} 3 & 1 & 0 \\ 0 & \frac{8}{3} & 1 \\ 0 & 0 & \frac{5}{8} \end{bmatrix} \begin{bmatrix} x \\ y \\ z \end{bmatrix} = \begin{bmatrix} 5 \\ \frac{25}{3} \\ \frac{15}{8} \end{bmatrix} \tag{2.35}$$

Applying back substitution, we get the solution as:

$$\frac{5}{8} z = \frac{15}{8} \qquad \therefore\ z = 3$$

$$\frac{8}{3} y + z = \frac{25}{3} \qquad \therefore\ y = 2$$

$$3x + y = 5 \qquad \therefore\ x = 1$$

2.6.2 Gauss Elimination with Forward Substitution

Here, we first transform $AX = B$ to $LX = B$, where L is the *lower triangular matrix*. Then the transformed system is solved through forward substitution (starting from the first equation and working forwards) to find the final solution.

Accordingly, Equation (2.1) needs to be converted to:

$$\begin{aligned} &a_{11}x_1 && = b_1 \\ &a_{21}x_1 + a_{22}x_2 && = b_2 \\ &a_{31}x_1 + a_{32}x_2 + a_{33}x_3 && = b_3 \\ &\quad\vdots && \\ &a_{n1}x_1 + a_{n2}x_2 + a_{n3}x_3 + \quad \cdots \quad + a_{n\,n}x_n && = b_n \end{aligned} \tag{2.36}$$

Notice that here we have omitted the superscript of all coefficients (a_{ij}) and constants (b_i).

Obviously,

$$L=\begin{bmatrix} a_{11} & 0 & 0 & \cdots & 0 \\ a_{21} & a_{22} & 0 & \cdots & 0 \\ a_{31} & a_{32} & a_{33} & \cdots & 0 \\ \cdot & \cdot & \cdot & & \cdot \\ \cdot & \cdot & \cdot & & \cdot \\ a_{n1} & a_{n2} & a_{n3} & \cdots & a_{nn} \end{bmatrix} \tag{2.37}$$

Solving the first equation followed by second, third and so on of Equation set (2.36), one obtains:

$$\begin{aligned} x_1 &= \frac{b_1}{a_{11}} \\ x_2 &= \frac{b_2 - a_{21}x_1}{a_{22}} \\ x_3 &= \frac{b_3 - a_{31}x_1 - a_{32}x_2}{a_{33}} \\ &\vdots \\ x_n &= \frac{b_n - a_{n\ n-1}x_{n-1} - a_{n\ n-2}x_{n-2} - \cdots - a_{n1}x_1}{a_{n\ n}} \\ &= \frac{b_n - \sum_{j=1}^{n-1} a_{n\ j}x_j}{a_{n\ n}} \end{aligned} \tag{2.38}$$

provided that $a_{ii} \neq 0$, where $i = 1, 2, \ldots\ldots, n$.

Since the unknowns are solved by forward substitutions, this method is referred to as the *forward substitution* approach.

Like the back substitution, in the forward substitution method, the B matrix between the original system (i.e., $AX = B$) and the transformed system (i.e., $LX = B$) is different. Thus, it is better to write $LX = D$ or so to avoid any confusion.

Example 2.10 Gauss elimination with forward substitution

Repeat Example 2.9 with

$$\begin{aligned} 3x + y &= 5 \\ x + 3y + z &= 10 \\ y + z &= 5 \end{aligned} \tag{2.34}$$

Find the solution by using the Gauss elimination with forward substitution.

Solution

In the first step, we target to eliminate z from the first two equations (notice that z is already not there in the first equation):

$$\begin{aligned} 3x+y&=5 \\ x+2y&=5 \qquad R_2 \to R_2 - R_3 \\ y+z&=5 \end{aligned}$$

Next, eliminate y from the first equation:

$$\begin{aligned} \frac{5}{2}x &= \frac{5}{2} \qquad R_1 \to R_1 - \frac{1}{2}R_2 \\ x+2y&=5 \\ y+z&=5 \end{aligned}$$

Finally, apply forward substitution and get the solution as:

$$\begin{aligned} \frac{5}{2}x &= \frac{5}{2} \qquad \therefore x=1 \\ x+2y&=5 \qquad \therefore y=2 \\ y+z&=5 \qquad \therefore z=3 \end{aligned}$$

2.6.3 Operational Counts in Gauss Elimination with Back Substitution

Recall that the operational count refers to the total number of divisions and multiplications involved in solving a system of equations. Let us now understand how the operational counting is done with taking a sample case of the Gauss elimination with backward sweep method. Here, the operational counting we do during the (i) conversion of the system $AX = B$ to $UX = C$, and (ii) back substitution. It is given in the following section.

- **Operational counting during conversion of $AX = B$ to $UX = C$**

Number of Divisions

Step 1: Dividing $a_{21}^{(1)}, a_{31}^{(1)}, \cdots a_{n1}^{(1)}$ by first pivot $a_{11}^{(1)}$: $n-1$

Step 2: Dividing $a_{32}^{(2)}, a_{42}^{(2)}, \cdots a_{n2}^{(2)}$ by second pivot $a_{22}^{(2)}$: $n-2$

.

.

.

Step $n-1$: Dividing $a_{n\ n-1}^{(n-1)}$ by $(n-1)$th pivot $a_{n-1\ n-1}^{(n-1)}$: 1

$$\textbf{\textit{Total}} = \sum(n-1) = \frac{n(n-1)}{2}$$

Number of Multiplications

Step 1: It involves the following row operations: $R_2 \to R_2 - \frac{a_{21}^{(1)}}{a_{11}^{(1)}} R_1, \ldots\ldots, R_n \to R_n - \frac{a_{n1}^{(1)}}{a_{11}^{(1)}} R_1.$

Accordingly, the number of multiplications involved in:

Second equation: n

Third equation: n

.

.

.

nth equation: n

***Sum of multiplications for total (n–1)* equations** $= n(n-1)$

Step 2: Sum of multiplications for total (n–2) equations $= (n-1)(n-2)$

.

.

.

Total $= \sum n(n-1) = \frac{n}{3}(n-1)(n+1)$

- **Operational counting during back substitution**

Number of Divisions

In Equation (2.32): 1

In Equation (2.33): $n-1$

Total $= n$

Number of Multiplications

Multiplications are involved in the term $\sum_{j=i+1}^{n} a_{ij}^{(i)} x_j$ of Equation (2.33) as:

for $i = n-1$: 1

for $i = n-2$: 2

.

.

.

for $i = 1$: $n-1$

Total $= \sum (n-1) = \frac{n(n-1)}{2}$

- **Grand total (operational counts in Gauss elimination with back substitution)**

Let us now add the operational counts obtained above during conversion and back substitution:

$$\text{Total number of divisions} = \frac{n(n-1)}{2} + n = \frac{n(n+1)}{2}$$

$$\text{Total number of multiplications} = \frac{n}{3}(n-1)(n+1) + \frac{n(n-1)}{2} = \frac{n(n-1)(2n+5)}{6}$$

$$\textbf{Grand total} \text{ (operational counts)} = \frac{n(n+1)}{2} + \frac{n(n-1)(2n+5)}{6} = \frac{n(n^2+3n-1)}{3}$$

$$\approx \frac{n^3}{3} \text{ (for large } n\text{)}$$

Remarks

1. As shown, among the two key steps, namely forward elimination and back substitution, the former step comprises the bulk of the computational effort, particularly for large systems of equations.
2. In a similar fashion, we can compute the total number of additions and subtractions involved in the backward sweep as:

   $$\frac{2n^3+3n^2-5n}{6}$$

   It is usually not counted because, as stated, the multiplication/division takes relatively more time on a computer than the addition/subtraction.
3. For a system of 100 × 100 matrix, Gauss elimination involves total 3.3×10^5 operations, whereas Cramer rule involves 10^{158} operations. Obviously, the Cramer rule is computationally much more expensive than the Gauss elimination.

   However, still the Gauss elimination requires a reasonably large computational effort. To improve this situation, we will further examine the other methods in the subsequent sections.
4. For the reduction of the round-off error, the pivoting strategy is introduced in Gauss elimination method, which will be discussed in the next phase.

2.7 Gauss Elimination with Pivoting

What is a pivot element?

In the Gauss elimination, the original system having the form of $AX = B$ [Equation (2.1)] is transformed into $UX = C$ [Equation (2.31)], in which, each of the non-zero diagonal elements, namely $a_{11}^{(1)}, a_{22}^{(2)}, a_{33}^{(3)}, \ldots, a_{nn}^{(n)}$, is called *pivot* or *pivot element*.

Let us understand this through the following example.

Example 2.11 Pivot element (revisit Example 2.9)

The original system includes:

$$3x + y = 5$$

$$x + 3y + z = 10$$

$$y + z = 5$$

Mark the pivot element in every step during the transformation to the equivalent upper triangular system.

Solution

For solving this problem by the use of Gauss elimination with back substitution, let us first form the augmented matrix as:

$$\text{pivot} \rightarrow \begin{bmatrix} \textcircled{3} & 1 & 0 & 5 \\ 1 & 3 & 1 & 10 \\ 0 & 1 & 1 & 5 \end{bmatrix}$$

In the first step, the first row is used to eliminate the elements in the first column below the diagonal. Accordingly, the first row is referred as the *pivot row* and the element $a_{11}^{(1)}$ (here, 3) as the *pivot element*. Performing a row operation, $R_2 \rightarrow R_2 - \frac{1}{3}R_1$, we get:

$$\begin{bmatrix} 3 & 1 & 0 & 5 \\ 0 & \boxed{\frac{8}{3}} & 1 & \frac{25}{3} \\ 0 & 1 & 1 & 5 \end{bmatrix} \leftarrow \text{pivot (row 2)}$$

In the next step, the second row is used to eliminate the elements in the second column that lie below the diagonal. Accordingly, here the second row is the pivotal row and the element $a_{22}^{(2)}$ (here, $\frac{8}{3}$) is the pivot element. With $R_3 \rightarrow R_3 - \frac{3}{8}R_2$, we have:

$$\begin{bmatrix} 3 & 1 & 0 & 5 \\ 0 & \frac{8}{3} & 1 & \frac{25}{3} \\ 0 & 0 & \boxed{\frac{5}{8}} & \frac{15}{8} \end{bmatrix}$$

There is no more operation required to execute since the system $AX = B$ got transformed into $UX = C$. The third pivot element is $a_{33}^{(3)}$ (here, $\frac{5}{8}$).

Then using the back substitution algorithm, we got the solution in Example 2.9 previously.

What is pivoting?

We have developed the Gauss elimination algorithm previously with back substitution (see Sub-section 2.6.1) having total $(n-1)$ steps. When any one of the pivot elements, $a_{11}^{(1)}, a_{22}^{(2)}, a_{33}^{(3)}, \ldots, a_{nn}^{(n)}$, gets vanished or becomes very small compared to the rest of the elements in that row, then row interchanging (or swapping) is recommended to be performed so that one can obtain a non-vanishing pivot or can avoid the multiplication by a large number. This strategy is known as the *pivoting*.

With this, we perform the following elementary operations in Gauss elimination with pivoting on *aug A*:

Interchanges: The order of two rows can be changed: $R_i \leftrightarrow R_j$

Scalling: A row can be multiplied by a non-zero constant: $R_i \rightarrow aR_i$ $(a \neq 0)$

Replacement: A row can be replaced by summing up that row with a multiple of any other row: $R_i \rightarrow R_i + aR_j$

Let us start our discussion on pivoting strategy with a simple example below.

Example 2.12 No pivoting required

Solve the following set of equations:

$$12x_1 - 5x_2 + x_3 = 8$$
$$3x_1 + 15x_2 - 9x_3 = 9$$
$$x_1 - 4x_2 + 10x_3 = 7$$

using the Gauss elimination (back substitution) method.

Solution

Notice that this system is *diagonally dominant* and thus, pivoting is not required at all. Let us confirm this by moving a couple of more steps forward.

Performing row operations, we have with

first elimination stage:

$$12x_1 - 5x_2 + x_3 = 8$$
$$\frac{65}{4}x_2 - \frac{37}{4}x_3 = 7 \qquad R_2 \rightarrow R_2 - \frac{1}{4}R_1$$
$$-\frac{43}{12}x_2 + \frac{119}{12}x_3 = \frac{19}{3} \qquad R_3 \rightarrow R_3 - \frac{1}{12}R_1$$

second elimination stage:

$$12x_1 - 5x_2 + x_3 = 8$$
$$\frac{65}{4}x_2 - \frac{37}{4}x_3 = 7$$
$$\frac{6144}{780}x_3 = \frac{1536}{195} \qquad R_3 \rightarrow R_3 + \frac{43}{195}R_2$$

Then using back substitution, we easily get the solution as:

$x_3 = 1$

$x_2 = 1$

$x_1 = 1$

Obviously, in solving this problem, we did not face any pivoting related issue.

Remark
There is no need of any pivoting if the matrix A is:

- Diagonally dominant[7] (shown in the above example)
- Real, symmetric and positive definite[7] (left for your exercise)

Types of pivoting
There are three types of pivoting strategy:

- Partial pivoting
- Scaled partial pivoting
- Complete or full pivoting

They all are detailed in the following sections with examples.

2.7.1 Partial Pivoting

Let us consider $a_{11}^{(1)} = 0$, with which, one cannot use first row to eliminate the elements in first column below the diagonal. In such a situation, we cannot continue the Gauss elimination straightaway and the necessity of pivoting arises. For this, in the first stage of elimination, it is necessary to find the largest element (in magnitude) in the first column. To make it first pivot element, the corresponding row of this element is then interchanged with the first row.

In similar fashion, the second elimination stage starts with searching the largest element in the second column among the $(n-1)$ elements (i.e., excluding the first element, $a_{12}^{(1)}$) as the second pivot. Then, one needs to interchange the second row with the row having the second pivot element. Obviously, this procedure is true for the Gauss elimination with back substitution and it ends as soon as the original system $AX = B$ gets the form $UX = C$. This strategy is referred to as the *partial pivoting* or *maximal column pivoting*.

To generalize the partial pivoting algorithm, let us consider any column p. We first verify the magnitude of all elements in that column p that lie on or below the diagonal. Then identify row k, in which, the element has the largest magnitude (absolute value),

$$\begin{aligned} \left|a_{kp}^{(p)}\right| &= \max\left\{ \left|a_{pp}^{(p)}\right|, \left|a_{p+1\,p}^{(p)}\right|, \ldots, \left|a_{n-1\,p}^{(p)}\right|, \left|a_{np}^{(p)}\right| \right\} \\ &= \max\left|a_{ip}^{(p)}\right| \qquad p \le i \le n \end{aligned} \tag{2.39}$$

and then interchange rows p and k if $k > p$.

7 Detailed in Appendix A2.2.2.

Remark

The Gauss elimination method does not work when the diagonal or pivot element $a_{pp}^{(p)} = 0$. In such a situation, we talked about the application of partial pivoting governed by Equation (2.39). Alternatively, one can find the first row k (below row p), where $a_{kp}^{(p)} \neq 0$ (instead of maximum $\left|a_{kp}^{(p)}\right|$ considered in partial pivoting where $k > p$); then interchanging row p and row k leads to obtain a non-zero pivot element. This pivoting is called the *trivial pivoting*. However, this strategy involves larger round-off error over the partial pivoting.

Computer uses fixed-precision arithmetic and thus, there is a possibility of having some error in each arithmetic operation. To reduce the propagation of this error, it is recommended to use the partial pivoting rather than the trivial pivoting.

Example 2.13 Partial pivoting

Solve the following set of equations:

$$\begin{aligned} x_1 - x_2 + 5x_3 &= 5 \\ 3x_1 - 3x_2 + 7x_3 &= 7 \\ x_1 + 2x_2 + 4x_3 &= 10 \end{aligned} \tag{2.40}$$

using the Gauss elimination (back substitution) with partial pivoting.

Solution

The system equations can be represented in matrix form as:

$$\begin{bmatrix} 1 & -1 & 5 \\ 3 & -3 & 7 \\ 1 & 2 & 4 \end{bmatrix} \begin{bmatrix} x_1 \\ x_2 \\ x_3 \end{bmatrix} = \begin{bmatrix} 5 \\ 7 \\ 10 \end{bmatrix} \tag{2.41}$$

Applying Gauss elimination technique, we have with

first elimination stage:

$$\begin{bmatrix} 1 & -1 & 5 \\ 0 & 0 & -8 \\ 0 & 3 & -1 \end{bmatrix} \begin{bmatrix} x_1 \\ x_2 \\ x_3 \end{bmatrix} = \begin{bmatrix} 5 \\ -8 \\ 5 \end{bmatrix} \qquad \begin{aligned} R_2 &\to R_2 - 3R_1 \\ R_3 &\to R_3 - R_1 \end{aligned} \tag{2.42}$$

second elimination stage:

Since $a_{22}^{(2)} = 0$, we cannot move forward with the Gauss elimination method alone and thus, the partial pivoting strategy needs to be applied. For this, let us identify the largest element (in magnitude) in the second column that lies on or below the diagonal. It is obviously 3 that corresponds to the third row. Making interchange between the second and the third row,

$$\begin{bmatrix} 1 & -1 & 5 \\ 0 & 3 & -1 \\ 0 & 0 & -8 \end{bmatrix} \begin{bmatrix} x_1 \\ x_2 \\ x_3 \end{bmatrix} = \begin{bmatrix} 5 \\ 5 \\ -8 \end{bmatrix}$$

Then solving with back substitution,

$$x_1 = 2$$
$$x_2 = 2$$
$$x_3 = 1$$

Remark

Row swapping gets stopped at any stage when the pivot and all elements below that pivot in a particular column are zero. This type of situation arises if the set of equations does not have unique solution.

Example 2.14 Effect of round-off error

Consider a system truly represented by the following set of equations:

$$\begin{aligned} x_1 + x_2 + x_3 &= 1 \\ x_1 + x_2 + 2x_3 &= 2 \\ x_1 + 2x_2 + 3x_3 &= 2 \end{aligned} \tag{2.43}$$

Perform row operations with the target of eliminating x_1 from the last two equations and get:

$$\begin{aligned} x_1 + x_2 + x_3 &= 1 & & \\ x_3 &= 1 & & R_2 \rightarrow R_2 - R_1 \\ x_2 + 2x_3 &= 1 & & R_3 \rightarrow R_3 - R_1 \end{aligned}$$

Reordering the above three equations in an upper triangular form, we get the solution as:

$$x_3 = 1$$
$$x_2 = -1$$
$$x_1 = 1$$

This is the solution of the actual system.

Now let us consider a little different scenario. Because of round-off error (caused by the finite precision in the computer), suppose the system Equation (2.43) gets a slightly different form as:

$$\begin{aligned} x_1 + x_2 + x_3 &= 1 \\ x_1 + (1+\varepsilon)x_2 + 2x_3 &= 2 \\ x_1 + 2x_2 + 3x_3 &= 2 \end{aligned} \tag{2.44}$$

where, ε is a very small number and it is of the order of round-off error. Solve this problem by Gauss elimination (back substitution) with and without pivoting, and compare their results (with reference to the solution obtained above as: $x_1 = 1$, $x_2 = -1$, $x_3 = 1$).

Solution

Eliminating x_1 from the last two equations,

$$\begin{aligned} x_1 + x_2 + x_3 &= 1 \\ \varepsilon x_2 + x_3 &= 1 \qquad R_2 \rightarrow R_2 - R_1 \\ x_2 + 2x_3 &= 1 \qquad R_3 \rightarrow R_3 - R_1 \end{aligned} \tag{2.45}$$

Without Pivoting

Performing Gauss elimination, we obtain from Equation (2.45),

$$\begin{aligned} x_1 + x_2 + x_3 &= 1 \\ \varepsilon x_2 + x_3 &= 1 \\ \left(2 - \frac{1}{\varepsilon}\right) x_3 &= 1 - \frac{1}{\varepsilon} \qquad R_3 \rightarrow R_3 - \frac{1}{\varepsilon} R_2 \end{aligned} \tag{2.46}$$

Using back substitution,

$$\begin{aligned} x_3 &= \frac{1 - \frac{1}{\varepsilon}}{2 - \frac{1}{\varepsilon}} \\ x_2 &= \frac{1}{\varepsilon}(1 - x_3) \\ x_1 &= 1 - x_2 - x_3 \end{aligned} \tag{2.47}$$

Let us consider $\varepsilon = 0.0001$, which is, say, of the order of round-off error. With this, we would get the solution from Equation (2.47) by using a hypothetical computer with 5 digits of accuracy as:

$$x_3 = 1.0001$$
$$x_2 = -1.0000$$
$$x_1 = 0.9999$$

Notice that it is very close to the solution obtained before for the actual system.

At this moment, it is interesting to explore that if we have 4 digits of precision, then we would obtain a completely different set of solution (for the same $\varepsilon = 0.0001$) from Equation set (2.47) as:

$$x_3 = 1.000$$
$$x_2 = 0.0$$
$$x_1 = 0.0$$

Obviously, it is far away from the given solution ($x_1 = 1, x_2 = -1, x_3 = 1$).

With Pivoting

Applying partial pivoting, we obtain with swapping the last two equations of Equation set (2.45):

$$x_1 + x_2 + x_3 = 1$$
$$x_2 + 2x_3 = 1$$
$$\varepsilon x_2 + x_3 = 1$$

Eliminating x_2 from last equation,

$$\begin{aligned} &x_1 + x_2 + x_3 = 1 \\ &x_2 + 2x_3 = 1 \\ &(1-2\varepsilon)x_3 = 1-\varepsilon \qquad R_3 \rightarrow R_3 - \varepsilon R_2 \end{aligned} \tag{2.48}$$

Solving,

$$\begin{aligned} x_3 &= \frac{1-\varepsilon}{1-2\varepsilon} \\ x_2 &= 1 - 2x_3 \\ x_1 &= 1 - x_2 - x_3 \end{aligned} \tag{2.49}$$

Considering $\varepsilon = 0.0001$ and 4 digits of precision, we would get:

$$x_3 = 1.000$$
$$x_2 = -1.000$$
$$x_1 = 1.000$$

It is basically the right solution.

Remark

For a hypothetical computer with 4 digits of accuracy, we got correct answer when we applied the partial pivoting in Gauss elimination and absolutely wrong answer for the same method without pivoting. Although, today such type of restriction having 4 digits of accuracy in scientific computing seems to be absurd, but the issue concerning the effect of round-off error on Gauss elimination is properly illustrated.

Here we can conclude that the pivoting is the only strategy to reduce the round-off error in Gauss elimination. In other words, this pivoting strategy provides for greater numerical accuracy.

2.7.2 Scaled Partial Pivoting

The scaled pivoting would be an improved version of partial pivoting strategy by means of reducing the effect of round-off error. In this scheme, the pivot element is scaled largest element in magnitude in its row.

Let us construct the scale vector:

$$S = [S_1, \ S_2, \ ..., \ S_n] \tag{2.50}$$

with finding the largest element in each row of matrix A as:

$$S_i = \max_{1 \le j \le n} |a_{ij}| \qquad 1 \le i \le n \tag{2.51}$$

Recall that i stands for row and j for column.

For the first pivot element, then we form the scaled pivot vector as:

$$R = \left[\left| \frac{a_{11}^{(1)}}{S_1} \right|, \ \left| \frac{a_{21}^{(1)}}{S_2} \right|, \ .. \ , \ \left| \frac{a_{n1}^{(1)}}{S_n} \right| \right] \tag{2.52}$$

The largest element in this vector corresponds to the pivot row. It means if $\left|\frac{a_{i1}}{S_i}\right|$ is the largest scaled pivot element, then i th row should be the pivot row.

Repeat the same procedure to all pivot elements with vectors S and R. The application of this scaled partial pivoting is shown in the next Example 2.15.

2.7.3 Full or Complete Pivoting

In this pivoting strategy, we look for the largest element (in magnitude) in matrix A and select it as the first pivot. This requires an interchange of the equations as well as the position of the variables. This full pivoting (also called as the *maximum* or the *complete* pivoting) algorithm performs with:

$$\left|a_{kl}^{(p)}\right| = \max \left|a_{ij}^{(p)}\right| \qquad p \le i, j \le n \tag{2.53}$$

Then interchange rows p and k, and columns p and l.

In an example below, we will compare this full pivoting with the other two pivoting techniques, namely partial and scaled partial pivoting.

Remarks

1. Full pivoting scheme is appropriate to deal with ill-conditioned systems.[8]
2. Full pivoting is most likely to provide the better performance than the partial pivoting and the scaled partial pivoting.
3. Since the full pivoting is a time-consuming task and more importantly, it involves relatively large computational complexity, the partial pivoting is often preferred.

---

8 It is detailed in Appendix A2.2.3.

Example 2.15 Comparing three pivoting strategies

Solve the following problem:

$$\begin{aligned} x_1+3x_2+2x_3&=-1\\ 2x_1-x_2+100x_3&=53\\ 3x_1+x_2+200x_3&=102 \end{aligned} \tag{2.54}$$

using the partial, scaled partial, and full pivoting method with three digits of precision (actual solution: $x_1 = 1, x_2 = -1, x_3 = 0.5$).

Solution

For this linear system, we have:

$$aug\ A=\begin{bmatrix} 1 & 3 & 2 & -1\\ 2 & -1 & 100 & 53\\ 3 & 1 & 200 & 102 \end{bmatrix}$$

Partial Pivoting

In the first column of *aug A*, 3 is the largest element. Accordingly, we interchange the first and third rows $(R_1 \leftrightarrow R_3)$ as:

$$\begin{bmatrix} 3 & 1 & 200 & 102\\ 2 & -1 & 100 & 53\\ 1 & 3 & 2 & -1 \end{bmatrix}$$

Performing Gauss elimination,

$$\begin{bmatrix} 3 & 1 & 200 & 102\\ 0 & -1.67 & -33.3 & -15\\ 0 & 2.67 & -64.7 & -35 \end{bmatrix} \qquad \begin{aligned} &R_2 \to R_2-0.667R_1 \left(\text{i.e., } R_2 \to R_2-\frac{2}{3}R_1\right)\\ &R_3 \to R_3-0.333R_1 \left(\text{i.e., } R_3 \to R_3-\frac{1}{3}R_1\right) \end{aligned}$$

In the second column (on or below the diagonal), 2.67 is the largest element in magnitude. Thus, further interchanging second and third rows $(R_2 \leftrightarrow R_3)$,

$$\begin{bmatrix} 3 & 1 & 200 & 102\\ 0 & 2.67 & -64.7 & -35\\ 0 & -1.67 & -33.3 & -15 \end{bmatrix}$$

It gives,

$$\begin{bmatrix} 3 & 1 & 200 & 102\\ 0 & 2.67 & -64.7 & -35\\ 0 & 0 & -73.8 & -36.9 \end{bmatrix} \qquad R_3 \to R_3+0.625R_2 \left(\text{i.e., } R_3 \to R_3+\frac{1.67}{2.67}R_2\right)$$

Back substitution yields:

$$x_3 = \frac{36.9}{73.8} = 0.5$$

$$x_2 = -0.992$$

$$x_1 = 0.997$$

This set of obtained solution has a slight difference with the actual solution (given as: $x_1 = 1$, $x_2 = -1$, $x_3 = 0.5$).

Scaled Partial Pivoting

Let us find the largest element in each row of matrix A (not of *aug A*). Highlighting them, we rewrite the augmented matrix as:

$$aug\ A = \begin{bmatrix} 1 & 3 & 2 & -1 \\ 2 & -1 & 100 & 53 \\ 3 & 1 & 200 & 102 \end{bmatrix} \begin{matrix} \to 3 \\ \to 100 \\ \to 200 \end{matrix}$$

The scale vector is [Equation (2.50)]:

$$S = [S_1,\ S_2,\ S_3] = [3,\ 100,\ 200]$$

and the scaled pivot vector is [Equation (2.52)]:

$$R = \left[\left|\frac{a_{11}^{(1)}}{S_1}\right|,\ \left|\frac{a_{21}^{(1)}}{S_2}\right|,\ \left|\frac{a_{31}^{(1)}}{S_3}\right| \right] = \left[\frac{1}{3},\ \frac{2}{100},\ \frac{3}{200} \right]$$

Notice that the first element, $\frac{1}{3}$ in vector R is the largest one in magnitude and thus, the first row is the pivot row. Consequently, there is no need of any row swapping.

On applying the row operations, we have:

$$aug\ A = \begin{bmatrix} 1 & 3 & 2 & -1 \\ 0 & -7 & 96 & 55 \\ 0 & -8 & 194 & 105 \end{bmatrix} \begin{matrix} \to 3 \\ \to 100 \\ \to 200 \end{matrix} \qquad \begin{matrix} \\ R_2 \to R_2 - 2R_1 \\ R_3 \to R_3 - 3R_1 \end{matrix}$$

Now,

$$R = \left[\left|\frac{a_{22}^{(2)}}{S_2}\right|,\ \left|\frac{a_{32}^{(2)}}{S_3}\right| \right] = \left[\frac{7}{100},\ \frac{8}{200} \right]$$

Again there is no need of row interchanging since $\frac{7}{100}$ is the largest one. The above augmented matrix subsequently gets the following form:

$$aug\ A = \begin{bmatrix} 1 & 3 & 2 & -1 \\ 0 & -7 & 96 & 55 \\ 0 & 0 & 84.3 & 42.1 \end{bmatrix} \qquad R_3 \to R_3 - 1.14R_2 \ \left(\text{i.e., } R_3 \to R_3 - \frac{8}{7}R_2 \right)$$

Solving,

$x_3 = 0.499$
$x_2 = -1.01$
$x_1 = 1.03$

This set of obtained solution has some difference with the actual solution.

Full Pivoting

In the augmented matrix of the given problem, the first column is related with variable x_1, second column with x_2 and third column with x_3. Accordingly, we write:

$$aug\ A = \begin{array}{c} \begin{array}{cccc} x_1 & x_2 & x_3 & \end{array} \\ \begin{bmatrix} 1 & 3 & 2 & -1 \\ 2 & -1 & 100 & 53 \\ 3 & 1 & 200 & 102 \end{bmatrix} \end{array}$$

Notice that the largest element in matrix A (not in $aug\ A$) is 200. To make it first pivot element, one needs to interchange rows 1 and 3 ($R_1 \leftrightarrow R_3$), and columns 1 and 3 ($C_1 \leftrightarrow C_3$). With this,

$$\begin{array}{c} \begin{array}{cccc} x_3 & x_2 & x_1 & \end{array} \\ \begin{bmatrix} 200 & 1 & 3 & 102 \\ 100 & -1 & 2 & 53 \\ 2 & 3 & 1 & -1 \end{bmatrix} \end{array}$$

Performing row operations:

$$\begin{array}{c} \begin{array}{cccc} x_3 & x_2 & x_1 & \end{array} \\ \begin{bmatrix} 200 & 1 & 3 & 102 \\ 0 & -1.5 & 0.5 & 2 \\ 0 & 2.99 & 0.97 & -2.02 \end{bmatrix} \end{array} \qquad \begin{array}{l} R_2 \to R_2 - 0.5R_1 \quad \left(\text{i.e., } R_2 \to R_2 - \dfrac{1}{2}R_1\right) \\ R_3 \to R_3 - 0.01R_1 \quad \left(\text{i.e., } R_3 \to R_3 - \dfrac{1}{100}R_1\right) \end{array}$$

Now, 2.99 is the largest among the elements in second and third rows of matrix A. Thus, we further move to interchange the rows as:

$$\begin{array}{c} \begin{array}{cccc} x_3 & x_2 & x_1 & \end{array} \\ \begin{bmatrix} 200 & 1 & 3 & 102 \\ 0 & 2.99 & 0.97 & -2.02 \\ 0 & -1.5 & 0.5 & 2 \end{bmatrix} \end{array}$$

Applying $R_3 \to R_3 + 0.502R_2$ (i.e., $R_3 \to R_3 + \dfrac{1.5}{2.99}R_2$),

$$\begin{array}{ccc} x_3 & x_2 & x_1 \end{array}$$
$$\begin{bmatrix} 200 & 1 & 3 & 102 \\ 0 & 2.99 & 0.97 & -2.02 \\ 0 & 0 & 0.987 & 0.987 \end{bmatrix}$$

Solving,
$x_1 = 1$
$x_2 = -1$
$x_3 = 0.5$
Notice that it is same with the given actual solution (i.e., $x_1 = 1, x_2 = -1, x_3 = 0.5$).

Remark
In this example, the full pivoting provides the best performance followed by the partial pivoting and then the scaled partial pivoting.

2.7.4 Finding Determinant of Matrix *A* by Gauss Elimination

Finding determinant by co-factor expansion is not a practical approach, particularly for large systems. In this regard, Gauss elimination provides an additional benefit by finding the determinant of matrix A in a simple and straightforward way. Basically, we look for computing det A in an easy and simple way to know, for example, whether unique solution exists or not for a given system. Now question is: how to find det$A$ using Gauss elimination?

The determinant of an upper triangular matrix, U is simply the product of its diagonal elements (u_{ii}) as:

$$\det U = a_{11}^{(1)} \times a_{22}^{(2)} \times a_{33}^{(3)} \times \ \dots \ \times a_{n\,n}^{(n)} = \prod_{i=1}^{n} u_{i\,i} \tag{2.55}$$

With this, for the problem discussed in Example 2.9, one can get based on Equation (2.35):

$$\det U = 3 \times \frac{8}{3} \times \frac{5}{8} = 5$$

This formula is also valid for lower triangular and diagonal matrix.

Transforming $AX = B$ to $UX = C$, one can directly find the determinant of matrix A as:

$$\det A = (-1)^p \det U = (-1)^p \prod_{i=1}^{n} u_{i\,i} \tag{2.56}$$

Here p is the number of row and column interchanges involved during transformation. Note that the sign of determinant changes with every row (and column) interchange.

Remarks

1. Notice that if there is no row (and column) swapping involved, one can directly use Equation (2.55) for finding det A.
2. In finding det$A$, this method (involved n^3 operations) reduces the computational cost for large systems ($n > 5$) over the approach using a co-factor expansion (involved $n!$ operations).

Example 2.16 Validating Equation (2.56)

For the following system:

$$\begin{aligned} 3x - 3y + 5z &= 14 \\ 6x - 6y + 7z &= 25 \\ x + y + z &= 0 \end{aligned} \tag{2.57}$$

find det A based on Equation (2.56).

Solution

Here,

$$A = \begin{bmatrix} 3 & -3 & 5 \\ 6 & -6 & 7 \\ 1 & 1 & 1 \end{bmatrix}$$

and so, $|A| = 3\times(-13) + 3\times(-1) + 5\times 12 = 18$. Now, let us find $|A|$ based on Equation (2.56) and verify whether it is 18 or not!

For this, we convert $AX = B$ to $UX = C$ as:

$$\begin{bmatrix} 3 & -3 & 5 \\ 0 & 0 & -3 \\ 0 & 2 & -\frac{2}{3} \end{bmatrix} \begin{bmatrix} x \\ y \\ z \end{bmatrix} = \begin{bmatrix} 14 \\ -3 \\ -\frac{14}{3} \end{bmatrix} \qquad \begin{aligned} R_2 &\to R_2 - 2R_1 \\ R_3 &\to R_3 - \frac{1}{3}R_1 \end{aligned} \tag{2.58}$$

Interchanging second and third row (i.e., $p = 1$),

$$\begin{bmatrix} 3 & -3 & 5 \\ 0 & 2 & -\frac{2}{3} \\ 0 & 0 & -3 \end{bmatrix} \begin{bmatrix} x \\ y \\ z \end{bmatrix} = \begin{bmatrix} 14 \\ -\frac{14}{3} \\ -3 \end{bmatrix}$$

Apply back substitution and get:

$$\begin{aligned} z &= 1 \\ y &= -2 \\ x &= 1 \end{aligned}$$

Obviously,

$$U = \begin{bmatrix} 3 & -3 & 5 \\ 0 & 2 & -\frac{2}{3} \\ 0 & 0 & -3 \end{bmatrix}$$

From Equation (2.55),

$$\det U = \prod_{i=1}^{n} u_{i\,i} = -18$$

Using Equation (2.56), we get (with $p = 1$):

$$\det A = (-1)^p \det U = 18$$

Notice that the same value (i.e., 18) we got before by the approach using co-factor expansion. Hence, our validation is complete.

An alternative way: Through column interchanging

Instead of interchanging rows, one can also interchange columns in solving the above problem. Accordingly, Equation (2.58) gets the following form with swapping the second and third column (i.e., $p = 1$):

$$\begin{bmatrix} 3 & 5 & -3 \\ 0 & -3 & 0 \\ 0 & -\frac{2}{3} & 2 \end{bmatrix} \begin{bmatrix} x \\ z \\ y \end{bmatrix} = \begin{bmatrix} 14 \\ -3 \\ -\frac{14}{3} \end{bmatrix}$$

Take care in redefining the unknown vector X so that the equations remain unaltered. Performing Gauss elimination further,

$$\begin{bmatrix} 3 & 5 & -3 \\ 0 & -3 & 0 \\ 0 & 0 & 2 \end{bmatrix} \begin{bmatrix} x \\ z \\ y \end{bmatrix} = \begin{bmatrix} 14 \\ -3 \\ -4 \end{bmatrix} \qquad R_3 \to R_3 - \frac{2}{9} R_2$$

Solving through back substitution,

$$x = 1$$

$$y = -2$$

$$z = 1$$

The same solution we got through row exchanging before.

Notice that here,

$$U = \begin{bmatrix} 3 & 5 & -3 \\ 0 & -3 & 0 \\ 0 & 0 & 2 \end{bmatrix}$$

and so, $\det U = \prod_{i=1}^{n} u_{i\,i} = -18$. With this, we get:

$$\det A = (-1)^p \det U = 18$$

Here, p = number of column interchange (= 1).

2.8 LU Decomposition or Factorization Method

In the Gauss elimination method that involves elimination followed by substitution, it is seen that most of the computational effort is attributed to the elimination step. Thus, attention needs to be paid to improve this step. In this light, there is a scheme, in which, the coefficient matrix, A of the system $AX = B$ is decomposed or factorized into the product of a lower triangular matrix, L and an upper triangular matrix, U as:

$$A = LU \tag{2.59}$$

Thus, it is referred to as the *LU decomposition* or *factorization* method.[9] In the above equation,

$$L = \begin{bmatrix} l_{11} & 0 & 0 & \cdots & 0 \\ l_{21} & l_{22} & 0 & \cdots & 0 \\ l_{31} & l_{32} & l_{33} & \cdots & 0 \\ \cdot & \cdot & \cdot & & \cdot \\ \cdot & \cdot & \cdot & & \cdot \\ l_{n1} & l_{n2} & l_{n3} & \cdots & l_{nn} \end{bmatrix} \tag{2.60}$$

$$U = \begin{bmatrix} u_{11} & u_{12} & u_{13} & \cdots & u_{1n} \\ 0 & u_{22} & u_{23} & \cdots & u_{2n} \\ 0 & 0 & u_{33} & \cdots & u_{3n} \\ \cdot & \cdot & \cdot & & \cdot \\ \cdot & \cdot & \cdot & & \cdot \\ 0 & 0 & 0 & \cdots & u_{nn} \end{bmatrix} \tag{2.61}$$

Plugging Equation (2.59) into the system equation,

$$AX = B$$

we get,

$$LUX = B \tag{2.62}$$

One can write this as the following two systems of equations:

9 Also called the *triangularization* method.

$$LY = B \tag{2.63a}$$

$$UX = Y \tag{2.63b}$$

Finding L and U, at first solve Equation (2.63a) to find Y by forward substitution. Then solve Equation (2.63b) to find the solution X by using backward sweep. This is given in algorithmic form below.

Algorithm

There are three steps involved in obtaining the solution of the linear system $AX = B$ as:

Step 1: Construct the matrices L and U such that

$$A = LU \tag{2.59}$$

Step 2: Solve $LY = B$ [Equation (2.63a)] for Y using forward substitution.

Step 3: Solve $UX = Y$ [Equation (2.63b)] for X using back substitution.

Types of LU Decomposition

We need to compute the triangular matrices, L and U, such that:

$$A = LU = \begin{bmatrix} l_{11} & 0 & 0 & \dots & 0 \\ l_{21} & l_{22} & 0 & \dots & 0 \\ l_{31} & l_{32} & l_{33} & \dots & 0 \\ \cdot & \cdot & \cdot & & \cdot \\ \cdot & \cdot & \cdot & & \cdot \\ l_{n1} & l_{n2} & l_{n3} & \dots & l_{n\,n} \end{bmatrix} \begin{bmatrix} u_{11} & u_{12} & u_{13} & \dots & u_{1n} \\ 0 & u_{22} & u_{23} & \dots & u_{2n} \\ 0 & 0 & u_{33} & \dots & u_{3n} \\ \cdot & \cdot & \cdot & & \cdot \\ \cdot & \cdot & \cdot & & \cdot \\ 0 & 0 & 0 & \dots & u_{n\,n} \end{bmatrix}$$

$$= \begin{bmatrix} l_{11}u_{11} & l_{11}u_{12} & l_{11}u_{13} & \cdots & l_{11}u_{1n} \\ l_{21}u_{11} & l_{21}u_{12}+l_{22}u_{22} & l_{21}u_{13}+l_{22}u_{23} & \cdots & l_{21}u_{1n}+l_{22}u_{2n} \\ l_{31}u_{11} & l_{31}u_{12}+l_{32}u_{22} & l_{31}u_{13}+l_{32}u_{23}+l_{33}u_{33} & \cdots & l_{31}u_{1n}+l_{32}u_{2n}+l_{33}u_{3n} \\ \cdot & \cdot & \cdot & & \cdot \\ \cdot & \cdot & \cdot & & \cdot \\ l_{n1}u_{11} & l_{n1}u_{12}+l_{n2}u_{22} & l_{n1}u_{13}+l_{n2}u_{23}+l_{n3}u_{33} & \cdots & l_{n1}u_{1n}+l_{n2}u_{2n}+l_{n3}u_{3n}+\dots+l_{n\,n}u_{nn} \end{bmatrix} \tag{2.64}$$

Comparing Equations (2.3) and (2.64),

$$l_{i1}u_{1j}+l_{i2}u_{2j}+l_{i3}u_{3j}+\dots+l_{i\,n}u_{n\,j}=a_{i\,j} \qquad 1\le i,j\le n \tag{2.65}$$

where,

$$l_{i\,j}=0 \qquad j>i \tag{2.66a}$$

$$u_{i\,j}=0 \qquad i>j \tag{2.66b}$$

Notice that there are total n^2 equations[10] in the form of Equation (2.65). Let us then count the total number of unknown variables as $(n^2 + n)$:

total elements in L matrix [Equation (2.60)]: $1+2+3+....+(n-1)+n=\frac{n(n+1)}{2}$

total elements in U matrix [Equation (2.61)]: $n+(n-1)+(n-2)+...+2+1=\frac{n(n+1)}{2}$

To have a unique solution, obviously one needs to predefine n variables. For this, there are two techniques we have:

i) Crout method[11]
It considers $u_{ii}=1$ for $1\leq i\leq n$.

ii) Doolittle method[12]
It makes $l_{ii}=1$ for $1\leq i\leq n$.

Both these methods are detailed below with their applications.

2.8.1 Crout Method

To discuss this scheme, let us start with:

$$l_{i1}u_{1j}+l_{i2}u_{2j}+l_{i3}u_{3j}+...+l_{in}u_{nj}=a_{ij} \qquad 1\leq i,j\leq n \tag{2.65}$$

In which,

$$l_{ij}=0 \qquad j>i \tag{2.66a}$$

$$u_{ij}=0 \qquad i>j \tag{2.66b}$$

As mentioned, the Crout method additionally considers:

$$u_{ii}=1 \qquad 1\leq i\leq n \tag{2.67}$$

All the elements of the coefficient matrix A are known. With this, the Crout method involves the following computations to solve a given system.

- *Compute 1st column of matrix L and 1st row of matrix U* (here, $u_{11} = 1$)

Based on Equation (2.64) with $u_{11} = 1$, the first column ($j = 1$) of matrix L is same with that of matrix A. Accordingly,

$$l_{i1}=a_{i1} \qquad 1\leq i\leq n \tag{2.68a}$$

Having this, we proceed to compute the first row of matrix U as:

$$u_{11}=1$$

$$u_{1j}=\frac{a_{1j}}{l_{11}} \qquad 2\leq j\leq n \tag{2.68b}$$

10 Each row has n equations and thus total n^2 equations for n rows.

11 Named after the American mathematician Prescott Durand Crout (1907–1984).

12 Named after the American mathematician Myrick Hascall Doolittle (1830–1911).

- *Compute 2nd column of matrix L and 2nd row of matrix U* (here, $u_{22} = 1$)

Similarly, we have for

Second column of L ($j = 2$):

$$l_{i2} = a_{i2} - l_{i1}u_{12} \qquad 2 \le i \le n \text{ (since } l_{12} = 0) \tag{2.68c}$$

Second row of U ($i = 2$):

$$u_{22} = 1$$

$$u_{2j} = \frac{a_{2j} - l_{21}u_{1j}}{l_{22}} \qquad 3 \le j \le n \text{ (since } u_{21} = 0 \text{ and } u_{22} = 1) \tag{2.68d}$$

- *Compute kth column of matrix L and kth row of matrix U* (here, $u_{kk} = 1$)

kth column of L ($j = k$):

$$l_{ik} = a_{ik} - \sum_{j=1}^{k-1} l_{ij}u_{jk} \qquad k \le i \le n \tag{2.68e}$$

kth row of U ($i = k$):

$$u_{kk} = 1$$

$$u_{kj} = \frac{a_{kj} - \sum_{m=1}^{k-1} l_{km}u_{mj}}{l_{kk}} \qquad k+1 \le j \le n \tag{2.68f}$$

This way, we can find all the elements of triangular matrices L and U in Equation (2.64). Next, we would like to solve a problem by using this methodology chalked out above for the Crout method.

Example 2.17 Crout method

Solve the following linear system:

$$\begin{aligned} 3x_1 - 2x_2 + x_3 &= 2 \\ x_1 + 3x_2 - x_3 &= 5 \\ 2x_1 + x_2 - 2x_3 &= -5 \end{aligned} \tag{2.69}$$

using the Crout method.

Solution

Factorizing the coefficient matrix A into the product of L and U, we can write it in the form of Equation (2.59) with using the Crout method as:

$$\begin{bmatrix} 3 & -2 & 1 \\ 1 & 3 & -1 \\ 2 & 1 & -2 \end{bmatrix} = \begin{bmatrix} l_{11} & 0 & 0 \\ l_{21} & l_{22} & 0 \\ l_{31} & l_{32} & l_{33} \end{bmatrix} . \begin{bmatrix} 1 & u_{12} & u_{13} \\ 0 & 1 & u_{23} \\ 0 & 0 & 1 \end{bmatrix}$$

$$= \begin{bmatrix} l_{11} & l_{11}u_{12} & l_{11}u_{13} \\ l_{21} & l_{21}u_{12}+l_{22} & l_{21}u_{13}+l_{22}u_{23} \\ l_{31} & l_{31}u_{12}+l_{32} & l_{31}u_{13}+l_{32}u_{23}+l_{33} \end{bmatrix} \tag{2.70}$$

Equating the terms on both sides,

$$l_{11}=3 \qquad l_{11}u_{12}=-2 \qquad l_{11}u_{13}=1$$
$$l_{21}=1 \qquad l_{21}u_{12}+l_{22}=3 \qquad l_{21}u_{13}+l_{22}u_{23}=-1$$
$$l_{31}=2 \qquad l_{31}u_{12}+l_{32}=1 \qquad l_{31}u_{13}+l_{32}u_{23}+l_{33}=-2$$

Obviously, there are 9 (= n^2) equations available for 12 (= $n^2 + n$) unknown variables.[13] For finding them, let us perform the following computations as discussed before.

- First column ($j = 1$)

Based on Equation (2.68a):

$$l_{11}=3 \qquad l_{21}=1 \qquad l_{31}=2$$

- First row ($i = 1$)

Using Equation (2.68b):

$$u_{11}=1 \qquad u_{12}=\frac{a_{12}}{l_{11}}=-\frac{2}{3} \qquad u_{13}=\frac{a_{13}}{l_{11}}=\frac{1}{3}$$

- Second column ($j = 2$)

Using Equation (2.68c):

$$l_{12}=0 \qquad l_{22}=a_{22}-l_{21}u_{12} = 3+\frac{2}{3}=\frac{11}{3} \qquad l_{32}=a_{32}-l_{31}u_{12} = 1+\frac{4}{3}=\frac{7}{3}$$

13 Actually 9 variables are unknown [see Equation (2.70)] because of considering $u_{ii} = 1$.

- Second row ($i = 2$)

Using Equation (2.68d):

$$u_{21} = 0 \qquad u_{22} = 1 \qquad u_{23} = \frac{a_{23} - l_{21}u_{13}}{l_{22}}$$

$$= \frac{-1 - \frac{1}{3}}{\frac{11}{3}} = -\frac{4}{11}$$

- Third column ($j = 3$)

Using Equation (2.68e):

$$l_{13} = 0 \qquad l_{23} = 0 \qquad l_{33} = a_{33} - l_{31}u_{13} - l_{32}u_{23}$$

$$= -2 - \frac{2}{3} + \frac{28}{33} = -\frac{20}{11}$$

- Third row ($i = 3$)

$$u_{31} = 0 \qquad u_{32} = 0 \qquad u_{33} = 1$$

It gives:

$$L = \begin{bmatrix} 3 & 0 & 0 \\ 1 & \frac{11}{3} & 0 \\ 2 & \frac{7}{3} & -\frac{20}{11} \end{bmatrix} \qquad U = \begin{bmatrix} 1 & -\frac{2}{3} & \frac{1}{3} \\ 0 & 1 & -\frac{4}{11} \\ 0 & 0 & 1 \end{bmatrix}$$

In the next step, we use Equation (2.63a) (i.e., $LY = B$) as:

$$\begin{bmatrix} 3 & 0 & 0 \\ 1 & \frac{11}{3} & 0 \\ 2 & \frac{7}{3} & -\frac{20}{11} \end{bmatrix} \begin{bmatrix} y_1 \\ y_2 \\ y_3 \end{bmatrix} = \begin{bmatrix} 2 \\ 5 \\ -5 \end{bmatrix}$$

Using forward substitution,

$$3y_1 = 2 \qquad \therefore y_1 = \frac{2}{3}$$

$$y_1 + \frac{11}{3}y_2 = 5 \qquad \therefore y_2 = \frac{13}{11}$$

$$2y_1 + \frac{7}{3}y_2 - \frac{20}{11}y_3 = -5 \quad \therefore y_3 = 5$$

In the final step, we use Equation (2.63b) (i.e., $UX = Y$) as:

$$\begin{bmatrix} 1 & -\frac{2}{3} & \frac{1}{3} \\ 0 & 1 & -\frac{4}{11} \\ 0 & 0 & 1 \end{bmatrix} \begin{bmatrix} x_1 \\ x_2 \\ x_3 \end{bmatrix} = \begin{bmatrix} \frac{2}{3} \\ \frac{13}{11} \\ 5 \end{bmatrix}$$

Using backward sweep, we get the final solution as:

$$x_3 = 5$$
$$x_2 = 3$$
$$x_1 = 1$$

2.8.2 Doolittle Method

Like the Crout method, we start the Doolittle method with:

$$l_{i1}u_{1j} + l_{i2}u_{2j} + l_{i3}u_{3j} + \ldots + l_{i\,n}u_{n\,j} = a_{i\,j} \qquad 1 \le i, j \le n \tag{2.65}$$

In which,

$$l_{i\,j} = 0 \qquad j > i \tag{2.66a}$$

$$u_{i\,j} = 0 \qquad i > j \tag{2.66b}$$

Additionally,

$$l_{i\,i} = 1 \qquad 1 \le i \le n \tag{2.71}$$

Having all the elements of matrix A known, the Doolittle method involves the following computations to solve a given system $AX = B$.

- *Compute 1st row of matrix U and 1st column of matrix L* (here, $l_{11} = 1$)

Based on Equation (2.64) with $l_{11} = 1$, the first row ($i = 1$) of matrix U is identical with that of A. Accordingly,

$$u_{1j} = a_{1j} \qquad 1 \le j \le n \tag{2.72a}$$

Having this, we proceed to compute the elements of first column of matrix L as:

$$l_{11} = 1$$

$$l_{i1} = \frac{a_{i1}}{u_{11}} \qquad 2 \le i \le n \tag{2.72b}$$

- *Compute 2nd row of matrix U and 2nd column of matrix L* (here, $l_{22} = 1$)

Second row of U ($i = 2$):

$$u_{2j} = a_{2j} - l_{21}u_{1j} \qquad 2 \le j \le n \tag{2.72c}$$

Second column of L ($j = 2$):

$$l_{22} = 1$$

$$l_{i2} = \frac{a_{i2} - l_{i1}u_{12}}{u_{22}} \qquad 3 \le i \le n \tag{2.72d}$$

- *Compute kth row of matrix U and kth column of matrix L* (here, $l_{kk} = 1$)

kth row of U ($i = k$):

$$u_{kj} = a_{kj} - \sum_{m=1}^{k-1} l_{km}u_{mj} \qquad k \le j \le n \tag{2.72e}$$

kth column of L ($j = k$):

$$l_{kk} = 1$$

$$l_{ik} = \frac{a_{ik} - \sum_{m=1}^{k-1} l_{im}u_{mk}}{u_{kk}} \qquad k+1 \le i \le n \tag{2.72f}$$

By this way, one can find all the elements of matrices L and U in Equation (2.64).

This Doolittle method is illustrated in the following example.

Example 2.18 Doolittle method

Repeat Example 2.17 that includes the following system equations:

$$3x_1 - 2x_2 + x_3 = 2$$
$$x_1 + 3x_2 - x_3 = 5$$
$$2x_1 + x_2 - 2x_3 = -5$$

Solve it by using the Doolittle method.

Solution

Decomposing the coefficient matrix A into the product of L and U, we get from Equation (2.59) with the use of Doolittle method as:

$$\begin{bmatrix} 3 & -2 & 1 \\ 1 & 3 & -1 \\ 2 & 1 & -2 \end{bmatrix} = \begin{bmatrix} 1 & 0 & 0 \\ l_{21} & 1 & 0 \\ l_{31} & l_{32} & 1 \end{bmatrix} \begin{bmatrix} u_{11} & u_{12} & u_{13} \\ 0 & u_{22} & u_{23} \\ 0 & 0 & u_{33} \end{bmatrix}$$
$$= \begin{bmatrix} u_{11} & u_{12} & u_{13} \\ l_{21}u_{11} & l_{21}u_{12} + u_{22} & l_{21}u_{13} + u_{23} \\ l_{31}u_{11} & l_{31}u_{12} + l_{32}u_{22} & l_{31}u_{13} + l_{32}u_{23} + u_{33} \end{bmatrix} \tag{2.73}$$

Equating the terms on both sides,

$$u_{11} = 3 \qquad u_{12} = -2 \qquad u_{13} = 1$$

$$l_{21}u_{11} = 1 \qquad l_{21}u_{12} + u_{22} = 3 \qquad l_{21}u_{13} + u_{23} = -1$$

$$l_{31}u_{11} = 2 \qquad l_{31}u_{12} + l_{32}u_{22} = 1 \qquad l_{31}u_{13} + l_{32}u_{23} + u_{33} = -2$$

Let us perform the following computations as discussed before.

- First row (i = 1)

Using Equation (2.72a):

$$u_{11} = 3 \qquad u_{12} = -2 \qquad u_{13} = 1$$

- First column (j = 1)

Based on Equation (2.72b):

$$l_{11} = 1 \qquad l_{21} = \frac{a_{21}}{u_{11}} = \frac{1}{3} \qquad l_{31} = \frac{a_{31}}{u_{11}} = \frac{2}{3}$$

- Second row (i = 2)

$$u_{21} = 0 \qquad u_{22} = a_{22} - l_{21}u_{12} = \frac{11}{3} \qquad u_{23} = a_{23} - l_{21}u_{13} = -\frac{4}{3}$$

- Second column (j = 2)

$$l_{12} = 0 \qquad l_{22} = 1 \qquad l_{32} = \frac{a_{32} - l_{31}u_{12}}{u_{22}} = \frac{7}{11}$$

- Third row (i = 3)

$$u_{31} = 0 \qquad u_{32} = 0 \qquad u_{33} = a_{33} - l_{31}u_{13} - l_{32}u_{23} = -\frac{20}{11}$$

- Third column (j = 3)

$$l_{13} = 0 \qquad l_{23} = 0 \qquad l_{33} = 1$$

It gives:

$$L = \begin{bmatrix} 1 & 0 & 0 \\ \frac{1}{3} & 1 & 0 \\ \frac{2}{3} & \frac{7}{11} & 1 \end{bmatrix} \qquad U = \begin{bmatrix} 3 & -2 & 1 \\ 0 & \frac{11}{3} & -\frac{4}{3} \\ 0 & 0 & -\frac{20}{11} \end{bmatrix}$$

In the next step, we use Equation (2.63a) (i.e., $LY = B$) as:

$$\begin{bmatrix} 1 & 0 & 0 \\ \frac{1}{3} & 1 & 0 \\ \frac{2}{3} & \frac{7}{11} & 1 \end{bmatrix} \begin{bmatrix} y_1 \\ y_2 \\ y_3 \end{bmatrix} = \begin{bmatrix} 2 \\ 5 \\ -5 \end{bmatrix}$$

Using the forward substitution,

$$y_1 = 2$$

$$\frac{1}{3}y_1 + y_2 = 5 \qquad \therefore y_2 = \frac{13}{3}$$

$$\frac{2}{3}y_1 + \frac{7}{11}y_2 + y_3 = -5 \qquad \therefore y_3 = -\frac{100}{11}$$

In the final step, we use Equation (2.63b) (i.e., $UX = Y$) as:

$$\begin{bmatrix} 3 & -2 & 1 \\ 0 & \frac{11}{3} & -\frac{4}{3} \\ 0 & 0 & -\frac{20}{11} \end{bmatrix} \begin{bmatrix} x_1 \\ x_2 \\ x_3 \end{bmatrix} = \begin{bmatrix} 2 \\ \frac{13}{3} \\ -\frac{100}{11} \end{bmatrix}$$

Using the backward sweep, we get the final solution as:

$$x_3 = 5$$

$$x_2 = 3$$

$$x_1 = 1$$

Remarks on LU Decomposition

1. The Gauss elimination with backward sweep involves $\sim n\,(n+m)^2$ operations to work on the $n\times(n+m)$ augmented matrix; whereas it is $\sim n^2(n+m)$ for the LU decomposition + (fore and aft) sweep. The number of operations in the LU decomposition is thus almost same with that in the Gauss elimination with backward sweep. For more details about it, you can consult the book by Mathews (2000).
2. In the factorization/decomposition phase, there is no operation involved in the right hand side (i.e., in the B vector) of the given system. Thus, the LU decomposition method may require a little less effort compared to the Gauss elimination that performs elimination of the matrix A, along with manipulations of the right hand side vector B. But at the same time it is worth noting that the LU decomposition requires both the forward and backward substitution, whereas the Gauss elimination involves any one (either backward or forward substitution) among them.

3. As seen in the last two solved examples, the LU decomposition does not require any pivoting strategy. This approach works fine as long as the pivot elements are not small or zero. Otherwise, the pivoting strategy needs to be employed with LU decomposition. For example, if $l_{22} = 0$ (or $u_{22} = 0$), we should do row interchanging (pivoting) to get the solution.

2.9 Cholesky Method

For the general form of coefficient matrix A, the solution methodology is discussed in the preceeding section with the Crout and Doolittle method. If this A matrix is symmetric positive definite,[14] there is an easier approach available, and that is called the *Cholesky*[15] or the *square root* method.

In this Cholesky method, matrix A is decomposed as:

$$A = LL^{\mathrm{T}} \tag{2.74}$$

As mentioned, L denotes the lower triangular matrix and L^T is its transpose. With this, the system ($AX = B$) gets the following form:

$$LL^{\mathrm{T}}X = B \tag{2.75}$$

which can be split into:

$$L^{\mathrm{T}}X = Y \tag{2.76a}$$

$$LY = B \tag{2.76b}$$

First compute Y from Equation (2.76b) by the forward substitution and then find X from Equation (2.76a) by the back substitution.

Note that for the lower triangular matrix L, we obtain its elements as:

$$l_{ii} = \left(a_{ii} - \sum_{j=1}^{i-1} l_{ij}^2 \right)^{\frac{1}{2}} \qquad i = 1,2,\ldots,n \tag{2.77a}$$

$$l_{ij} = \frac{1}{l_{jj}} \left(a_{ij} - \sum_{k=1}^{j-1} l_{jk} l_{ik} \right) \qquad i = j+1,\ j+2,\ldots,n \ \&\ j = 1,2,\ldots,n \tag{2.77b}$$

$$l_{ij} = 0 \qquad i < j \tag{2.77c}$$

Remarks

1. The Cholesky decomposition can also be made for matrix A as:

$$A = UU^{\mathrm{T}} \tag{2.78}$$

---

14 Detailed in Appendix A2.2.2.

15 Named after the French military officer and mathematician Andre-Louis Cholesky (1875–1918).

in which, U is the upper triangular matrix and U^T is its transpose. For the ith row:

$$u_{ii} = \left(a_{ii} - \sum_{k=1}^{i-1} u_{ki}^2 \right)^{\frac{1}{2}}$$

$$u_{ij} = \frac{1}{u_{ii}} \left(a_{ij} - \sum_{k=1}^{i-1} u_{ki} u_{kj} \right) \qquad j = i+1,\ i+2, \ldots, n$$

2. This method involves relatively low operational counts ($\approx n^3/6$ for large n). But remember, the Cholesky method only deals with systems having positive definite matrix.

Example 2.19 Cholesky method

Solve the following system of linear equations:

$$2x_1 - x_2 = 3$$
$$-x_1 + 2x_2 - x_3 = -3$$
$$-x_2 + 2x_3 = 5$$

using the Cholesky method.

Solution

It is shown in Example A2.2 (Appendix A2.2.2) that the matrix

$$A = \begin{bmatrix} 2 & -1 & 0 \\ -1 & 2 & -1 \\ 0 & -1 & 2 \end{bmatrix}$$

is a positive definite matrix.

Using the Cholesky decomposition method with

$$A = LL^T$$

we have,

$$\begin{bmatrix} 2 & -1 & 0 \\ -1 & 2 & -1 \\ 0 & -1 & 2 \end{bmatrix} = \begin{bmatrix} l_{11} & 0 & 0 \\ l_{21} & l_{22} & 0 \\ l_{31} & l_{32} & l_{33} \end{bmatrix} \begin{bmatrix} l_{11} & l_{21} & l_{31} \\ 0 & l_{22} & l_{32} \\ 0 & 0 & l_{33} \end{bmatrix}$$

Comparing both sides:

$$l_{11}^2 = 2 \qquad \therefore l_{11} = \sqrt{2}$$

$$l_{11}l_{21} = -1 \qquad \therefore l_{21} = -\frac{1}{\sqrt{2}}$$

$$l_{11}l_{31} = 0 \qquad \therefore l_{31} = 0$$

$$l_{21}^2 + l_{22}^2 = 2 \qquad \therefore l_{22} = \sqrt{1.5}$$

$$l_{21}l_{31} + l_{22}l_{32} = -1 \qquad \therefore l_{32} = -\frac{1}{\sqrt{1.5}}$$

$$l_{31}^2 + l_{32}^2 + l_{33}^2 = 2 \qquad \therefore l_{33} = \frac{2}{\sqrt{3}}$$

Alternatively, one can find all the above elements of L matrix using Equation (2.77). Anyway, it gives:

$$L = \begin{bmatrix} \sqrt{2} & 0 & 0 \\ -\frac{1}{\sqrt{2}} & \sqrt{1.5} & 0 \\ 0 & -\frac{1}{\sqrt{1.5}} & \frac{2}{\sqrt{3}} \end{bmatrix}$$

In the next step, we use Equation (2.76b) (i.e., $LY = B$):

$$\begin{bmatrix} \sqrt{2} & 0 & 0 \\ -\frac{1}{\sqrt{2}} & \sqrt{1.5} & 0 \\ 0 & -\frac{1}{\sqrt{1.5}} & \frac{2}{\sqrt{3}} \end{bmatrix} \begin{bmatrix} y_1 \\ y_2 \\ y_3 \end{bmatrix} = \begin{bmatrix} 3 \\ -3 \\ 5 \end{bmatrix}$$

Using forward substitution,

$$\sqrt{2}\,y_1 = 3 \qquad \therefore y_1 = \frac{3}{\sqrt{2}}$$

$$-\frac{1}{\sqrt{2}}y_1 + \sqrt{1.5}\,y_2 = -3 \qquad \therefore y_2 = -\frac{3}{2\sqrt{1.5}}$$

$$-\frac{1}{\sqrt{1.5}}y_2 + \frac{2}{\sqrt{3}}y_3 = 5 \qquad \therefore y_3 = 2\sqrt{3}$$

In the final step, we use Equation (2.76a) (i.e., $L^{T}X = Y$) as:

$$\begin{bmatrix} \sqrt{2} & -\frac{1}{\sqrt{2}} & 0 \\ 0 & \sqrt{1.5} & -\frac{1}{\sqrt{1.5}} \\ 0 & 0 & \frac{2}{\sqrt{3}} \end{bmatrix} \begin{bmatrix} x_1 \\ x_2 \\ x_3 \end{bmatrix} = \begin{bmatrix} \frac{3}{\sqrt{2}} \\ -\frac{3}{2\sqrt{1.5}} \\ 2\sqrt{3} \end{bmatrix}$$

Using back substitution, we get the final solution as:

$$\frac{2}{\sqrt{3}}x_3 = 2\sqrt{3} \qquad \therefore x_3 = 3$$

$$\sqrt{1.5}x_2 - \frac{1}{\sqrt{1.5}}x_3 = -\frac{3}{2\sqrt{1.5}} \qquad \therefore x_2 = 1$$

$$\sqrt{2}x_1 - \frac{1}{\sqrt{2}}x_2 = \frac{3}{\sqrt{2}} \qquad \therefore x_1 = 2$$

2.10 Thomas Algorithm or Tridiagonal Matrix Algorithm

There are many systems, for which, the coefficient matrix A contains non-zero elements only at the diagonal, upper diagonal and lower diagonal. For solving this type of tridiagonal systems of linear equations, the *Thomas algorithm*[16] is quite popular. It is a simplified form of the Gauss elimination and this approach requires relatively less computational effort since most of the matrix elements are already zero.

Let us consider a tridiagonal system of n unknowns:

$$a_i x_{i-1} + b_i x_i + c_i x_{i+1} = d_i \qquad i = 1, 2, \ldots, n \tag{2.79}$$

in which, $a_1 = c_n = 0$. In matrix form:

$$AX = \begin{bmatrix} b_1 & c_1 & 0 & 0 & \ldots & 0 & 0 & 0 \\ a_2 & b_2 & c_2 & 0 & \ldots & 0 & 0 & 0 \\ 0 & a_3 & b_3 & c_3 & \ldots & 0 & 0 & 0 \\ . & . & . & . & . & . & . & . \\ . & . & . & . & . & . & . & . \\ 0 & 0 & 0 & 0 & \ldots & a_{n-1} & b_{n-1} & c_{n-1} \\ 0 & 0 & 0 & 0 & \ldots & 0 & a_n & b_n \end{bmatrix} \begin{bmatrix} x_1 \\ x_2 \\ x_3 \\ . \\ . \\ x_{n-1} \\ x_n \end{bmatrix} = \begin{bmatrix} d_1 \\ d_2 \\ d_3 \\ . \\ . \\ d_{n-1} \\ d_n \end{bmatrix} \tag{2.80}$$

16 Named after the British physicist and applied mathematician Llewellyn Hilleth Thomas (1903–1992).

Here, the tridiagonal matrix A is called the *Jacobi matrix*.

This system [Equation (2.80)] can be transformed into an upper triangular form with all diagonal elements equal to unity. Having this, let us define:

$$x_i = P_i x_{i+1} + Q_i \tag{2.81a}$$

$$x_{i-1} = P_{i-1} x_i + Q_{i-1} \tag{2.81b}$$

Inserting Equation (2.81b) into (2.79),

$$a_i\left(P_{i-1}x_i + Q_{i-1}\right) + b_i x_i + c_i x_{i+1} = d_i$$

which gives:

$$x_i = -\frac{c_i}{b_i + a_i P_{i-1}} x_{i+1} + \frac{d_i - a_i Q_{i-1}}{b_i + a_i P_{i-1}} \tag{2.82}$$

Comparing Equations (2.81a) and (2.82),

$$P_i = -\frac{c_i}{b_i + a_i P_{i-1}} \tag{2.83a}$$

$$i = 1, 2, 3, \ldots$$

$$Q_i = \frac{d_i - a_i Q_{i-1}}{b_i + a_i P_{i-1}} \tag{2.83b}$$

To find the constants P and Q, these two *recurrence relations* can be used. But for this, one needs initial values, P_0 and Q_0. Considering $i = 1$, we get from Equation (2.79):

$$a_1 x_0 + b_1 x_1 + c_1 x_2 = d_1$$

Since $a_1 = 0$,

$$x_1 = -\frac{c_1}{b_1} x_2 + \frac{d_1}{b_1} \tag{2.84}$$

Comparing this Equation (2.84) with (2.82) for $i = 1$:

$$P_0 = Q_0 = 0 \tag{2.85}$$

Now we can determine all the constants P and Q from Equation (2.83).

As stated, Equation (2.81a) represents an upper triangular system. With this, we can get the solution x_i using backward sweep. At the starting (i.e., $i = n$), Equation (2.81a) yields:

$$x_n = P_n x_{n+1} + Q_n$$

Since $P_n = 0$ [from Equation (2.83a) with $C_n = 0$], we get:

$$x_n = Q_n \tag{2.86}$$

Likewise, compute $x_{n-1}, x_{n-2}, \ldots, x_1$.

Remarks

1. Tridiagonal system of linear equations is quite common in staged operations[17] in process engineering. This type of equations is also encountered when the finite difference technique is applied on the boundary value problems (discussed in Chapter 5 of this book).
2. Number of operations involved in Thomas algorithm is in the order of n (actually $5n - 8$), whereas it is of the order of n^3 for the Gauss elimination with backward sweep. It reveals that this Thomas algorithm leads to reduce the computational load and thus, the round-off error significantly.

Example 2.20 Thomas algorithm

Solve the following system equations:

$$x_1 + x_2 = 2$$

$$3x_1 - 2x_2 + x_3 = 2$$

$$2x_2 + 3x_3 - x_4 = 4$$

$$x_3 + 3x_4 = 4$$

using the Thomas algorithm.

Solution

System equations can be represented in the following matrix form:

$$\begin{bmatrix} 1 & 1 & 0 & 0 \\ 3 & -2 & 1 & 0 \\ 0 & 2 & 3 & -1 \\ 0 & 0 & 1 & 3 \end{bmatrix} \begin{bmatrix} x_1 \\ x_2 \\ x_3 \\ x_4 \end{bmatrix} = \begin{bmatrix} 2 \\ 2 \\ 4 \\ 4 \end{bmatrix}$$

Comparing with Equation (2.80):

$$\begin{array}{llll} a_1 = 0 & b_1 = 1 & c_1 = 1 & d_1 = 2 \\ a_2 = 3 & b_2 = -2 & c_2 = 1 & d_2 = 2 \\ a_3 = 2 & b_3 - 3 & c_3 = 1 & d_3 - 4 \\ a_4 = 1 & b_4 = 3 & c_4 = 0 & d_4 = 4 \end{array}$$

---

17 Involved in, for example, absorption and stripping column.

Using Equation (2.83), we get P and Q (with $P_0 = Q_0 = 0$) as:

$$P_1 = -\frac{c_1}{b_1 + a_1 P_0} = -1 \qquad Q_1 = \frac{d_1 - a_1 Q_0}{b_1 + a_1 P_0} = 2$$

$$P_2 = -\frac{c_2}{b_2 + a_2 P_1} = \frac{1}{5} \qquad Q_2 = \frac{d_2 - a_2 Q_1}{b_2 + a_2 P_1} = \frac{4}{5}$$

$$P_3 = -\frac{c_3}{b_3 + a_3 P_2} = \frac{5}{17} \qquad Q_3 = \frac{d_3 - a_3 Q_2}{b_3 + a_3 P_2} = \frac{12}{17}$$

$$P_4 = -\frac{c_4}{b_4 + a_4 P_3} = 0 \qquad Q_4 = \frac{d_4 - a_4 Q_3}{b_4 + a_4 P_3} = 1$$

To get the solution, first use Equation (2.86):

$$x_4 = Q_4 = 1$$

Then Equation (2.81a) gives:

$$x_3 = P_3 x_4 + Q_3 = 1$$

$$x_2 = P_2 x_3 + Q_2 = 1$$

$$x_1 = P_1 x_2 + Q_1 = 1$$

So the final solution is: $x_1 = x_2 = x_3 = x_4 = 1$.

2.11 Gauss–Jordan Method

Gauss–Jordan method[18] is an extended version of the Gauss elimination. Here attempt is made to transform the coefficient matrix A to a diagonal matrix rather than a triangular matrix. For this, elimination of elements is performed in the *aug A* matrix below as well as above the pivot element such that one can get the solution without performing the forward or back substitution.

This Gauss–Jordan method is illustrated in the following example.

Example 2.21. Gauss–Jordan method

Solve the following system equations:

$$3x_1 - 3x_2 + 5x_3 = 14$$

$$6x_1 - 6x_2 + 7x_3 = 25$$

$$x_1 + x_2 + x_3 = 0$$

using the Gauss–Jordan method.

18 Named after the German mathematician Carl Friedrich Gauss (1777–1855) and German geodesist Wilhelm Jordan (1842–1899).

Solution

For this system, we previously got in Example 2.16 the following upper triangular form:

$$\begin{bmatrix} 3 & -3 & 5 \\ 0 & 2 & -\frac{2}{3} \\ 0 & 0 & -3 \end{bmatrix} \begin{bmatrix} x_1 \\ x_2 \\ x_3 \end{bmatrix} = \begin{bmatrix} 14 \\ -\frac{14}{3} \\ -3 \end{bmatrix}$$

It gives:

$$aug\ A = \begin{bmatrix} 3 & -3 & 5 & 14 \\ 0 & 2 & -\frac{2}{3} & -\frac{14}{3} \\ 0 & 0 & -3 & -3 \end{bmatrix}$$

Till this point, the Gauss elimination and Gauss–Jordan schemes perform the same computation.

In the next phase, the Gauss–Jordan method attempts to vanish the non-diagonal elements in A matrix. Accordingly, we perform elementary row operations:

$$\begin{bmatrix} 3 & 0 & 4 & 7 \\ 0 & 2 & 0 & -4 \\ 0 & 0 & -3 & -3 \end{bmatrix} \qquad \begin{matrix} R_1 \rightarrow R_1 + 1.5R_2 \\ R_2 \rightarrow R_2 - \frac{2}{9}R_3 \end{matrix}$$

Again,

$$\begin{bmatrix} 3 & 0 & 0 & 3 \\ 0 & 2 & 0 & -4 \\ 0 & 0 & -3 & -3 \end{bmatrix} \qquad R_1 \rightarrow R_1 + \frac{4}{3}R_3$$

Notice that the coefficient matrix A has got its diagonal form above. Without using back substitution, we directly get:

$$3x_1 = 3 \qquad \therefore x_1 = 1$$

$$2x_2 = -4 \qquad \therefore x_2 = -2$$

$$-3x_3 = -3 \qquad \therefore x_3 = 1$$

Remarks

1. In many textbooks,[19] the Gauss–Jordan method is described with the following conversion for the system $AX = B$:

$$\left[A \mid B\right] \rightarrow \left[I \mid E\right]$$

19 For example, see the book by Gupta (2003).

in which, I is the identity matrix of the same order with A matrix. Here, $\left[A \mid B\right]$ is the augmented matrix of A and $\left[I \mid E\right]$ is its transformed equivalent form.

It reveals that the diagonal elements in the preceding Example 2.21 need to be further normalized to unity. We ended there with the following matrix:

$$aug\ A = \begin{bmatrix} 3 & 0 & 0 & 3 \\ 0 & 2 & 0 & -4 \\ 0 & 0 & -3 & -3 \end{bmatrix}$$

Normalizing the first, second and third rows by dividing them with 3, 2 and −3, respectively, we get:

$$\begin{bmatrix} 1 & 0 & 0 & 1 \\ 0 & 1 & 0 & -2 \\ 0 & 0 & 1 & 1 \end{bmatrix} \equiv \left[I \mid E\right]$$

From this, one can directly get the solution as: $x_1 = 1, x_2 = -2, x_3 = 1$.

2. The CPU time required for the Gauss–Jordan method is about 1.5 times of that required for the Gauss elimination. Thus, it is not so attractive scheme for solving $AX = B$.

   However, this Gauss–Jordan method is useful for other purposes, including A^{-1} determination through:

$$\left[A \mid I\right] \rightarrow \left[I \mid A^{-1}\right]$$

3. There are a couple of linear systems, for which, all non-zero coefficients (in A matrix) are unity. For this, one can see the worked-out Case Study 2.4 that deals with traffic flow.
4. Like the Gauss elimination, the Gauss–Jordan technique may require partial pivoting (i.e., row interchanging).

2.12 Operational Counts of Direct Methods: A Comparison

Various direct methods are discussed and employed in the preceeding sections to solve the linear systems of equations. Because of the fixed-precision arithmetic used in the computer, there is a possibility for loss of significant digits in each arithmetic operation. This loss is quite remarkable particularly for large systems (i.e., for large n), which may lead to erroneous results. Let us now compare the direct methods discussed so far in terms of their operational counts (approximate) for large n in Table 2.2.

TABLE 2.2 Comparative operational counts of direct methods

Method	Operational counts
Cramer rule	$n!$
Gauss–Jordan	$\frac{n^3}{2}$
Gauss elimination	$\frac{n^3}{3}$
LU decomposition (Crout and Doolittle)	$\frac{n^3}{3}$
Cholesky method (only positive definite matrix)	$\frac{n^3}{6}$
Thomas algorithm (only tridiagonal system)	n

Indirect or Iterative Methods

We have learned so far how to solve the linear systems with the use of direct methods, like Gauss elimination, LU decomposition, and Thomas algorithm. These techniques are easy to implement and involve a fixed/finite number of operations to find an exact solution (if exists) subject to round-off error. However, as indicated, these methods may lead to erroneous result, particularly for large systems, due to the loss of significant digits. This apart, as the size of the system grows, the computational cost of direct methods increases. Thus the direct methods are not suitable for n more than about 25 that is quite common in engineering applications.

To overcome these problems concerning round-off error and computational effort (for large systems), there is a different class of methods, called the *iterative* or *indirect* methods. For them, we need to assume an initial guess of the solution and then update it in every iteration. This way, improve the solution value successively and terminate the procedure as soon as we reach sufficiently close to the solution. But what is frustrating is that the iterative method may not converge to the solution (this issue is elaborated later).

The indirect methods can provide more accurate solution than for example, the Gauss elimination method (a direct method) because of their ability to overcome the round-off errors by subsequent iterations.

Importantly, the indirect methods may also reduce the computational cost over the direct methods. This is because the iterative methods only involve matrix–vector multiplications. For a full matrix, each iteration scales as $\sim n^2$, indicating a significant savings achieved by the indirect methods (with a reasonably small number of iterations) over the Gauss elimination. Because of these reasons, the iterative methods get preference for large engineering systems.

Next, we would like to study the three iterative approaches as:

- Jacobi method
- Gauss–Seidel method
- Relaxation method

2.13 Jacobi Method or Method of Simultaneous Displacement

Let us start with the following system:

$$\begin{aligned} &a_{11}x_1+a_{12}x_2+\dots\dots\dots+a_{1n}x_n=b_1\\ &a_{21}x_1+a_{22}x_2+\dots\dots\dots+a_{2n}x_n=b_2\\ &\vdots\\ &a_{n1}x_1+a_{n2}x_2+\dots\dots\dots+a_{n\,n}x_n=b_n \end{aligned} \tag{2.1}$$

Simply rearranging the above equations,

$$\begin{aligned} x_1&=\frac{b_1-a_{12}x_2-\dots\dots\dots-a_{1n}x_n}{a_{11}}\\ x_2&=\frac{b_2-a_{21}x_1-a_{23}x_3-\dots\dots\dots-a_{2n}x_n}{a_{22}}\\ &\vdots\\ x_n&=\frac{b_n-a_{n1}x_1-a_{n2}x_2-\dots\dots\dots-a_{n\;n-1}x_{n-1}}{a_{n\,n}} \end{aligned} \tag{2.87}$$

With this, for a typical $(k+1)$th iteration, Jacobi method[20] yields:

$$\begin{aligned} x_1^{(k+1)}&=\frac{1}{a_{11}}\left[b_1-\left(a_{12}x_2^{(k)}+a_{13}x_3^{(k)}+\dots\dots\dots+a_{1n}x_n^{(k)}\right)\right]\\ x_2^{(k+1)}&=\frac{1}{a_{22}}\left[b_2-\left(a_{21}x_1^{(k)}+a_{23}x_3^{(k)}+\dots\dots\dots+a_{2n}x_n^{(k)}\right)\right]\\ &\vdots\\ x_n^{(k+1)}&=\frac{1}{a_{n\,n}}\left[b_n-\left(a_{n1}x_1^{(k)}+a_{n2}x_2^{(k)}+\dots\dots\dots+a_{n\;n-1}x_{n-1}^{(k)}\right)\right] \end{aligned} \tag{2.88}$$

In general,

$$x_i^{(k+1)}=\frac{1}{a_{i\,i}}\left[b_i-\sum_{\substack{j=1\\ j\neq i}}^{n}a_{i\,j}x_j^{(k)}\right] \tag{2.89}$$

where, $k=0,1,2,3,\dots$, $i=1,2,3,\dots,n$ and $a_{i\,i}\neq 0$. This is the final form of the Jacobi method.

20 Named after the German mathematician Carl Gustav Jacob Jacobi (1804–1851).

In the first iteration step, we calculate $x_i^{(1)}$, which requires an initial guess:

$$X^{(0)} = \begin{bmatrix} x_1^{(0)} \\ x_2^{(0)} \\ \cdot \\ \cdot \\ x_n^{(0)} \end{bmatrix}$$

One can set all of these entries to zero if there is no better information about X.

Continue the iterations until the solution satisfies the following stopping criteria:

$$\left|X^{(k+1)} - X^{(k)}\right| \le \text{tol} \tag{2.90a}$$

$$\left|AX^{(k+1)} - B\right| \le \text{tol} \tag{2.90b}$$

where, 'tol' denotes the desired tolerance. Equation (2.90a) is frequently used when the iterations are converging. It is also recommended to use Equation (2.90b) at least once at the end of the calculation, if not at every iteration.

Remark

It becomes obvious that the Jacobi method performs computation for all elements simultaneously and thus, it is also called the *method of simultaneous displacement.*

Example 2.22 Jacobi method

Solve the following system:

$$3x_1 - 2x_2 + x_3 = 2$$

$$x_1 + 3x_2 - x_3 = 3$$

$$x_1 - 2x_2 + 3x_3 = 2$$

with the given: $x_1^{(0)} = x_2^{(0)} = x_3^{(0)} = 0$ and $\text{tol} = 10^{-3}$.

Solution

Rearranging the given system equations:

$$x_1 = \frac{1}{3}\left(2 + 2x_2 - x_3\right)$$

$$x_2 = \frac{1}{3}\left(3 - x_1 + x_3\right)$$

$$x_3 = \frac{1}{3}\left(2 - x_1 + 2x_2\right)$$

For iteration 1 ($k = 0$):

$$x_1^{(1)} = \frac{1}{3}(2 + 2x_2^{(0)} - x_3^{(0)}) = \frac{2}{3} = 0.6667$$

$$x_2^{(1)} = \frac{1}{3}(3 - x_1^{(0)} + x_3^{(0)}) = 1$$

$$x_3^{(1)} = \frac{1}{3}(2 - x_1^{(0)} + 2x_2^{(0)}) = \frac{2}{3} = 0.6667$$

Likewise, we continue the calculation and the produced results are listed in Table 2.3.

Table 2.3 Results of Jacobi iterations

Iteration	x_1	x_2	x_3
1	0.6667	1.0	0.6667
2	1.1111	1.0	1.1111
3	0.9630	1.0	0.9630
4	1.0123	1.0	1.0123
5	0.9959	1.0	0.9959
6	1.0014	1.0	1.0014
7	0.9995	1.0	0.9995
8	1.0002	1.0	1.0002

Now, let us check the convergence. From Equation (2.90a):

$$\left|x_1^{(8)} - x_1^{(7)}\right| = 0.7\times10^{-3} < 10^{-3}\ (= \text{tol})$$

$$\left|x_2^{(8)} - x_2^{(7)}\right| = 0 < 10^{-3}$$

$$\left|x_3^{(8)} - x_3^{(7)}\right| = 0.7\times10^{-3} < 10^{-3}$$

Notice that the solution (x_1 = 1.0002, x_2 = 1.0, x_3 = 1.0002) satisfies the first stopping criterion. Although it is shown only for the last step, but this criterion is originally checked in every iteration step, based on which, we stop at $k = 7$.

We further check with Equation (2.90b):

$$\left|3x_1 - 2x_2 + x_3 - 2\right| = 0.8\times10^{-3} < 10^{-3}\ \ (= \text{tol})$$

$$\left|x_1 + 3x_2 - x_3 - 3\right| = 0 < 10^{-3}$$

$$\left|x_1 - 2x_2 + 3x_3 - 2\right| = 0.8\times10^{-3} < 10^{-3}$$

which validates our second tolerance criterion as well. So the final solution is:

x_1 = 1.0002
x_2 = 1.0
x_3 = 1.0002

Remark

If the system is strictly diagonally dominant [Equation (2.107)], then the convergence is guaranteed. This criterion is sufficient but not necessary for convergence (this is further elaborated in Section 2.16).

Here, the given system is diagonally dominant and it converges to its solution (true solution is: $x_1 = x_2 = x_3 = 1$).

Example 2.23 Divergence of Jacobi method

Repeat Example 2.22 with rearranging the system equations as:

$$x_1 + 3x_2 - x_3 = 3$$
$$3x_1 - 2x_2 + x_3 = 2$$
$$x_1 - 2x_2 + 3x_3 = 2$$

Given: $x_1^{(0)} = x_2^{(0)} = x_3^{(0)} = 0$ and $\text{tol} = 10^{-3}$.

Solution

Using the Jacobi method:

$$x_1 = 3 - 3x_2 + x_3$$
$$x_2 = \frac{1}{2}(3x_1 + x_3 - 2)$$
$$x_3 = \frac{1}{3}(2 - x_1 + 2x_2)$$

With the initial guess, $x_1^{(0)} = x_2^{(0)} = x_3^{(0)} = 0$, we produce a sequence of approximations, $X^{(1)}, X^{(2)}, \ldots, X^{(6)}$, in Table 2.4.

TABLE 2.4 Results of Jacobi iterations

Iteration	x_1	x_2	x_3
1	3.0	−1.0	0.6667
2	6.6667	3.8333	−1.0
3	−9.4999	8.50	0.9999
4	-21.5001	−14.7499	9.4999
5	56.7496	−28.5002	-1.9999
6	86.5007	83.1244	-37.250

Remark

Notice that this example system is *not* diagonally dominant and it diverges from the solution.

2.14 Gauss–Seidel Method or Method of Successive Displacement

To improve the convergence rate of the iterations, the Jacobi method is extended to an iterative approach, called the *Gauss–Seidel approach*[21]. In this technique, X is computed with the use of the most recent estimates. Let us now discuss how it is done.

For solving the system $AX = B$ [Equation (2.1)], Equation (2.88) is modified to:

$$
\begin{aligned}
x_1^{(k+1)} &= \frac{1}{a_{11}}\left[b_1 - \left(a_{12}x_2^{(k)} + a_{13}x_3^{(k)} + \dots\dots + a_{1n}x_n^{(k)}\right)\right] \\
x_2^{(k+1)} &= \frac{1}{a_{22}}\left[b_2 - \left(a_{21}x_1^{(k+1)} + a_{23}x_3^{(k)} + \dots\dots + a_{2n}x_n^{(k)}\right)\right] \\
&\vdots \\
x_n^{(k+1)} &= \frac{1}{a_{n\,n}}\left[b_n - \left(a_{n1}x_1^{(k+1)} + a_{n2}x_2^{(k+1)} + \dots\dots + a_{n\;n-1}x_{n-1}^{(k+1)}\right)\right]
\end{aligned} \tag{2.91}
$$

This set of equations can be written in consolidated form as:

$$
x_i^{(k+1)} = \frac{1}{a_{i\,i}}\left[b_i - \sum_{j=1}^{i-1} a_{i\,j}x_j^{(k+1)} - \sum_{j=i+1}^{n} a_{i\,j}x_j^{(k)}\right] \tag{2.92}
$$

where, $k = 0, 1, 2, 3, \ldots$ and $i = 1, 2, 3, \ldots, n$. This is the final form of the Gauss–Seidel method.[22]

Differences between Jacobi and Gauss–Seidel method

Although the Jacobi and Gauss–Seidel methods look extremely similar in terms of their formulas, the following differences exist between them:

1. Gauss–Seidel method computes $x^{(k+1)}$ using the elements of $X^{(k+1)}$ that have already been computed and those of $X^{(k)}$, which are not available for $(k + 1)$th iteration. Hence, unlike the Jacobi method, the Gauss–Seidel requires a single storage vector, in which, the elements get exchanged with their respective new values as soon as they are available. This is the advantage of the Gauss–Seidel algorithm.
2. As indicated above, successive displacement of x-elements is involved in the Gauss–Seidel approach and thus, it is also called the *method of successive displacement*. On the other hand, as stated before, the Jacobi method performs computation for all elements simultaneously and thus, that is often called the *method of simulataneous displacement.*
3. Gauss–Seidel provides a faster convergence rate over the Jacobi method without any additional effort (see also by comparing the solutions between Examples 2.22 and 2.24).

21 Named after the German mathematicians Carl Friedrich Gauss (1777–1855) and Philipp Ludwig von Seidel (1821–1896).

22 This method is similar in spirit with the simple fixed-point iterative method employed to solve the non-linear algebraic equations (see Appendix 3A).

At this point, it is worth emphasizing that if the coefficient matrix A is strictly diagonally dominant [Equation (2.107)], both the Jacobi and Gauss–Seidel methods converge irrespective of the initial guess $X^{(0)}$.

Example 2.24 Gauss–Seidel method

Solve Example 2.22 having the following equations:

$$3x_1 - 2x_2 + x_3 = 2$$
$$x_1 + 3x_2 - x_3 = 3$$
$$x_1 - 2x_2 + 3x_3 = 2$$

with the use of Gauss–Seidel method. Given: $x_1^{(0)} = x_2^{(0)} = x_3^{(0)} = 0$ and $\text{tol} = 10^{-3}$.

Solution

Using the Gauss–Seidel approach [Equation (2.91)],

$$x_1^{(k+1)} = \frac{1}{3}(2 + 2x_2^{(k)} - x_3^{(k)})$$

$$x_2^{(k+1)} = \frac{1}{3}(3 - x_1^{(k+1)} + x_3^{(k)})$$

$$x_3^{(k+1)} = \frac{1}{3}(2 - x_1^{(k+1)} + 2x_2^{(k+1)})$$

With $k = 0, 1, 2, 3, \ldots$, we get the solution as shown in Table 2.5 against each iteration.

TABLE 2.5 Results of Gauss–Seidel iterations

Iteration	x_1	x_2	x_3
1	0.6667	0.7778	0.9630
2	0.8642	1.0329	1.0672
3	0.9995	1.0226	1.0152
4	1.01	1.0017	0.9978
5	1.0019	0.9986	0.9984
6	0.9996	0.9996	0.9999
7	0.9998	1.0000	1.0000

Final solution is obtained within the tolerance limit (10^{-3}) as:

$$x_1 = 0.9998$$
$$x_2 = 1.0000$$
$$x_3 = 1.0000$$

Remark

Notice that the Gauss–Seidel method converges in 7 iterations and for the same sample system, Jacobi method requires 8 iterations (see Example 2.22). So, there is an improvement achieved by the former approach over the latter.

2.15 Relaxation Method

It is observed above that the Gauss–Seidel approach improves the convergence rate over the Jacobi method. To accelerate that rate further with better flexibility, there is an approach called the *relaxation method.* This is basically a generalization of the Jacobi and Gauss–Seidel method made by introducing an additional degree of freedom, namely *relaxation factor.*

Let us discuss this iterative relaxation method to solve the linear system represented by Equation (2.1). We can rewrite Equation (2.87) as:

$$
\begin{aligned}
x_1 &= x_1 - \frac{a_{11}x_1}{a_{11}} + \frac{1}{a_{11}}\left[b_1 - (a_{12}x_2 + a_{13}x_3 + \dots\dots + a_{1n}x_n)\right] \\
x_2 &= x_2 - \frac{a_{22}x_2}{a_{22}} + \frac{1}{a_{22}}\left[b_2 - (a_{21}x_1 + a_{23}x_3 + \dots\dots + a_{2n}x_n)\right] \\
&\vdots \\
x_n &= x_n - \frac{a_{n\,n}x_n}{a_{n\,n}} + \frac{1}{a_{n\,n}}\left[b_n - (a_{n1}x_1 + a_{n2}x_2 + \dots\dots + a_{n\;n-1}x_{n-1})\right]
\end{aligned}
\tag{2.93}
$$

Rearranging,

$$
\begin{aligned}
x_1 &= x_1 + \frac{1}{a_{11}}\left[b_1 - (a_{11}x_1 + a_{12}x_2 + a_{13}x_3 + \dots\dots + a_{1n}x_n)\right] \\
x_2 &= x_2 + \frac{1}{a_{22}}\left[b_2 - (a_{21}x_1 + a_{22}x_2 + a_{23}x_3 + \dots\dots + a_{2n}x_n)\right] \\
&\vdots \\
x_n &= x_n + \frac{1}{a_{n\,n}}\left[b_n - (a_{n1}x_1 + a_{n2}x_2 + a_{n3}x_3 + \dots\dots + a_{n\;n-1}x_{n-1} + a_{n\,n}x_n)\right]
\end{aligned}
\tag{2.94}
$$

Defining the *residual* (r) as:

$$
\begin{aligned}
r_1 &= b_1 - (a_{11}x_1 + a_{12}x_2 + a_{13}x_3 + \dots\dots + a_{1n}x_n) \\
r_2 &= b_2 - (a_{21}x_1 + a_{22}x_2 + a_{23}x_3 + \dots\dots + a_{2n}x_n) \\
&\vdots \\
r_n &= b_n - (a_{n1}x_1 + a_{n2}x_2 + a_{n3}x_3 + \dots\dots + a_{n\,n}x_n)
\end{aligned}
\tag{2.95}
$$

For exact solution, these residuals should be zero:

$$
r_i = b_i - \sum_{j=1}^{n} a_{i\,j}x_j = 0 \tag{2.96}
$$

where $1 \le i \le n$. With this background, attempt is made to speed up the iteration process by multiplying these residual terms with a factor, w (i.e., relaxation factor). Accordingly, Equation (2.94) gets the following form:

$$
\begin{aligned}
x_1 &= x_1 + \frac{w}{a_{11}}\left[b_1 - (a_{11}x_1 + a_{12}x_2 + a_{13}x_3 + \dots\dots + a_{1n}x_n)\right] \\
x_2 &= x_2 + \frac{w}{a_{22}}\left[b_2 - (a_{21}x_1 + a_{22}x_2 + a_{23}x_3 + \dots\dots + a_{2n}x_n)\right] \\
&\vdots \\
x_n &= x_n + \frac{w}{a_{n\,n}}\left[b_n - (a_{n1}x_1 + a_{n2}x_2 + a_{n3}x_3 + \dots\dots + a_{n\;n-1}x_{n-1} + a_{n\,n}x_n)\right]
\end{aligned} \tag{2.97}
$$

Rearranging,

$$
\begin{aligned}
x_1 &= (1-w)x_1 + \frac{w}{a_{11}}\left[b_1 - (a_{12}x_2 + a_{13}x_3 + \dots\dots + a_{1n}x_n)\right] \\
x_2 &= (1-w)x_2 + \frac{w}{a_{22}}\left[b_2 - (a_{21}x_1 + a_{23}x_3 + \dots\dots + a_{2n}x_n)\right] \\
&\vdots \\
x_n &= (1-w)x_n + \frac{w}{a_{n\,n}}\left[b_n - (a_{n1}x_1 + a_{n2}x_2 + a_{n3}x_3 + \dots\dots + a_{n\;n-1}x_{n-1})\right]
\end{aligned} \tag{2.98}
$$

- Based on Jacobi method (simultaneous displacement):

$$
x_i^{(k+1)} = (1-w)x_i^{(k)} + \frac{w}{a_{i\,i}}\left[b_i - \sum_{\substack{j=1 \\ j\neq i}}^{n} a_{i\,j}x_j^{(k)}\right] \tag{2.99}
$$

where, $k = 0, 1, 2, 3, \ldots$ and $i = 1, 2, 3, \ldots, n$.

- Based on Gauss–Seidel method (successive displacement):

$$
x_i^{(k+1)} = (1-w)x_i^{(k)} + \frac{w}{a_{i\,i}}\left[b_i - \sum_{j=1}^{i-1} a_{i\,j}x_j^{(k+1)} - \sum_{j=i+1}^{n} a_{i\,j}x_j^{(k)}\right] \tag{2.100}
$$

where, $k = 0, 1, 2, 3, \ldots$ and $i = 1, 2, 3, \ldots, n$.

- Types of relaxation method

Equations (2.99) and (2.100) represent the relaxation method with:

- $0 < w < 1$ called the *under-relaxation* method
- $w > 1$ called the *over-relaxation* method

Remarks

1. When Equation (2.100) is used with over-relaxation, it is often called *successive over-relaxation* (SOR) method.
2. The formula given in Equation (2.100) gets preference over that in Equation (2.99) for solving the linear systems. This is because the successive displacement (Gauss–Seidel) provides faster convergence than the simultaneous displacement (Jacobi).

3. When $w = 1$, the relaxation method represented by Equations (2.99) and (2.100) reduces to the Jacobi [Equation (2.89)] and Gauss–Seidel method [Equation (2.92)], respectively.

Example 2.25 Successive relaxation method

Solve Example 2.24 by the use of the successive relaxation method [i.e., Equation (2.100)],

$$3x_1 - 2x_2 + x_3 = 2$$
$$x_1 + 3x_2 - x_3 = 3$$
$$x_1 - 2x_2 + 3x_3 = 2$$

with the given: $x_1^{(0)} = x_2^{(0)} = x_3^{(0)} = 0$ and $\text{tol} = 10^{-3}$ for the four different w values, namely 0.8, 0.9, 1.0 and 1.1.

Solution

Based on Equation (2.98) [or (2.100)],

$$x_1^{(k+1)} = (1-w)x_1^{(k)} + \frac{w}{3}(2 + 2x_2^{(k)} - x_3^{(k)})$$

$$x_2^{(k+1)} = (1-w)x_2^{(k)} + \frac{w}{3}(3 - x_1^{(k+1)} + x_3^{(k)})$$

$$x_3^{(k+1)} = (1-w)x_3^{(k)} + \frac{w}{3}(2 - x_1^{(k+1)} + 2x_2^{(k+1)})$$

Varying $k = 0,1,2,\ldots$, we produce the solution for four different w values as shown in Table 2.6.

TABLE 2.6 Results of successive relaxation method

(a) with $w = 0.8$

Iteration	x_1	x_2	x_3
1	0.53333	0.65778	0.74193
2	0.79297	0.91794	0.95983
3	0.92554	0.99273	1.00795
4	0.97911	1.00624	1.01048
5	0.99635	1.00502	1.00574
6	1.00041	1.00242	1.00233
7	1.00075	1.00091	1.00075
8	1.00043	1.00026	1.00018

(b) with $w = 0.9$

Iteration	x_1	x_2	x_3
1	0.6	0.72	0.852
2	0.8364	0.97668	1.02029
3	0.96356	1.01469	1.02177
4	0.99864	1.00841	1.00763
5	1.00262	1.00234	1.00138
6	1.00125	1.00027	0.99993
7	1.00031	0.99991	0.99985

(c) with $w = 1.0$

Iteration	x_1	x_2	x_3
1	0.66667	0.77778	0.96296
2	0.8642	1.03292	1.06722
3	0.99954	1.02256	1.01519
4	1.00997	1.00174	0.99783
5	1.00188	0.99865	0.99847
6	0.99961	0.99962	0.99988
7	0.99979	1.00003	1.00009

(d) with $w = 1.1$

Iteration	x_1	x_2	x_3
1	0.73333	0.83111	1.07393
2	0.87571	1.08957	1.10386
3	1.04003	1.01445	0.98553
4	1.0119	0.98889	0.98893
5	0.99472	0.99899	1.0023
6	0.99894	1.00133	1.00113
7	1.00067	1.00004	0.99967
8	1.00008	0.99984	0.99989

Remarks

1. Here we got the best performance when the relaxation factor, w is about 0.9 and 1.0. In some cases, the optimal value of w is possible to determine analytically.[23]
2. Successive relaxation method with $w = 1$ leads to the Gauss–Seidel method and thus, they produce the same results (verify by comparing with the results of Example 2.24).

2.16 Convergence of Iterative Methods

Convergence is a major concern for the iterative methods (i.e., Jacobi, Gauss–Seidel and relaxation). As indicated, if these iterative methods converge, we prefer them since they involve lower computational effort and higher accuracy compared to the direct methods.

23 For details, see the book by Lambers and Sumner (2019).

An iterative approach used to determine the solution of a linear system, $AX = B$ takes the following form:

$$X^{(k+1)} = PX^{(k)} + Q \tag{2.101}$$

Let X^* be the exact solution and thus,

$$X^* = PX^* + Q \tag{2.102}$$

Subtracting Equation (2.102) from (2.101),

$$X^{(k+1)} - X^* = P\left(X^{(k)} - X^*\right) \tag{2.103}$$

Since the true error[24] is:

$$\mathbf{e}^{(k)} = X^{(k)} - X^*$$

one can rewrite Equation (2.103) as:

$$\mathbf{e}^{(k)} = P\mathbf{e}^{(k-1)} \tag{2.104}$$

Accordingly,

$$\begin{aligned} \mathbf{e}^{(1)} &= P\mathbf{e}^{(0)} \\ \mathbf{e}^{(2)} &= P\mathbf{e}^{(1)} = P^2\mathbf{e}^{(0)} \\ &\vdots \\ \mathbf{e}^{(k)} &= P\mathbf{e}^{(k-1)} = P^k\mathbf{e}^{(0)} \end{aligned} \tag{2.105}$$

Using the properties of matrix norms (see Appendix A2.2.2):

$$\left\|\mathbf{e}^{(k)}\right\| \le \left\|P^k\right\| \; \left\|\mathbf{e}^{(0)}\right\|$$

It gives:

$$\left\|\mathbf{e}^{(k)}\right\| \le \left\|P\right\|^k \; \left\|\mathbf{e}^{(0)}\right\| \tag{2.106}$$

Hence, an iterative method is guaranteed to converge if $\|P\| < 1$. It reveals that $\mathbf{e}^{(k)} \to 0$ (i.e., $\left\|\mathbf{e}^{(k)}\right\| \to 0$) as $k \to \infty$. It should be noted that similar formulation for iterative convergence methods of non-linear systems will be discussed in next Chapter 3. But remember, the ability of these linear methods to converge is much more predictable than for the methods of non-linear systems.

Remark

Before ending this discussion, let us note that there is a condition sometimes we encounter in practice for a system of $AX = B$, whose A matrix is (strictly) diagonally dominant. It means,

$$\left|a_{i\,i}\right| > \sum_{\substack{j=1 \\ j \ne i}}^{n} \left|a_{i\,j}\right| \tag{2.107}$$

24 In Chapter 1, k is used as a suffix [see Equation (1.9)].

for $i=1, 2, 3, ..., n$. With this, the Jacobi and Gauss–Seidel methods are guaranteed to converge to the solution. These iterative approaches may also converge even if the matrix A is not diagonally dominant, revealing that Equation (2.107) is a sufficient, but not necessary condition.

Recall that in Example 2.22, the above condition [Equation (2.107)] is satisfied and thus, the convergence is achieved there.

2.17 Solved Case Studies

So far we have learned various elegant methods and their applications to a wide variety of mathematical systems of linear equations. Now we would like to solve a few case studies that are frequently encountered in engineering disciplines.

Case Study 2.1 Gauss elimination with backward sweep (3-reactors system)

A flowsheet consisting of three chemical reactors is shown in Figure 2.4. Here, the following notations are used:

F = volumetric flow rate (m^3/sec)
C = concentration of reactant A (kg/m^3)
V = volume of liquid in the reactor (m^3)

In all three reactors, a first-order liquid phase chemical reaction:

$$A \rightarrow P$$

takes place. The rate of disappearance of reactant A is thus given as:

$$(-r) = kC$$

where, the reaction rate constant, k = 0.1 sec^{-1}. With this given information,

i) develop the component A balance equations for all three reactors operated at steady state
ii) find three concentrations of component A (C_1, C_2 and C_3).

Solution

i) For reactant A, we can rewrite Equation (1.1) at steady state as:

$$\begin{pmatrix}\text{Rate of}\\ \text{input of } A\end{pmatrix} = \begin{pmatrix}\text{Rate of}\\ \text{output of } A\end{pmatrix} + \begin{pmatrix}\text{Rate of}\\ \text{depletion of } A\end{pmatrix}$$

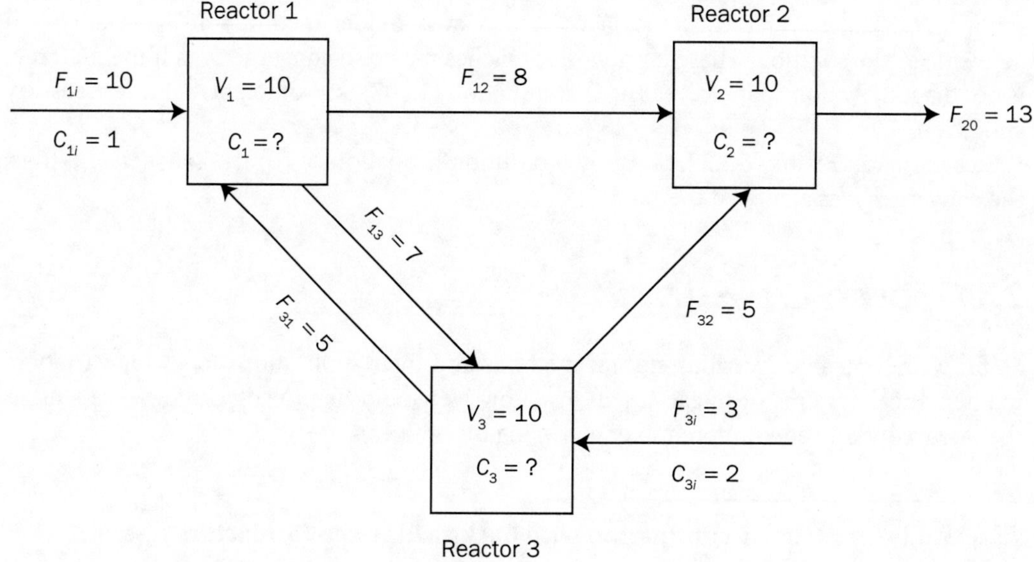

FIGURE 2.4 A system of three reactors

Here, the unit of all three terms is kg of A/sec. Note that the rate of depletion of component $A = (-r)V$. With this, we make A balance for the three reactors as follows:

***Reactor* 1**

$$F_{1i}C_{1i} + F_{31}C_3 = F_{12}C_1 + F_{13}C_1 + 0.1C_1V_1$$

It gives:

$$16C_1 - 5C_3 = 10 \tag{2.108a}$$

***Reactor* 2**

$$F_{12}C_1 + F_{32}C_3 = F_{20}C_2 + 0.1C_2V_2$$

It yields:

$$8C_1 - 14C_2 + 5C_3 = 0 \tag{2.108b}$$

***Reactor* 3**

$$F_{3i}C_{3i} + F_{13}C_1 = F_{31}C_3 + F_{32}C_3 + 0.1C_3V_3$$

Or,

$$-7C_1 + 11C_3 = 6 \tag{2.108c}$$

Derivation of component A balance equations is complete.

ii) The steady state mass balance equations for the 3-reactors system are developed above as:

$$16C_1 - 5C_3 = 10 \tag{2.108a}$$

$$8C_1 - 14C_2 + 5C_3 = 0 \tag{2.108b}$$

$$-7C_1 + 11C_3 = 6 \tag{2.108c}$$

To solve this set of linear Equations (2.108), we form:

$$aug\ A = \begin{bmatrix} 16 & 0 & -5 & 10 \\ 8 & -14 & 5 & 0 \\ -7 & 0 & 11 & 6 \end{bmatrix}$$

Performing forward elimination:

$$\begin{bmatrix} 16 & 0 & -5 & 10 \\ 0 & -14 & \frac{15}{2} & -5 \\ 0 & 0 & \frac{141}{16} & \frac{83}{8} \end{bmatrix} \qquad \begin{matrix} R_2 \rightarrow R_2 - \frac{1}{2}R_1 \\ R_3 \rightarrow R_3 + \frac{7}{16}R_1 \end{matrix}$$

With backward sweep, we get the final concentrations as:

$C_1 = 0.9929$ kg/m^3
$C_2 = 0.9878$ kg/m^3
$C_3 = 1.1773$ kg/m^3

Case Study 2.2 An electrical network

Consider a simple electrical network shown in Figure 2.5. Before going to develop the model and solving that model, let us define all the elements associated in this circuit.

Here all the surrounding thick straight lines denote the wires. As shown, J_1 and J_2 are the two junction points or nodes. The symbol having two parallel vertical bars represents the battery, in which, the longer one is the positive terminal and shorter one the negative terminal. This battery typically provides voltage (in volt (V)) and as a result, the current, denoted by I (in ampere (A)), flows through the wires. The zigzag symbol shown in the network represents the resistor that typically resists or opposes the flow of current, which results in voltage drop. This resistance (R) is measured in ohm (Ω) and the corresponding value of R for all 4 resistors is shown in the figure. Moreover, the supply of voltage by the two batteries is also given there.

Here, our goal is to find the currents (I in A) flowed through the wires with developing required system of equations. For solving those equations, you are asked to employ an iterative Gauss–Seidel method. If it fails to converge, use the Gauss elimination approach.

Note: The solution is as follows: $I_1 = -\frac{5}{18}$, $I_2 = \frac{35}{9}$ and $I_3 = \frac{25}{6}$ A.

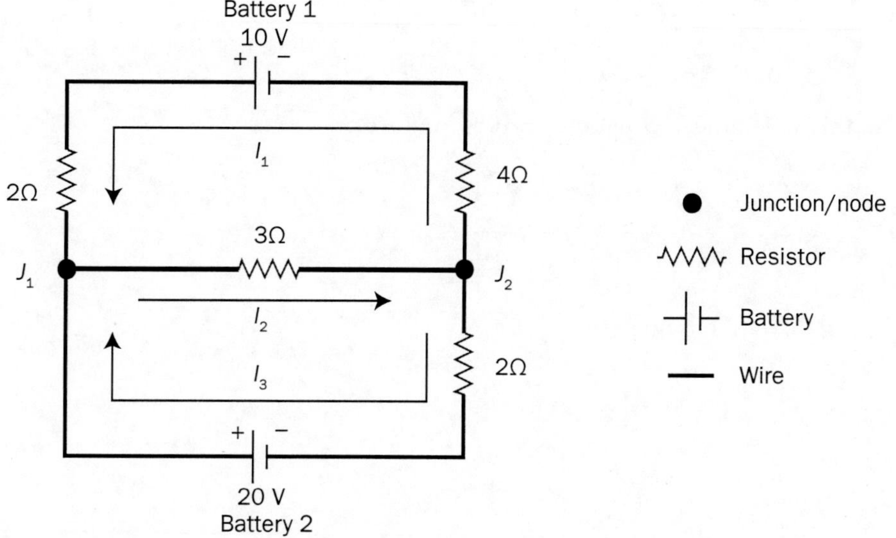

FIGURE **2.5** An electrical network

Solution

As shown in Figure 2.5, the current I_1 flows through the upper branch, I_2 through the middle branch and I_3 through the lower branch. The direction of flow of all these three currents as shown in the figure is chosen arbitrarily. However, after computing the values of these three currents, we have the scope to verify whether our assumed flow directions are correct or wrong. Let us wait for that.

For finding I_1, I_2 and I_3, obviously we need total three equations. For obtaining those equations, we will use the Kirchhoff's law.

Kirchhoff's first law (for junction)

Kirchhoff's first law applies to currents at a junction in a circuit and thus, it is often called as *Kirchhoff's current law*. This law states that for a junction in an electrical circuit, the sum of currents flowing into the junction is equal to the sum of currents flowing out of that junction.

Applying this law, we get for:

Junction J_1

$$I_1 + I_3 = I_2$$

Junction J_2

$$I_2 = I_1 + I_3$$

Notice that both the equations are same. Thus, we get one algebraic equation by the application of the Kirchhoff's first law as:

$$I_1 - I_2 + I_3 = 0 \tag{2.109a}$$

We need two more equations to make the system completely specified (i.e., degrees of freedom = 0). For this, let us apply the Kirchhoff's second law.

Kirchhoff's second law (for loop)

It states that the sum of potential differences across a closed circuit must be equal to zero. This law, which is also known as *Kirchhoff's voltage law*, can further be inferred from the following statement: in a closed-loop, the amount of voltage gains is equal to the amount of voltage drops. It means:

Voltage gains – Voltage drops = 0

In the example system, the voltage gains for battery 1 and battery 2 are 10 and 20 V, respectively. Whereas, the voltage drop is associated with the resistors that we encounter in a closed circuit and it is calculated by the Ohm's law[25] as:

Voltage drop = $I.R$

Next we apply the Kirchhoff's second law on the two loops and it is discussed below.

Upper loop

For developing the balance equation, let us start from the location of battery 1, which is shown as *starting point* in Figure 2.6. Then gradually move across the three resistors one after another associated in the upper loop (having resistance of 2, 3 and subsequently 4 Ω), and finally come back to the initial location (i.e., battery 1). It is depicted in Figure 2.6, based on which, the Kirchhoff's second law yields:

$$10-2I_1-3I_2-4I_1=0$$

Rearranging,

$$6I_1+3I_2=10 \qquad (2.109b)$$

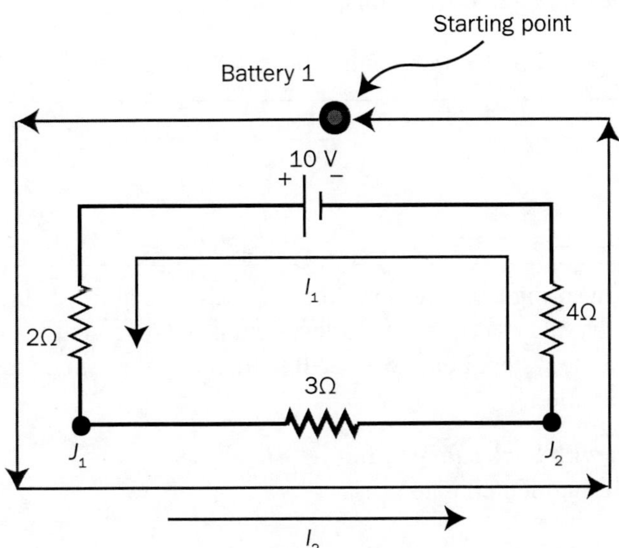

FIGURE 2.6 Applying Kirchhoff's second law on upper loop

25 The current through a resistor equals the voltage across it divided by its resistance.

Lower loop

Similarly, for the lower loop, we get:

$$20-3I_2-2I_3=0$$

It yields,

$$3I_2+2I_3=20 \tag{2.109c}$$

So, we finally get a system of three linear algebraic equations:

$I_1-I_2+I_3=0$ from Kirchhoff's first law (2.109a)

$6I_1+3I_2=10$ from Kirchhoff's second law (upper loop) (2.109b)

$3I_2+2I_3=20$ from Kirchhoff's second law (lower loop) (2.109c)

Now we would like to solve this set of linear Equations (2.109) by an iterative method (Gauss–Seidel) as asked.

Gauss–Seidel method

Using the Gauss–Seidel technique [Equation (2.91)], one can rewrite the set of Equations (2.109) as:

$$I_1^{(k+1)}=I_2^{(k)}-I_3^{(k)}$$

$$I_2^{(k+1)}=\frac{1}{3}(10-6I_1^{(k+1)})$$

$$I_3^{(k+1)}=\frac{1}{2}(20-3I_2^{(k+1)})$$

Supposing $I_1^{(0)}=I_2^{(0)}=I_3^{(0)}=0$ and $\text{tol}=10^{-4}$, we perform the computation with varying $k=0,1,2,3,...$ as shown in Table 2.7 against each iteration.

TABLE 2.7 Results of Gauss–Seidel iterations

Iteration	I_1	I_2	I_3
1	0.0	3.33333	5.00000
2	–1.66667	6.66667	0.0
3	6.66667	–10.00001	25.00001
4	–35.00002	73.33337	–100.00006

Clearly, we are diverging from the solution. The reason is quite obvious from the model Equations (2.109) that the electrical network system is *not* diagonally dominant.

To solve the example system Equations (2.109), then we move to employ a direct method, namely Gauss elimination.

Gauss elimination with back substitution

Let us first form the augmented matrix as:

$$aug\ A=\begin{bmatrix}1 & -1 & 1 & 0\\ 6 & 3 & 0 & 10\\ 0 & 3 & 2 & 20\end{bmatrix}$$

Performing forward elimination ($R_2 \rightarrow R_2 - 6R_1$ followed by $R_3 \rightarrow R_3 - \frac{1}{3}R_2$):

$$\begin{bmatrix} 1 & -1 & 1 & 0 \\ 0 & 9 & -6 & 10 \\ 0 & 0 & 4 & \frac{50}{3} \end{bmatrix}$$

With backward sweep, we get the final currents (tol $= 10^{-5}$) as:

$$I_1 = -\frac{5}{18} = -0.27778 \text{ A}$$

$$I_2 = \frac{35}{9} = 3.88889 \text{ A}$$

$$I_3 = \frac{25}{6} = 4.16667 \text{ A}$$

Remarks

1. Since computed I_1 is negative, so its actual direction of flow must be opposite to that shown in Figure 2.5. Correcting it, we get the modified Figure 2.7.
2. Voltage gain for a battery is positive when we move from negative to positive terminal during the model development by the use of Kirchhoff's second law. Accordingly, for the network with that shown in Figure 2.8(a),

   Voltage gain = – 10 V

   Similarly, for Figure 2.8(b),

   Voltage gain = 10 V

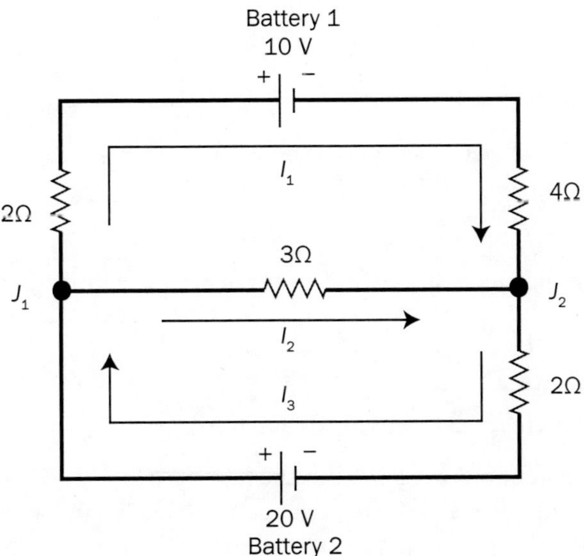

FIGURE 2.7 The electrical network after correcting the flow direction of I_1

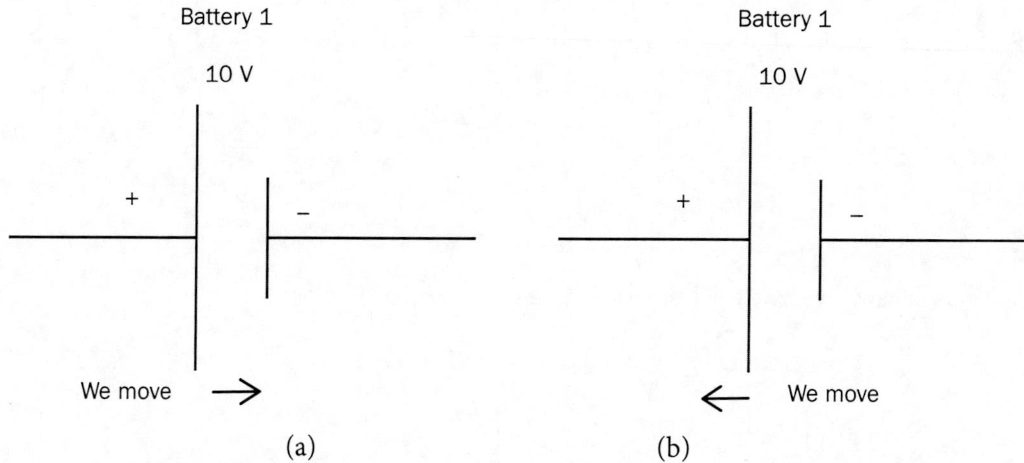

Figure 2.8 Moving from (a) positive to negative terminal (voltage gain = –10 V), and (b) negative to positive terminal (voltage gain = 10 V)

3. As far as voltage drop in a resistor is concerned, it is positive as long as the direction of our movement is same with the direction of current flow. It becomes clear from the following two examples shown in Figure 2.9.

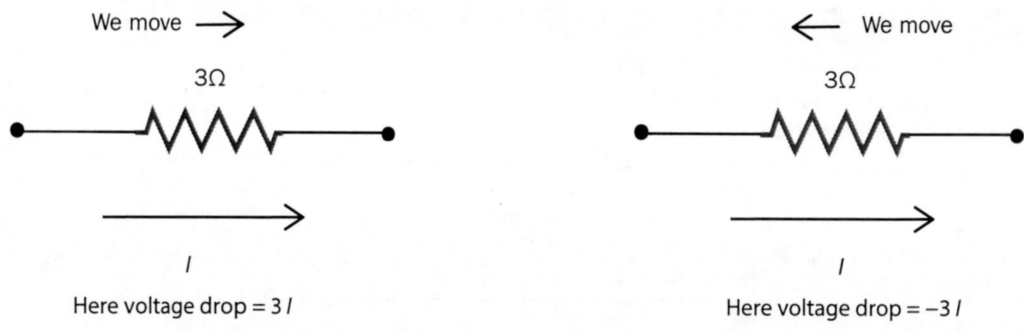

Figure 2.9 Calculating voltage drop by the use of Ohm's law with showing its sign that varies with the direction of our movement

Notice that both the above Remarks 2 and 3 are concerned with Kirchhoff's second law.

Case Study 2.3 An electrical network (revisited)

With the basic understanding we gained in the last Case Study 2.2, you are now asked to find the current flowing through the electrical network shown in Figure 2.7.

Solution

Junction J_2

The Kirchhoff's first law yields for junction J_2:

$$I_1 + I_2 - I_3 = 0 \tag{2.110a}$$

Upper loop

Next, we would like to apply the Kirchhoff's second law. Based on Figure 2.10, we get for upper loop:

$$-10 = 4I_1 - 3I_2 + 2I_1$$

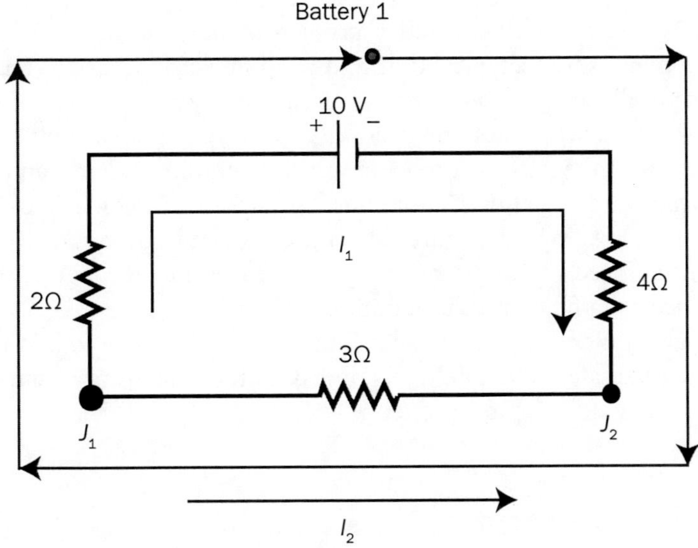

FIGURE **2.10** Applying Kirchhoff's second law on upper loop

It gives,

$$-6I_1 + 3I_2 = 10 \tag{2.110b}$$

Lower loop

If we move in the same direction with I_3, the lower loop yields:

$$20 = 3\,I_2 + 2\,I_3$$

It gives,

$$3I_2 + 2I_3 = 20 \tag{2.110c}$$

So, we get the following system of linear equations:

$$I_1 + I_2 - I_3 = 0 \tag{2.110a}$$

$$-6I_1 + 3I_2 = 10 \tag{2.110b}$$

$$3I_2 + 2I_3 = 20 \tag{2.110c}$$

Solving Equations (2.110) by the use of Gauss elimination with back substitution, we finally obtain:

$I_1 = 0.27778$ A
$I_2 = 3.88889$ A
$I_3 = 4.16667$ A

Clearly, these computed values satisfy the modeling Equations (2.110) with a tolerance of 10^{-5}.

Case Study 2.4 Traffic flow (network of one-way streets)

Next we would like to solve another practical problem of the linear algebraic equation system. In this interesting case study, a typical traffic flow in a network of various one-way streets is considered and it is illustrated in Figure 2.11. As shown, there are total four intersections or junction points (also called nodes), namely J_1, J_2, J_3 and J_4. Note that all the numbers given in this figure represent the flow of number of vehicles per minute. For example, there are 75 vehicles heading towards junction J_1 per minute. Similarly, 100 vehicles are exiting junction J_2 per minute. Here, arrows in the figure show the direction of traffic flow.

Traffic engineers need the information of this vehicle flow through each street of this network to device a traffic control program, particularly for fixing the time duration for traffic lights, speed limit of the vehicles, etc. This is quite important to minimize the delays and improve the road safety. Definitely, this time duration should vary from time to time in a day, from weekdays to weekend, and so on.

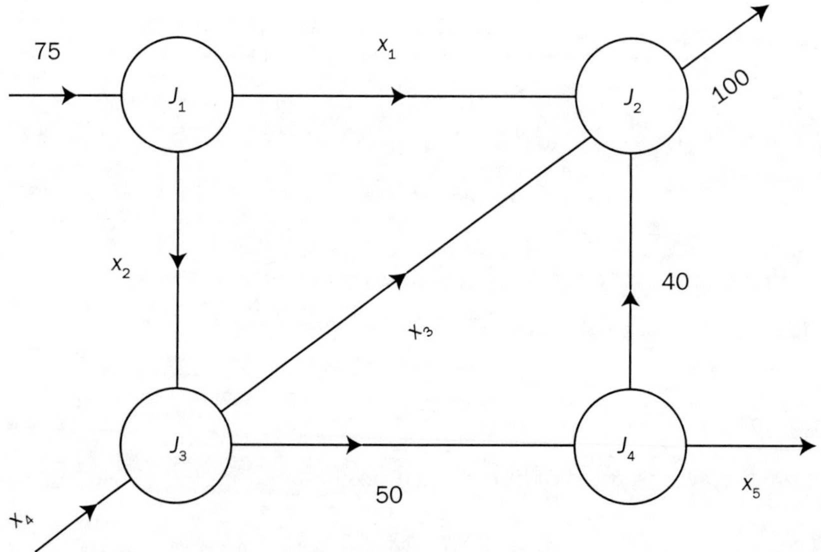

FIGURE 2.11 Traffic flow in a network

Some time ago, these traffic flow data were collected directly from the spot by counting the flow of vehicles through all connecting streets. Unfortunately, these data were lost thereafter for a couple of streets, for which, we have used the unknown variables from x_1 to x_5 in Figure 2.11. Based on the information available in our hand as shown in that figure, we are asked to find these 5 unknowns x_1 through x_5 (where, x = number of vehicles/min) by using the Gauss–Jordan elimination method.

Solution

At first, we would like to make the flow balance to develop the relevant model equations required for finding x_1 to x_5. Before doing this job for the entire network (based on total balance) as well as for each node/junction (component balance) of that network, let us write the governing flow balance equation as:

$$\text{Total flow in} = \text{Total flow out} \tag{2.111}$$

There may be a case that a couple of vehicles are parked in a Mall (shopping center) located in between two junctions for say, an hour. Again, a motor accident may occur in between two junctions and those damaged vehicles are remained there for a reasonably long time. Such cases we are not directly dealing here and with this assumption,[26] Equation (2.111) is used below.

Flow balance (entire network)

Equation (2.111) yields:

$$75 + x_4 = 100 + x_5 \tag{2.112a}$$

Flow balance (for individual junction)

Applying Equation (2.111), further we get for:

Junction J_1

$$75 = x_1 + x_2 \tag{2.112b}$$

Junction J_2

$$x_1 + x_3 + 40 = 100 \tag{2.112c}$$

Junction J_3

$$x_2 + x_4 = x_3 + 50 \tag{2.112d}$$

Junction J_4

$$50 = 40 + x_5 \tag{2.112e}$$

Rearranging, we get the following set of 5 respective equations:

$$x_4 - x_5 = 25 \tag{2.113a}$$

$$x_1 + x_2 = 75 \tag{2.113b}$$

26 It is not very unrealistic to assume that (i) the number of vehicles that entered a mall is approximately equal to the number of vehicles that left the mall, particularly at busy hours (due to parking space constraint), (ii) happening of road accident is rare in this network.

$$x_1 + x_3 = 60 \tag{2.113c}$$

$$x_2 - x_3 + x_4 = 50 \tag{2.113d}$$

$$x_5 = 10 \tag{2.113e}$$

Obviously, it represents a system of linear equations, in which, notice that all the coefficients are here 1.

To solve this linear system, let us form the augmented matrix as:

$$aug\ A = \begin{bmatrix} 0 & 0 & 0 & 1 & -1 & 25 \\ 1 & 1 & 0 & 0 & 0 & 75 \\ 1 & 0 & 1 & 0 & 0 & 60 \\ 0 & 1 & -1 & 1 & 0 & 50 \\ 0 & 0 & 0 & 0 & 1 & 10 \end{bmatrix} \tag{2.114}$$

Notice that it is a sparse matrix, in which, most of its entries are zero. Moreover, as indicated, the remaining all elements are 1, and we are somehow understanding why one would prefer to use the Gauss–Jordan method.

Here, we also follow the Gauss–Jordan algorithm. Accordingly, attempt is made to have all the diagonal elements as 1, and to vanish the non-diagonal elements in A matrix.

Performing the following row operation, $R_2 \to R_2 - R_3$:

$$aug\ A = \begin{bmatrix} 0 & 0 & 0 & 1 & -1 & 25 \\ 0 & 1 & -1 & 0 & 0 & 15 \\ 1 & 0 & 1 & 0 & 0 & 60 \\ 0 & 1 & -1 & 1 & 0 & 50 \\ 0 & 0 & 0 & 0 & 1 & 10 \end{bmatrix}$$

Subsequently, we conduct $R_4 \to R_4 - R_2$ followed by $R_1 \to R_1 - R_4$ and get:

$$aug\ A = \begin{bmatrix} 0 & 0 & 0 & 0 & -1 & -10 \\ 0 & 1 & -1 & 0 & 0 & 15 \\ 1 & 0 & 1 & 0 & 0 & 60 \\ 0 & 0 & 0 & 1 & 0 & 35 \\ 0 & 0 & 0 & 0 & 1 & 10 \end{bmatrix}$$

Again, performing $R_1 \to R_1 + R_5$:

$$aug\ A = \begin{bmatrix} 0 & 0 & 0 & 0 & 0 & 0 \\ 0 & 1 & -1 & 0 & 0 & 15 \\ 1 & 0 & 1 & 0 & 0 & 60 \\ 0 & 0 & 0 & 1 & 0 & 35 \\ 0 & 0 & 0 & 0 & 1 & 10 \end{bmatrix}$$

Rearranging the rows, we finally get:

$$aug\ A = \begin{bmatrix} x_1 & x_2 & x_3 & x_4 & x_5 & \\ 1 & 0 & 1 & 0 & 0 & 60 \\ 0 & 1 & -1 & 0 & 0 & 15 \\ 0 & 0 & 0 & 1 & 0 & 35 \\ 0 & 0 & 0 & 0 & 1 & 10 \\ 0 & 0 & 0 & 0 & 0 & 0 \end{bmatrix} \tag{2.115}$$

Notice that except the third column, in all other remaining columns (of A matrix) we got zero entries above and below the diagonal elements (i.e., 1). The variable associated with this third column, x_3 is considered as the *free* variable (that could have many possible values and this will become clear a little later) and the rest four variables, x_1, x_2, x_4 and x_5 are treated as dependent variables.

Anyway, on the completion of the Gauss–Jordan algorithm, we convert the matrix rows [in Equation (2.115)] into equations as:

$$x_1 = 60 - x_3 \text{ (based on 1st row)}$$
$$x_2 = x_3 + 15 \text{ (based on 2nd row)}$$
$$x_3 = \text{free variable}$$
$$x_4 = 35 \text{ (based on 3rd row)}$$
$$x_5 = 10 \text{ (based on 4th row)}$$

Basically, this is the solution obtained for the given traffic flow system by using the Gauss–Jordan method. Accordingly, we reproduce Figure 2.12 inserting our solution.

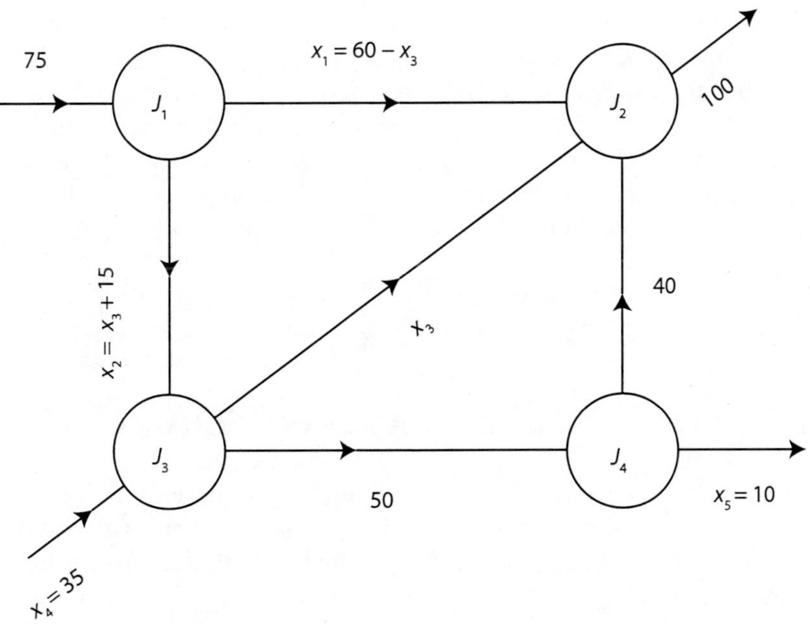

FIGURE 2.12 Traffic flow in a network after finding unknowns (i.e., x_1 to x_5)

Remarks

1. It is obvious that the variable x_3 is remained undermined. It reveals that the given data for the traffic flow system are not enough to find the all 5 unknowns, namely x_1 to x_5.
2. In this situation, one can have infinitely many solutions, depending on the value someone is going to adopt for the free variable x_3.
3. To make this traffic flow system completely specified, the vehicle flow information for one more street [i.e., x_3 that should be integer and ≤ 60 (otherwise, the one-way street rule gets violated)] needs to be pre-specified.

2.18 Summary and Concluding Remarks

This chapter covers several elegant methods for solving the systems of linear algebraic equations. A wide variety of problems are considered and solved using suitable solution techniques, among the direct and indirect methods. In Chapters 5 and 6 of this book, there are several case studies which are modeled and then discretized for simulation. Applying the finite difference method on them results in various sets of algebraic equations; for linear cases, we will use these techniques there.

General Recommendations

If there is no convergence problem, it is recommended to use the iterative methods for higher accuracy (i.e., reduced round-off error), although these methods may require large number of iterations to get the solution. With the advent of fast and inexpensive digital computers, today, the computation load is not a major concern until and unless it is prohibitively large. This apart, for the systems, of which, matrix A is sparse (i.e., most of the elements are zero), the direct methods can cause '*fill-in*' (i.e., introducing new non-zero elements in the matrix), and thereby reducing efficiency.

But remember, there is a limited class of systems that satisfies the convergence criterion. For them, it is recommended to use the direct methods. Unfortunately, for higher-order systems, these methods involve large number of arithmetic computations, thereby large round-off error. To reduce this error, one can employ the pivoting strategy with the direct method (e.g., Gauss elimination with back substitution + partial pivoting).

Appendix 2A

A2.1 Linear versus Non-linear Algebraic Equation

A *single variable* function that can form a straight line in two dimensions is said to be *linear*. It is basically a polynomial of degree 1. Similarly, a function of *two variables* is said to be *linear* when it can form a plane in three dimensions. As indicated, the power of variables should be one.

Examples

Examples are given for a couple of cases in the following section:

Linear single variable function

For single variable case, we have:

$$f(x) = 3x + 5$$

It is linear since the independent variable, x appears only with degree one.

Non-linear single variable function

$$f(x) = 3x^2 + 5$$

Here, x^2 is a non-linear (quadratic) term.

Linear double variable function

For double variable case, both x and y are unknown variable. With this, we have the following examples,

$$f(x, y) = x + 2y \qquad \text{... linear function}$$

$$\begin{cases} 5x + y = 6 \\ 3x + 2y = 9 \end{cases} \qquad \text{... system of linear equations}$$

Non-linear double variable function

It may include the following examples,

$$f(x, y) = x^2 e^{-y} \qquad \text{... non-linear function}$$

$$\begin{cases} x^2 + 3y = 5 \\ 3x + 2y = 9 \end{cases} \qquad \text{... non-linear system of equations}$$

The first equation is non-linear (and second one is linear) and thus, it is overall non-linear.

A2.2 Some Basics of Vector and Matrix Algebra

A2.2.1 Vector

A real n-dimensional vector X is an ordered set of n real numbers. It is commonly written in the following coordinate form:

$$X = (x_1, x_2, \ldots, x_n) \tag{A2.1}$$

We can also write it as:

$$[X] = [x_1 \quad x_2 \quad \cdots \quad x_n]$$

The numbers, namely $x_1, x_2, \ldots, x_n$, are the components of X. The set comprising all n-dimensional vectors is referred to as *n-dimensional space.* A vector is said to be *position vector* when it is employed to represent a point or position in space. On the other hand, it is called *displacement vector* when that vector represents a movement between the two points in space.

Let us consider another vector Y with:

$$Y = (y_1, y_2, \ldots, y_n) \tag{A2.2}$$

and then the vectors have the following properties.

Vector Properties

i) $X = Y$ if and only if $x_i = y_i$ for $i = 1, 2, \ldots, n$

ii) $X + Y = (x_1 + y_1, x_2 + y_2, \ldots, x_n + y_n)$

iii) $-X = (-x_1, -x_2, \ldots, -x_n)$

iv) $Y - X = (y_1 - x_1, y_2 - x_2, \ldots, y_n - x_n)$

v) $Y - X = Y + (-X)$

vi) Scalar multiplication
$cX = (cx_1, cx_2, \ldots, cx_n)$ where c is a real number (scalar)

vii) Linear combination
$cX + dY = (cx_1 + dy_1, cx_2 + dy_2, \ldots, cx_n + dy_n)$

viii) Dot product
$X \bullet Y = (x_1 y_1 + x_2 y_2 + \ldots + x_n y_n)$

Vector Norms

A *norm* is a real-valued function that gives a measure of the size (or length) of multi-component mathematical entries such as vectors or matrices. There are three most commonly used vector norms, which include:

i) Absolute norm (L_1 norm)

$$\|X\|_1 = \sum_{i=1}^{n} |x_i| \tag{A2.3}$$

ii) Euclidean norm (L_2 norm)

$$\|X\|_2 = \left(\sum_{i=1}^{n} x_i^2 \right)^{\frac{1}{2}} \tag{A2.4}$$

iii) Maximum norm (L_∞ norm)

$$\|X\|_\infty = \max_{1 \le i \le n} |x_i| \tag{A2.5}$$

This is also known as the *uniform vector* norm.

A2.2.2 Matrix

A matrix is defined as a rectangular array of numbers, which is arranged in rows and columns in a systematic manner. A matrix A consisting of m number of rows and n number of columns is called an $m \times n$ (read 'm by n') matrix. With this, we can write:

$$A = (a_{ij})_{m\times n} \qquad \text{for } 1 \le i \le m$$
$$1 \le j \le n$$

Here, 'A' represents a matrix and 'a' the elements of that matrix A. As stated, a_{ij} is basically a number in location (i, j) (i.e., stored in ith row and jth column of the array).

Recalling A Few Useful Definitions

The square matrix A is said to be

- Non-singular if $|A| \neq 0$
- Orthogonal if $A^{-1} = A^T$
- Symmetric if $A = A^T$
- Skew symmetric if $A = -A^T$
- Null if $a_{ij} = 0,\ i, j = 1,\ 2,\ \ldots,\ n$
- Unit if $a_{ij} = 0\ (i \neq j)$ and $a_{ii} = 1$ for $i = 1,\ 2,\ \ldots,\ n$
- Diagonal if $a_{ij} = 0,\ i \neq j$
- Lower triangular if $a_{ij} = 0,\ j > i$
- Upper triangular if $a_{ij} = 0,\ i > j$
- Tridiagonal if $a_{ij} = 0$ for $|i - j| > 1$
- Diagonally dominant if $|\,a_{ii}\,| \ge \sum_{\substack{j=1\\ i\neq j}}^{n} |\,a_{ij}\,|,\ i = 1,\ 2,\ \ldots,\ n$
- Strictly diagonally dominant if $|\,a_{ii}\,| > \sum_{\substack{j=1\\ i\neq j}}^{n} |\,a_{ij}\,|,\ i = 1,\ 2,\ \ldots,\ n$
- Banded matrix if $a_{ij} = 0$ for $j - i > p$ and $i - j > q$ with bandwidth (w) of $p + q + 1$ (further elaborated later)

In the above discussion, we have used the following notations:

A = square matrix of order n
A^T = transpose of A (also denoted by A')
A^{-1} = inverse of A

Hermitian Matrix

A square matrix A with complex entries is said to be a Hermitian matrix[27] (denoted by A^H or A^*) if:

$$A = (\bar{A})^T \tag{A2.6}$$

27 Named after the French mathematician Charles Hermite (1822–1901).

Or,

$$A' = \bar{A} \qquad \text{(A2.7)}$$

where,

$\bar{A}$ = complex conjugate of A or conjugate matrix of A

Example A2.1 Hermitian matrix

Prove that the following matrix, A is a Hermitian matrix:

$$A = \begin{bmatrix} 2 & 4-5i \\ 4+5i & -3 \end{bmatrix}$$

Solution

For the given A matrix, we have:

$$A' = \begin{bmatrix} 2 & 4+5i \\ 4-5i & -3 \end{bmatrix}$$

and

$$\bar{A} = \begin{bmatrix} 2 & 4+5i \\ 4-5i & -3 \end{bmatrix}$$

Obviously, $A' = \bar{A}$ and thus matrix A is a Hermitian matrix.

Unitary and Normal Matrix

The square matrix A is said to be unitary if:

$$A^{-1} = (\bar{A})^{\mathrm{T}}$$

and normal if:

$$AA^{*} = A^{*}A$$

Positive Definite Matrix

A real $n \times n$ symmetric matrix A is said to be *symmetric positive definite*, or simply *positive definite* matrix if:

i) $X^{*}AX > 0$ for any vector $X \neq \mathbf{0}$ and $X^{*} = (\bar{X})^{T}$

ii) $X^{*}AX = 0$ when $X = \mathbf{0}$

where,

X = column vector with elements x_i, $i = 1, 2, ..., n$

X^{T} = row vector with elements x_i, $i = 1, 2, ..., n$

$\bar{X}$ = complex conjugate of X

We should note that only a symmetric matrix is considered positive definite if it satisfies the condition given above. A non-symmetric matrix A can never be called positive definite even if it satisfies $X^*AX > 0$ for any non-zero vector X.

If A is a positive definite matrix, then it has the following properties:

- A is non-singular
- All diagonal elements of A are positive
- Largest entry of A lies on the diagonal
- All eigenvalues of A are positive

Example A2.2 Positive definite matrix

Show that the matrix

$$A = \begin{bmatrix} 2 & -1 & 0 \\ -1 & 2 & -1 \\ 0 & -1 & 2 \end{bmatrix}$$

is positive definite.

Solution

For a non-zero vector X with entries x_1, x_2 and x_3, we have:

$$X^*AX = (\overline{X})^T AX = \begin{bmatrix} \overline{x}_1 & \overline{x}_2 & \overline{x}_3 \end{bmatrix} \begin{bmatrix} 2 & -1 & 0 \\ -1 & 2 & -1 \\ 0 & -1 & 2 \end{bmatrix} \begin{bmatrix} x_1 \\ x_2 \\ x_3 \end{bmatrix}$$

$$= [2\overline{x}_1 - \overline{x}_2 \quad -\overline{x}_1 + 2\overline{x}_2 - \overline{x}_3 \quad -\overline{x}_2 + 2\overline{x}_3] \begin{bmatrix} x_1 \\ x_2 \\ x_3 \end{bmatrix}$$

$$= 2x_1\overline{x}_1 - x_1\overline{x}_2 - \overline{x}_1x_2 + 2\overline{x}_2x_2 - x_2\overline{x}_3 - \overline{x}_2x_3 + 2x_3\overline{x}_3$$

Let

$$x_1 = \alpha_1 + i\beta_1 \qquad x_2 = \alpha_2 + i\beta_2 \qquad x_3 = \alpha_3 + i\beta_3$$

Thus,

$$\overline{x}_1 = \alpha_1 - i\beta_1 \qquad \overline{x}_2 = \alpha_2 - i\beta_2 \qquad \overline{x}_3 = \alpha_3 - i\beta_3$$

Accordingly,

$$X^*AX = 2(\alpha_1^2 + \beta_1^2) - 2(\alpha_1\alpha_2 + \beta_1\beta_2) + 2(\alpha_2^2 + \beta_2^2) - 2(\alpha_2\alpha_3 + \beta_2\beta_3) + 2(\alpha_3^2 + \beta_3^2)$$

$$= (\alpha_1 - \alpha_2)^2 + (\beta_1 - \beta_2)^2 + (\alpha_2 - \alpha_3)^2 + (\beta_2 - \beta_3)^2 + (\alpha_1^2 + \beta_1^2 + \alpha_3^2 + \beta_3^2)$$

$\equiv$ non-negative (i.e., > 0)

[zero only if $x_1 = x_2 = x_3 = 0$ (i.e., X is a zero vector)]

Hence, A is positive definite.

Remark

If there is no complex part in the entries of vector X, that means:

$$x_1 = \bar{x}_1 = \alpha_1 \qquad x_2 = \bar{x}_2 = \alpha_2 \qquad x_3 = \bar{x}_3 = \alpha_3$$

then one can use the following condition for positive definite matrix:

$$X^T AX > 0 \text{ for any vector } X \neq \mathbf{0}$$

Banded Matrix

An $n \times n$ matrix A is said to have *upper bandwidth p* if

$$a_{ij} = 0 \qquad \text{for } j - i > p$$

Similarly, this matrix A has *lower bandwidth q* if

$$a_{ij} = 0 \qquad \text{for } i - j > q$$

A matrix having upper bandwidth of p and lower bandwidth of q is said to have *bandwidth* $w = p + q + 1$. An $n \times n$ matrix A is said to be *banded* if $w < 2n - 1$.

When $p = q = 1$, the matrix is called *tridiagonal.* Upper triangular and lower triangular matrices are the special cases of banded matrices.

Remark

Banded structure of a matrix leads to secure a significant computational savings. It is possible by storing only the values of the non-zero entries and their locations, rather than every entry of a large matrix A. In MATLAB, there is a provision to store such entries of the coefficient matrices that are *sparse.*[28]

Example A2.3 Banded matrix

Consider the following matrix:

$$A = \begin{bmatrix} -2 & 3 & 0 & 0 & 0 \\ 3 & 1 & 5 & 0 & 0 \\ 0 & -7 & 8 & 6 & 0 \\ 0 & 0 & 3 & 5 & 8 \\ 0 & 0 & 0 & -3 & 7 \end{bmatrix}$$

Find the bandwidth of this matrix.

Solution

In this example, the upper bandwidth $p = 1$ since $a_{ij} = 0$ whenever $j - i > 1$; see for example, $a_{13} = a_{24} = a_{35} = 0$. Similarly, $q = 1$ since $a_{ij} = 0$ whenever $i - j > 1$; see for example, $a_{31} = a_{42} = a_{53} = 0$. So the total bandwidth is: $w = p + q + 1 = 3$. Notice that A is truly a tridiagonal matrix and thus, it has $p = q = 1$.

---

28 Which means that most of the entries in the matrix are zero.

Remark

Let us make a little modification in the above A matrix:

$$A = \begin{bmatrix} -2 & 3 & 0 & 0 & 0 \\ 3 & 1 & 5 & 0 & 0 \\ 0 & -7 & 8 & 6 & 0 \\ 0 & 0 & 3 & 5 & 8 \\ 0 & 1 & 2 & -3 & 7 \end{bmatrix}$$

For this, it is easy to show $p = 1$ and $q = 3$ (don't think $q = 1$ because of $a_{31} = a_{41} = a_{42} = a_{51} = 0$; but remember $a_{52} \neq 0$). Thus, the total bandwidth is 5.

Matrix Norms

The matrix norm, $\|A\|$, is a non-negative number that satisfies the following properties:

i) $\|A\| > 0$ if $A \neq \mathbf{0}$ and $\|\mathbf{0}\| = 0$
ii) $\|cA\| = |c| \ \|A\|$ where c is an arbitrary complex number
iii) $\|A + B\| \leq \|A\| + \|B\|$
iv) $\|AB\| \leq \|A\| \ \|B\|$

Most common norms for the square $n \times n$ matrix, A include:

i) L_1 norm or *column-sum* norm or *maximum absolute column-sum* norm

$$\|A\|_1 = \max_{1 \leq j \leq n} \sum_{i=1}^{n} |a_{ij}| \tag{A2.8}$$

ii) L_∞ norm or *maximum absolute row-sum* norm

$$\|A\|_\infty = \max_{1 \leq i \leq n} \sum_{j=1}^{n} |a_{ij}| \tag{A2.9}$$

iii) L_2 norm or *Hilbert* norm or *Spectral* norm

$$\|A\|_2 = \sqrt{\lambda_{\max}} \tag{A2.10}$$

Here, $\lambda_{\max}$ is the largest eigenvalue of A^+A, in which A^+ is the adjoint (transpose for real matrices) of A.

iv) *Euclidean* or *Frobenius* norm (analogous to L_2 norm of a vector)

$$\|A\|_e = \|A\|_F = \left(\sum_{i=1}^{n} \sum_{j=1}^{n} a_{ij}^2 \right)^{\frac{1}{2}} \tag{A2.11}$$

A2.2.3 Ill Conditioning

Consider a linear system,

$$AX = B$$

in which, a small perturbation in either A or B leads to a large change in the solution X $(= A^{-1}B)$. Such systems are often called *ill-conditioned* systems, for which, the round-off errors can become a serious problem.

In practice, the values of process parameters, which constitute matrix A and B, are not exactly known; there is always some uncertainty remained. Indeed, determining parameter values somewhat involves either direct or indirect measurements, which are never precise. With this, let us note a few circumstances when ill-conditioning typically occurs:

- When matrix A is *nearly singular* (i.e., $|A|$ is close to zero)
- In a system of two equations, when two lines are nearly parallel
- In a system of three equations, when three planes are nearly parallel

---

Example A2.4 Ill-conditioning

Let us take a simple example to understand the ill-conditioning. For the following system of two equations,

$$\begin{bmatrix} 1 & 2 \\ 0.38 & 0.75 \end{bmatrix} \begin{bmatrix} x_1 \\ x_2 \end{bmatrix} = \begin{bmatrix} 3 \\ 1.14 \end{bmatrix}$$

the exact solution is:

$x_1 = 3$

$x_2 = 0$

Because of an error in measurement, suppose we get a little different value of a_{21} as 0.39 (instead of 0.38). Accordingly, the system is modified to:

$$\begin{bmatrix} 1 & 2 \\ 0.39 & 0.75 \end{bmatrix} \begin{bmatrix} x_1 \\ x_2 \end{bmatrix} = \begin{bmatrix} 3 \\ 1.14 \end{bmatrix}$$

The exact solution of this new system is:

$x_1 = 1$

$x_2 = 1$

which differs significantly from the solution obtained before.

Condition Number

Prior to numerical solution of a system, one needs to know whether it is *ill-conditioned* or *well-conditioned.* This is because for ill-conditioned systems, some specific numerical methods are there (e.g., Gauss elimination with full pivoting); Gauss elimination (without pivoting) fails to deal with such systems. Now to identify the ill-conditioning of a system, the concept of *condition number* of matrix A came into picture and it is elaborated below.

Let us start with the matrix norm that should be consistent with the vector norm. For any vector X and matrix A,

$$\|AX\| \le \|A\| \ \|X\| \tag{A2.12}$$

It is also shown before that the norm of the product of two matrices is less than or equal to the product of the norms of the individual matrices:

$$\|AC\| \le \|A\| \ \|C\| \tag{A2.13}$$

With this background, let us consider two different cases below.

Case 1: Small perturbation in B (in the system $AX = B$)
Let us perturb B by a small quantity ΔB, which leads to change the solution X by ΔX. Accordingly,

$$A(X + \Delta X) = B + \Delta B \tag{A2.14}$$

Since $AX = B$,

$$A\Delta X = \Delta B \tag{A2.15}$$

It gives:

$$\Delta X = A^{-1}\Delta B \tag{A2.16}$$

Clearly it tells us the change occurred in solution X against a change in B.

From Equation (A2.13):

$$\|\Delta X\| \le \|A^{-1}\| \ \|\Delta B\| \tag{A2.17}$$

and

$$\|B\| \le \|A\| \ \|X\| \tag{A2.18}$$

Inverting Equation (A2.18),

$$\frac{1}{\|B\|} \ge \frac{1}{\|A\| \ \|X\|}$$

and from Equation (A2.17),

$$\|\Delta B\| \ge \frac{\|\Delta X\|}{\|A^{-1}\|}$$

Combining,

$$\frac{\|\Delta B\|}{\|B\|} \ge \frac{\|\Delta X\|}{\|X\|}\left(\frac{1}{\|A\| \ \|A^{-1}\|}\right) \tag{A2.19}$$

Rearranging,

$$\frac{\|\Delta X\|}{\|X\|} \le \text{Cond } (A)\frac{\|\Delta B\|}{\|B\|} \tag{A2.20}$$

where the *condition number*,

$$\text{Cond }(A) = \|A\| \ \|A^{-1}\| \tag{A2.21}$$

Case 2: Small perturbation in A (in the system $AX = B$)
Like Equation (A2.14),

$$(A + \Delta A)(X + \Delta X) = B \tag{A2.22}$$

Since $AX = B$,

$$\Delta A(X + \Delta X) = -A\Delta X$$

It gives:

$$-\Delta X = A^{-1}\Delta A(X + \Delta X) \tag{A2.23}$$

Using Equation (A2.13),

$$\|\Delta X\| \le \|A^{-1}\| \ \|\Delta A\| \ \|X + \Delta X\|$$

Rearranging,

$$\frac{\|\Delta X\|}{\|X + \Delta X\|} \le \|A^{-1}\| \ \|\Delta A\| \tag{A2.24}$$

Multiplying right hand side by $\dfrac{\|A\|}{\|A\|}$,

$$\frac{\|\Delta X\|}{\|X + \Delta X\|} \le \text{Cond }(A)\frac{\|\Delta A\|}{\|A\|} \tag{A2.25}$$

Remarks

1. The system $AX = B$ is said to be *ill-conditioned* if Cond (A) is large; with this, small relative changes in A or B (due to say, round-off error) will lead to produce large relative changes in X.
2. If the condition number is close to unity, then the system $AX = B$ is *well-conditioned*. In such a situation, one can have a better flexibility in choosing a suitable numerical solution method.

Example A2.5 Condition number

Find the condition number of a system with

$$A = \begin{bmatrix} 1 & 2 \\ 2 & -1 \end{bmatrix}$$

by using the Euclidean norm.

Solution

For this,

$$|A| = -5$$

$$A^{-1} = \frac{\text{adj } A}{|A|} = -\frac{1}{5}\begin{bmatrix} -1 & -2 \\ -2 & 1 \end{bmatrix}$$

$$= \begin{bmatrix} 0.2 & 0.4 \\ 0.4 & -0.2 \end{bmatrix}$$

Using the Euclidean norm:

$$\|A\| = \sqrt{1+4+4+1} = \sqrt{10}$$

$$\|A^{-1}\| = \sqrt{0.04+0.16+0.16+0.04} = \sqrt{0.4} = \frac{2}{\sqrt{10}}$$

So, the condition number is obtained as:

$$\text{Cond } (A) = \|A\| \; \|A^{-1}\| = \sqrt{10} \times \frac{2}{\sqrt{10}} = 2$$

Condition number is small and somewhat close to 1. Thus, the given system is considered well-conditioned.

Exercise

Review Questions

2.1 There are two vectors, $X = (2,\ 3,\ 4,\ -5)$ and $Y = (5,\ 1,\ -2,\ 4)$. Find:
- Sum, $X + Y$
- Difference, $X - Y$
- Length, $\|X\|$
- Dot product, $X \bullet Y$
- Displacement from X to Y, $Y - X$
- Distance from X to Y, $\|Y - X\|$

2.2 Discuss the different types of solutions and their conditions for the systems of algebraic equations.

2.3 Why the '*naive*' term is used with Gauss elimination?

2.4 Why we go for pivoting with Gauss elimination and other direct methods?

2.5 What is trivial pivoting? Does it involve more computational load than partial pivoting?

2.6 Why full pivoting leads to perform better than partial pivoting?

2.7 What are the different types of triangular systems of linear equations?

2.8 Show that for triangularizing an $n \times n$ matrix in LU decomposition method, the number of multiplications and divisions required is $(n^3 - n)/3$. Use the Gauss elimination approach to get the form $A = LU$.

2.9 Estimate the operational counts as $(5n\text{–}8)$ for the Thomas algorithm.

2.10 List the relative merits and demerits of all the direct and indirect methods.

2.11 For a system $AX = B$, we perturb both A and B with the quantities of ΔA and ΔB respectively. It gives:

$$(A + \Delta A)\bar{X} = B + \Delta B$$

Here, X is the exact solution and $\bar{X}$ the approximate solution. With this, show that:

$$\frac{\|\bar{X} - X\|}{\|X\|} \leq \frac{\text{Cond }(A)}{\left(1 - \|A^{-1}\Delta A\|\right)}\left[\frac{\|\Delta B\|}{\|B\|} + \frac{\|\Delta A\|}{\|A\|}\right]$$

where, Cond (A) = condition number of matrix $A = \|A\| \; \|A^{-1}\|$.

Practice Problems

2.12 Find the eigenvalues of:

$$A = \begin{bmatrix} 1 & 3 \\ -1 & 2 \end{bmatrix}$$

2.13 Find the eigenvalues of:

$$A = \begin{bmatrix} 3 & 2 & -1 \\ 1 & 3 & 2 \\ 2 & 1 & -3 \end{bmatrix}$$

2.14 Find the eigenvalues of:

$$A = \begin{bmatrix} 2 & 0 & 0 \\ 0 & 3 & 4 \\ 0 & 4 & 9 \end{bmatrix}$$

2.15 Find the condition number of the following system:

$$\begin{bmatrix} 2.1 & 1.8 \\ 6.2 & 5.3 \end{bmatrix}\begin{bmatrix} x_1 \\ x_2 \end{bmatrix} = \begin{bmatrix} 2.0 \\ 6.3 \end{bmatrix}$$

2.16 Find the inverse of the matrix:

$$A = \begin{bmatrix} 1 & 1 & 1 \\ 1 & 1 & 2 \\ 1 & 2 & 3 \end{bmatrix}$$

2.17 Find the inverse of the following matrix:

$$A = \begin{bmatrix} 3 & 5 & -7 \\ 6 & 4 & 8 \\ 5 & -8 & 2 \end{bmatrix}$$

using the Gauss elimination method with transforming $AX = B$ to $UX = C$.

2.18 The Wood and Berry distillation column is modeled by:

$$\begin{bmatrix} x_D \\ x_B \end{bmatrix} = \begin{bmatrix} 12.8 & -18.9 \\ 6.6 & -19.4 \end{bmatrix}\begin{bmatrix} R \\ V_B \end{bmatrix}$$

where,

x_D = top product (distillate) composition (mole fraction)

x_B = bottoms composition (mole fraction)

R = reflux rate

V_B = vapor boil-up rate

Find R and V_B when we would like to maintain the product purity at 99 mol%.

2.19 Show that the following matrix is symmetric positive definite:

$$A = \begin{bmatrix} 12 & 4 & -1 \\ 4 & 7 & 1 \\ -1 & 1 & 6 \end{bmatrix}$$

2.20 Show that the following matrix is positive definite:

$$A = \begin{bmatrix} 9 & -3 & 3 & 9 \\ -3 & 17 & -1 & -7 \\ 3 & -1 & 17 & 15 \\ 9 & -7 & 15 & 44 \end{bmatrix}$$

2.21 Find the condition number for the matrix:

$$A = \begin{bmatrix} 4 & 7 \\ 2 & 6 \end{bmatrix}$$

using the

i) 1-norm

ii) Euclidean norm

2.22 Find the condition number for the matrix:

$$A = \begin{bmatrix} 1 & 1 & 3 \\ 1 & 3 & -3 \\ 2 & -4 & 5 \end{bmatrix}$$

using the

i) 1-norm

ii) Euclidean norm

2.23 Compute the L_1 norm for the following matrix:

$$A = \begin{bmatrix} 1 & 2 & 3 \\ -2 & 1 & 4 \\ 3 & -2 & -1 \end{bmatrix}$$

2.24 Compute the L_∞ norm for the following matrix:

$$A = \begin{bmatrix} 15 & -20 & 25 \\ 25 & 20 & -15 \\ 15 & -25 & 20 \end{bmatrix}$$

2.25 Compute the L_1, L_2 and L_∞ norm for the following vector:

$$X = \begin{bmatrix} 5 & 10 & -15 & -25 \end{bmatrix}$$

2.26 Solve the following problem using the matrix inversion method:

$$1.538z_1 - 0.756z_2 + 0.035z_3 = 4.331$$
$$1.004z_1 + 3.752z_2 - 2.183z_3 = -15.80$$
$$2.231z_1 - 0.376z_2 + 3.925z_3 = 13.21$$

(**Ans:** $z_1 = 1.212$, $z_2 = -3.153$, $z_3 = 2.375$).

2.27 Solve the following problem:

$$\begin{bmatrix} -5 & 4 & 0 & 0 \\ 0.5 & -3 & 1.5 & 0 \\ 0 & 0.75 & -3 & 1.25 \\ 0 & 0 & 0.833 & -3 \end{bmatrix} \begin{bmatrix} x_1 \\ x_2 \\ x_3 \\ x_4 \end{bmatrix} = - \begin{bmatrix} 20 \\ 20 \\ 20 \\ 370 \end{bmatrix}$$

(**Ans:** $X = [48.29 \quad 55.37 \quad 81.30 \quad 145.91]^T$).

2.28 Solve the following system of equations:

$$x_1 + 3x_2 - 5x_3 = -1$$
$$9x_1 + 15x_2 - 16x_3 = 8$$
$$31x_1 + 22x_2 - 19x_3 = 34$$

using the Cramer rule (**Ans:** $x_1 = x_2 = x_3 = 1$).

2.29 Solve the following linear system:

$$4x_1 - x_2 = 75$$
$$-x_1 + 4x_2 + x_3 = 100$$
$$-x_2 + 4x_3 - x_4 = 100$$
$$-x_3 + 4x_4 = 75$$

using the:

i) Cramer rule
ii) Matrix inversion method

2.30 Solve the following equations:

$$x_1 - x_2 + 5x_3 = 5$$
$$3x_1 - 3x_2 + 7x_3 = 7$$
$$x_1 + 2x_2 + 4x_3 = 10$$

using the Gauss elimination (back substitution) with and without pivoting.

2.31 Solve the following problem using the Gauss elimination:

$$1.133x_1 + 5.281x_2 = 6.414$$
$$24.14x_1 - 1.210x_2 = 22.93$$

to show the difference between trivial and partial pivoting by the use of four digits of precision.

2.32 Solve the following problem:

$$\begin{bmatrix} 0.0001 & 1 \\ 1 & 1 \end{bmatrix} \begin{bmatrix} x_1 \\ x_2 \end{bmatrix} = \begin{bmatrix} 1 \\ 2 \end{bmatrix}$$

by the Gauss elimination (back substitution) with and without pivoting. Use a hypothetical computer with 4 digits of precision.

2.33 Solve the following system:

$$3.129z_1 - 2.731z_2 - 1.317z_3 + 0.031z_4 = 0.871$$
$$0.113z_1 + 4.368z_2 - 3.013z_3 + 1.673z_4 = 12.29$$
$$1.910z_1 - 2.372z_2 + 5.018z_3 - 3.124z_4 = -15.19$$
$$0.337z_1 + 2.730z_2 + 1.039z_3 - 7.013z_4 = -12.69$$

using the Gauss elimination method
(**Ans:** $z_1 = 0.5$, $z_2 = 1.0$, $z_3 = -1.5$, $z_4 = 2.0$).

2.34 Use Gauss elimination to construct the LU decomposition of the following matrix:

$$A = \begin{bmatrix} 1 & 2 & 3 \\ -1 & 2 & 6 \\ 1 & -2 & 8 \end{bmatrix}$$

2.35 Solve the following system:

$$5x_1 + 2x_2 + x_3 = 22$$
$$3x_1 + 7x_2 - x_3 = -27$$
$$x_1 + 2x_2 + 3x_3 = 16$$

using the (i) Crout method, and (ii) Doolittle method
(**Ans:** $x_1 = 5$, $x_2 = -5$, $x_3 = 7$).

2.36 Solve the following system equations:

$$7x_1 + x_2 - x_3 = 23$$
$$x_1 - 9x_2 + 7x_3 = -25$$
$$3x_1 - 5x_2 + 9x_3 = 19$$

using the (i) Crout method, and (ii) Doolittle method
(**Ans:** $x_1 = 3$, $x_2 = 7$, $x_3 = 5$).

2.37 Solve the following system of equations with LU decomposition (+ pivoting):

$$3x_1 - 2x_2 + x_3 = -10$$
$$2x_1 + 6x_2 - 4x_3 = 44$$
$$-8x_1 - 2x_2 + 5x_3 = -26$$

(**Ans:** $x_1 = 1.2121$, $x_2 = 6.4848$, $x_3 = -0.6667$).

2.38 Solve the following system:

$$3x_1 + 2x_2 + x_3 + x_4 = 4$$
$$2x_1 - 3x_2 + x_3 + 2x_4 = -9$$
$$x_1 + 2x_2 + 5x_3 - 2x_4 = 4$$
$$2x_1 - x_2 - x_3 - 3x_4 = 7$$

using the (i) Crout method, and (ii) Doolittle method
(**Ans:** $x_1 = 1$, $x_2 = 2$, $x_3 = -1$, $x_4 = -2$).

2.39 Consider the following system:

$$\begin{aligned} 3x_1 - x_2 + x_3 &= 8 \\ -x_1 + 3x_2 + x_3 &= 4 \\ x_1 + x_2 + 2x_3 &= 7 \end{aligned}$$

i) show that matrix A is symmetric positive definite.
ii) solve the problem with the Cholesky method with $A = LL^T$.
iii) solve the problem with the Cholesky method with $A = UU^T$.

(**Ans:** $x_1 = 3$, $x_2 = 2$, $x_3 = 1$).

2.40 Solve the following system:

$$\begin{aligned} 12z_1 + 4z_2 - z_3 &= 122 \\ 4z_1 + 7z_2 + z_3 &= 105 \\ -z_1 + z_2 + 6z_3 &= 61 \end{aligned}$$

using the Cholesky method
(**Ans:** $z_1 = 8$, $z_2 = 9$, $z_3 = 10$).

2.41 Solve the following system of linear equations:

$$\begin{aligned} 2x_1 - x_2 &= 3 \\ -x_1 + 2x_2 - x_3 &= -3 \\ -x_2 + 2x_3 &= 5 \end{aligned}$$

by performing the Cholesky decomposition as $A = UU^T$
(**Ans:** $x_1 = 2$, $x_2 = 1$, $x_3 = 3$).

2.42 For a cooling fin system [see Case Study 5.1 (Chapter 5)], we get the following set of equations:

$$\begin{bmatrix} -3 & 1 & 0 & 0 & 0 \\ 1 & -3 & 1 & 0 & 0 \\ 0 & 1 & -3 & 1 & 0 \\ 0 & 0 & 1 & -3 & 1 \\ 0 & 0 & -1 & 0 & 1 \end{bmatrix} \begin{bmatrix} \theta_1 \\ \theta_2 \\ \theta_3 \\ \theta_4 \\ \theta_5 \end{bmatrix} = \begin{bmatrix} -1 \\ 0 \\ 0 \\ 0 \\ 0 \end{bmatrix}$$

Find θ using a suitable numerical technique
(**Ans:** $\theta_1 = 0.3830$, $\theta_2 = 0.1489$, $\theta_3 = 0.0638$, $\theta_4 = 0.0425$, $\theta_5 = 0.0638$).

2.43 Solve the following tridiagonal system using the Thomas algorithm:

$$\begin{bmatrix} 2 & -1 & 0 & 0 & 0 \\ -1 & 2 & -1 & 0 & 0 \\ 0 & -1 & 2 & -1 & 0 \\ 0 & 0 & -1 & 2 & -1 \\ 0 & 0 & 0 & -1 & 2 \end{bmatrix} \begin{bmatrix} u_1 \\ u_2 \\ u_3 \\ u_4 \\ u_5 \end{bmatrix} = k^2 \begin{bmatrix} 2 \\ 2 \\ 2 \\ 2 \\ 2 \end{bmatrix}$$

with $k = \dfrac{1}{6}$

(**Ans:** $u_1 = 0.138889$, $u_2 = 0.222222$, $u_3 = 0.25$, $u_4 = 0.222222$, $u_5 = 0.138889$).

2.44 Solve the following linear set of equations:

$$4u_1 - u_2 - u_3 = 0$$
$$u_1 - 4u_2 + u_4 = 0$$
$$-u_1 + 4u_3 - u_4 = 6$$
$$-u_2 - u_3 + 4u_4 = 3$$

This system is obtained in Example 6.2 (Chapter 6)
(**Ans**: $u_1 = 0.625$, $u_2 = 0.5$, $u_3 = 2.0$, $u_4 = 1.375$).

2.45 Solve the following tridiagonal system with the use of Thomas algorithm:

$$\begin{bmatrix} a & -1 & 0 \\ -1 & a & -1 \\ 0 & -1 & a \end{bmatrix} \begin{bmatrix} x_1 \\ x_2 \\ x_3 \end{bmatrix} = \begin{bmatrix} 60b+180 \\ 60b+130 \\ 60b+180 \end{bmatrix}$$

where, $a = 3$ and $b = -1$. This system is obtained in Case Study 6.4 (Chapter 6).

2.46 Find the inverse of matrix A:

$$A = \begin{bmatrix} 1 & -2 & 3 \\ 2 & 1 & -3 \\ 1 & 2 & -1 \end{bmatrix}$$

using the Gauss–Jordan method with

$$[A \mid I] \to [I \mid A^{-1}]$$

2.47 Using the Gauss–Jordan method, find A^{-1} for:

$$A = \begin{bmatrix} 1 & 2 & 1 \\ 4 & 3 & -1 \\ 3 & 2 & 1 \end{bmatrix}$$

2.48 Consider the following system:

$$5x_1 + 2x_2 + x_3 = 9$$
$$2x_1 - 3x_2 - 2x_3 = 5$$
$$x_1 + 2x_2 - 3x_3 = -3$$

i) Transform the augumented matrix using the Gauss–Jordan method

$$aug\ A \equiv [A \mid B \mid I] \to [I \mid \underline{B} \mid A^{-1}]$$

and find $\underline{B}$ and A^{-1}.

ii) Solve the given system using the Gauss–Jordan method.

2.49 Solve the following tridiagonal system of linear equations using the Thomas algorithm obtained in Case study 6.8 of this book:

$$\begin{bmatrix} -(4\beta+1) & 4\beta & 0 & 0 \\ \frac{\beta}{2} & -(2\beta+1) & \frac{3\beta}{2} & 0 \\ 0 & \frac{3\beta}{4} & -(2\beta+1) & \frac{5\beta}{4} \\ 0 & 0 & \frac{5\beta}{6} & -(2\beta+1) \end{bmatrix} \begin{bmatrix} x_1 \\ x_2 \\ x_3 \\ x_4 \end{bmatrix} = \begin{bmatrix} -50 \\ -50 \\ -50 \\ -50-175\beta \end{bmatrix}$$

with $\beta = 2.3$.

2.50 Solve the following system of algebraic equations for $a=-2$:

$$\begin{bmatrix} 1 & 0 & 0 & 0 & 0 & 0 & 0 & 0 & 0 \\ 1 & a & 1 & 0 & 0 & 0 & 0 & 0 & 0 \\ 0 & 1 & a & 1 & 0 & 0 & 0 & 0 & 0 \\ 0 & 0 & 1 & a & 1 & 0 & 0 & 0 & 0 \\ 0 & 0 & 0 & 1 & a & 1 & 0 & 0 & 0 \\ 0 & 0 & 0 & 0 & 1 & a & 1 & 0 & 0 \\ 0 & 0 & 0 & 0 & 0 & 1 & a & 1 & 0 \\ 0 & 0 & 0 & 0 & 0 & 0 & 1 & a & 1 \\ 0 & 0 & 0 & 0 & 0 & 0 & 0 & 0 & 1 \end{bmatrix} \begin{bmatrix} x_1 \\ x_2 \\ x_3 \\ x_4 \\ x_5 \\ x_6 \\ x_7 \\ x_8 \\ x_9 \end{bmatrix} = \begin{bmatrix} 0 \\ 0 \\ 0 \\ 0 \\ 0 \\ 0 \\ 0 \\ 0 \\ 1 \end{bmatrix}$$

2.51 Solve the following system equations using the Gauss–Jordan method:

$$105x_1+59x_2-37x_3=-28$$
$$x_1-213x_2+527x_3=1268$$
$$78x_1+37x_2-320x_3=-599$$

(Ans: $x_1=1,\ x_2=-1,\ x_3=2$).

2.52 Solve the following linear system of equations using the Gauss–Jordan method:

$$3.152x_1-5.378x_2+7.015x_3=13.90$$
$$2.730x_1+5.881x_2-6.318x_3=-4.0$$
$$1.116x_1-7.307x_2+3.216x_3=-4.72$$

(Ans: $x_1 = 1.116,\ x_2 = 2.215,\ x_3 = 3.178$).

2.53 Consider the following system of equations:

$$4x_1-x_2+x_3=6$$
$$4x_1-8x_2+x_3=-22$$
$$-2x_1+x_2+5x_3=10$$

Assume the initial guess values as: $x_1^{(0)}=1$, $x_2^{(0)}=2$ and $x_3^{(0)}=3$. Find the solution (tol = 10^{-6}) using

i) Jacobi method
ii) Gauss–Seidel method

(Ans: $x_1=2,\ x_2=4,\ x_3=2$).

2.54 Rearrange the system equations of Problem 2.53 as:

$$-2x_1+x_2+5x_3=10$$
$$4x_1-x_2+x_3=6$$
$$4x_1-8x_2+x_3=-22$$

Does the solution converge when we use the Jacobi method with $x_1^{(0)}=1$, $x_2^{(0)}=2$ and $x_3^{(0)}=3$?

2.55 Rearrange the system equations of Problem 2.53 as:

$$-2x_1+x_2+5x_3=10$$
$$4x_1-8x_2+x_3=-22$$
$$4x_1-x_2+x_3=6$$

Given: $x_1^{(0)}=1$, $x_2^{(0)}=2$ and $x_3^{(0)}=3$. Check the convergence for:

i) Jacobi method

ii) Gauss–Seidel method

2.56 Solve the following linear system of equations using the Gauss–Seidel method:

$$10.045x_1-3.284x_2+5.313x_3=20.515$$
$$2.156x_1+9.324x_2-6.139x_3=2.336$$
$$3.574x_1+2.136x_2-8.213x_3=-18.003$$

Given: $x_1^{(0)}=x_2^{(0)}=x_3^{(0)}=2$ and $tol=10^{-3}$

(**Ans**: x_1 = 1.0433, x_2 = 2.1132, x_3 = 3.1956).

2.57 Consider a two-dimensional body shown in Figure 2.13 for steady state heat conduction. For this, the model equations include:

$$4T_1=50+T_2+T_3+200$$
$$4T_2=T_1+50+T_4+200$$
$$4T_3=50+T_4+50+T_1$$
$$4T_4=T_3+50+50+T_2$$

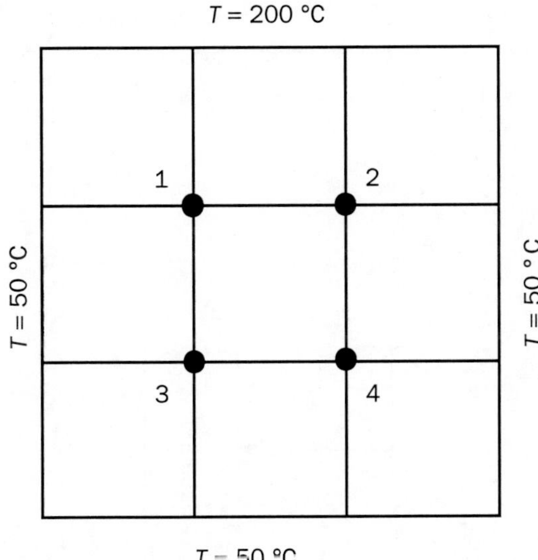

Figure 2.13 Two-dimensional body

Find temperature T at 4 nodal points using the successive relaxation method with: $T_1^{(0)}=T_2^{(0)}=T_3^{(0)}=T_4^{(0)}=50$ °C, $tol=10^{-3}$ and $w=1.1$.

2.58 Solve the following system:

$$10x_1-5x_2+3x_3+x_4=-5$$
$$2x_1+9x_2-4x_3+2x_4=18$$
$$3x_1-2x_2+7x_3-x_4=-5$$
$$8x_1-9x_2-10x_3+32x_4=12$$

using the successive relaxation method with assuming initial guess values for $tol = 10^{-4}$, and $w = 0.9$ and 1.2.

2.59 Solve the following set of equations:

$$1.138x_1 - 0.0124x_2 + x_3 = 0.1012$$
$$2.746x_1 + 3.736x_2 - 1.189x_3 = 12.4558$$
$$3.214x_1 - 4.380x_2 + 5.782x_3 = -12.4187$$

using the successive relaxation method with $x_1^{(0)} = x_2^{(0)} = x_3^{(0)} = 1$, $tol = 10^{-3}$ and $w = 0.92$.

2.60 Solve the following set of equations:

$$9.165x + 3.212y - 5.638z = 16.138$$
$$9.299x - 15.378y + 5.124z = 16.138$$
$$2.134x + 5.613y - 10.328z = 16.138$$

using the successive relaxation method with $x^{(0)} = y^{(0)} = z^{(0)} = 2$, $tol = 10^{-4}$ and $w = 1.1$.

2.61 The separation of a ternary system, consisting of species *A*, *B* and *C*, is carried out by the use of two separators as shown in Figure 2.14.

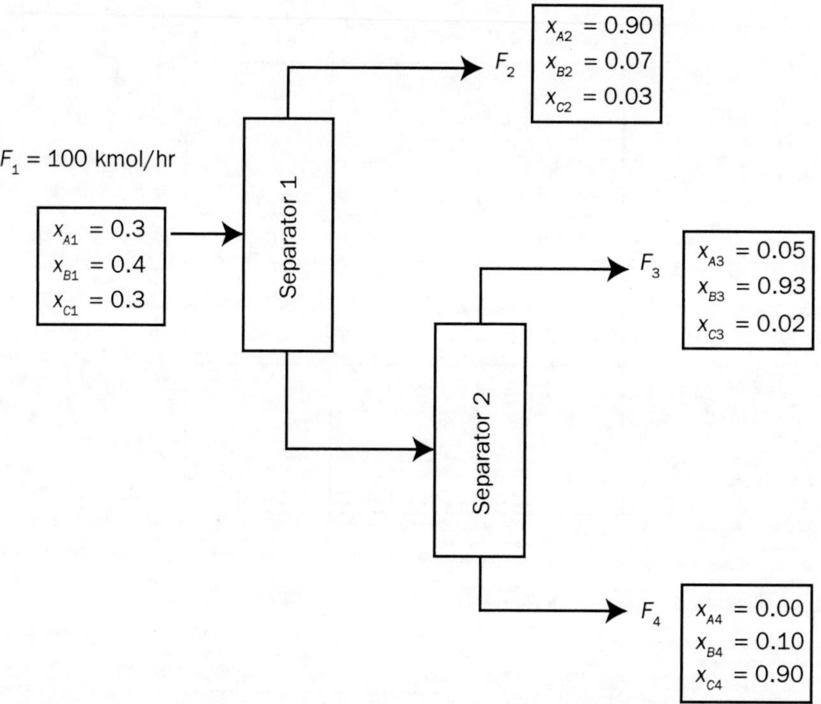

FIGURE 2.14 Process configuration for separating a ternary system (F = flow rate and x = mole fraction)

(a) Write three component mole balance equations based on:

$$F_1 x_{A1} = F_2 x_{A2} + F_3 x_{A3} + F_4 x_{A4}$$
$$F_1 x_{B1} = F_2 x_{B2} + F_3 x_{B3} + F_4 x_{B4}$$
$$F_1 x_{C1} = F_2 x_{C2} + F_3 x_{C3} + F_4 x_{C4}$$

(b) Solve the resulting equations using the Gauss elimination method with:
i) Partial pivoting
ii) Scaled partial pivoting
iii) Full pivoting

2.62 There are three streams (F_1, F_2 and F_3) that are getting blended in a mixer shown in Figure 2.15. Each stream contains three species, namely *A*, *B* and *C*.

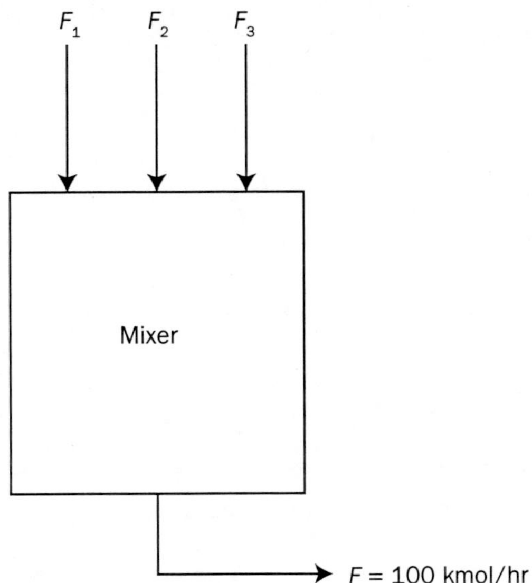

Figure 2.15 Schematic of a mixer

The composition (mole fraction) of all three inlet streams is listed in Table 2.8.

Table 2.8 Composition of all inlet streams

Stream	Species *A*	Species *B*	Species *C*
F_1	0.3	0.3	0.4
F_2	0.1	0.2	0.7
F_3	0.8	0.15	0.05

Both the flowrates F_2 and F_3 are to be twice the flow rate F_1.

(a) Develop the total mole balance equation and find the flow rates by hand.

Hint: $F = 100 = F_1 + F_2 + F_3 = F_1 + 2F_1 + 2F_1 = 20 + 40 + 40$ (i.e., $F_1 = 20$, $F_2 = F_3 = 40$).

(b) Develop three component mole balance equations and find the composition of exit stream.

(c) Having the composition of exit stream known, solve the developed component mole balance equations when only F is known (i.e., F_1, F_2 and F_3 are unknown) by using the Gauss–Seidel method (tol = 10^{-3}), and find the flow rate of all three inlet streams.

2.63 Consider a network of one-way roads demonstrated in Figure 2.16. All the numbers in this figure represent the number of vehicles flowed per hour. Find the four unknowns, x_1 to x_4, using the Gauss–Jordan method and make comments, if any.

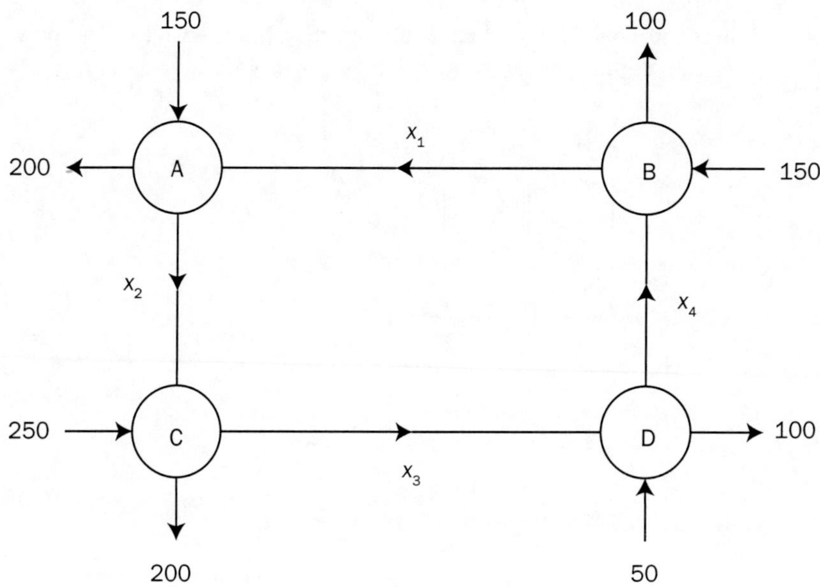

FIGURE **2.16** A rectangular network of roads

2.64 Consider a traffic circle illustrated in Figure 2.17.

i) Use the Gauss–Jordan method to find the traffic flows, x_1 to x_5 (number of vehicles/min), for which, the data are not available.

ii) Any comment on the type of solution (unique solution/no solution/infinitely many solutions)?

iii) If x_3 is additionally given in this problem as 50, find the rest unknown vehicle flow rates using the same Gauss–Jordan method. Any comment on the type of solution (unique solution/no solution/infinitely many solutions)?

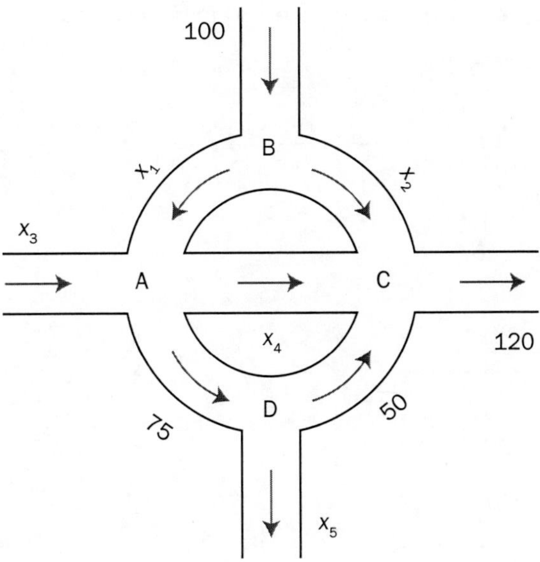

FIGURE 2.17 A circular network of streets

2.65 An electrical network system is shown in Figure 2.18. Find the currents, I_1 to I_6 using a suitable direct as well as indirect method.

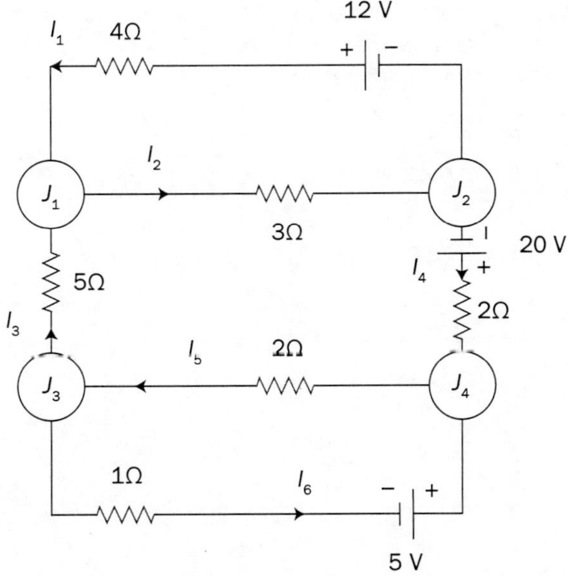

FIGURE 2.18 An electrical network system

2.66 For the electrical network system detailed in Figure 2.19, find the loop currents, I_1, I_2 and I_3.

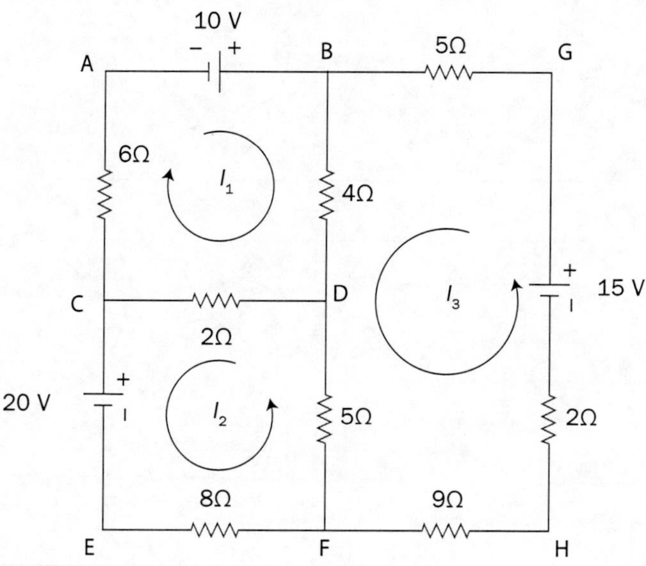

FIGURE 2.19 An electrical network system

2.67 The separation of a quaternary mixture, consisting of species *A*, *B*, *C* and *D*, takes place in an integrated process shown in Figure 2.20.

i) Develop the model equations based on the component mole balance at steady state condition. The steady state compositions (mole fractions) for all four product streams are shown in the figure.

ii) Find the product flow rates (D_1, B_1, D_2 and B_2) by solving the model equations developed in Part (a) by using the:
   - Gauss elimination method
   - Gauss–Seidel method

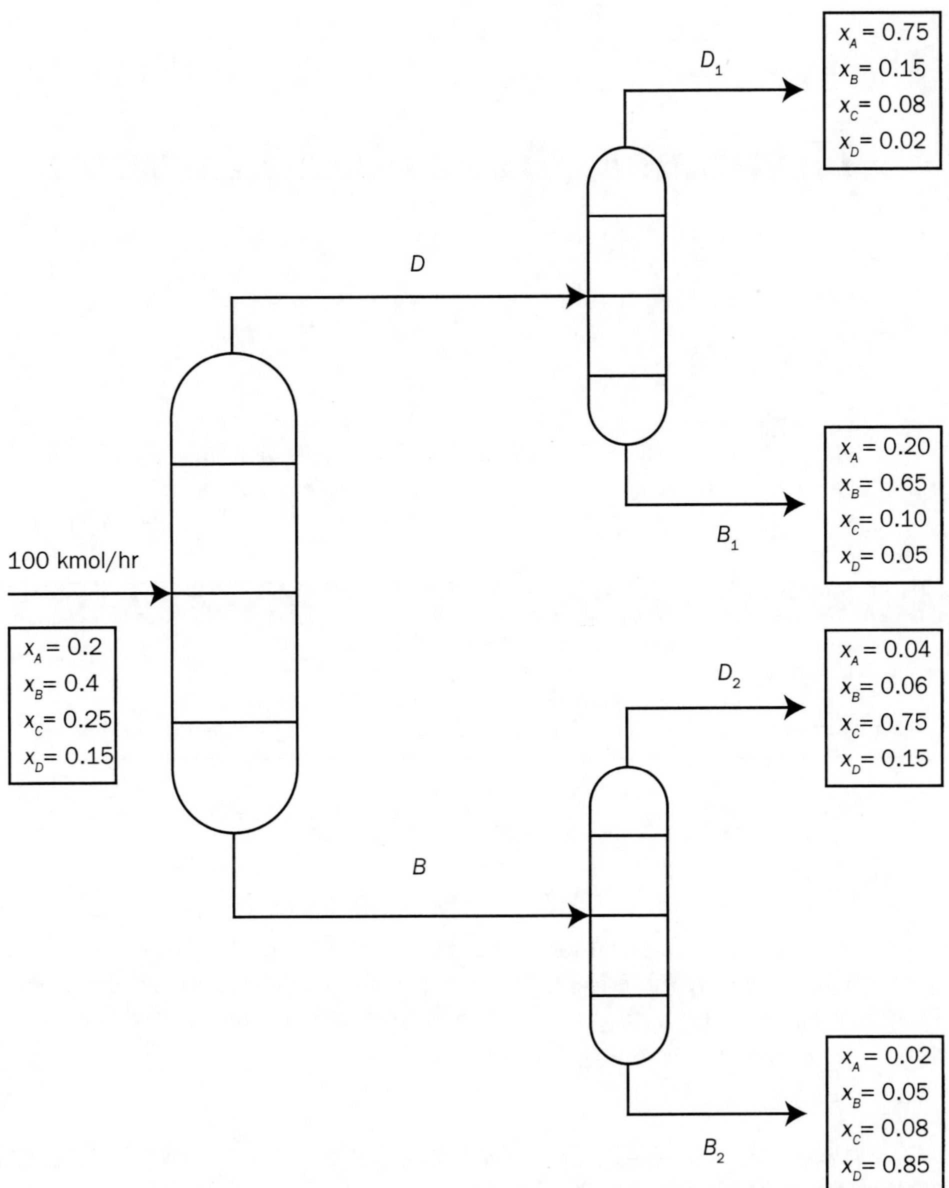

Figure 2.20 Schematic of the integrated process for the separation of a quaternary mixture

References

Gupta, S. K. (2003). *Numerical Methods for Engineers*, 1st ed. (reprinted version), New Delhi: New Age International (P) Ltd.

Lambers, J. V. and Sumner, A. C. (2019). *Explorations in Numerical Analysis*, 1st ed., Singapore: World Scientific.

Mathews, J. H. (2000). *Numerical Methods for Mathematics, Science, and Engineering*, 2nd ed., New Delhi: Prentice-Hall.

3

Non-linear Algebraic Equation

"Calculating does not equal mathematics. It's a subsection of it. In years gone by it was the limiting factor, but computers now allow you to make the whole of mathematics more intellectual."
– Conrad Wolfram

Key Learning Objectives

- Learning how to solve systems of non-linear algebraic equations
- Knowing the bracketing, open, and polynomial-based numerical methods
- Learning how to develop the numerical algorithm for different systems
- Simulating a couple of industry-relevant processes with choosing a suitable algorithm

3.1 Introduction

This chapter covers the numerical solution of the systems of non-linear algebraic equations. For this, there are several numerical techniques available, which are often called the *iterative convergence methods*. Prior to discussing these methods, let us start with a non-linear function of a single variable represented by:

$$y = f(x) = 0 \tag{3.1}$$

One can solve it graphically by producing a plot of y versus x. As shown in Figure 3.1a, the curve intersects the x-axis at point x^*, which is called the *root* (or zero) of the function $f(x)$, since it satisfies Equation (3.1). It is true that a non-linear algebraic equation may have multiple solutions or roots (e.g., x_1^*, x_2^* and x_3^*) as shown in Figure 3.1b. In fact, we all know that an nth order polynomial has n solutions.

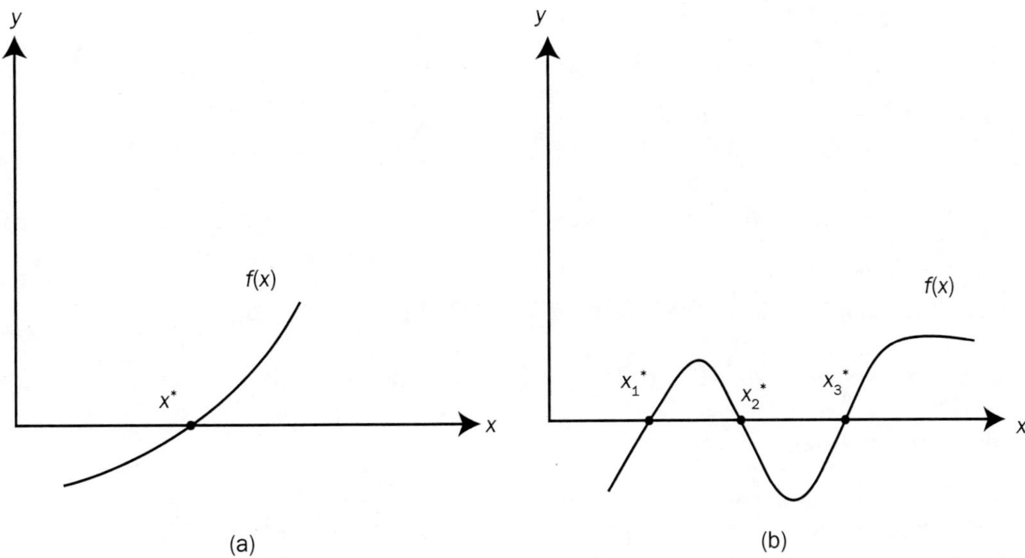

FIGURE 3.1 Graphical solution to $f(x) = 0$: (a) single solution (x^*), and (b) multiple solutions (x_1^*, x_2^* and x_3^*)

This *graphical* method is useful for rough estimates of the root and because of the lack of precision, it has very limited application. However, one can use this rough estimate as an initial guess. Alternatively, one can use the *trial-and-error* approach, which starts with assuming the initial guess value of x and evaluates whether $f(x) = 0$. If it is not satisfied (that is almost always true), re-assume another guess for x and again evaluate $f(x)$ to check the improvement in result, if any. This is repeated until $f(x)$ reaches sufficiently close to zero.

It is somewhat obvious that finding root through the graphical or trial-and-error method is insufficient and inadequate for scientific and engineering problems. Thus the necessity of numerical methods arises. Next, we will pay our attention to know various numerical methods and how to solve the non-linear systems by judiciously using them.

3.2 Iterative Convergence Methods

Classifying Numerical Methods

There are typically three classes of iterative convergence methods available for numerical solution. They include:

- Bracketing method[1]
  - Bisection method
  - False Position method

1 It is called so because this type of method starts with two initial guesses that *bracket* the root (i.e., they are on either side of the root) and this bracketing continues till the end.

- Open method[2]
  - Secant method
  - Newton–Raphson method
- Polynomial-based method[3]
  - Muller method
  - Chebyshev method

All these numerical methods are discussed in the following sections with their applications.

3.3 Bisection or Interval Halving Method

The *Bisection method* typically involves the following computational steps in sequence.

Computational steps

Step 1: Identify two guess values of x (say, x_k and x_{k+1}) at the starting such that one where the function $f(x)$ is negative (<0) and another where $f(x)$ is positive (>0). This leads to:

$$f(x_k)f(x_{k+1})<0$$

Step 2: Determine the midpoint and then find corresponding $f(x)$ at that point. The midpoint is obtained from:

$$x_{k+2}=\frac{1}{2}(x_k+x_{k+1})$$

Then the corresponding function $f(x_{k+2})$ needs to be evaluated.

Step 3: In this step, our job is to replace one of the two x values (i.e., either x_k or x_{k+1}) by the value of x at the midpoint (i.e., x_{k+2}). For this, one should follow as:

if the value of $f(x_k)$ has the same sign with that of $f(x_{k+2})$, then replace x_k by x_{k+2} (i.e., $x_k = x_{k+2}$), else replace x_{k+1} by x_{k+2} (i.e., $x_{k+1} = x_{k+2}$).

Step 4: Check for convergence. If it is not converged, go back to Step 2.

Note that in this numerical algorithm, the iterations proceed with $k = 0, 1, 2, \ldots$ and so on.

Since we bisect the x-interval in each iteration step, it is called the *Bisection method.* Figure 3.2 describes this numerical technique pictorially and the flowchart based on the above four steps is developed in Figure 3.3. This iterative method continues until the size of the interval shrinks below a convergence tolerance level. When it goes lower than a specified tolerance limit (denoted by 'tol'), we get an approximate value of the root as the solution.

This Bisection method is also called the *interval halving* approach, since in each iteration, it can halve the size of the interval. Further, seeing Figure 3.4, it becomes clear to us that why it falls under the class of the *bracketing* method. In fact, we could make such a plot between $f(x)$ versus x and get a decent guess for x (i.e., x_0 and x_1) from bracketing.

2 This method requires one or more initial guesses and there is no such exercise involved in bracketing the root (i.e., remained *open*).

3 This method is derived through polynomial approximation.

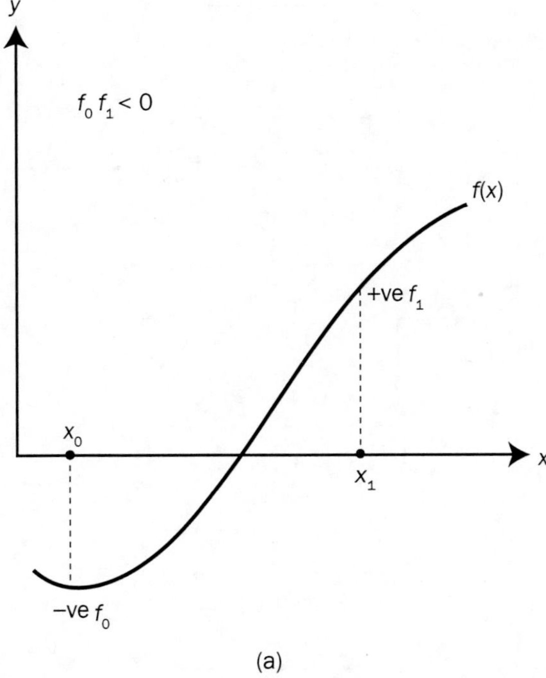

(a)

Figure 3.2a Representing computational Step 1 of the Bisection method (how to identify two guess values of x, namely x_0 and x_1) [here, $f_k = f(x_k)$].

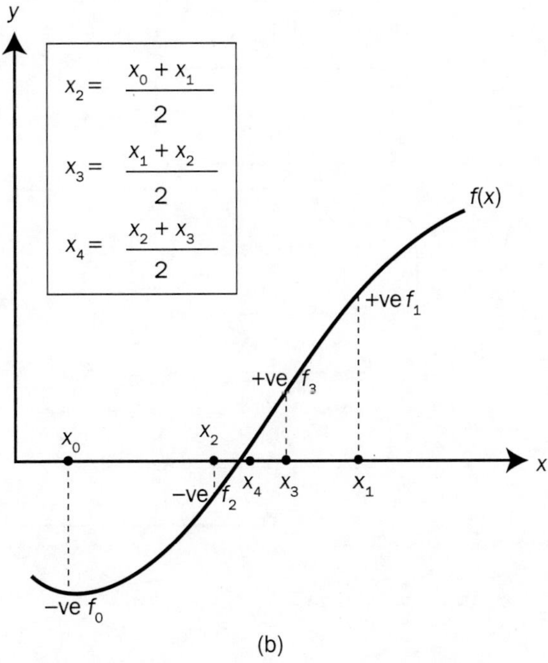

(b)

Figure 3.2b Graphical representation of the Bisection method as explained above with four computational steps.

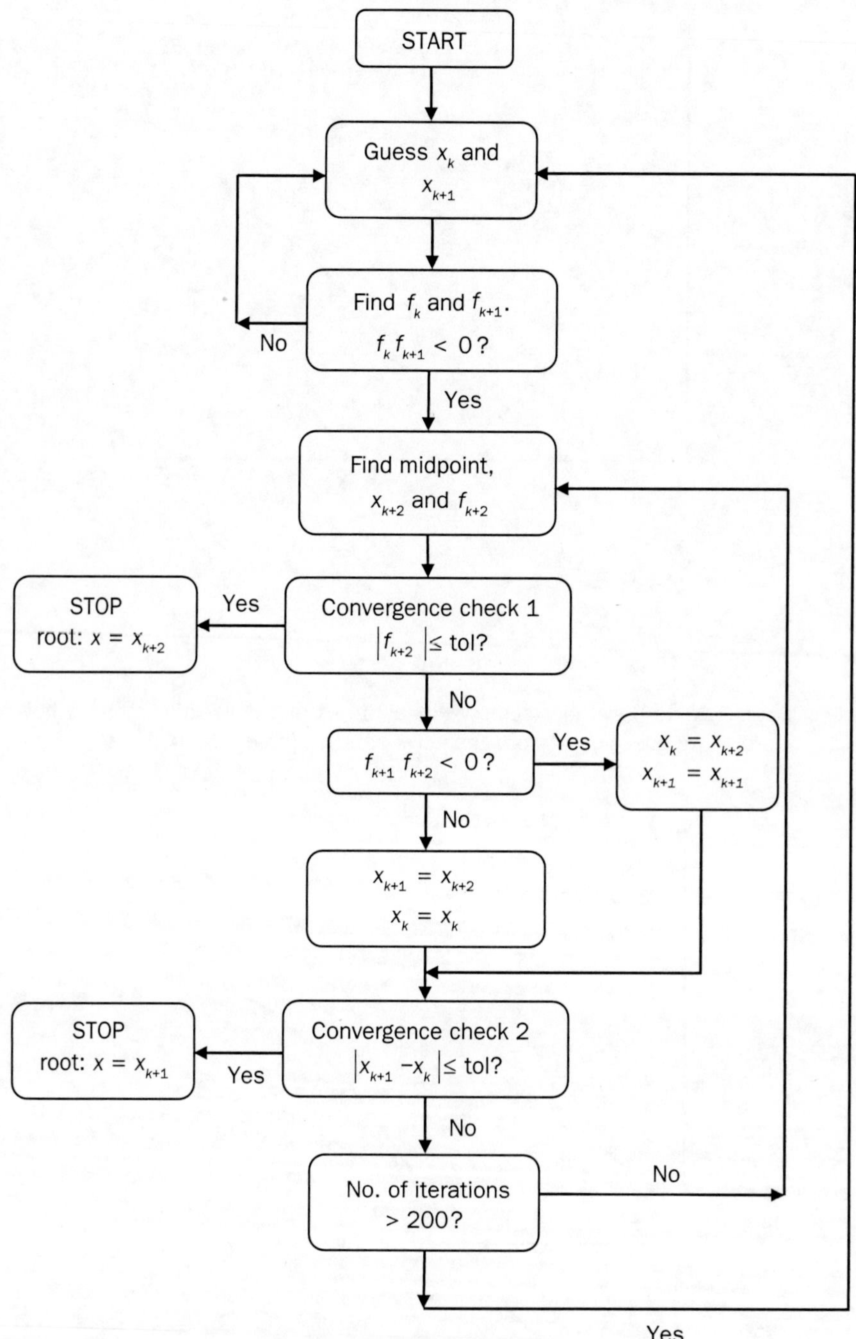

FIGURE 3.3 Flowchart for the Bisection method[4]

4 It is shown that here we stop our iteration if either of the two convergence checks is satisfied. However, one may use both of them as followed in a few of the cases later.

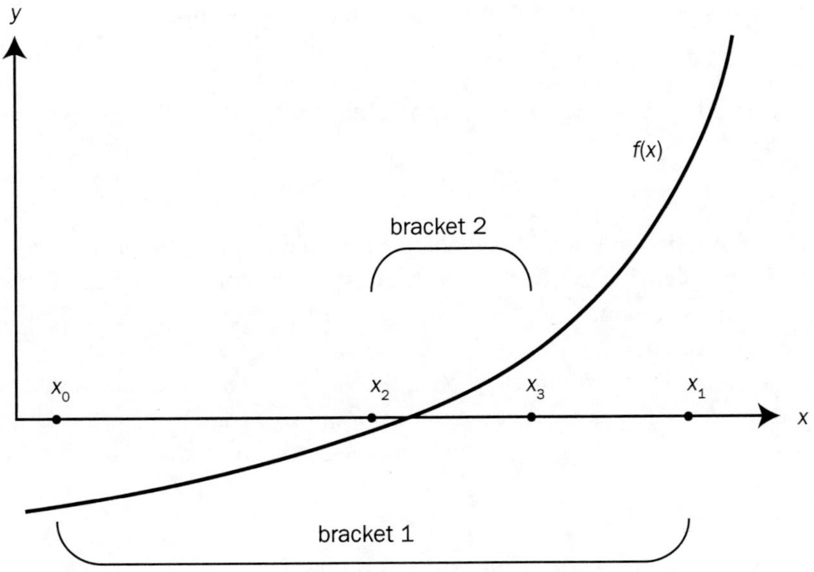

FIGURE 3.4 Brackets (successive) in the Bisection method

Remarks

1. As shown, the Bisection method locates a root by repeatedly narrowing the distance between the two guesses. If an interval contains a root, this simple numerical approach never fails.

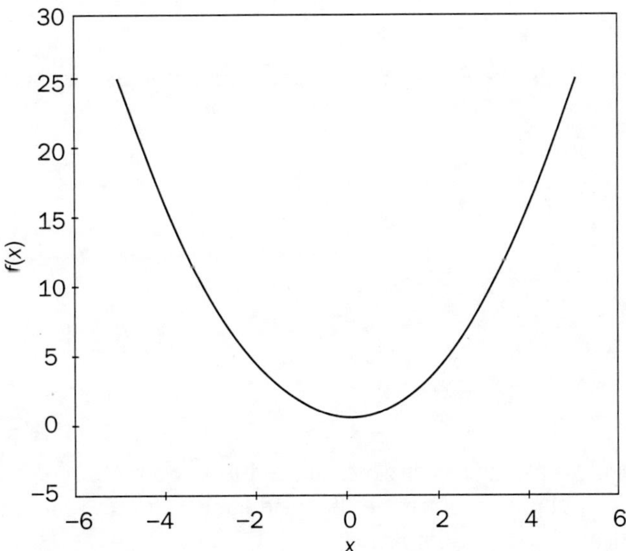

FIGURE 3.5 Illustration of $f(x) = x^2$, whose root is: $x = 0$

At this point of time, let us take an interesting example,

$f(x) = x^2$

illustrated in Figure 3.5. For this, one cannot identify the two guess values that lead to satisfy:

$f(x_k)\, f(x_{k+1}) < 0$

Thus, Bisection method is not suitable for such an exceptional system.

2. Major problem of the Bisection method is its slow convergence rate. This is because this method retains many old estimates. This issue is addressed in the next part with an example.
3. It is not easily extendable to multi-variable systems. For example, for an n-dimensional system, this technique requires n different initial brackets. Then in each bracketed region, the interval needs to be narrowed down, which is not a trivial task.

3.3.1 Linear Convergence: Bisection Method

The Bisection method is continued iteratively until the following stopping condition is satisfied:

$$|E_{k+1}| = |x_{k+1} - x_k| \le \text{tol}$$

Here, E is the error between two successive iterates. Recall that 'tol' denotes the pre-determined tolerance threshold.

As mentioned, the line segment (with an initial length, $|E_1| = |x_1 - x_0|$) is divided into halves at each iteration. Accordingly, we have:

$$\text{at } k = 1 \qquad |E_2| = \frac{|E_1|}{2}$$

where, as indicated, E_1 is the error based on two guess values, x_k and x_{k+1} (with $k = 0$). Similarly,

$$\text{at } k = 2 \qquad |E_3| = \frac{|E_2|}{2} = \frac{|E_1|}{2^2}$$

$$\text{at } k = k \qquad |E_{k+1}| = \frac{|E_k|}{2} = \frac{|E_1|}{2^k} \tag{3.2}$$

Based on Equation (A3.11)[5], we get:

$$\alpha = 0.5$$

$$m = 1$$

This indicates the *linear rate of convergence* of the Bisection method.

Note that this formulation will further be used later in finding the number of iterations required for the Bisection method *a priori*.

5 See Appendix A3.2.

Example 3.1 Bisection method

Find the root of the following non-linear algebraic equation,

$$f(x) = 5x - \exp(-x^2) + \sin x = 0$$

using the initial guess values of 0 and 0.5. Acceptable tolerance limit is given as 10^{-6}.

Solution

Using the computer-assisted algorithm developed above for Bisection method, we produce the results in Table 3.1 through iteration and obtain the root x as 0.162445 within the given tolerance.[6] Note that here x is in radian and absolute error is $|x_{k+1} - x_k|$.

Further, let us do the second check with:

$$|f(x^*)| = |f(0.162445)| = 2.5 \times 10^{-7}$$

Obviously, it satisfies the desired tolerance limit (i.e., 10^{-6}) as well.

Remark

The Bisection method converges in 19 iterations, whereas the Secant method finds root in 4 iterations (shown later in Example 3.2 for the same equation). It clearly reveals that the Bisection method provides a slower convergence compared to the Secant method.

TABLE 3.1 Numerical results of the Bisection method

Iteration (k)	x_k	x_{k+1}	Absolute error
0	0.000000	0.500000	0.500000
1	0.000000	0.250000	0.250000
2	0.125000	0.250000	0.125000
3	0.125000	0.187500	0.062500
4	0.156250	0.187500	0.031250
5	0.156250	0.171875	0.015625
6	0.156250	0.164062	0.007812
7	0.160156	0.164062	0.003906
8	0.162109	0.164062	0.001953
9	0.162109	0.163086	0.000977
10	0.162109	0.162598	0.000489
11	0.162354	0.162598	0.000244
12	0.162354	0.162476	0.000122
13	0.162415	0.162476	0.000061
14	0.162415	0.162445	0.000030
15	0.162430	0.162445	0.000015

Contd

6 Note that for a fair comparison between all convergence methods, the desired tolerance in terms of absolute error [Equation (1.19)] is considered as a stopping criterion. The true error [Equation (1.9)] is not considered here since we do not know the exact solution, x^* *a priori*.

Contd

Iteration (k)	x_k	x_{k+1}	Absolute error
16	0.162437	0.162445	0.000008
17	0.162441	0.162445	0.000004
18	0.162443	0.162445	0.000002
19	0.162444	0.162445	0.000001

3.3.2 Finding Number of Iterations: Bisection Method

The number of iterations required in the Bisection method to satisfy the desired tolerance criterion can be estimated *a priori*. For this, let us rewrite Equation (3.2) for the *n*th iteration as:

$$|E_n| = \frac{|E_0|}{2^n}$$

Notice that here E_0 is the difference between the two initial guess values. We stop our computation as soon as we meet:

$$\frac{|E_0|}{2^n} \leq \text{tol}$$

Taking logarithm and rearranging,

$$n \geq \frac{\log|E_0| - \log(\text{tol})}{\log 2}$$

This is the final formula that can be used irrespective of the form of function, f to find the number of iterations *a priori*.

Applying to Example 3.1

Here,

$$|E_0| = 0.5$$

$$\text{tol} = 10^{-6}$$

So,

$$n \geq \frac{\log(0.5) + 6}{\log 2} = 18.93 \approx 19$$

So we need at the most 19 iterations (same is noticed in Example 3.1) to obtain the approximate solution accurate to 10^{-6}.

Remarks

1. This procedure of finding the number of iterations is valid only when we use the stopping criterion in terms of the absolute error, $|E|$.
2. A few more cases are listed in Table 3.2 for the two initial guess values of 0 and 1.

TABLE 3.2 Number of iterations (= n) in the Bisection method

tol	n
10^{-2}	7
10^{-3}	10
10^{-4}	14
10^{-5}	17
10^{-6}	20
10^{-7}	24
10^{-8}	27

3.4 Secant Method

The Secant method can improve the slow convergence involved in the Bisection method. Rather than bisecting the interval to have the next approximation to the root, the Secant method involves the construction of a secant line and finds its x-intercept as the next root estimate.

Let us adopt Equation (3.1) for finding its root. For this, we guess x_k and x_{k+1} as the two approximations to the root. Then, as stated, construct a straight line by joining the two points (x_k, f_k) and (x_{k+1}, f_{k+1}). This line is called the *secant* (or *chord*) of the function f, and thus the name of the method (i.e., the Secant or Chord method).

The slope of the straight line has the following form:

$$m = \frac{f_{k+1} - f_k}{x_{k+1} - x_k} \tag{3.3}$$

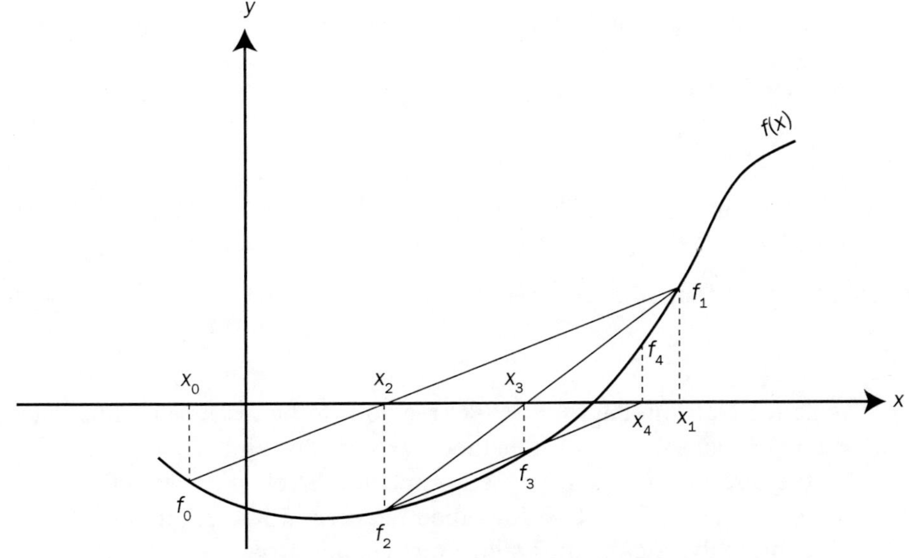

FIGURE 3.6 Graphical representation of the Secant method

The point of intersection of this straight line with the x-axis is considered as the next approximation to the root. To determine it (i.e., x_{k+2}), one needs to form an equation of straight line as:

$$f_{k+2}-f_{k+1}=m(x_{k+2}-x_{k+1})=\frac{f_{k+1}-f_k}{x_{k+1}-x_k}(x_{k+2}-x_{k+1})$$

To find its x-intercept, let $y = f_{k+2} = 0$. Accordingly, we get:

$$x_{k+2}=x_{k+1}-\frac{x_{k+1}-x_k}{f_{k+1}-f_k}f_{k+1} \tag{3.4a}$$

Rearranging,

$$x_{k+2}=\frac{x_k f_{k+1}-x_{k+1}f_k}{f_{k+1}-f_k} \tag{3.4b}$$

This is the final form of the *Secant* or *Chord method.* This technique keeps on iterating until the guess comes sufficiently close to the root. Figure 3.6 illustrates this numerical algorithm graphically.

Computational steps

In the following, the steps involved in the Secant method are presented sequentially:

Step 1: Guess x_k and x_{k+1}, and find the corresponding f_k and f_{k+1}.

Step 2: Based on the following points, (x_k, f_k) and (x_{k+1}, f_{k+1}), find the slope from:

$$m=\frac{f_{k+1}-f_k}{x_{k+1}-x_k} \tag{3.3}$$

Step 3: Find x_{k+2} from Equation (3.4a), which is written as:

$$x_{k+2}=x_{k+1}-\frac{f_{k+1}}{m}$$

and then compute f_{k+2}.

Step 4: Check for convergence. If it is not converged, go back to Step 2 and repeat the calculations.

As indicated, the iterations in the above Secant algorithm proceed with $k = 0, 1, 2, \ldots$ and so on.

Remarks

1. Unlike the Newton–Raphson (N–R) method (to be discussed a little later) that requires one guess, the Secant method starts with two initial guess values.
2. Like the Bisection method, the Secant method involves one function evaluation per iteration; whereas, the N–R method requires two function evaluations (f and f') that make the computation sometimes complicated.

3. The Secant method retains the most recent two estimates and thus, it would provide faster convergence rate than the bracketing method (i.e., Bisection and False Position).
4. After reaching the most recent location [say, (x_{k+2}, f_{k+2})], the Secant method sees back the last location [i.e., (x_{k+1}, f_{k+1})] to make a straight line for finding the next root (i.e., x_{k+3}). In other words, the Secant method requires the recent point and the last point for finding x_{k+3}. However, there is no such exercise involved in the N–R method that requires only the recent point [say, (x_{k+2}, f_{k+2})] for finding x_{k+3}. Hence, the Secant method is likely to provide a comparatively slow response than the N–R. It will further be reflected in the subsequent discussion.

3.4.1 Super Linear Convergence: Secant Method

Recall the *true error* defined in Equation (1.9) as:

$$e_k = x_k - x^*$$

which gives,

$$x_k = x^* + e_k$$

Similarly,

$$x_{k-1} = x^* + e_{k-1}$$

$$x_{k+1} = x^* + e_{k+1}$$

Recall that x^* is the root of Equation (3.1).

Based on the Secant method [Equation (3.4a)],

$$x_{k+1} = x_k - \frac{x_k - x_{k-1}}{f_k - f_{k-1}} f_k \tag{3.5a}$$

which gives,

$$e_{k+1} = \frac{e_{k-1} f_k - e_k f_{k-1}}{f_k - f_{k-1}} \tag{3.5b}$$

According to the mean value theorem,

$$f'(\xi_k) = \frac{f(x_k) - f(x^*)}{x_k - x^*}$$

in which, ξ is some value in the range $x_k < \xi < x^*$.

Since $f(x^*) = 0$ and $e_k = x_k - x^*$, we get:

$$f'(\xi_k) = \frac{f_k}{e_k}$$

This gives,

$$f_k = e_k f'(\xi_k)$$

Similarly,

$$f_{k-1}=e_{k-1}f'(\xi_{k-1})$$

From Equation (3.5b),

$$e_{k+1}=e_k e_{k-1}\frac{f'(\xi_k)-f'(\xi_{k-1})}{f_k-f_{k-1}}$$

So,

$$e_{k+1}\propto e_k e_{k-1} \tag{3.5c}$$

Based on Equation (A3.10)[7] [i.e., $e_{k+1}=\alpha e_k^m$],

$$e_{k+1}\propto e_k^m$$

So,

$$e_k\propto e_{k-1}^m \tag{3.5d}$$

From Equation (3.5c),

$$e_k^m\propto e_{k-1}^m\, e_{k-1}$$

This means,

$$e_k^m\propto e_{k-1}^{m+1}$$

So,

$$e_k\propto e_{k-1}^{(m+1)/m}$$

Comparing this with Equation (3.5d),

$$m=\frac{m+1}{m}$$

Rearranging and then solving it,

$$m=\frac{1\pm\sqrt{5}}{2}$$

Since m is never negative, so $m = 1.618$.

It leads to faster convergence than linear convergence [i.e., $m = 1$ (observed in the Bisection method)] and the rate of convergence of the Secant method is thus called *super linear.*

Example 3.2 Secant method

Repeat the problem of Example 3.1 to find the root using the Secant method with the same initial guess values (0 and 0.5) and tolerance limit (10^{-6}).

---

7 See Appendix A3.2.

Solution

Secant method is detailed above, using which we obtain the results shown in Table 3.3. The root is obtained within 4 iterations (first estimated value is x_2) as 0.162445.

TABLE 3.3 Numerical results of the Secant method

Iteration (k)	x_k	x_{k+1}	Absolute error
0	0.000000	0.500000	0.5
1	0.500000	0.156220	0.34378
2	0.156220	0.162237	0.006017
3	0.162237	0.162445	0.000208
4	0.162445	0.162445	0.000000

3.5 False Position Method

If the approximations are computed such that:

$$f_k f_{k+1} < 0$$

is satisfied, then the numerical technique represented by Equation (3.4a) [or Equation (3.4b)] is referred as the *False Position* or the *Regula Falsi* method.[8] It reveals that this approach combines the concept of the Bisection and Secant methods. Figure 3.7 illustrates this technique.

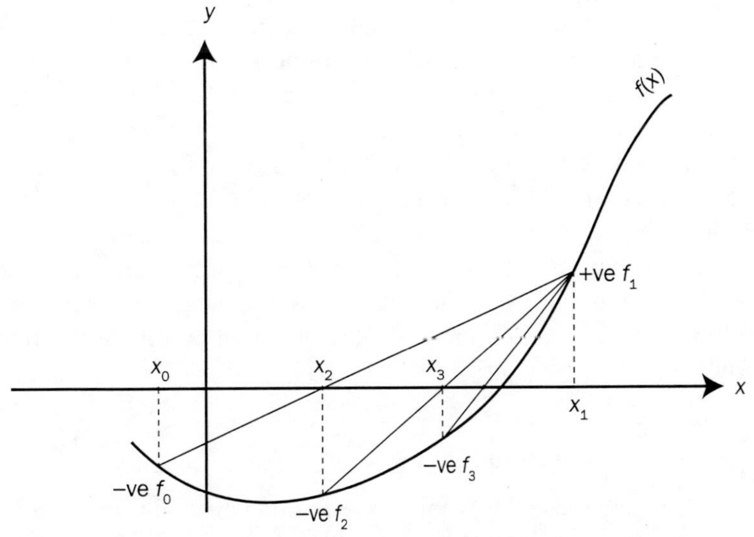

FIGURE 3.7 Graphical representation of the False Position method

8 This method involves the trial-and-error approach by using the test (i.e., *false*) values for x such that f_k and f_{k+1} are of opposite sign. For this reason, perhaps it is called as *False Position* (in Latin, *Regula Falsi*) method.

Like the Secant method, the False Position scheme constructs a straight line to approximate the function in the local region of interest. The key difference between these two convergence methods is that the Secant method keeps the two most recent estimates, whereas the False Position method retains the most recent estimate and the next recent one which satisfies $f_k f_{k+1} < 0$.

Computational steps

The following steps can be followed to solve Equation (3.1) by the use of the False Position approach:

Step 1: Guess x_k and x_{k+1} such that one where $f_k < 0$ and another where $f_{k+1} > 0$. This same step is followed in the Bisection method.

Step 2: Find x_{k+2} from Equation (3.4a):

$$x_{k+2} = x_{k+1} - \frac{x_{k+1} - x_k}{f_{k+1} - f_k} f_{k+1} \tag{3.4a}$$

and then determine the corresponding f_{k+2}. This step is followed in the Secant method.

Step 3: Verify the sign of f_{k+2}. If $f(x_k)$ has the same sign with $f(x_{k+2})$, replace x_k by x_{k+2}, else x_{k+1} will be replaced. This is again followed in the Bisection method.

Step 4: Check for convergence. If it is not converged, go back to Step 2 and repeat the computation.

Remarks

1. Since the False Position approach may retain a reasonably old reference point to maintain an opposite sign bracket around the root, it would provide a slower convergence rate over the Secant method but faster rate than the Bisection technique. Indeed, when the function $f(x)$ is convex,[9] the Regula Falsi method has linear convergence (i.e., order of convergence (m) = 1, which is same with Bisection method) and if $f(x)$ is not convex, m is at most 1.618 (same with Secant method).
2. Like the Bisection method, the False Position approach always converges. In few cases, the open method (i.e., Secant and N–R) diverges, whereas under the same conditions, the bracketing method (i.e., Bisection and False Position) converges slowly.

Example 3.3 False Position method

Repeat Example 3.1 to find the root using the False Position method with the same initial guess values (0 and 0.5) and tolerance limit (10^{-6}).

9 A function f is said to be convex when the chord, joining any two points of a curve, always lies above that curve.

Solution

Following the computational steps of the False Position method outlined above, we produce the results in Table 3.4. The root is obtained within 4 iterations (first estimated value is x_2) as 0.162445.

TABLE 3.4 Numerical results of the False Position method

Iteration (*k*)	x_k	x_{k+1}	Absolute error	$\lvert f_{k+1} \rvert$
0	0.000000	0.500000	0.500000	2.200625
1	0.500000	0.156219	0.343781	0.039212
2	0.500000	0.162237	0.337763	0.001311
3	0.500000	0.162438	0.337562	0.0000444
4	0.500000	0.162445	0.337555	2.5×10^{-7}

Remarks

1. Obviously, the False Position method (4 iterations)[10] secures a faster rate of convergence than the Bisection method (19 iterations).
2. Another interesting finding in this example system is that even after converging to the actual root (= 0.162445), the absolute error is reasonably large and it seems unacceptable. However, here $\lvert f_{k+1} \rvert$ satisfies the tolerance limit and thus, we accept the root as the final solution.
3. If one eventually knows the true solution (x^*) a *priori*, then it is obvious that with respect to the absolute true error ($\lvert e_k \rvert$), the Bisection method converges in 14 iterations, whereas the Secant method takes 3 iterations and the False Position method requires 4 iterations.

3.6 Newton–Raphson Method

The Newton–Raphson (N–R) method[11] is perhaps the most popular and commonly used technique to solve the systems of non-linear algebraic equations. This iterative convergence method is often called as *Newton's method.* It is easy to develop by the use of Taylor series expansion of $f(x)$ as:

$$f(x+\Delta x) = f(x) + \frac{\partial f(x)}{\partial x}\frac{\Delta x}{1\,!} + \frac{\partial^2 f(x)}{\partial x^2}\frac{(\Delta x)^2}{2\,!} + \frac{\partial^3 f(x)}{\partial x^3}\frac{(\Delta x)^3}{3\,!} + \ldots\ldots\ldots = 0 \qquad (3.6)$$

Neglecting second and all subsequent higher order terms, we get:

$$f(x) + \frac{\partial f(x)}{\partial x}\Delta x = 0 \qquad (3.7)$$

10 Basically, from fourth iteration onward, there is no change in the values of x_k and x_{k+1}. Now one could raise a question that *how you know what is happening after 4th iteration*? For that, we need to proceed for at least one more iteration (i.e., total five iterations). Accordingly, one may consider total 5 iterations (instead of 4) involved in the False Position method for performance comparison with the other iterative convergence approaches when absolute error (≤ tol) is placed as the only stopping criterion.

11 Named after Sir Isaac Newton (1643–1727) and English mathematician Joseph Raphson (1648–1715).

Rewriting,

$$\Delta x = \frac{-f(x)}{f'(x)} \tag{3.8}$$

in which, $f'(x) = \frac{\partial f(x)}{\partial x}$.

To compute x at iteration $k + 1$, we define:

$$\Delta x_{k+1} = x_{k+1} - x_k \tag{3.9}$$

Considering $f_k = f(x_k)$, Equation (3.8) yields:

$$\Delta x_{k+1} = \frac{-f_k}{f'_k} \tag{3.10}$$

Combining Equations (3.9) and (3.10), we have:

$$x_{k+1} = x_k - \frac{f_k}{f'_k} \tag{3.11}$$

This is the final form of the *Newton–Raphson* convergence method for a single-variable system.

Equation (3.11) of the N–R method can also be derived from the Secant Equation (3.5a) as well. For this, let us simply make the following finite difference approximation:

$$f'_k \approx \frac{f_k - f_{k-1}}{x_k - x_{k-1}} \tag{3.12}$$

The Newton's method is pictorially illustrated in Figure 3.8.

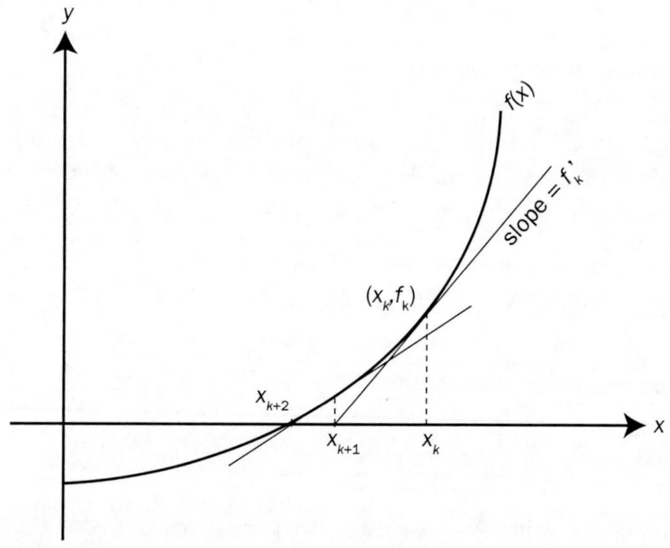

FIGURE 3.8 Graphical representation of the Newton–Raphson method

Computational steps

The computer-assisted N–R algorithm is developed with the following sequential steps:

Step 1: Assume x_k.

Step 2: Find f_k and then its derivative f'_k.

Step 3: Calculate x_{k+1} from Equation (3.11).

Step 4: Check for convergence. If it does not converge, go back to Step 2.

Note that finding a derivative (f'_k) analytically (exact result) is sometimes cumbersome. In such a case, one can prefer numerical differentiation (approximate result) by the use of say, explicit Euler method (detailed in Chapter 4).

3.6.1 Extension of N–R Method to Multi-variable Systems

It is quite simple to extend the Newton–Raphson technique to multi-variable systems. For this, let us represent a multi-variable system as:

$$F(X) = \mathbf{0} \tag{3.13}$$

This form of equation comprises a set of n equations with n variables, namely $x_1, x_2, \ldots\ldots, x_n$ as:

$$\begin{bmatrix} f_1(x_1, x_2, \ldots\ldots\ldots x_n) \\ f_2(x_1, x_2, \ldots\ldots\ldots x_n) \\ \cdot \\ \cdot \\ f_n(x_1, x_2, \ldots\ldots\ldots x_n) \end{bmatrix} = \begin{bmatrix} 0 \\ 0 \\ \cdot \\ \cdot \\ 0 \end{bmatrix} \tag{3.14}$$

Using Taylor series for each f_i (with neglecting the second and all subsequent higher-order terms), we get:

$$f_i(X + \Delta X) = f_i(X) + \sum_{j=1}^{n} \frac{\partial f_i(X)}{\partial x_j} \Delta x_j = 0 \tag{3.15}$$

In matrix form,

$$F(X) + \mathrm{J}\Delta X = \mathbf{0} \tag{3.16}$$

in which, the *Jacobian* matrix has the following form:

$$\mathrm{J} = \begin{bmatrix} \frac{\partial f_1}{\partial x_1} & \frac{\partial f_1}{\partial x_2} & \cdot & \cdot & \frac{\partial f_1}{\partial x_n} \\ \cdot & \cdot & \cdot & \cdot & \cdot \\ \cdot & \cdot & \cdot & \cdot & \cdot \\ \frac{\partial f_n}{\partial x_1} & \frac{\partial f_n}{\partial x_2} & \cdot & \cdot & \frac{\partial f_n}{\partial x_n} \end{bmatrix}$$

Rearranging Equation (3.16),

$$\Delta X = -J^{-1} F(X) \tag{3.17}$$

For X at iteration $k + 1$,

$$X_{k+1} - X_k = -J_k^{-1}\ F(X_k) \tag{3.18}$$

It gives,

$$X_{k+1} = X_k - J_k^{-1}\ F_k \tag{3.19}$$

This is the final form of the N–R method for multi-variable systems, in which X_k is basically a vector of x-values at iteration k.

It is now obvious that how this Equation (3.19) gets the form of Equation (3.11) for a single variable system. Here, the Jacobian is, in essence, analogous to the derivative of a multi-variate function.

Remarks

1. The N–R method has a reasonably good speed of convergence, leading to a reduced number of iterations. This point is elaborated below.
2. As shown, it is straightforward to extend this technique from single to multi-variable systems.
3. This method involves the evaluation of the derivative of a function at every iteration; doing it analytically is not always an easy task. Thus, as stated before, this job we can alternatively do through numerical route (finite difference approximation). At this point, it is interesting to note that this is perhaps more important to have a reduced number of iterations than the additional work required for computation.
4. The N–R method might face divergence problem, particularly when the function is strongly non-linear and the initial guess is reasonably poor. Issues related to convergence of this N–R method are briefly discussed in Appendix 3A.

Example 3.4 Newton–Raphson method

Revisit the problem discussed in Example 3.1, and find the root using the N–R method with an initial guess value of 0.5 against the tolerance limit of 10^{-6}.

Solution

Using the Newton–Raphson method presented before, the results are produced in Table 3.5. Obviously, the root x is obtained in 4 iterations as 0.162445 within the given tolerance.

Table 3.5 Numerical results of the Newton–Raphson method

Iteration	x_k	x_{k+1}	Absolute error
1 (k = 0)	0.500000	0.169396	0.330604
2 (k = 1)	0.169396	0.162451	0.006945
3 (k = 2)	0.162451	0.162445	0.000006
4 (k = 3)	0.162445	0.162445	0.000000

Remarks

1. This example system eventually *does not* confirm that the N–R method provides faster convergence (converged in 4 iterations) compared to the False Position (converged in 4 iterations) and Secant method (converged in 4 iterations). But there is a difference with the False Position method if we consider its convergence in 5 iterations (see note 10 provided under Example 3.3). However, this is not the general picture (rather quite uncommon!) as shown below with another example.
2. To get a different picture on the comparative rate of convergence between the Secant, False Position and N–R method, one can solve the following problem (Jain et al., 1995):

   $$xe^x - \cos x = 0$$

   which has the root 0.51775736. It is observed that with respect to the stopping criterion $|f_{k+1}|$, the N–R method (with $x_0 = 1$) converges in 5 iterations, whereas the Secant method ($x_0 = 0$ and $x_1 = 1$) gets the root in 7 iterations and the False Position method ($x_0 = 0$ and $x_1 = 1$) fails to converge even in 10 iterations. With this, one can arrange them according to their performance (rate of convergence) as: N–R > Secant > False Position.

3.6.2 A Little More About Newton–Raphson

Because of its wide application as an iterative convergence technique, we are curious to get a little more insight into this Newton–Raphson method. In this connection, let us note that apart from its straightforward extension from single variable to multi-variable systems, the N–R method features the followings:

A. It is very sensitive to initial guess.
B. It shows divergence when the root is close to the inflection point.
C. It provides quadratic convergence when the root is close to the solution.

Further, this N–R method is used in finding the:

D. Optimum of a function, and
E. Complex roots of a function.

Now, we would like to briefly discuss these issues one by one with examples.

Example 3.5 Newton–Raphson method: Poor initial guess

Find the root of the following function,

$$f(x) = x - \tan x$$

with a tolerance limit of 10^{-5}.

Solution

Differentiating,

$$f'(x) = 1 - \sec^2 x$$

Accordingly, we get from Equation (3.11):

$$x_{k+1} = x_k - \frac{f_k}{f'_k}$$

$$= x_k - \frac{x_k - \tan x_k}{1 - \sec^2 x_k}$$

1. **Initial guess** 1 ($x_0 = 5$): We produce the following set of results in Table 3.6 that clearly indicates the divergence of N–R method with an initial guess of 5.

Table 3.6 Numerical results with a poor initial guess for the N–R method

Iteration	x_k	x_{k+1}	Absolute error
1	5.000000	5.733339	0.733339
2	5.733339	22.62787	16.894531
3	22.62787	62.65731	40.02944
4	62.65731	2083.468	2020.8107
5	2083.468	6695.371	4611.903
6	6695.371	18929.3	12233.929
7	18929.3	21750.86	2821.56

2. **Initial guess 2** ($x_0 = 4.6$): We change our initial guess to 4.6. With this, the solution converges to the root $x = 4.493409$ as shown in Table 3.7.

Table 3.7 Numerical results with a good initial guess for the N–R method

Iteration	x_k	x_{k+1}	Absolute error
1	4.600000	4.545732	0.054268
2	4.545732	4.506146	0.039586
3	4.506146	4.494172	0.011974
4	4.494172	4.493412	0.000760
5	4.493412	4.493409	0.000003
6	4.493409	4.493409	0.00000

Second check

$|f(x^*)| = |f(4.493409)| = 9.2\times10^{-6}$ (i.e., within desired tolerance limit).

Remarks

1. It becomes obvious from this example that the poor guess of a root of the non-linear function leads to divergence in the N–R method. However, this divergence is not so common for such a little difference between the actual x (i.e., 4.493409) and guessed x (i.e., 5).
2. If we have knowledge about the behavior of a function or the graph of that function is available, we may choose a good initial guess. Else, the following approach is recommended.

3. If a system is very sensitive to initial guess, one can use the Bisection method for the first few iterations and subsequently employ the rapidly convergent N–R method.

Example 3.6 Newton–Raphson method: Poor initial guess (cycling phenomenon)

Let us study an interesting phenomenon, *cycling*, that occurs when the initial guess is poor in the N–R method for a few cases. For this, we adopt (Mathews, 2000):

$$f(x) = x^3 - x - 3 = 0$$

which has the root of 1.671699881.

Solution

We assume a poor guess of $x_0 = 0$. Accordingly, we get the following cyclic sequences:

$x_1 = -3.000000$ $x_2 = -1.961538$ $x_3 = -1.147176$ $x_4 = -0.006579$

$x_5 = -3.000389$ $x_6 = -1.961818$ $x_7 = -1.147430$ $x_8 = -0.007256$

Notice that there is a cycle with $x_{k+4} \approx x_k$ for $k = 0,1,2\ldots.$

However, if we start with a guess value close to the root (say, $x_0 = 2$), we comfortably converge to the solution (= 1.671699881) in 4 iterations.

Example 3.7 Newton–Raphson method: Divergence near inflection point

Find the root of the following non-linear function,

$$f(x) = (x-1)^3 + x$$

with $x_0 = 1.8$ and tolerance limit = 10^{-6}.

Solution

Equation (3.11) yields,

$$x_{k+1} = x_k - \frac{(x_k - 1)^3 + x_k}{3\,(x_k - 1)^2 + 1}$$

With this, we get the results in Table 3.8 when $x_0 = 1.8$ and tolerance limit – 10^{-6}.

TABLE 3.8 Numerical results of the N–R method with divergence near inflection point

Iteration	x_k	x_{k+1}	Absolute error
1	1.8000000	1.0082191	0.7917809
2	1.0082191	0.0002037	1.0080154
3	0.0002037	0.2500763	0.2498726
4	0.2500763	0.3139616	0.0638853
5	0.3139616	0.3176604	0.0036988
6	0.3176604	0.3176721	1.17×10^{-5}
7	0.3176721	0.3176721	0.000000

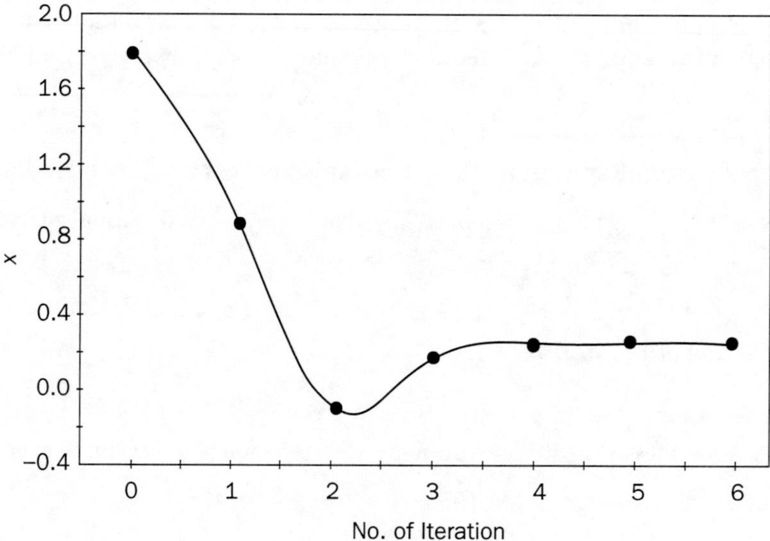

FIGURE 3.9 Divergence near inflection point

Second check

$$|f(x^*)| = |f(0.3176721)| = 2.3 \times 10^{-7} \text{ (i.e., within desired tolerance limit)}$$

Remarks

1. Here we observe that when x reaches 1.0082191 (see the first row in Table 3.8), which is close to the inflection point (that is at $x = 1$ obtained by considering $f''(x) = 0$), it starts diverging and then finally converges to the actual solution (i.e., 0.3176721). This is also depicted in Figure 3.9.
2. Divergence near the inflection point can be observed more prominently if you make a little modification in the above example,

   $$f(x) = (x-1)^3 + 0.512$$

   for which, the inflection is at $x = 1$.

   Basically with $x = 1$, $f' = 3(x-1)^2 = 0$ and it leads to the next estimate as: $x_{k+1} = x_k - \infty = -\infty$. However, in this example, x_1 is obtained 0.925926 when the guess value is $x_0 = 1.6$. Since x_1 is not exactly 1, we get $x_2 = -30.1534$ (not $-\infty$). After having divergence near the inflection point, it finally converges (in 15 iterations when tolerance = 10^{-6}) to the root, $x^* = 0.2$.

Example 3.8 Newton–Raphson method: Quadratic convergence

Find the root of the following non-linear function,

$$f(x) = e^{-x} - x$$

with $x_0 = 0.5$ and tolerance limit $= 10^{-6}$.

Solution

Using the N–R method,

$$x_{k+1} = x_k - \frac{\exp(-x_k) - x_k}{-\exp(-x_k) - 1}$$

Starting with $x_0 = 0.5$, we finally converge to the root $x = 0.567143$ as shown in Table 3.9.

TABLE 3.9 Numerical results of the N–R method with quadratic convergence

Iteration	x_k	x_{k+1}	Absolute error
1	0.500000	0.566311	6.6311×10^{-2}
2	0.566311	0.567143	8.32×10^{-4}
3	0.567143	0.567143	0.000

Second check

$$|f(x^*)| = |f(0.567143)| = 4.6\times10^{-7} \text{ (i.e., within desired tolerance limit)}$$

Remark

In this example, we observe that when the root is close to the solution, the N–R method provides quadratic convergence (reflected in the *absolute error* column in Table 3.9).

In the following, we would like to know the theory involved behind the quadratic convergence of the N–R method.

3.6.3 Quadratic Convergence: Newton–Raphson Method

From Equation (3.11),

$$x_{k+1} = x_k - \frac{f(x_k)}{f'(x_k)}$$

Let x^* be the root and so, $f(x^*)=0$. The above equation can be written as:

$$x^* - x_{k+1} = x^* - x_k + \frac{f(x_k)}{f'(x_k)}$$

That is,

$$x^* - x_{k+1} = x^* - x_k - \frac{f(x^*) - f(x_k)}{f'(x_k)}$$

A three-term Taylor series expansion,

$$f(x^*) = f(x_k) + f'(x_k)(x^* - x_k) + f''(\xi)\frac{(x^* - x_k)^2}{2}$$

where, ξ is some value in the range $x_k < \xi < x^*$.

Rearranging,

$$\frac{f(x^*)-f(x_k)}{f'(x_k)}=x^*-x_k+\frac{f''(\xi)(x^*-x_k)^2}{2f'(x_k)}$$

That is,

$$\frac{f(x^*)-f(x_k)}{f'(x_k)}-(x^*-x_k)=\frac{f''(\xi)(x^*-x_k)^2}{2f'(x_k)}$$

The left hand side is derived above as $x_{k+1}-x^*$. Thus, we get:

$$x_{k+1}-x^*=\frac{f''(\xi)}{2f'(x_k)}\left(x_k-x^*\right)^2$$

That is,

$$e_{k+1}=\frac{f''(\xi)}{2f'(x_k)}\ e_k^2$$

where, the true error $e=x-x^*$.

As we converge, $x_k\to x^*$ and e_{k+1} becomes proportional to the square of e_k. Also, based on Equation (A3.10), we get the order of convergence (m) = 2. This is called the *quadratic convergence*, indicating that in each iteration the numerical solution would improve by two decimal places.

Example 3.9 Newton–Raphson method: Finding minimum of a function

Here, our goal is to use the N–R method to approximate the minimum (or optimum) of a function, f when this f is differentiable. This can be accomplished by determining the critical point (x^c) at which

$$f'(x)=0$$

For this, let us take an example of a function:

$$f(x)=x+2\sin x$$

Now we need to find x^c at which $f'(x=x^c)=0$ with a desired tolerance of 10^{-5} and the approximate value of $f(x^c)$.

Solution

To solve,

$$f'(x)=0$$

we use the N–R method with

$$x_{k+1}=x_k-\frac{f'_k}{f''_k} \qquad \text{[from Equation (3.11)]}$$

For the sample function,

$$f'_k = 1 + 2\cos x_k$$
$$f''_k = -2\sin x_k$$

Adopting an initial guess of $x_0 = -2$, we produce the results as documented in Table 3.10.

Table 3.10 Numerical results of the N–R method used for minimizing the function $f(x) = x + 2\sin x$

Iteration	x_k	x_{k+1}	Absolute error
1	−2.0	−2.0922176	0.0922176
2	−2.0922176	−2.0943938	0.0021762
3	−2.0943938	−2.0943951	1.3e-6 (<10^{-5})

Obviously, we obtain $x^c = -2.0943951$ at which

$$f(x^c) = -3.8264459$$
$$f'(x^c) = 0.0$$
$$f''(x^c) = 1.7320508 > 0 \text{ (i.e., minimization of } f)$$

Remark

In the similar fashion, one can use the Secant method to make $f'(x) = 0$. For this, rewrite Equation (3.4a) as:

$$x_{k+1} = x_k - \frac{x_k - x_{k-1}}{f'_k - f'_{k-1}} f'_k$$

Notice that like N–R, it does not require second derivatives of f.

A short note on complex roots

For a polynomial having n number of complex roots, one can also employ the Newton–Raphson method with the initial guess, x_0 being a complex number. For this, Equation (3.19) can similarly be used with X_k and F_k being complex numbers or functions. In the following worked out example, we will discuss the solution of an equation that has three complex roots.

Example 3.10 Newton–Raphson method: Complex roots

Solve the following complex equation using the N–R method (tol = 10^{-4}) (Gupta, 2019):

$$f(z) = -z^3 + 2iz + 3i = 0$$

where the variable z is complex.

Solution

Let us assume $z = x + iy$. Substituting it, we have:

$$f(z) = f(x+iy) = -(x+iy)^3 + 2i\ (x+iy) + 3i = 0$$

It gives:

$$-(x^3+3ix^2y-3xy^2-iy^3)+2ix-2y+3i=0$$

On separating the real and imaginary parts:

$$(-x^3+3xy^2-2y)+i\ (-3x^2y+y^3+2x+3)=0$$

It is fairly true that the given function, $f(z)$, becomes zero if and only if the real and imaginary parts both vanish separately. Since,

$$f(z) = f(x+iy) = f_1(x, y)+if_2(x, y) = 0$$

for the example system, we accordingly have:

$$f_1(x, y)=-x^3+3xy^2-2y=0$$

$$f_2(x, y)=y^3-3x^2y+2x+3=0$$

This is obviously a two-variable system, for which, we can use the N–R formula [i.e., Equation (3.19)]. Accordingly, we construct the Jacobian matrix as:

$$J=\begin{bmatrix} \dfrac{\partial f_1}{\partial x} & \dfrac{\partial f_1}{\partial y} \\ \dfrac{\partial f_2}{\partial x} & \dfrac{\partial f_2}{\partial y} \end{bmatrix}$$

$$=\begin{bmatrix} -3x^2+3y^2 & 6xy-2 \\ -6xy+2 & -3x^2+3y^2 \end{bmatrix}$$

Next, for the two resulting cubic equations, assume three different sets of initial approximations, (x_0, y_0) and obtain:[12]

$$x_1 = -1.00638 \qquad y_1 = 0.337588$$

$$x_2 = 1.48412 \qquad y_2 = 1.11040$$

$$x_3 = -0.477743 \qquad y_3 = -1.44799$$

So the three complex roots are:

$$z_1=-1.00638+i\ 0.337588$$

$$z_2=1.48412+i\ 1.11040$$

$$z_3=-0.477743-i\ 1.44799$$

---

12 A real initial guess never produces a complex root.

Most of the iterative convergence methods we have discussed so far have simply approximated the non-linear function, $f(x)$ in the neighborhood of the root by a straight line. However, one can expect a better matching to the actual curve when the non-linear $f(x)$ is approximated in the neighborhood of the root by a quadratic polynomial. This is followed in the next two iterative approaches, namely Muller and Chebyshev methods.

3.7 Muller Method

Muller iterative convergence approach[13] is formulated based on a polynomial and thus, as stated before, it is categorized as a polynomial-based approach. Considering the following second-degree (quadratic) equation:

$$p(x)=a_0x^2+a_1x+a_2=0 \tag{3.20}$$

in which, $a_0(\neq 0), a_1$ and a_2 are the three parameters. Here we guess the three values for the unknown x as: x_{k-2}, x_{k-1} and x_k. Our goal is to find the root x^* of $f(x)=0$.

Let us use the following conditions to find a_0, a_1 and a_2:

$$p_{k-2}=f_{k-2}=a_0x_{k-2}^2+a_1x_{k-2}+a_2 \tag{3.21a}$$

$$p_{k-1}=f_{k-1}=a_0x_{k-1}^2+a_1x_{k-1}+a_2 \tag{3.21b}$$

$$p_k=f_k=a_0x_k^2+a_1x_k+a_2 \tag{3.21c}$$

Inserting a_0, a_1 and a_2 in Equation (3.20), and then rearranging,

$$p(x)=\frac{(x-x_{k-1})(x-x_k)}{(x_{k-2}-x_{k-1})(x_{k-2}-x_k)}f_{k-2}+\frac{(x-x_{k-2})(x-x_k)}{(x_{k-1}-x_{k-2})(x_{k-1}-x_k)}f_{k-1}$$
$$+\frac{(x-x_{k-2})(x-x_{k-1})}{(x_k-x_{k-2})(x_k-x_{k-1})}f_k=0 \tag{3.22}$$

Representing Equation (3.22) in the following form:

$$\frac{g(g+g_k)}{g_{k-1}(g_{k-1}+g_k)}f_{k-2}-\frac{g(g+g_k+g_{k-1})}{g_kg_{k-1}}f_{k-1}+\frac{(g+g_k)(g+g_k+g_{k-1})}{g_k(g_k+g_{k-1})}f_k=0 \tag{3.23}$$

in which,

$$g=x-x_k$$
$$g_k=x_k-x_{k-1}$$
$$g_{k-1}=x_{k-1}-x_{k-2}$$

13 Named after American mathematician and computer scientist David Eugene Muller (1924–2008).

Denoting,

$$\lambda = g/g_k$$
$$\lambda_k = g_k/g_{k-1}$$
$$\delta_k = 1 + \lambda_k$$

Equation (3.23) yields:

$$\lambda^2 c_k + \lambda h_k + \delta_k f_k = 0 \tag{3.24}$$

in which,

$$h_k = \lambda_k^2 f_{k-2} - \delta_k^2 f_{k-1} + (\lambda_k + \delta_k) f_k$$
$$c_k = \lambda_k (\lambda_k f_{k-2} - \delta_k f_{k-1} + f_k)$$

Dividing Equation (3.24) by λ^2,

$$\frac{\delta_k f_k}{\lambda^2} + \frac{h_k}{\lambda} + c_k = 0 \tag{3.25}$$

Notice that this is a quadratic equation in terms of $1/\lambda$. Solving it,

$$\lambda = \frac{2\delta_k f_k}{-h_k \pm \sqrt{(h_k^2 - 4\delta_k f_k c_k)}} = \lambda_{k+1} \tag{3.26}$$

Sign in the denominator of Equation (3.26) is chosen such that λ_{k+1} has the smallest absolute value.

Since $\lambda = g/g_k$, we have:

$$\lambda_{k+1} = \frac{x - x_k}{x_k - x_{k-1}} \tag{3.27}$$

It gives,

$$x = x_k + \lambda_{k+1}(x_k - x_{k-1}) \tag{3.28}$$

Replacing x by x_{k+1},

$$x_{k+1} = x_k + \lambda_{k+1}(x_k - x_{k-1}) \tag{3.29}$$

This is the final form of the *Muller* method, which is pictorially demonstrated in Figure 3.10. Note that the next approximation x_{k+1} is obtained in this method as the zero of the second-degree curve passing through the following three points: (x_{k-2}, f_{k-2}), (x_{k-1}, f_{k-1}) and (x_k, f_k).

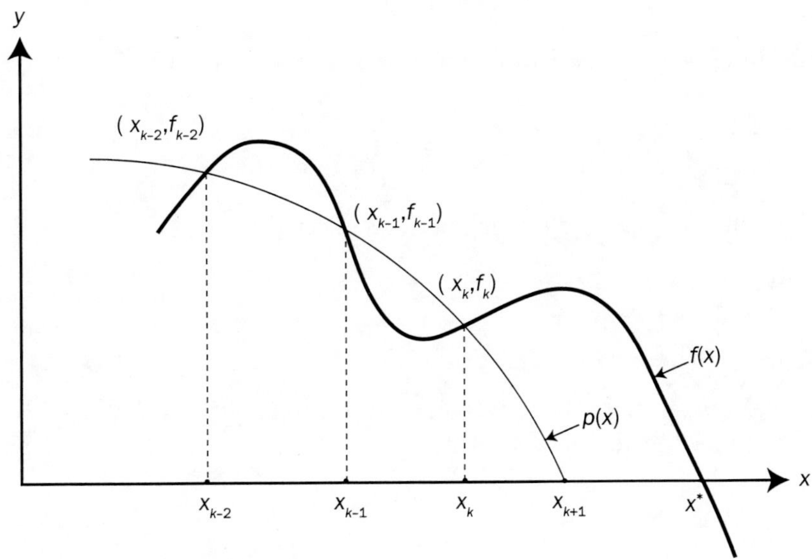

FIGURE 3.10 Graphical representation of the Muller method

Computational steps

The steps involved in the Muller method for finding the root are briefed below.

Step 1: Assume x_{k-2}, x_{k-1} and x_k.

Step 2: Calculate the corresponding f_{k-2}, f_{k-1} and f_k.

Step 3: Check the convergence.

If $|f_k| \leq \text{tol}$

go to Step 9, else to next step.

Step 4: Find g_k and g_{k-1} from:

$$g_k = x_k - x_{k-1}$$
$$g_{k-1} = x_{k-1} - x_{k-2}$$

Step 5: Find λ_k and δ_k from:

$$\lambda_k = g_k / g_{k-1}$$
$$\delta_k = 1 + \lambda_k$$

Step 6: Calculate h_k and c_k from:

$$h_k = \lambda_k^2 f_{k-2} - \delta_k^2 f_{k-1} + (\lambda_k + \delta_k) f_k$$
$$c_k = \lambda_k(\lambda_k f_{k-2} - \delta_k f_{k-1} + f_k)$$

Step 7: Find λ_{k+1}.

For this, first calculate λ from Equation (3.26):

$$\lambda = \frac{2\delta_k f_k}{-h_k \pm \sqrt{(h_k^2 - 4\delta_k f_k c_k)}} \tag{3.26}$$

The λ that has the smallest magnitude is selected as the λ_{k+1}.

Step 8: Update x.
From Equation (3.29),

$$x_{k+1} = x_k + \lambda_{k+1}(x_k - x_{k-1}) \tag{3.29}$$

One can again check the convergence here using Equation (1.19).
To continue the iteration, go to Step 2.

Step 9: Stop.

Remarks

1. Computational complexity involved in the Muller approach is more than that in the N–R method. However, the positive side is that it involves one function evaluation per iteration, whereas it is 2 for N–R.
2. Muller method uses a second-degree polynomial (i.e., quadratic equation), while the Newton–Raphson uses a straight line (linear equation). Hence for curvilinear functions, the Muller method is likely to secure a faster rate of convergence than N–R method.
3. On the other hand, to find x_{k+1}, the Muller method needs three last estimates (i.e., x_k, x_{k-1} and x_{k-2}), whereas the N–R method needs only one estimate (i.e., x_k). In this sense, the latter method seems to provide faster response but may have more chances of divergence.
4. It can be shown that the order of convergence of Muller method is about 1.84. It is observed before as 1 for the Bisection method, 1.618 for the Secant method and 2 for the N–R method. Thus overall, the N–R method is likely to achieve theoretically a better progress in each iteration compared to the Muller method followed by the Secant and then the Bisection approach.

Example 3.11 Muller method

Repeat the following non-linear algebraic equation,

$$f(x) = 5x - \exp(-x^2) + \sin x = 0$$

to find its root using the Muller method with the initial guess values of 0, 0.25 and 0.5. Acceptable tolerance limit is given as 10^{-6}.

Solution

Using the computer-assisted Muller algorithm presented above, we produce the results in Table 3.11. Here, we consider the three given assumed values as: $x_{k-2} = 0.0$, $x_{k-1} = 0.25$ and $x_k = 0.5$. With this, we get the solution (=0.162445) in 4 iterations.

Table 3.11 Numerical results of the Muller method

Iteration	x_k	x_{k+1}	Absolute error
0	0.00000	0.25000	0.25000
0	0.25000	0.50000	0.25000
1 (k = 0)	0.50000	0.162012	0.337988
2 (k = 1)	0.162012	0.1624438	0.0004318
3 (k = 2)	0.1624438	0.162445	1.2e-6
4 (k = 3)	0.162445	0.162445	0.0000

3.8 Chebyshev Method

The Chebyshev method[14] is also a polynomial (of degree two) based approach. Here, to find the coefficients, namely a_0, a_1 and a_2 involved in Equation (3.20), we use the following condition:

$$f_k = a_0 x_k^2 + a_1 x_k + a_2 \tag{3.21c}$$

Accordingly,

$$f'_k = 2a_0 x_k + a_1 \tag{3.30}$$

$$f''_k = 2a_0 \tag{3.31}$$

Finding a_0, a_1 and a_2 from the above three equations, we substitute them in Equation (3.20) and get:

$$f_k + (x - x_k) f'_k + \frac{1}{2}(x - x_k)^2 f''_k = 0 \tag{3.32}$$

Notice that this is the Taylor series expansion of $f(x)$ at $x = x_k$ obtained with neglecting the third- and higher-order terms.

Rearranging Equation (3.32),

$$x_{k+1} = x_k - \frac{f_k}{f'_k} - \frac{1}{2}(x_{k+1} - x_k)^2 \frac{f''_k}{f'_k}$$

Using Equation (3.11) (i.e., N–R method),

$$x_{k+1} = x_k - \frac{f_k}{f'_k} - \frac{1}{2}\frac{f_k^2}{(f'_k)^3} f''_k \tag{3.33}$$

This is the final form of the *Chebyshev method.*

14 Named after Russian mathematician Pafnuty Chebyshev (1821–1894).

Computational steps

The implementation of this advanced Chebyshev method goes through the following sequential steps:

Step 1: Assume x_k.

Step 2: Find f_k, f'_k and f''_k.

Step 3: Calculate x_{k+1} from Equation (3.33).

Step 4: Check for convergence. If it does not converge, go back to Step 2 and repeat the computation.

Remarks

1. Like the N–R approach, this Chebyshev method requires only one guess value for x (i.e., x_k).
2. The order of convergence of Chebyshev method is 3. Thus, it would show the best performance among all the iterative convergence methods discussed in this chapter. This is also reflected through the sample function (i.e., $f(x) = 5x - \exp(-x^2) + \sin x = 0$) solved with the use of all convergence methods for a systematic comparison.
3. It involves three evaluations (i.e., f_k, f'_k and f''_k) in each iteration, whereas in N–R method, we need f_k and f'_k.
4. To avoid the evaluation of second derivative, f''_k in this Chebyshev method, let us make a modification by rewriting Equation (3.32) as:

$$x_{k+1} - x_k = -\frac{f_k}{f'_k + \frac{1}{2}(x_{k+1} - x_k) f''_k}$$

$$\approx -\frac{f_k}{f'\left[x_k + \frac{1}{2}(x_{k+1} - x_k)\right]}$$

Using Equation (3.11) (i.e., N–R formula),

$$x_{k+1} = x_k - \frac{f_k}{f'\left[x_k - \frac{1}{2}\frac{f_k}{f'_k}\right]} \tag{3.34}$$

This is often called as the *multi-point iteration* method. Notice that although this scheme involves three evaluations (i.e., f_k, f'_k and f') as in Chebyshev method (i.e., f_k, f'_k and f''_k), but there is no function with the second derivative.

Example 3.12 Chebyshev method

Repeat the following non-linear algebraic equation,

$$f(x) = 5x - \exp(-x^2) + \sin x = 0$$

to find its root using the Chebyshev method with an initial guess value of 0.5. Acceptable tolerance limit is given as 10^{-6}.

Solution

For the example system,

$$f_k = 5x_k - \exp(-x_k^2) + \sin x_k$$

$$f'_k = 5 + 2x_k \exp(-x_k^2) + \cos x_k$$

$$f''_k = 2\exp(-x_k^2)(1 - 2x_k^2) - \sin x_k$$

Using the Chebyshev algorithm developed before, we produce the results with $x_0 = 0.5$ in Table 3.12.

Table 3.12 Numerical results of the Chebyshev method

Iteration	x_k	x_{k+1}	Absolute error
1	0.50000	0.1669384	0.3330616
2	0.1669384	0.1624450	0.0044934
3	0.1624450	0.1624450	0.0000000

Notice that here we get the solution (= 0.162445) in 3 iterations.

Remark

This Chebyshev method converges in 3 iterations, showing the fastest convergence among all the numerical methods implemented before on the same example system. The reason becomes obvious if one just compares the order of convergence between those methods.

3.9 Efficiency of Iterative Method

So far we have paid our attention in learning various numerical methods to solve the systems of non-linear algebraic equations. Those methods are also systematically compared in terms of their performance (i.e., rate of convergence) and complexity level (i.e., number of evaluations), among others. Now we would like to combine them and form a single indicator called the *efficiency index* for an overall comparison.

The efficiency of an iterative convergence method depends on its order of convergence (m) and the total number of function and derivative evaluations (p) at each iteration step. The efficiency index (η) is accordingly defined by:

$$\eta = m^{1/p} \tag{3.35}$$

This is the final formula for efficiency estimation. Note that an iterative method is more efficient if the value of this efficiency index is larger.

Application to Bisection method

In Bisection method, there is one function evaluation involved in each iteration and so, $p = 1$. Again, m is derived before as 1. Accordingly, the efficiency index is obtained by using the above formula as $\eta = 1$.

Comparing iterative methods

Let us now compare the efficiency of different iterative methods. Table 3.13 lists them, along with the order of convergence (m) and number of function evaluations (p).

TABLE 3.13 Comparing efficiency

Method	m	p	η
Bisection	1	1	1
Secant	1.618	1	1.62
N–R	2	2	1.41
Muller	1.84	1	1.84
Chebyshev	3	3	1.44

Remark

Based on this analysis, we can conclude that the Muller method is the most efficient approach followed by the Secant, Chebyshev, N–R and then Bisection method.

A Short Note: General Recommendations

- If the rapid evaluation of $f(x)$ is not an issue, one can prefer the Bisection method since it never fails. However, slow convergence rate is a major cause of concerns.
- To improve the speed of convergence, one can employ the N–R and Chebyshev method when the function $f(x)$ is smooth enough, and f, f' and f'' are evaluated quite easily.
- The Secant method is a better choice than the N–R or Chebyshev method when the evaluation of f' and f'' is not an easy task. The Secant method also provides better performance than the Regula Falsi method. But the only problem is that the Secant method sometimes does not converge.
- If convergence problem is there, it is recommended to use the Bisection method for the first few iterations followed by the rapidly convergent N–R method for the rest iterations.
- If the function is curvilinear, it is wise to adopt a polynomial-based approach (i.e., Muller or Chebyshev method). Among them, the Muller method involves complexity, whereas the Chebyshev method requires three evaluations (i.e., f, f' and f'') per iteration.

3.10 Solved Examples and Case Studies

Now we would like to solve a wide variety of model problems, including a few industrially relevant case studies, by the use of the iterative convergence methods discussed before for the systems of non-linear algebraic equations.

Example 3.13 Secant method

Apply the Secant method to find the inverse of a number given as:

$$\alpha = \frac{1}{x} \tag{3.36}$$ [15]

where, $\alpha = 5$. Two initial guess values include: $x_0 = 0.1$ and $x_1 = 0.5$, and the desired tolerance is 10^{-6}.

Solution

According to the Secant method [from Equation (3.5a)],

$$x_{k+1} = \frac{x_{k-1} f_k - x_k f_{k-1}}{f_k - f_{k-1}}$$

which gives,

$$x_{k+1} = \frac{x_{k-1}\left(5 - \frac{1}{x_k}\right) - x_k\left(5 - \frac{1}{x_{k-1}}\right)}{\left(5 - \frac{1}{x_k}\right) - \left(5 - \frac{1}{x_{k-1}}\right)}$$

Simplifying,

$$x_{k+1} = x_k + x_{k-1} - 5 x_k x_{k-1}$$

It can be rewritten as:

$$x_{k+2} = x_{k+1} + x_k - 5 x_{k+1} x_k \tag{3.37}$$

Based on this equation and the given guess values ($x_0 = 0.1$ and $x_1 = 0.5$), we produce the results in Table 3.14 against each iteration.

Table 3.14 Numerical solutions of $\alpha = \frac{1}{x}$

Iteration (k)	x_k	x_{k+1}	Absolute error $[= \lvert x_{k+1} - x_k \rvert]$
0	0.1	0.5	0.4
1	0.5	0.35	0.15
2	0.35	−0.025	0.375
3	−0.025	0.36875	0.39375
4	0.36875	0.38984	0.02109

Contd

---

15 This equation represents a linear system and getting its analytical solution is quite easy and straightforward. However, we are starting this Section 3.10 with this linear system just to see the simple application of a root-finding method.

Contd

Iteration (k)	x_k	x_{k+1}	Absolute error $[=\|x_{k+1}-x_k\|]$
5	0.38984	0.03982	0.35002
6	0.03982	0.35204	0.31222
7	0.35204	0.32177	0.03027
8	0.32177	0.10743	0.21434
9	0.10743	0.25636	0.14893
10	0.25636	0.22608	0.03028
11	0.22608	0.19265	0.03343
12	0.19265	0.200958	0.008308
13	0.200958	0.2000352	0.000923
14	0.2000352	0.1999998	3.54e-5
15	0.1999998	0.2	2e-7

So, the inverse of α (=5) is obtained as 0.2.

Case Study 3.1 Spherical dome

Consider a spherical dome, which is a portion of a sphere cut off by a plane, shown in Figure 3.11. Its volume is given as:

$$V=\frac{\pi h^2}{3}(3r-h) \tag{3.38}$$

with

$$V=6\text{m}^3$$

$$r=3\text{m}$$

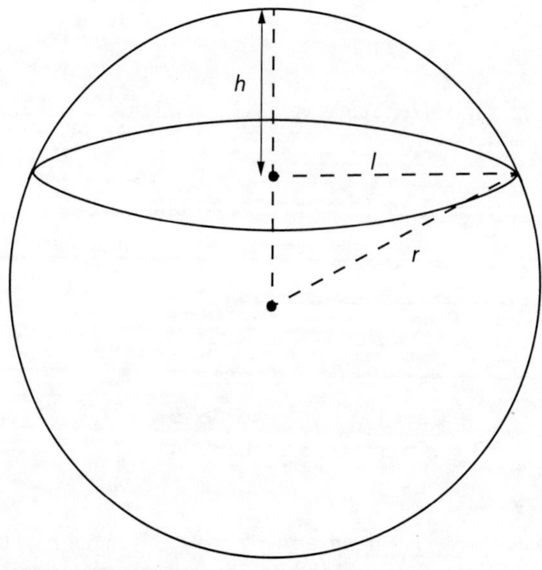

FIGURE 3.11 A spherical dome

Find the height (h) using the:

i) N–R method ($h_0 = 1$)
ii) Bisection method ($h_0 = 1$ and $h_1 = 0.5$)

Desired tolerance is 10^{-4}.

Solution
Substituting $V = 6$ and $r = 3$ in Equation (3.38) and rearranging, one obtains:

$$h^3 - 9h^2 + 5.727 = 0$$

Now, we need to solve,

$$f(h) = h^3 - 9h^2 + 5.727 = 0 \tag{3.39}$$

Its derivative is:

$$f'(h) = 3h^2 - 18h \tag{3.40}$$

i) N–R method
For the N–R method ($h_0 = 1$), we do the following calculations:

Iteration 1
Using Equation (3.11),

$$\begin{aligned} h_1 &= h_0 - \frac{f(h_0)}{f'(h_0)} \\ &= 1 - \frac{-2.273}{-15} \\ &= 0.8485 \end{aligned}$$

It gives an error (absolute) of 0.1515, which is higher than the given tolerance limit (= 10^{-4}) and thus, we go for second iteration.

Iteration 2
Similarly,

$$\begin{aligned} h_2 &= h_1 - \frac{f(h_1)}{f'(h_1)} \\ &= 0.8485 - \frac{-0.1417}{-13.1131} \\ &= 0.8377 \end{aligned}$$

Obviously, error = 0.0108 and we repeat the calculation in the third iteration.

Iteration 3

$$h_3 = h_2 - \frac{f(h_2)}{f'(h_2)}$$
$$= 0.8377 - \frac{-0.0008229}{-12.9734}$$
$$= 0.8376$$

Here, error = 0.0001, which satisfies the tolerance limit of 10^{-4} and we stop our iteration. In Table 3.15, all the results obtained above are documented.

TABLE 3.15 Numerical solutions (N–R method) of a spherical dome

Iteration	h_k	h_{k+1}	Absolute error
1	1.0	0.8485	0.1515
2	0.8485	0.8377	0.0108
3	0.8377	0.8376	10^{-4}

The height is finally obtained by the use of the N–R method in 3 iterations as 0.8376 m.

ii) Bisection method

For the Bisection method ($h_0 = 1$ and $h_1 = 0.5$), we do the following calculations:

Iteration 1

For $h_0 = 1, f(h_0) = -2.273$

and for $h_1 = 0.5, f(h_1) = 3.602$

Obviously, $f(h_0)f(h_1)<0$ and thus the midpoint is:

$$h_2 = \frac{1}{2}(1+0.5)$$
$$= 0.75$$

which gives,

$$f(h_2) = 1.0864$$

Since $f(h_2)$ has the same sign with $f(h_1)$, h_1 should be replaced by h_2. It gives an error $(=|h_2 - h_0|)$ of 0.25 that is much higher than the tolerance limit ($=10^{-4}$).

Iteration 2

It starts with $h_0 = 1$ and $h_2 = 0.75$. We have seen above: $f(h_0)f(h_2) < 0$. Thus, the midpoint is:

$$h_3 = \frac{1}{2}(1+0.75)$$
$$= 0.875$$

It gives,

$f(h_3) = -0.4937$

Since $f(h_3)$ has the same sign with $f(h_0)$, h_0 should be replaced by h_3. It gives an error of 0.125.

We continue the iterations until the stopping criterion is satisfied. The produced results are listed in Table 3.16. Notice that we execute 13 iterations to get the final solution.

Table 3.16 Numerical solutions (Bisection method) of a spherical dome

Iteration (k)	h_k	h_{k+1}	Absolute error
0	1.0	0.5	0.5
1	1.0	0.75	0.25
2	0.75	0.875	0.125
3	0.875	0.8125	0.0625
4	0.8125	0.8437	0.0312
5	0.8437	0.8281	0.0156
6	0.8437	0.8359	0.0078
7	0.8359	0.8398	0.0039
8	0.8359	0.8378	0.0019
9	0.8378	0.8368	0.001
10	0.8378	0.8373	0.0005
11	0.8378	0.8375	0.0003
12	0.8378	0.8376	0.0002
13	0.8376	0.8377	10^{-4}

Case Study 3.2 Spherical dome (revisited): Multiple roots or solutions

Repeat the spherical dome problem (Case Study 3.1) that is further asked to solve by the N–R method with:

i) $h_0 = -1$
ii) $h_0 = 9$

Desired tolerance is 10^{-4}.

Solution

i) Using the N–R method with $h_0 = -1$, we get the results as shown in Table 3.17.

Table 3.17 Numerical solutions (N–R method with $h_0 = -1$) of a spherical dome

Iteration	h_k	h_{k+1}	Absolute error
1	–1.0	–0.7965	0.2035
2	–0.7965	–0.7664	0.0301
3	–0.7664	–0.7658	0.0006
4	–0.7658	–0.7658	0.000

The height is obtained as –0.7658 m. Since the height never be negative, so it is not acceptable.

ii) Using the N–R method with $h_0 = 9$, we further get the solution in each iteration as shown in Table 3.18.

Table 3.18 Numerical solutions (N–R method with $h_0 = 9$) of a spherical dome

Iteration	h_k	h_{k+1}	Absolute error
1	9.0	8.9293	0.0707
2	8.9293	8.9281	0.0012
3	8.9281	8.9281	0.000

Here, the height is obtained as 8.9281 m that is surprisingly greater than the diameter (= 6 m). So, it is also not feasible.

Remark

Solving the model equation (cubic) of a spherical dome [i.e., Equation (3.39)], we get three real roots as: 0.8376, 8.9281 and –0.7658 as briefed in Table 3.19.

Table 3.19 Numerical solution of a spherical dome with different starting values used in the N–R method

Starting value	Root (height in m)	Feasible?
1.0	0.8376	Yes
–1.0	–0.7658	No
9.0	8.9281	No

Among the multiple solutions or roots, $h = 0.8376$ m is a feasible solution because of the reason stated above and the final height is thus 0.8376 m.

Case Study 3.3 Finding compressibility factor from an equation of state (EoS) model

We all know,

$$PV = nRT \tag{3.41}$$

This represents the *ideal gas equation of state*[16] model and it is further written as:

$$Pv = RT \tag{3.42}$$

Here,

P = absolute pressure of the gas
V = volume of the gas
v = specific volume of the gas
n = number of moles of the gas
R = universal gas constant
T = absolute temperature of the gas

16 An equation that relates the pressure, temperature and specific volume of a substance is referred as an equation of state.

Real gases do not follow the ideal gas law [that is represented by Equation (3.42)]. The deviation at given temperature and pressure is accounted for by introducing a correction factor, called *compressibility factor* (Z). Accordingly, Equation (3.42) is modified for real gas to:

$$Pv = ZRT \tag{3.43}$$

For ideal gases, obviously $Z = 1$. This compressibility factor in a couple of EoS models (e.g., Peng–Robinson (PR) and Soave–Redlich–Kwong (SRK)) is calculated based on the following form of cubic equation:

$$Z^3 - AZ^2 + BZ - C = 0 \tag{3.44}$$

where, A, B and C are dependent on phase composition, temperature and pressure.

Problem statement

Here, we assume that A, B and C are constant for simplicity. With this, the following form of equation is considered:

$$Z^3 - 2Z^2 + 475.24Z - 950.48 = 0 \tag{3.45}$$

Solve it by using the N–R method with $Z_0 = 1$ and desired tolerance = 10^{-5}.

Solution

In this example,

$$f(Z) = Z^3 - 2\,Z^2 + 475.24\,Z - 950.48 = 0$$

Its derivative is then,

$$f'(Z) = 3Z^2 - 4\,Z + 475.24$$

Using the N–R method [Equation (3.11)],

$$Z_{k+1} = Z_k - \frac{Z_k^3 - 2Z_k^2 + 475.24\,Z_k - 950.48}{3Z_k^2 - 4Z_k + 475.24}$$

Based on this form of N–R equation, we finally get one root for compressibility factor (i.e., $Z_1 = 2$) among the three roots. Next, we need to determine the other two roots, namely Z_2 and Z_3. For this, one can factor it out (analytically) of the quadratic equation as shown below.

Having Z_1, one can rewrite the cubic Equation (3.45) as:

$$(Z - Z_1)(aZ^2 + bZ + c) = 0 \tag{3.46}$$

which gives,

$$aZ^3 + (b - 2a)Z^2 + (c - 2b)\,Z - 2c = 0 \tag{3.47}$$

Comparing Equations (3.45) and (3.47),

$$a = 1 \qquad b - 2a = -2 \qquad c - 2b = 475.24$$

$$\Rightarrow b = 0 \qquad \Rightarrow c = 475.24$$

Accordingly, Equation (3.46) yields:

$$(Z-2)(Z^2+475.24)=0$$

From this,

$$(Z^2+475.24)=0$$

$$\Rightarrow \quad Z=\pm i(21.8)$$

Hence, the three roots of the cubic Equation (3.45) are obtained for Z as:

$$Z_1=2$$

$$Z_2=i(21.8)$$

$$Z_3=-i(21.8)$$

Note that the complex roots have no physical significance for compressibility factor and thus, we finally adopt $Z = 2$.

Remarks

1. If we produce a plot of $f(Z)$ vs. Z, we will see for this example that the produced graph crosses the x-axis only once at $Z = 2$. It indicates that the function $f(Z)$ has a single real root and other 2 roots are complex.
2. We all know that a cubic equation has three roots. They are:
   - either 1 real and 2 conjugate imaginary roots
   - or 3 real roots

   For compressibility factor (Z), the real root is adopted in the former case since as stated, the complex roots have no physical meaning. For the latter case with 3 real roots, follow the guideline (Reid et al., 1977) to know how to select Z among them.

Case Study 3.4 Overhead condenser of a distillation column

An overhead condenser is typically equipped with the distillation column to condense its top vapor stream. As shown in Figure 3.12, it is basically a double pipe heat exchanger, which totally condenses the saturated top vapor (inner stream) by the use of cooling water (outer stream) flowed counter-currently.

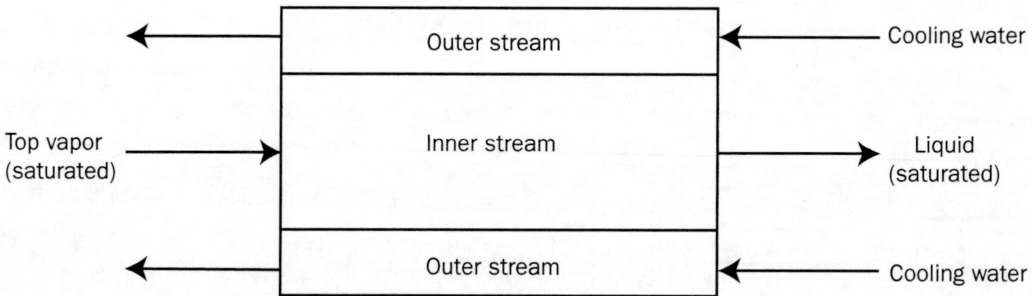

FIGURE 3.12 An overhead total condenser (double pipe heat exchanger)

Model Equations

The heat balance yields an expression of condenser duty (Q_c) as:

$$Q_c = U_c A_c \Delta T_{LMTDc} = m_{cw} C_{pcw} \Delta T_{cw} \tag{3.48}$$

in which, the log-mean temperature difference (LMTD) is defined as:

$$\Delta T_{LMTDc} = \frac{(T_T - T_{cwi}) - (T_T - T_{cwo})}{\ln \dfrac{(T_T - T_{cwi})}{(T_T - T_{cwo})}} \tag{3.49}$$

and the temperature difference in the outer stream (i.e., cooling water),

$$\Delta T_{cw} = T_{cwo} - T_{cwi} \tag{3.50}$$

Here,

T_{cwi} = cooling water (i.e., outer stream) inlet temperature

T_{cwo} = cooling water outlet temperature

U_c = overall heat transfer coefficient

T_T = saturated temperature of top vapor (i.e., inner stream)

m_{cw} = cooling water flow rate

C_{pcw} = specific heat capacity of cooling water

A_c = heat transfer area

Problem statement

For this counter-current double pipe heat exchanger, the following information is available:

T_{cwi} = 90 °F

U_c = 150 Btu/hr.ft^2.°F

T_T = 178 °F

m_{cw} = 35000 lb/hr

C_{pcw} = 1 Btu/lb.°F

A_c = 65 ft^2

Develop the computer-assisted algorithm to determine the coolant outlet temperature (T_{cwo}) using the N–R method [guessed $T_{cwo}(t = 0) = 100$ °F] within a tolerance of 10^{-6}. Compare the numerical solution with analytical result.

Solution

Inserting Equations (3.49) and (3.50) into Equation (3.48),

$$U_c A_c \frac{(T_T - T_{cwi}) - (T_T - T_{cwo})}{\ln \dfrac{(T_T - T_{cwi})}{(T_T - T_{cwo})}} = m_{cw} C_{pcw} (T_{cwo} - T_{cwi})$$

This gives,

$$\frac{U_c A_c}{m_{cw} C_{pcw}} \frac{(T_{cwo} - T_{cwi})}{\ln \frac{(T_T - T_{cwi})}{(T_T - T_{cwo})}} - (T_{cwo} - T_{cwi}) = 0$$

Rearranging,

$$\frac{U_c A_c}{m_{cw} C_{pcw}} - \ln \frac{(T_T - T_{cwi})}{(T_T - T_{cwo})} = 0 = f(T_{cwo})$$

Using given values,

$$\alpha A_c - \ln \frac{88}{178 - T_{cwo}} = 0 = f(T_{cwo}) \tag{3.51}$$

where, $\alpha = \frac{U_c}{m_{cw} C_{pcw}}$ = 0.004285714. To solve this equation, the N–R method is asked to employ.

Differentiating Equation (3.51) with respect to T_{cwo},

$$f'(T_{cwo}) = -\frac{1}{178 - T_{cwo}} \tag{3.52}$$

With this, we use the N–R formula,

$$T_{new} = T_{old} - \frac{f}{f'}$$

and get the results as shown in Table 3.20 for an initial guess $T_{cwo}(t = 0) = 100$ °F.

Table 3.20 Cooling water outlet temperature at each iteration of the N–R method

Iteration (k)	T_{cwo} (°F)
0	100.00
1	112.319580763474
2	111.402370251138
3	111.395936944706
4	111.395936638187

In the fourth iteration, we get the final T_{cwo} as 111.395936 °F. The analytical solution is easily obtained from Equation (3.51) as 111.395949 °F.

Case Study 3.5 Finding bubble point temperature: N–R method

Bubble point calculations are typically involved in distillation columns. In this, along with bubble point temperature (T), vapor phase component composition (y_i) is also computed based on the following known information:

- liquid phase component composition (x_i)
- total operating pressure (P_t)

Antoine constants (A_i, B_i and C_i) are definitely known. An iterative convergence technique is typically used for bubble point calculations. For the N–R method, the computer-assisted algorithm is formulated below in step-wise fashion for a particular time step.

Step 1: Guess *bubble point temperature* (T). For this, one can start with:

$$T = \Sigma x_i T_{BPi} \tag{3.53}$$

where, T_{BPi} refers to the boiling point temperature of component *i*.

Step 2: Find vapor pressure for each component *i* (P_i^S) from the Antoine equation:

$$P_i^S = \exp\left[A_i - \frac{B_i}{T + C_i}\right] \tag{3.54}$$

For first iteration, take T from Step 1 and for subsequent iterations, take from Step 6 calculated at the previous iterative step. As stated, Antoine constants are known (see the book by Reid et al., 1977 for various components).

Step 3: Compute vapor phase composition (y_i). For this, let us derive the formula for y_i below.

The partial pressure of component *i* (P_i) is determined from

Dalton's law (for ideal vapor phase):

$$P_i = P_t y_i \tag{3.55}$$

Raoult's law (for ideal liquid phase):

$$P_i = P_i^S x_i \tag{3.56}$$

At equilibrium,

$$P_i = P_t y_i = P_i^S x_i \tag{3.57}$$

It leads to:

$$y_i = \frac{P_i^S x_i}{P_t} \tag{3.58}$$

This formula is used to compute y_i, for which, we need to adopt P_i^S from Step 2, and both x_i and P_t are specified.

Step 4: Check the convergence based on the following condition:

$$\left|\Sigma y_i - 1\right| \le \text{tol}$$

If it is satisfied, go to Step 7 with having the answer for:

- y_i (obtained in Step 3)
- T [guessed in Step 1 (for first iteration) or obtained in Step 6 at the last iterative step (for subsequent iterations)]

If it is not satisfied, go to next step.

Step 5: Let us assume:

$$f(T) = \Sigma y_i - 1$$

$$= \frac{1}{P_t} \Sigma x_i \exp\left(A_i - \frac{B_i}{T + C_i}\right) - 1 \quad (3.59)$$

So, its derivative has the following form:

$$f'(T) = \frac{df(T)}{dT} = \Sigma y_i \frac{B_i}{(T + C_i)^2} \quad (3.60)$$

Step 6: Update temperature. Using the N–R method:

$$T_{new} = T_{old} - \frac{f(T_{old})}{f'(T_{old})} \quad (3.61)$$

Both f and f' are already calculated in Step 5. T_{old} is to be taken from:

- Step 1 (for first iteration)
- Step 6 (for subsequent iterations calculated in last iterative step)

To continue the computation, go back to Step 2.

Step 7: Stop.

At this point, it should be noted that there is as such no need to further impose the stopping criterion on temperature in Step 6; it is there on y_i in Step 4.

Problem statement

Find the bubble point temperature and vapor phase composition for a binary system, consisting of benzene (suffix '*B*') and toluene (suffix '*T*'), based on the specified information documented below. The Antoine constants are given in Table 3.21.

Liquid phase composition (x_B/x_T)	0.3/0.7 (mol fract)
Total pressure (P_t)	760 mm Hg

TABLE **3.21** Antoine constants (vapor pressure in mm Hg and temperature in K)

Component	Boiling point (°C)	A_i	B_i	C_i
Benzene	80.1	15.9008	2788.51	–52.36
Toluene	110.6	16.0137	3096.52	–53.67

Use the N–R method with a desired tolerance of 10^{-6}.

Solution

To solve this problem, let us follow the computer-assisted algorithm formulated above with 7 sequential steps.

Step 1: The guessed temperature is [from Equation (3.53)]:

$$T = 0.3 \times 80.1 + 0.7 \times 110.6$$
$$= 101.45°\text{C}$$
$$= 374.45 \text{ K}$$

Step 2: Vapor pressure calculation.
Using Equation (3.54),

$$P_B^S = 1398.63 \text{ mm Hg}$$

$$P_T^S = 578.60 \text{ mm Hg}$$

Step 3: Finding vapor phase composition.
From Equation (3.58),

$$y_B = 0.55209$$

$$y_T = 0.53292$$

Step 4: Convergence check.
Based on the calculated y_i in the last Step 3,

$$\left|\Sigma y_i - 1\right| = 0.08501 > \text{tol}\left(=10^{-6}\right)$$

Since it is larger than the desired tolerance, go to next Step 5.

Step 5: Find $f(T)$ and its derivative, $f'(T)$.
From Equation (3.59),

$$f(T) = 0.08501$$

Again, from Equation (3.60),

$$f'(T) = y_B \frac{B_B}{(T+C_B)^2} + y_T \frac{B_T}{(T+C_T)^2}$$
$$= 0.0308767$$

Step 6: Update temperature.
From Equation (3.61),

$$T_{\text{new}} = 374.45 - \frac{0.08501}{0.0308767}$$
$$= 371.6968 \text{ K}$$

For next iteration, go to Step 2.

Using the above iterative steps, we get the results documented in Table 3.22 from the developed computer program.

TABLE 3.22 Bubble point temperature at each iteration of the N–R method

Iteration (k)	T (K)
0	374.45
1	371.696809342001
2	371.608038683759
3	371.607948866496
4	371.607948866404

In the fourth iteration, we get the final T that satisfies the desired tolerance of 10^{-6}.

Case Study 3.6 Finding bubble point temperature: Muller method

Repeat the last Case Study 3.5 to find the bubble point temperature of the same binary benzene/toluene system by using a different method, namely Muller iterative convergence method. For this, let us first develop the computational steps. Then we will follow those steps to find the bubble point.

Step 1: Guess the bubble point temperature.

We know that the Muller method requires three guess values, namely T_{k-2}, T_{k-1} and T_k. For this, let us first assume the middle one (T_{k-1}) based on Equation (3.53) as:

$$T_{k-1} = \Sigma x_i T_{BPi} \tag{3.62}$$

Then, let's consider:

$$T_{k-2} < T_{k-1}$$

$$T_k > T_{k-1}$$

Roughly speaking, one may adopt ±5% of T_{k-1} as T_k and T_{k-2}, respectively.

Step 2: Find vapor pressure for each component i (P_i^S) from the Antoine equation [Equation (3.54)] at all three temperatures.

Step 3: Compute f_{k-2}, f_{k-1} and f_k at temperature T_{k-2}, T_{k-1} and T_k, respectively, based on Equation (3.59):

$$f = \frac{1}{P_t}\Sigma x_i P_i^S - 1 = \frac{1}{P_t}\Sigma x_i \exp\left(A_i - \frac{B_i}{T+C_i}\right) - 1$$

Step 4: Compute vapor phase composition (y_i) from Equation (3.58).

Step 5: Check convergence based on the following condition:

$$\left|\Sigma y_i - 1\right| \le \text{tol}$$

Here, y_i calculated at T_k can be used. If the above condition is satisfied, go to Step 11, else go to next step.

Step 6: Find g_k and g_{k-1} from:

$$g_k = T_k - T_{k-1}$$

$$g_{k-1} = T_{k-1} - T_{k-2}$$

Step 7: Find λ_k and δ_k from:

$$\lambda_k = g_k / g_{k-1}$$
$$\delta_k = 1 + \lambda_k$$

Step 8: Calculate h_k and c_k from:

$$h_k = \lambda_k^2 f_{k-2} - \delta_k^2 f_{k-1} + (\lambda_k + \delta_k) f_k$$
$$c_k = \lambda_k (\lambda_k f_{k-2} - \delta_k f_{k-1} + f_k)$$

Step 9: Find λ_{k+1}.
First calculate λ from Equation (3.26):

$$\lambda = \frac{2\delta_k f_k}{-h_k \pm \sqrt{(h_k^2 - 4\delta_k f_k c_k)}} \tag{3.26}$$

The λ that is smallest in magnitude is selected as the λ_{k+1}.

Step 10: Update T.
From Equation (3.29),

$$T_{k+1} = T_k + \lambda_{k+1}(T_k - T_{k-1}) \tag{3.63}$$

To continue the iteration, go to Step 2.

Step 11: Stop.

Following these computational steps, we solve the same bubble point temperature problem considered in Case Study 3.5 and get the results presented in Table 3.23. Here, we assume the following three initial temperature values:

$$T_k = 374.45 + 20 = 394.45 \text{ K}$$
$$T_{k-1} = \sum x_i T_{BPi} = 374.45 \text{ K}$$
$$T_{k-2} = 374.45 - 20 = 354.45 \text{ K}$$

Table 3.23 Bubble point temperature in each iteration of the Muller method

Iteration (k)	T (K)
0	354.45/374.45/394.45
1	371.673802785525
2	371.607681849931
3	371.607948891282
4	371.607948866404

Like the N–R method, the Muller iterative convergence approach obtains the solution in the fourth iteration only with satisfying the desired tolerance of 10^{-6}.

Case Study 3.7 Finding dew point temperature

Compute the dew point temperature and liquid phase composition for the same binary system considered in Case Study 3.5, which consists of benzene (suffix '*B*') and toluene (suffix '*T*'), based on the known information provided below:

Vapor phase composition (y_B/y_T)	0.4/0.6 (mol fract)
Total pressure (P_t)	760 mm Hg

Adopt Antoine constants (vapor pressure in mm Hg and temperature in K) from Table 3.21. Use the N–R method with tolerance = 10^{-6}.

Solution

To find the dew point temperature by using the N–R root finding method, the computations involved in each time step are shown below in sequence.

Step 1: Guess dew point temperature as:

$$T = 0.4 \times 80.1 + 0.6 \times 110.6$$
$$= 98.40\,^0\mathrm{C}$$
$$= 371.40\ \mathrm{K}$$

Step 2: Compute vapor pressure.
Using Antoine Equation (3.54),

$$P_B^S = 1287.54 \text{ mm Hg}$$
$$P_T^S = 527.39 \text{ mm Hg}$$

Step 3: Finding liquid phase composition.
At equilibrium, we know (for ideal mixture)

$$P_t y_i = P_i^S x_i \tag{3.57}$$

which gives:

$$x_i = \frac{P_t y_i}{P_i^S} \tag{3.64}$$

Accordingly,
$x_B = 0.2361$

$x_T = 0.8646$

Step 4: Convergence check.

$$\left|\sum x_i - 1\right| = 0.1007 > \mathrm{tol}\left(= 10^{-6}\right)$$

Step 5: Find $f(T)$ and its derivative, $f'(T)$.

Let us consider,

$$f(T) = \Sigma x_i - 1$$
$$= P_t \Sigma y_i \exp\left(\frac{B_i}{T+C_i} - A_i\right) - 1 \quad (3.65)$$

Its derivative is:

$$f'(T) = \frac{df(T)}{dT} = -\Sigma x_i \frac{B_i}{(T+C_i)^2} \quad (3.66)$$

With this,

$$f(T) = 0.1007$$
$$f'(T) = -0.032988$$

Step 6: Update dew point temperature.
Using the N–R method [Equation (3.11)],

$$T_{new} = T_{old} - \frac{f(T)}{f'(T)}$$

we get,

$$T_{new} = 374.45 \text{ K}$$

For the next iteration, go to Step 2. You continue this work and it is left for your exercise to find the final value of the dew point temperature.

Case Study 3.8 Equilibrium flash calculation

A flash drum is schematically shown in Figure 3.13 that operates at a certain pressure (P_t) and temperature (T).

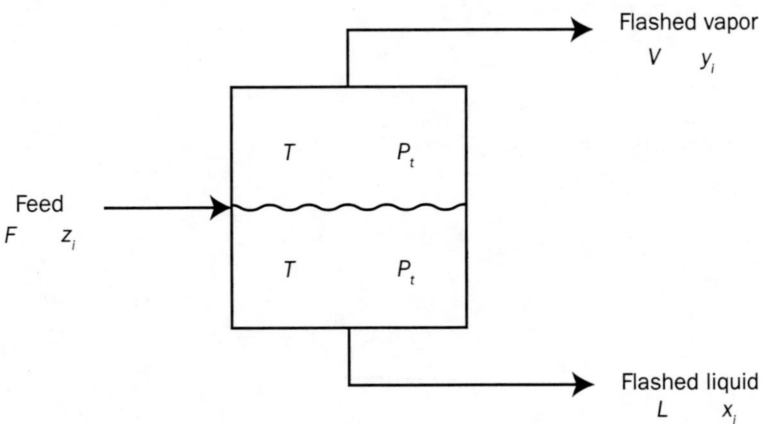

FIGURE 3.13 Flash drum

Here,

F = feed flow rate

z_i = feed composition (mol fract)

L = flashed liquid flow rate

x_i = liquid composition (mol fract)

V = flashed vapor flow rate

y_i = vapor composition (mol fract)

Deriving model (steady state)

Total mole balance

$$F = L + V \tag{3.67}$$

Component i balance

$$F\, z_i = L\, x_i + V\, y_i \tag{3.68}$$

Equilibrium relation

$$y_i = K_i\, x_i \tag{3.69}$$

where, K is the vapor liquid equilibrium coefficient.

Summation

$$\sum x_i = 1 \tag{3.70a}$$

$$\sum y_i = 1 \tag{3.70b}$$

Solution methodology

Substituting Equation (3.69) into Equation (3.68),

$$F\, z_i = L\, x_i + V\, K_i x_i$$

Further inserting Equation (3.67) and simplifying,

$$x_i = \frac{F\, z_i}{F + V(K_i - 1)} \tag{3.71}$$

Let us fix the stopping criterion as:

$$\left|\sum x_i - 1\right| \le \text{tol} \tag{3.72}$$

Recall that 'tol' denotes the tolerance limit.

Here we consider,

$$f(V) = \sum x_i - 1 \tag{3.73a}$$

That means,

$$f(V) = \sum \frac{F\, z_i}{F + V(K_i - 1)} - 1 \tag{3.73b}$$

Thus,

$$f'(V) = \sum \frac{F\, z_i(1-K_i)}{[F+V(K_i-1)]^2} \tag{3.74}$$

With this, let us list the known and unknown terms typically involved in the flash calculation:

Knowns: T, P_t, F, z_i

Unknowns: L, x_i, V, y_i

To find these unknown terms, one needs to use an iterative convergence technique. For the N–R method, the following sequential steps can be followed.

Step 1: Compute equilibrium coefficient, K_i (for ideal mixture).

From Equation (3.69),

$$K_i = \frac{y_i}{x_i}$$

That gives,

$$K_i = \frac{P_i^S}{P_t}$$

Use the Antoine Equation (3.54) to find the vapor pressure (P_i^S) based on known flash temperature (T); flash pressure (P_t) is also specified here.

Step 2: Assume V ($0 \le V \le F$).

Step 3: Compute L from Equation (3.67).

Step 4: Calculate x_i from Equation (3.71).

Step 5: Check the convergence [Equation (3.72)]. If it is satisfied, go to Step 7 followed by Step 8, else go to next Step 6.

Step 6: Update V.

For this, compute $f(V)$ and $f'(V)$ from Equations (3.73a) and (3.74), respectively. Then, apply the N–R method,

$$V_{new} = V_{old} - \frac{f(V)}{f'(V)} \tag{3.75}$$

To continue the calculation, go to Step 3.

Step 7: Compute y_i.

For this, use Equation (3.69).

Step 8: Stop.

Problem statement

Perform the flash calculations for a quaternary mixture with the following given information, including those in Table 3.24 (Smith and VanNess, 1987).

$P_t = 1$ atm

$T = 60$ °C

$F = 100$ kmol/hr

Table 3.24 Given composition and vapor liquid equilibrium coefficient for the feed mixture

Component	z_i	K_i (at 1 atm and 60°C)
n-hexane	0.25	1.694
ethanol	0.40	0.636
methyl cyclopentane	0.20	1.668
benzene	0.15	1.070

Using the N–R method, find L, V, x_i and y_i. For this, start with $V_0 = 100$ kmol/hr ($= F$) and desired tolerance is 10^{-5}.

Solution

To solve this flash drum problem, we will follow the computer-assisted steps discussed before in order.

Step 1: The equilibrium coefficient (K_i) for all four components is specified and thus, their calculation is not involved here.

Step 2: Guessed value of V is 100 (i.e., $V_0 = F$).

Step 3: Calculate L.

From Equation (3.67), we get:

$L = 0$

Step 4: Determine x_i.

From Equation (3.71),

$x_1 = 0.1476$

$x_2 = 0.6289$

$x_3 = 0.1199$

$x_4 = 0.1402$

Step 5: Convergence check.

From Equation (3.72),

$\left|\sum x_i - 1\right| = 0.0366 > \text{tol } (=10^{-5})$

Since it is not satisfied, let us go to the next step.

Step 6: Update V.

Equations (3.73a) and (3.74) yield, respectively:

$$f(V) = \sum x_i - 1 = 0.0366$$

$$f'(V) = \sum \frac{F\, z_i(1-K_i)}{[F+V(K_i-1)]^2} = 0.002423$$

From Equation (3.75),

$$V_{\text{new}} = 100 - \frac{0.0366}{0.002423} = 84.895$$

To continue the iteration, go to Step 3.

Using the above iteration scheme, we develop the computer program and produce the results in Table 3.25. Notice that we get the final flashed vapor outflow rate (V) of 81.626285 kmol/hr at fifth iteration that leads to satisfy the desired tolerance of 10^{-5}.

Based on this V, you can now easily determine the final liquid flow rate (L) [from Equation (3.67)], and liquid and vapor composition [x_i from Equation (3.71) and y_i from Equation (3.69)] that are left for you.

TABLE 3.25 Flash calculations using the N–R method

Iteration (k)	V (kmol/hr)
0	100.00
1	84.8943401744555
2	81.7578581638538
3	81.6265121172687
4	81.6262849764718
5	81.6262849757929

Example 3.14 A multi-variable (two-variable) system

Solve the following system of equations:

$$f_1\,(x, y) = x^2 - 2y - 8 = 0$$

$$f_2\,(x, y) = -x^2 - y^2 + 16 = 0$$

using the N–R technique with the initial guess: $x_0 = 1$ and $y_0 = 0.5$. The desired tolerance is 10^{-5}.

Solution

Recall the general form of the Newton–Raphson method for multi-variable systems:

$$X_{k+1} = X_k - J_k^{-1}\; F_k \tag{3.19}$$

Note that X_k is a vector of x-values at iteration k. Accordingly, we get for the example system:

$$\begin{bmatrix} x \\ y \end{bmatrix}_{k+1} = \begin{bmatrix} x \\ y \end{bmatrix}_k - \begin{bmatrix} 2x & -2 \\ -2x & -2y \end{bmatrix}_k^{-1} \begin{bmatrix} x^2 - 2y - 8 \\ -x^2 - y^2 + 16 \end{bmatrix}_k$$

The inverse can readily be obtained from the following definition:

$$A^{-1} = \frac{\text{adj } A}{|A|} \tag{2.12}$$

Using the initial guess values, $x_0 = 1$ and $y_0 = 0.5$, we get a set of results documented in Table 3.26.

Table 3.26 Numerical solutions of the example multi-variable system

Iteration (*k*)	*x*	*y*
0	1	0.5
1	7.249999	2.750000
2	4.462930	2.075000
3	3.576078	2.000915
4	3.465855	2.000000
5	3.464102	2.000000
6	3.464102	2.000000

The final roots are obtained as:

$x = 3.464102$

$y = 2.0$

Example 3.15 A five-variable system

Solve the following multi-variable system:

$$\theta_0 - 2\theta_1 + \theta_2 + \alpha^2 \beta \sin\theta_1 = \alpha^2 f_1$$
$$\theta_1 - 2\theta_2 + \theta_3 + \alpha^2 \beta \sin\theta_2 = \alpha^2 f_2$$
$$\theta_2 - 2\theta_3 + \theta_4 + \alpha^2 \beta \sin\theta_3 = \alpha^2 f_3$$
$$\theta_3 - 2\theta_4 + \theta_5 + \alpha^2 \beta \sin\theta_4 = \alpha^2 f_4$$
$$\theta_4 - 2\theta_5 + \theta_6 + \alpha^2 \beta \sin\theta_5 = \alpha^2 f_5$$

where, $\alpha = 0.001$, $\beta = 10$, $\theta_0 = 0.7$ and $\theta_6 = 0.5$. Initial guess values for the N–R method include:

$\theta_1(0) = 0.1$

$\theta_2(0) = 0.2$

$\theta_3(0) = 0.3$

$\theta_4(0) = 0.4$

$\theta_5(0) = 0.5$

and the tolerance limit is 10^{-6}. Note that similar set of system equations will arise in Case Study 5.7 of this book (Chapter 5).

Solution

Applying the multi-variable N–R method [Equation (3.19)], we have:

$$\begin{bmatrix}\theta_1\\\theta_2\\\theta_3\\\theta_4\\\theta_5\end{bmatrix}_{k+1}=\begin{bmatrix}\theta_1\\\theta_2\\\theta_3\\\theta_4\\\theta_5\end{bmatrix}_{k}-\frac{1}{\alpha^2}\begin{bmatrix}-2+\alpha^2\beta\cos\theta_1 & 1 & 0 & 0 & 0\\ 1 & -2+\alpha^2\beta\cos\theta_2 & 1 & 0 & 0\\ 0 & 1 & -2+\alpha^2\beta\cos\theta_3 & 1 & 0\\ 0 & 0 & 1 & -2+\alpha^2\beta\cos\theta_4 & 1\\ 0 & 0 & 0 & 1 & -2+\alpha^2\beta\cos\theta_5\end{bmatrix}_k^{-1}$$

$$\frac{1}{\alpha^2}\begin{bmatrix}\theta_0-2\theta_1+\theta_2+\alpha^2\beta\sin\theta_1\\ \theta_1-2\theta_2+\theta_3+\alpha^2\beta\sin\theta_2\\ \theta_2-2\theta_3+\theta_4+\alpha^2\beta\sin\theta_3\\ \theta_3-2\theta_4+\theta_5+\alpha^2\beta\sin\theta_4\\ \theta_4-2\theta_5+\theta_6+\alpha^2\beta\sin\theta_5\end{bmatrix}_k$$

Solving this, we get the results in Table 3.27.

Table 3.27 Numerical solutions of the five-variable system

Iteration (k)	θ_1	θ_2	θ_3	θ_4	θ_5
0	0.1	0.2	0.3	0.4	0.5
1	0.66668133702	0.63335668357	0.60002595389	0.56668927115	0.53334719521
2	0.66668122568	0.63335626755	0.60002539089	0.56668886760	0.53334697589
3	0.66668122568	0.63335626755	0.60002539089	0.56668886760	0.53334697589

So the final roots include:

$\theta_1 = 0.66668122568$

$\theta_2 = 0.63335626755$

$\theta_3 = 0.60002539089$

$\theta_4 = 0.56668886760$

$\theta_5 = 0.53334697589$

Example 3.16 A six-variable system

Solve the following multi-variable system:

$$2C_1^2 + 30C_1 - 12C_2 = 18$$
$$-2C_2^2 + 9C_1 - 12C_2 + 3C_3 = 0$$
$$-2C_3^2 + 9C_2 - 12C_3 + 3C_4 = 0$$
$$-2C_4^2 + 9C_3 - 12C_4 + 3C_5 = 0$$
$$-2C_5^2 + 9C_4 - 12C_5 + 3C_6 = 0$$
$$-2C_6^2 + 12C_5 - 12C_6 = 0$$

using the N–R method with the following initial guess values:

$$C_1(0) = 5$$
$$C_2(0) = 1$$
$$C_3(0) = 10$$
$$C_4(0) = 5$$
$$C_5(0) = 1$$
$$C_6(0) = 5$$

The desired tolerance limit is 10^{-6}.

Solution

Based on Equation (3.19), we have:

$$\begin{bmatrix} C_1 \\ C_2 \\ C_3 \\ C_4 \\ C_5 \\ C_6 \end{bmatrix}_{k+1} = \begin{bmatrix} C_1 \\ C_2 \\ C_3 \\ C_4 \\ C_5 \\ C_6 \end{bmatrix}_{k} - \begin{bmatrix} 4C_1+30 & -12 & 0 & 0 & 0 & 0 \\ 9 & -4C_2-12 & 3 & 0 & 0 & 0 \\ 0 & 9 & -4C_3-12 & 3 & 0 & 0 \\ 0 & 0 & 9 & -4C_4-12 & 3 & 0 \\ 0 & 0 & 0 & 9 & -4C_5-12 & 3 \\ 0 & 0 & 0 & 0 & 12 & -4C_6-12 \end{bmatrix}_k^{-1} \begin{bmatrix} 2C_1^2+30C_1-12C_2-18 \\ -2C_2^2+9C_1-12C_2+3C_3 \\ -2C_3^2+9C_2-12C_3+3C_4 \\ -2C_4^2+9C_3-12C_4+3C_5 \\ -2C_5^2+9C_4-12C_5+3C_6 \\ -2C_6^2+12C_5-12C_6 \end{bmatrix}_k$$

Solving it, the computer program produces the following set of results (Table 3.28).

Table 3.28 Numerical solutions of the six-variable system

Iteration (k)	C_1	C_2	C_3	C_4	C_5	C_6
0	5.00	1.00	10.00	5.00	1.00	5.00
1	1.83383713272	1.97421854507	4.36084098612	3.00142663826	2.26558135252	2.41209045043
2	1.02425311305	1.12623592111	1.79841420152	1.59369731017	1.35166817698	1.28676662058
3	0.86168945529	0.77357042702	0.82523839800	0.80288079276	0.74862353437	0.71703333145
4	0.83206037481	0.69539208818	0.60369279827	0.53883681029	0.49135304202	0.46572825992
5	0.83013538363	0.69019195978	0.58792028457	0.51137287286	0.45556264310	0.42563617040
6	0.83012220980	0.69015600501	0.58780093061	0.51107565385	0.45503198051	0.42493684186
7	0.83012220844	0.69015600127	0.58780091717	0.51107561034	0.45503187619	0.42493667928
8	0.83012220844	0.69015600127	0.58780091717	0.51107561034	0.45503187619	0.42493667928

Final solution is obtained as:

$C_1 = 0.83012220$

$C_2 = 0.69015600$

$C_3 = 0.58780092$

$C_4 = 0.51107561$

$C_5 = 0.45503187$

$C_6 = 0.42493668$

3.11 Summary and Concluding Remarks

In this chapter, we learned a couple of elegant iterative techniques, which can be employed for solving a wide variety of process engineering problems, among other general systems. Detailing comparative merits and demerits of those techniques with examples, a general recommendation is made so that one can judiciously choose an appropriate method for a non-linear system. Finally, this chapter ends with solving a few interesting model problems.

Appendix 3A

A3.1 Condition for Convergence

A3.1.1 Single Variable Successive Substitutions (Fixed-point method)

Let us first briefly know about the simple fixed-point method. For this, we start with Equation (3.1),

$$f(x) = 0$$

To derive the iteration formula, we can rewrite the function as $f(x) = x - g(x)$. Accordingly,

$$f(x) = x - g(x) = 0 \tag{A3.1}$$

At each step, x_{k+1} can be computed from the old value $g(x_k)$. With this, based on an initial guess x_0, we find x_1 from:

$$x_1 = g(x_0)$$

Similarly, $x_2 = g(x_1)$ (A3.2)

$$\vdots$$

$$x_{k+1} = g(x_k)$$

This is the *fixed-point iteration* formula. It is also called the *one-point iteration* or *successive substitution*. Recall that the same philosophy we saw in the iterative methods formulated to solve the systems of linear algebraic equations in Chapter 2.

Next we move to formulate the condition for convergence of this fixed-point method. Let us assume x^* be the solution and thus, $f(x^*) = 0$ with:

$$x^* = g(x^*) \tag{A3.3}$$

This root x^* is also called the *fixed point* because inserting x^* into $g(x)$ gives back x^* by the above equation.

We can write then:

$$x_{k+1} - x^* = g(x_k) - g(x^*)$$

or,

$$x_{k+1} - x^* = \frac{g(x_k) - g(x^*)}{x_k - x^*} \ (x_k - x^*) \tag{A3.4}$$

Using the mean value theorem of calculus, one can say that there exists at least one value ξ_k somewhere in between x_k and x^* [i.e., $\xi_k \in (x_k, x^*)$] such that

$$x_{k+1} - x^* = g'(\xi_k) \ (x_k - x^*) \tag{A3.5}$$

Recall that the true error is $e_k = x_k - x^*$. Accordingly,

$$e_{k+1} = g'(\xi_k) \ e_k \tag{A3.6}$$

Taking absolute values,

$$|e_{k+1}| = |g'(\xi_k)| \ |e_k| \qquad \xi_k \in (x_k, x^*) \tag{A3.7}$$

If the initial guess x_0 is chosen within an interval about the solution x^*, then the iteration will converge if

$$|g'(x)| < 1 \tag{A3.8}$$

for all x in that interval.

To elaborate this issue, let us consider $|g'(x)| = 0.1$. Accordingly,

$$e_1 = 0.1 \, e_0$$

Similarly,

$$e_2 = 0.1 \ e_1 = 0.01 \ e_0$$

$$e_3 = 0.1 \ e_2 = 0.001 \ e_0$$

.

.

and so on.

Clearly, the error is converging toward zero. Thus,

A. For convergence

$$|g'(x)| < 1$$

which means, $-1 < g' < 1$.

Again the nature of convergence can be categorized with:

i) $-1 < g' < 0$ (oscillatory convergence)
ii) $0 < g' < 1$ (error decreases monotonically)

B. For divergence

$$|g'(x)| > 1$$

which means, $g' > 1$ and $g' < -1$.

C. If $|g'(x)| = 1$, it is inconclusive.

Remarks

1. As shown, there is a linear decay (contraction) of the error and ultimately:

   $\lim_{k\to\infty} e_k = 0$

   This is often known as the *contraction mapping*.
2. Equation (A3.8) is the *sufficient condition* for convergence. It means, if this condition is satisfied, the convergence is assured; if it is not satisfied, the system may or may not converge.
3. This convergence condition is elegant and really helpful. However, it is practically not useful since we do not know the solution, x^* *a priori*. Thus, we do the convergence check by typically satisfying any one or more of the following criteria:

i) $|x_{k+1} - x_k| \le \text{tol}$

ii) $\left|\dfrac{x_{k+1} - x_k}{x_{k+1}}\right| \le \text{tol}$

iii) $|f(x_k)| \le \text{tol}$

These three criteria and their use are detailed previously in Chapter 1.

A3.1.2 Multi-variable Successive Substitutions

For multi-variable systems, it is quite straightforward to extend the convergence condition to:

$$\sum_{j=1}^{n} \left|\frac{\partial g_i}{\partial x_j}\right| < 1 \qquad \text{(A3.9)}$$

Like Equation (A3.8), Equation (A3.9) is the *sufficient condition* for convergence.

Application to Newton–Raphson method

Using Equation (A3.2),

$$x_{k+1} = g(x_k)$$

we rewrite Equation (3.11) as:

$$g(x_k) = x_k - \frac{f(x_k)}{f'(x_k)}$$

Notice that this $g(x_k)$ is the N–R fixed-point function.

Differentiating,

$$g'(x_k) = 1 - \frac{(f'(x_k))^2 - f(x_k)f''(x_k)}{(f'(x_k))^2}$$

$$= \frac{f(x_k)f''(x_k)}{(f'(x_k))^2}$$

At the solution x^*,

$$g'(x^*) = \frac{f(x^*)f''(x^*)}{\left(f'(x^*)\right)^2}$$

Since $f(x)=0$ at the root x^*, then necessarily

$$g'(x^*)=0$$

as long as $f'(x^*) \neq 0$. So by virtue of the continuity of the function g′, there will necessarily be a neighborhood of the root x^* such that $|g'(x)|<1$. This reveals that the N–R method will converge to the solution if the initial guess is sufficiently close to that solution.

Example A3.1 Convergence characteristics of a single variable system

Analyze the convergence characteristics of the following example system:

$$f(x)=x^2-5x+4=0$$

Solution

As stated, a system [Equation (A3.2)] definitely converges (sufficient condition) to the root x^* if one starts from x_0 and if $|g'(x)|<1$ over the entire domain of x covered by the iterates x_k.

One can rearrange the above equation as:

$$x=x^2-4x+4=g(x)$$

Differentiating,

$$g'(x)=2x-4$$

For convergence,

$$|2x-4|<1$$

which leads to:

$$\frac{3}{2}<x<\frac{5}{2}$$

At this point it is interesting to note that both the roots (i.e., x^* = 1 and 4) themselves lie outside of the interval we found out above that guarantees convergence. Obviously, what range of x we got (i.e., $\frac{3}{2}<x<\frac{5}{2}$) is basically a conservative requirement.

Remarks

1. The condition for convergence obtained above cannot explain the example case definitively. As mentioned, this is merely a sufficient condition and a local result. Thus, it is repeated that if $|g'|<1$ is satisfied, then the convergence is assured; if not, we may or may not converge.

2. We can also rearrange the given system equation with a few other ways as, for example:
   - $x=(5x-4)^{1/2}$
   - $x=\dfrac{x^2+4}{5}$

   With these forms, one can also perform the similar analysis based on contraction mapping made above.

Example A3.2 Convergence characteristics of a two-variable system

Consider the following two-variable system:

$$f_1(x_1, x_2)=2x_1^3+8x_1-4x_2-2=0$$

$$f_2(x_1, x_2)=x_2^2+3x_2-4x_1-6=0$$

The exact solution is: $[x_1 \quad x_2]^T=[1 \quad 2]^T$. Employing the N–R technique with the initial guess values of $x_{10}=x_{20}=1$, solve the problem and discuss the convergence characteristics with reference to Equation (A3.9).

Solution

For this example, we have:

$$\begin{bmatrix} x_1 \\ x_2 \end{bmatrix}_{k+1}=\begin{bmatrix} x_1 \\ x_2 \end{bmatrix}_k-\begin{bmatrix} 6x_1^2+8 & -4 \\ -4 & 2x_2+3 \end{bmatrix}_k^{-1}\begin{bmatrix} 2x_1^3+8x_1-4x_2-2 \\ x_2^2+3x_2-4x_1-6 \end{bmatrix}_k$$

Using the initial guess values, $x_{10}=x_{20}=1$, we get a set of results documented in Table A3.1.

TABLE A3.1 Numerical solutions of the two-variable system

Iteration (k)	x_1	x_2
0	1.00	1.00
1	1.074074	2.259258
2	1.005495	2.0118636
3	1.000022	2.000033

The final roots are obtained as: x_1 = 1.000022 and x_2 = 2.000033.

To study the convergence characteristics of this system, let us arrange the system equations in the following format:

$$x_{1\,k+1}=2x_{1k}^3+9x_{1k}-4x_{2k}-2\equiv g_1(x_{1k}, x_{2k})$$

$$x_{2\,k+1}=x_{2k}^2+4x_{2k}-4x_{1k}-6\equiv g_2(x_{1k}, x_{2k})$$

For Equation (A3.9), we find:

$$\left|\frac{\partial g_1}{\partial x_1}\right| + \left|\frac{\partial g_1}{\partial x_2}\right| = |\ 6x_1^2+9\ | + |\ -4\ |$$

$$\left|\frac{\partial g_2}{\partial x_1}\right| + \left|\frac{\partial g_2}{\partial x_2}\right| = |\ -4\ | + |\ 2x_2+4\ |$$

For convergence, these values should be less than one in the entire path, starting from point $[1 \quad 1]^T$ to the root (unknown) $[1.000022 \quad 2.000033]^T$. Let us find them at the beginning (i.e., $[x_{10} \quad x_{20}]^T = [1 \quad 1]^T$),

$$\sum_{j=1}^{2}\left|\frac{\partial g_1}{\partial x_j}\right| = 19$$

$$\sum_{j=1}^{2}\left|\frac{\partial g_2}{\partial x_j}\right| = 10$$

Obviously, the sufficient condition [i.e., Equation (A3.9)] is violated immediately at the beginning and thus, this example system may or may not converge. But actually, it is shown in Table A3.1 that the system converges to the solution.

A3.2 Order of Convergence

Definition

An iterative numerical approach is said to be of *order m* or has the *rate of convergence m*, if m is the largest positive real number, for which, there exists a finite constant α ($\neq 0$) such that:

$$\lim_{n\to\infty}\frac{|e_{n+1}|}{|(e_n)^m|} = \alpha$$

Obviously, m (that is a positive real number) is referred as the *order of convergence* of e_n, and α as the *asymptotic error constant* that usually depends on the derivatives of f at $x = x^*$.

The above condition leads to:

$$|e_{n+1}| = \alpha |e_n|^m \tag{A3.10}$$

for large n. With this, if

$m = 1$, the numerical method has *linear* rate of convergence

$m = 2$, the numerical method has *quadratic* rate of convergence

$m = 3$, the numerical method has *cubic* rate of convergence

Further note that

$$\lim_{n\to\infty}\frac{|x_{n+1}-x_n|}{|x_n-x^*|}=\lim_{n\to\infty}\frac{|x_{n+1}-x^*|}{|x_n-x^*|}+\lim_{n\to\infty}\frac{|x^*-x_n|}{|x_n-x^*|}=0+1=1$$

It means, when n is very large,

$$|e_n|=|x_n-x^*|\approx|x_{n+1}-x_n|=|E_{n+1}|$$

Accordingly, Equation (A3.10) can be written as:

$$|E_{n+1}|=\alpha|E_n|^m \tag{A3.11}$$

Example A3.3 Rate of convergence

Show that the following sequence (converges to $\sqrt{c}$) has quadratic rate of convergence:

$$x_{n+1}=\frac{1}{2}x_n\left(1+\frac{c}{x_n^2}\right) \tag{A3.12}$$

Solution

The error defined in Equation (1.9) as:

$$e_n=x_n-x^*$$

Accordingly,

$$x_n=e_n+x^*$$

$$x_{n+1}=e_{n+1}+x^*$$

where, $x^*=\sqrt{c}=\lim_{n\to\infty}x_n$.

Plugging in Equation (A3.12):

$$e_{n+1}+x^*=\frac{1}{2}(e_n+x^*)\left[1+\frac{(x^*)^2}{(e_n+x^*)^2}\right]$$

$$=\frac{1}{2}(e_n+x^*)\left[1+\left(1+\frac{e_n}{x^*}\right)^{-2}\right]$$

$$=\frac{1}{2}(e_n+x^*)\left[2-\frac{2e_n}{x^*}+\frac{3e_n^2}{(x^*)^2}-......\right]$$

It gives:

$$e_{n+1}=\frac{1}{2x^*}e_n^2+o(e_n^3)$$

Comparing with Equation (A3.10), we get:

$$m = 2$$

$$\alpha = \frac{1}{2x^*}$$

Hence, it has quadratic rate of convergence.

EXERCISE

Review Questions

3.1 Find the order of convergence of the:
  i) Regula Falsi method
  ii) Muller method
  iii) Chebyshev method

3.2 Tabulate the major differences between the Bisection, False Position, Secant, Newton–Raphson, Muller and Chebyshev methods with their relative merits and demerits in terms of:
  – number of initial approximation(s)
  – possibility of divergence
  – order of convergence
  – computational effort
  – suitability, etc.

3.3 Show the graphical representation of the Chebyshev method.

3.4 Is there any example when the N–R method can provide a better performance than the Chebyshev method?

3.5 Find the efficiency index (η) of the multi-point iteration method [Equation (3.34)] based on the following formula:

$$\eta = m^{1/p} \tag{3.35}$$

Recall that m denotes the order of convergence and p the total number of function and derivative evaluations.

3.6 How reasonable the evaluation of an iterative convergence approach is through the efficiency index? Is there anything important missing from this correlation?

Practice Problems

3.7 Find the roots of the following non-linear equations:
  i) $x^3 - x - 5 = 0$
  ii) $2x^3 - 2.5x - 6 = 0$

by the use of the Bisection and False Position method.

3.8 Solve the following equation:

$$f(x) = x^3 - x$$

using the N–R method with three guess values (0.4472, 0.4475, 0.4480). Did you get three roots with those guess values? Any remark?

3.9 Solve the following form of equation:

$$\cos x = 3x$$

using the Secant method.

3.10 Find the root of the following non-linear equation:

$$x^3 - \sin(x^2) + 1 = 0$$

with $x_0 = 0.5$ (**Ans:** – 0.7649723).

3.11 (a) Solve the following equation:

$$f(x) = x^8 - 1$$

using the Bisection and False Position method with the initial guesses of 0 and 1.3.

(b) Which method shows better convergence rate and why?

(c) Do we face any problem when we go to apply the N–R method with an initial guess of 0.5?

3.12 Find the root of the following equation using the N–R and Muller method:

$$f(x) = x^2 - e^{-x} + \sin x = 0$$

3.13 Find the root of the following equation:

$$f(x) = \cos x - xe^{2x} = 0$$

by the use of the Chebyshev method with an initial guess value of 1.

3.14 Using a suitable root finding method, estimate the positive root of:

$$e^x = 1 + \frac{x}{1!} + \frac{x^2}{2!} + \frac{x^3}{3!} e^{0.3x}$$

with tol = 10^{-5}.

3.15 The following form of equation is derived for a spherical storage tank:

$$f(h) = h^3 - 9h^2 + 3.82 = 0$$

Find the height of liquid in the tank (h) using the False Position method.

3.16 Find the inverse of a number α (= 5) with

$$f(x) = \alpha - \frac{1}{x} = 0$$

For this,

i) use the Secant method with $x_0 = 0.5$, $x_1 = 1.0$ and desired tolerance = 10^{-6}.

ii) use the N–R method with $x_0 = 0.5$ and desired tolerance = 10^{-6}.

3.17 Find the root of the following non-linear function:

$$f(x) = (x-1)^3 + 0.512$$

with $x_0 = 1.6$ and tolerance limit $= 10^{-6}$. Any specific observation?

3.18 Discuss the convergence characteristics of the following system with reference to Equation (A3.8):

$$f(x) = x^2 + 1.65x - 9.4 = 0$$

The exact solution is: $[x_1 \quad x_2]^T = [2.35 \quad -4]^T$.

3.19 Solve the following equation:

$$Z = 1 + 2(5-Z)^2 \exp\left(1-\frac{1}{Z}\right)$$

using the N–R method with $Z_0 = 1.1$.

3.20 Solve the following system equation:

$$f(C) = C - 0.01 + 0.002\left[C - \left(\frac{0.01-C}{0.0848}\right)^{4.3}\right] = 0$$

using the N–R method with $C_0 = 0.01$.

3.21 Solve the following form of a hypothetical equation:

$$f = \frac{0.25}{\left[\log\left(\frac{\varepsilon}{3.7d_i} + \frac{5.74}{\mathrm{Re}^{0.9}\sqrt{f}}\right)\right]^2}$$

Here,

f = friction factor

ε = roughness = 0.001 mm

d_i = inside diameter = 5 mm

Re = Reynolds number = 5000

With this, find the friction factor using the:

i) N–R method
ii) Bisection method
iii) Muller method
iv) Chebyshev method

3.22 Repeat the above problem (use the same data given for ε, d_i and Re) to find the Darcy friction factor (f) by the use of the following Colebrook – White correlation that is applicable for a conduit flowing completely full of fluid at Re > 4000:

$$\frac{1}{\sqrt{f}} = -2\log\left(\frac{\varepsilon}{3.7d_i} + \frac{2.51}{\mathrm{Re}\sqrt{f}}\right)$$

Use the Bisection and False Position method with a tolerance of 10^{-6}.

3.23 In a rectangular open channel, the velocity of water (m/s) is estimated by using the Manning equation as:

$$\upsilon = \frac{\sqrt{S}}{\varepsilon}\left(\frac{wH}{w+2H}\right)^{2/3}$$

Using a suitable root finding method, estimate the depth H (in m) when:

channel slope, $S = 0.0003$
width, $w = 15$ m
roughness factor, $\varepsilon = 0.022$
$\upsilon = 0.48H$

3.24 Find the minimum of:

$$f(x) = \cos x - \sin x$$

using the (tolerance = 10^{-5})

i) Secant method with $x_0 = 1$ and $x_1 = 2.5$.

ii) N–R method with $x_0 = 3$.

3.25 Find the minimum of:

$$f(x) = \frac{x^3 - 2x^2 - 8}{x - 1}$$

using the N–R method (tolerance = 10^{-4}).

3.26 Consider the following water dissociation reaction occurred at very high temperature (~ 3000–4000 K at 1 atm):

$$H_2O \Leftrightarrow H_2 + \frac{1}{2} O_2$$

The equilibrium constant:

$$K = \frac{x}{1 - x}\sqrt{\frac{2P_t}{2 + x}}$$

Find the root x (mole fraction) for the total pressure, $P_t = 2.5$ atm and $K = 0.1$.

3.27 Find the root x of the following equation:

$$\frac{1}{y} = \sqrt{\frac{1}{R^2} + \left(xC - \frac{1}{xL}\right)^2}$$

for $y = 120$, $R = 220$, $L = 0.8$ and $C = 0.6 \times 10^{-6}$.

3.28 For a chemical reactor, the mole balance at steady state yields:

$$\frac{F}{V} C_i = \frac{F}{V} C + k\, C^{3.5}$$

Given:

$$\frac{F}{V} = 0.1 \text{ min}^{-1}$$

$$C_i = 1 \text{ lbmol/ft}^3$$

$$k = 0.04 \ \frac{\text{ft}^{7.5}}{\text{lbmol}^{2.5}\text{min}}$$

Find concentration C using a suitable iterative convergence method.

3.29 An exothermic zero-order reaction takes place in a continuous stirred-tank reactor. The steady state model in terms of the dimensionless reactor temperature (x) is given as:

$$0.42204 \exp\left(\frac{x}{1 + 0.05x}\right) = 1.3x$$

Find the two roots of this equation using the:

– Secant method

– N–R method

(**Ans**: $x = 0.55946$ and 2.0).

3.30 The fall velocity of a jumper (υ) is modeled as:

$$f(m) = \sqrt{\frac{gm}{c_D}} \tanh\left(\sqrt{\frac{gc_D}{m}}\, t\right) - \upsilon(t)$$

Using the N–R method, find the mass of the jumper (m) to have a velocity of 40 m/sec after 5 sec of free fall.

Given:

Drag coefficient, c_D = 0.25 kg/m

Acceleration due to gravity, g = 9.81 m/sec^2

and m_0 = 100 kg.

3.31 The concentration of microorganism in a pond is described by:

$$x = 5e^{-t} + 1.2e^{-0.3t}$$

Find the time required to have x = 1.3. For this, use the *accelerated Newton–Raphson method* (for single variable) given as:

$$x_{k+1} = x_k - \frac{g(x_k)}{g'(x_k)} \tag{3.76a}$$

where,

$$g(x) = \frac{f(x)}{f'(x)} \tag{3.76b}$$

3.32 Using the accelerated N–R method [Equation (3.76)], solve:

i) $f(x) = e^{-x} - x = 0$

with x_0 = 0.5.

ii) $f(x) = 5x - \exp(-x^2) + \sin x = 0$

with x_0 = 0.5. Is there any improvement in rate of convergence (number of iterations) over the classical N–R method used for the same system equation in Example 3.4?

3.33 Solve the following cubic polynomial that has complex roots by using the N–R method:

$$f(z) = z^3 + 7z + 13 = 0$$

3.34 Find three roots of the following cubic equation:[17]

$$\frac{\gamma^3}{4} - \gamma^2 + 3\gamma - 6 = 0$$

3.35 Solve the following equation that has three complex roots by using the N–R method:

$$f(z) = i\,z^3 + 4z^2 - 3ie^z = 0$$

3.36 Solve the following quartic polynomial that has complex roots by using the N–R method:

$$f(z) = z^4 - 3z^3 + 10z^2 - 4z + 5 = 0$$

3.37 Find one real and two complex conjugate roots for:

$$f(z) = z^3 - z^2 + 428.49z - 428.49 = 0$$

3.38 Consider a non-linear system represented by:

$$f(z) = z^3 + 2z^2 - z + 5$$

i) Graph the function and report the number of roots (real or imaginary or both) this system has.

ii) Find the roots by using the N–R method.

17 This form of equation is solved in analyzing the numerical stability of the fourth-order Runge–Kutta method (see Chapter 4).

3.39 Find the drag coefficient (c) by solving the following non-linear form of equation:

$$f(c) = \frac{667.38}{c}[1 - \exp(-0.146843c)] - 40 = 0$$

with a desired tolerance of 10^{-4} (**Ans**: 14.7802).

3.40 The van der Waals equation has the following form:

$$\left(P + \frac{a}{v^2}\right)(v - b) = RT$$

where,

Pressure (P) = 30 atm
Temperature (T) = 320 K
Parameters: a = 3.5 L^2.atm/mol^2
b = 0.04 L/mol
Universal gas constant (R) = 0.082057 L.atm/mol.K

Find the molar volume (v) using the *accelerated N–R method* [Equation (3.76)] with the guess value of 0.5 L/mol.

3.41 The Redlich–Kwong (RK) equation of state is given as:

$$P = \frac{RT}{v - b} - \frac{a}{v(v + b)\sqrt{T}}$$

The following information are given:

Pressure (P) = 10^7 Pa
Temperature (T) = 342 K
Parameters (in SI unit): a = 6.5
b = 3 × 10^{-5}
Universal gas constant (R) = 8.314 Pa.m^3/mol.K

Find the molar volume v (in m^3/mol) using the Secant and N–R method.

3.42 Show that the following sequence (converges to $\sqrt{c}$) provides quadratic rate of convergence:

$$x_{n+1} = \frac{1}{2}x_n\left(3 - \frac{x_n^2}{c}\right)$$

3.43 In a binary mixture of benzene and toluene, the partial pressure of benzene is given as 350 mm Hg and that of toluene is 410 mm Hg. The Antoine equation [Equation (3.54)] is used to approximate the vapor pressure as:

$$P_i^S = \exp\left[A_i - \frac{B_i}{T + C_i}\right] \tag{3.54}$$

Antoine constants are given in Table 3.21.

Using the N–R method, you are asked to determine the:

i) equilibrium temperature
ii) liquid phase composition (mol fract)

3.44 Considering liquid phase non-ideality, it is straightforward to derive an expression for the liquid phase composition (x) of a binary system as:

$$x_1 = \frac{P_t - \gamma_2 P_2^S}{\gamma_1 P_1^S - \gamma_2 P_2^S} \tag{3.77}$$

Here, γ denotes the activity coefficient.

We would like to solve the same benzene/toluene system considered in the last exercise Problem 3.43. The Antoine constants are provided there. For activity coefficients, let us use the Margules model highlighted below (Dorfman and Daoutidis, 2017):

$$\ln\gamma_1 = \alpha_1 x_2^2 + \beta_1 x_2^3$$

$$\ln\gamma_2 = \alpha_2 x_1^2 + \beta_2 x_1^3$$

where,

$$\alpha_i = A + 3(-1)^{i+1}B$$

$$\beta_i = 4(-1)^i B$$

with

$$A = 0.9$$

$$B = 0.4$$

Find the liquid phase composition, x (mol fract) using the N–R method of convergence at three different temperatures, 80, 90 and 100 °C, when P_t = 760 mm Hg.

3.45 Consider a 1:1 shell and tube heat exchanger, which acts as a bottom reboiler (partial) in a distillation column. As shown in Figure 3.14, hot oil passes through the shell side and the bottom liquid (saturated) passes co-currently through the tube side.

The heat balance yields:

$$Q_r = U_r A_r \Delta T_{LMTDr} = m_o C_{po} \Delta T_o$$

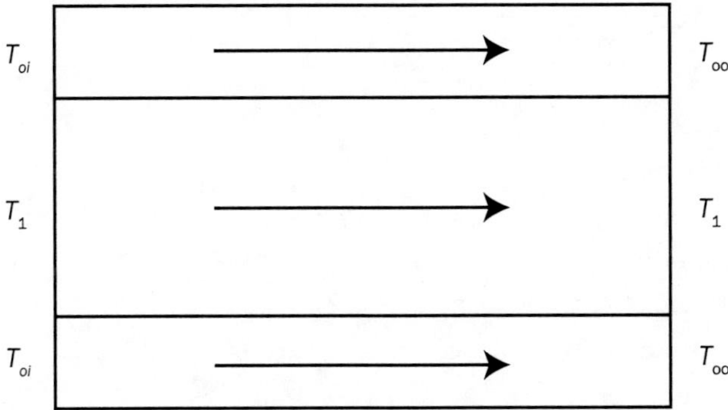

FIGURE 3.14 A 1:1 shell and tube heat exchanger

where,

$$\Delta T_{LMTDr} = \frac{(T_1 - T_{oi}) - (T_1 - T_{oo})}{\ln\dfrac{T_1 - T_{oi}}{T_1 - T_{oo}}}$$

$$\Delta T_o = T_{oo} - T_{oi}$$

$$T_{oo} = T_{oi}^{\alpha}$$

Given:

A_r = heat transfer area = 200 ft²

T_1 = temperature of bottom liquid = 230 °F

T_{oi} = inlet temperature of hot oil = 280 °F

m_o = hot oil flow rate = 28500 lb/hr

C_{po} = specific heat capacity of hot oil = 0.78 Btu/lb.°F

U_r = overall heat transfer coefficient = 250 Btu/hr.ft^2.°F

Using the N–R and Secant method, determine α with tol = 10^{-5}.

3.46 Compute the bubble point temperature (in °C) and vapor phase composition (in mol fract) of a binary ethanol/water system, which is ideal in vapor phase and non-ideal in liquid phase. Accordingly, one can use:

Modified Raoult's law (for non-ideal liquid phase):

$$P_i = \gamma_i P_i^S x_i \tag{3.78}$$

Dalton's law (for ideal vapor phase):

$$P_i = P_t y_i \tag{3.55}$$

Thus, we have at equilibrium:

$$y_i = \frac{\gamma_i P_i^S x_i}{P_t} \tag{3.79}$$

where γ_i is the activity coefficient of species i that is asked to determine from (Jana, 2018):

$$\ln\gamma_1 = -\ln(x_1 + \Lambda_{12}x_2) + x_2\left[\frac{\Lambda_{12}}{x_1 + \Lambda_{12}x_2} - \frac{\Lambda_{21}}{x_2 + \Lambda_{21}x_1}\right] \tag{3.80}$$

$$\ln\gamma_2 = -\ln(x_2 + \Lambda_{21}x_1) - x_1\left[\frac{\Lambda_{12}}{x_1 + \Lambda_{12}x_2} - \frac{\Lambda_{21}}{x_2 + \Lambda_{21}x_1}\right] \tag{3.81}$$

with

$\Lambda_{12} = 0.20916399$

$\Lambda_{21} = 0.82284181$

Here, Equations (3.80) and (3.81) represent the Wilson activity coefficient model.

To conduct the first three iterations of the N–R method, use the following information, including the Antoine constants given in Table 3.29:

Liquid phase composition (ethanol/water) 0.6/0.4 (mol fract)

Operating pressure (P_t) 1 atm

Table 3.29 Antoine constants (vapor pressure in mm Hg and temperature in K)

Component	Boiling point (°C)	A_i	B_i	C_i
Ethanol	78.37	18.5242	3578.91	–50.50
Water	100	18.3036	3816.44	–46.13

3.47 Repeat the above problem to find the dew point temperature and liquid phase composition using the N–R method with:

Vapor phase composition (ethanol/water) 0.6/0.4 (mol fract)

Operating pressure (P_t) 1 atm

3.48 There is a ternary mixture, consisting of 1 mol% methane, 59 mol% ethane and 40 mol% *n*-butane. Find the bubble point temperature at 51 psia using the Antoine constants given in Table 3.30.

TABLE **3.30** Antoine constants (vapor pressure in mm Hg and temperature in K)

Component	Boiling point (K)	A_i	B_i	C_i
Methane	111.5	15.2243	597.84	−7.16
Ethane	184	15.6637	1511.42	−17.16
n-butane	272	15.6782	2154.90	−34.42

3.49 Solve the following system of equations:

$$y_1^2 + y_1 y_2 = 10$$
$$3y_1 y_2^2 + y_2 = 60$$

by the use of the N–R method with:

$$y_{10} = 2$$
$$y_{20} = 4$$

3.50 Solve the following system of equations:

$$2y_1^2 - 5y_2^2 = 3$$
$$3y_1^3 + 2y_2^2 = 26$$

by the use of the N–R method with:

$$y_{10} = 1.1$$
$$y_{20} = 1$$

3.51 Solve the following system equations using the N–R method:

$$x^2 + y^2 = 17$$
$$2x^{\frac{1}{3}} + y^{\frac{1}{2}} = 4$$

with $x(0) = 1$, $y(0) = 2$ and tolerance = 10^{-4}.

3.52 Solve the following problem:

$$y_1 - 4y_1^2 - y_1 y_2 = 0$$
$$2y_2 - y_2^2 + 3y_1 y_2 = 0$$

by the use of the N–R method with:

i) $y_{10} = 1.0$ and $y_{20} = 1.1$
ii) $y_{10} = -1.0$ and $y_{20} = -1.1$

3.53 Solve the following system of equations:

$$\ln(x^2 + y) + y = 1$$
$$\sqrt{x} + xy = 0$$

correct to 5 decimal places.

3.54 Consider the following two-variable system:

$$f_1(x_1, x_2) = x_1^3 - 20.63x_1 - 4.21x_2 + 7 = 0$$
$$f_2(x_1, x_2) = x_2^2 + 7.28x_2 - 5.16x_1 + 18 = 0$$

The exact solution is: $[x_1 \quad x_2]^T = [1 \quad -3]^T$. Using the N–R technique with the initial guess values of $x_{10} = x_{20} = 0.5$, solve the problem and discuss the convergence characteristics with reference to Equation (A3.9).

3.55 Analyze the convergence characteristics of the following two-variable system with the use of Equation (A3.9):

$$x^2 - 2y - 8 = 0$$

$$-x^2 - y^2 + 16 = 0$$

Initial guess values are: $x_0 = 1$ and $y_0 = 0.5$.

3.56 Solve the following problem:

$$f_1(x_1, x_2) = \frac{5 + x_1 + x_2}{(20 - x_1)(50 - 2x_1 - x_2)^2} - 0.0005$$

$$f_2(x_1, x_2) = \frac{5 + x_1 + x_2}{(10 - x_2)(50 - 2x_1 - x_2)} - 0.03$$

by the use of the N–R method with $x_{10} = x_{20} = 2$.

3.57 Consider the following non-linear two-variable system:

$$x^2 + y^2 = 4$$

$$e^x + y = 1$$

i) Graph the functions and report the number of root(s) this system has.

ii) Find the root(s) by using the N–R method.

3.58 Consider the following system of two non-linear algebraic equations:

$$f(x, y) = 0$$

$$g(x, y) = 0$$

If (h, k) be the error in the initial guess (x_0, y_0), we have the roots as:

$$x^* = x_0 + h$$

$$y^* = y_0 + k$$

It yields:

$$f(x_0 + h, y_0 + k) = 0$$

$$g(x_0 + h, y_0 + k) = 0$$

i) Using the Taylor series expansion (with neglecting second- and higher-order terms), prove that:

$$h = \left(\frac{g f_y - f g_y}{f_x g_y - g_x f_y} \right)_{(x_0, y_0)}$$

$$k = \left(\frac{f g_x - g f_x}{f_x g_y - g_x f_y} \right)_{(x_0, y_0)}$$

ii) Assuming the next approximation to be close to the root, we have:

$$x_1 = x_0 + h$$

$$y_1 = y_0 + k$$

Accordingly, we can write:

$$x_{i+1} = x_i + \left(\frac{g\, f_y - f\, g_y}{f_x\, g_y - g_x\, f_y} \right)_{(x_i,\, y_i)} \qquad i = 0, 1, 2 \cdots$$

$$y_{i+1} = y_i + \left(\frac{f\, g_x - g\, f_x}{f_x\, g_y - g_x\, f_y} \right)_{(x_i,\, y_i)}$$

With this, the *accelerated Newton–Raphson* method is developed as:

$$x_{i+1} = x_i + \left(\frac{g\, f_y - f\, g_y}{f_x\, g_y - g_x\, f_y} \right)_{(x_i,\, y_i)} \qquad i = 0, 1, 2 \cdots$$

$$y_{i+1} = y_i + \left(\frac{f\, g_x - g\, f_x}{f_x\, g_y - g_x\, f_y} \right)_{(x_{i+1},\, y_i)}$$

Using this formula, solve the system equations given in Problems 3.50 and 3.51.

3.59 Solve the following set of equations using the Chebyshev method:

$$-3y_1 + y_2 + y_1 y_2 + 1 = 0$$
$$y_1 - 2y_2 + y_3 - y_1 y_2 + y_2 y_3 = 0$$
$$y_2 - 1.8y_3 - y_2 y_3 + 0.2 = 0$$

This set of equations is obtained in Example 5.4 of this book (Chapter 5). Guess the initial values as per your choice.

3.60 Perform the flash calculation[18] for a ternary mixture with the following given information:

Ternary system: *o*-xylene/*m*-xylene/*p*-xylene

P_t = 760 mm Hg

T = 141 °C

F = 100 kmol/hr

For the following form of the Antoine equation:

$$\log P_i^S = A_i + \frac{B_i}{T + C_i}$$

the coefficients are listed in Table 3.31. The feed composition (z_i) is also included in the same table.

Table 3.31 Antoine constants (vapor pressure in mm Hg and temperature in °C)

Component	Boiling point (°C)	z_i	A_i	B_i	C_i
o-xylene	144.4	0.3	6.99891	1474.679	213.69
m-xylene	139.1	0.3	7.00908	1462.266	215.11
p-xylene	138.4	0.4	6.99052	1453.430	215.31

Find L, V, x_i and y_i using the N–R method with a desired tolerance of 10^{-5}.

18 Same notations used here as in Case Study 3.8.

3.61 A multi-variable bioreactor model has the following form at steady state:

$$\frac{dx}{dt} = -D\,x + \mu\,x = 0$$

$$\frac{ds}{dt} = D(s_f - s) - \frac{\mu\,x}{Y} = 0$$

$$\frac{dp}{dt} = -D\,p + (\alpha\mu + \beta)x = 0$$

with (Monod kinetics)

$$\mu = \mu_{max}\frac{s}{k_s + s}$$

Given:

μ_{max} = 0.5 hr^{-1}

k_s = 0.25 g/L

s_f = 5 g/L

D = 0.1 hr^{-1}

Y = 0.4 g/g

α = 2.2 g/g

β = 0.2 hr^{-1}

(a) Find the concentration of substrate (*s*), biomass (*x*) and product (*p*) in g/L.

(b) Analyze the convergence characteristics with using Equation (A3.9).

3.62 Solve the following set of quadratic equations using the N–R method:

$$2y_1^2 + 30y_1 - 12y_2 - 18 = 0$$
$$2y_2^2 - 9y_1 + 12y_2 - 3y_3 = 0$$
$$2y_3^2 - 9y_2 + 12y_3 - 3y_4 = 0$$
$$2y_4^2 - 9y_3 + 12y_4 - 3y_5 = 0$$
$$2y_5^2 - 9y_4 + 12y_5 - 3y_6 = 0$$
$$2y_6^2 - 12y_5 + 12y_6 = 0$$

Ans

y_1 = 0.830122

y_2 = 0.690156

y_3 = 0.587801

y_4 = 0.511075

y_5 = 0.455031

y_6 = 0.424936

3.63 Solve the following problem using the N–R method with $\alpha = 1$ and $k = 0.25$:

$$-4x_1 - \alpha\,k^2x_1^2 + x_2 + x_4 + 2 = 0$$
$$x_1 - 4x_2 - \alpha\,k^2x_2^2 + x_3 + x_5 + 1 = 0$$
$$x_2 - 4x_3 - \alpha\,k^2x_3^2 + x_6 + 2 = 0$$
$$x_1 - 4x_4 - \alpha\,k^2x_4^2 + x_5 + x_7 + 1 = 0$$
$$x_2 + x_4 - 4x_5 - \alpha\,k^2x_5^2 + x_6 + x_8 = 0$$
$$x_3 + x_5 - 4x_6 - \alpha\,k^2x_6^2 + x_9 + 1 = 0$$
$$x_4 - 4x_7 - \alpha\,k^2x_7^2 + x_8 + 2 = 0$$
$$x_5 + x_7 - 4x_8 - \alpha\,k^2x_8^2 + x_9 + 1 = 0$$
$$x_6 + x_8 - 4x_9 - \alpha\,k^2x_9^2 + 2 = 0$$

Assume the required initial guess values. Note that these system equations are originated in an elliptic PDE problem (see Example 6.3 in Chapter 6).

3.64 Solve the following non-linear system of equations by the use of the N–R method:

$$-2x_1-k^2\phi^2x_1\exp\left(-\frac{\alpha}{x_4}\right)+x_2=0=f_1$$

$$x_1-2x_2-k^2\phi^2x_2\exp\left(-\frac{\alpha}{x_5}\right)+x_3=0=f_2$$

$$x_2-2x_3-k^2\phi^2x_3\exp\left(-\frac{\alpha}{x_6}\right)+1=0=f_3$$

$$1-2x_4-k^2\gamma x_1\exp\left(-\frac{\alpha}{x_4}\right)+x_5=0=f_4$$

$$x_4-2x_5-k^2\gamma x_2\exp\left(-\frac{\alpha}{x_5}\right)+x_6=0=f_5$$

$$x_5-2x_6-k^2\gamma x_3\exp\left(-\frac{\alpha}{x_6}\right)+1=0=f_6$$

with

$k = 0.25$

$\alpha = 0.01$

$\phi^2 = 1$

$\gamma = 15$

Assume the required initial guess values. Note that these equations are obtained when the ODEs of a boundary value problem (see Case Study 5.6) are discretized.

3.65 Solve the following set of equations and find u_1 to u_5 by using the N–R method with a tolerance limit of 10^{-6}:

$$2u_1^3-u_2=2k^2$$

$$-u_1+2u_2^2-u_3=2k^2$$

$$-u_2+2u_3-u_4^3=2k^2$$

$$-u_3^2+2u_4-u_5^3=2k^2$$

$$-u_4+2u_5^4=2k^2$$

where, $k = 0.1735$.

References

Dorfman, K. D. and Daoutidis, P. (2017). *Numerical Methods with Chemical Engineering Applications*, 1st ed., Cambridge: Cambridge University Press.

Gupta, R. K. (2019). *Numerical Methods: Fundamentals and Applications*, 1st ed., Cambridge: Cambridge University Press.

Jain, M. K., Iyengar, S. R. K., and Jain, R. K. (1995). *Numerical Methods for Scientific and Engineering Computation*, 3rd ed., New Delhi: New Age International.

Jana, A. K. (2018). *Chemical Process Modelling and Computer Simulation*, 3rd ed., New Delhi: Prentice-Hall.

Mathews, J. H. (2000). *Numerical Methods for Mathematics, Science, and Engineering*, 2nd ed., New Delhi: Prentice-Hall.

Reid, R. C., Prausnitz, J. M., and Sherwood, T. K. (1977). *The Properties of Gases and Liquids*, 3rd ed., New York: McGraw-Hill Book Company.

Smith, J. M. and VanNess, H. C. (1987). *Introduction to Chemical Engineering Thermodynamics*, 4th ed., New York: McGraw-Hill Book Company.

Part III

Systems of Differential Equations

An equation, which contains the derivatives of one or more dependent variables with respect to one or more independent variables, is called a *differential equation* (DE).

There are two types of differential equations:

- Ordinary differential equation (ODE)
- Partial differential equation (PDE)

Based on the numerical methods formulated and used to solve the ODEs, we further classify the ODE problems into two categories:

- Initial value problem (IVP)
- Boundary value problem (BVP)

In Chapters 4 and 5, we will cover the numerical solution of the ODE-IVP and ODE-BVP systems, respectively. The numerical schemes for the PDE systems will be discussed subsequently in Chapter 6.

4

Ordinary Differential Equation: Initial Value Problem (ODE-IVP)

"Among all of the mathematical disciplines the theory of differential equations is the most important... It furnishes the explanation of all those elementary manifestations of nature which involve time." – Sophus Lie

Key Learning Objectives

- Learning several numerical integration methods for initial value problems
- Analyzing numerical stability of those methods
- Example systems are solved to understand some important characteristics of the numerical methods
- Learning how to choose a suitable method for a specific system

4.1 Introduction

We will discuss here a couple of elegant and advanced numerical methods, and their use to solve the systems of first-order ordinary differential equations (ODEs).[1] At the beginning, let us first review a few basic concepts.

Order of an ODE

The order of the highest derivative appeared in a differential equation is typically termed as the *order* of that equation. For example,

$$p_1(x)\frac{d^2y}{dx^2}+p_2(x)\frac{dy}{dx}+p_3(x)y=p_4(x) \tag{4.1}$$

is a second-order ODE since the highest derivative that appears in it is second-order. Note that here, y is the dependent variable and x the independent variable.

1 An ODE is defined as a differential equation having one or more functions of one independent variable and the ordinary derivatives of those functions.

Degree of an ODE

The *degree* of a differential equation is the exponent/power of the highest order derivative in that equation. With this, Equation (4.1) has degree 1 since the exponent of the highest order derivative is 1.

Transforming Higher-Order ODE to a Set of First-Order ODEs

As stated, here we will learn how to solve the first-order ordinary differential equations (ODEs) numerically. For any higher-order ODE, however one can easily transform it to a set of first-order ODEs and that we are going to discuss. To understand this point, let us take an example system represented by a simple second-order ODE:

$$p(x)\frac{d^2y}{dx^2}+q(x)\frac{dy}{dx}=r(x) \tag{4.2a}$$

Let us consider,

$$y(x)=y_1(x)$$

$$\frac{dy_1}{dx}=y_2(x)$$

Accordingly, the second-order Equation (4.2a) gets the form of a set of two first-order ODEs:

$$\begin{cases} \dfrac{dy_1}{dx}=y_2(x) \\ \\ \dfrac{dy_2}{dx}=\dfrac{r(x)-q(x)y_2(x)}{p(x)} \end{cases}$$

In a similar fashion, one can reduce an nth-order ODE to:

$$\frac{dy_k(x)}{dx}=f_k(x,\ y_1,\ y_2,\dots\dots y_n) \tag{4.2b}$$

where, $k=1, 2, \dots\dots n$. This issue is further elaborated below with a second example.

Example 4.1 Transforming a third-order ODE to a set of first-order ODEs

Consider the following system of equation:

$$y'''(t)=y'(t)y(t)+5t\ y''(t) \tag{4.3a}$$

This third-order equation requires three initial conditions, given as:

$$y(0)=a_1$$

$$y'(0)=a_2$$

$$y''(0)=a_3$$

With this, transform Equation (4.3a) into a set of first-order ODEs.

Solution

Let us introduce,

$$y_1(t) = y(t)$$
$$y_2(t) = y'(t)$$
$$y_3(t) = y''(t)$$

Then Equation (4.3a) yields the following set of equations:

$$y'_1(t) = y_2(t)$$
$$y'_2(t) = y_3(t)$$
$$y'_3(t) = y_1(t)y_2(t) + 5t\ y_3(t)$$

This is the transformed set of first-order ODEs. In order to solve this set of equations, one can use the given initial conditions, which will be discussed a little later.

A little different approach of transformation

For Equation (4.3a), additionally one can also define,

$$y_4(t) = t$$

so that $y_4'(t) = 1$ and $y_4(t_0) = t_0$.

Based on Equation (4.2b), we get the following set of first-order ODEs:

$$y'(t) = f\left(y(t)\right) \quad [\text{here } x(=t) \text{ is also defined by } y(=y_4)] \tag{4.3b}$$

with

$$y' = \begin{bmatrix} y'_1 \\ y'_2 \\ y'_3 \\ y'_4 \end{bmatrix}$$

$$f(y) = \begin{bmatrix} y_2 \\ y_3 \\ y_1y_2 + 5y_3y_4 \\ 1 \end{bmatrix} \quad \text{and} \quad y(t_0) = \begin{bmatrix} a_1 \\ a_2 \\ a_3 \\ t_0 \end{bmatrix}.$$

Equation (4.3b) is said to be *autonomous* because it does not depend explicitly on time.

4.2 Classifying Ordinary Differential Equations

The ODEs can be classified on different bases. They include:

- Linear ODE versus non-linear ODE
(on the basis of linearity)
- Homogeneous ODE versus non-homogeneous ODE
(on the basis of homogeneity)

Note that the ODEs can also be classified based on the *order*.

4.2.1 Linear ODE versus Non-linear ODE

An ODE is said to be *linear* if the following conditions are satisfied:

- dependent variable y or its any derivative (e.g., y', y'' or y''') is not squared or cubed or more in power and their product (e.g., yy') does not appear
- coefficients[2] are function of independent variable and/or constant (i.e., the coefficients should not depend on y or its any derivative)

With this, the definition of *non-linear* ODE is quite obvious. Let us understand them through the following examples:

- $\frac{d^2y}{dx^2}+a\frac{dy}{dx}+by=c$ ..... linear ODE

(the power of y and its two derivatives is 1, and a, b and c are all constants)

- $\left(\frac{d^2y}{dx^2}\right)^2+a\frac{dy}{dx}+by=c$ ..... non-linear ODE

(the power of y'' is 2, although a, b and c are constants)

- $\frac{d^2y}{dx^2}+y\frac{dy}{dx}+by=c$ ..... non-linear ODE

(the power of y and its two derivatives is 1, but the coefficient of y' is y (i.e., a function of dependent variable))

4.2.2 Homogeneous ODE versus Non-Homogeneous ODE

A differential equation is said to be *homogeneous* if all terms in it contain the dependent variable y or its derivative; else, it is *non-homogeneous*. With this, Equation (4.1) is said to be:

- Homogeneous if $p_4(x) = 0$
- Non-homogeneous if $p_4(x) \neq 0$

Accordingly,

- $p_1(x)\frac{d^2y}{dx^2}+p_2(x)\frac{dy}{dx}+p_3(x)y=0$ .... homogeneous

2 The dictionary meaning of *coefficient* is as: 'a numerical or constant quantity placed before and multiplying the variable in an algebraic/differential expression.'

- $p_1(x)\dfrac{d^2y}{dx^2}+p_2(x)\dfrac{dy}{dx}+p_3(x)y=p_4(x)$ .... non-homogeneous

4.3 Classifying Systems of Ordinary Differential Equations

Based on the auxiliary conditions associated, the systems of ODEs are classified into:

- Initial value problem (IVP)
- Boundary value problem (BVP)

Let us briefly elaborate them with highlighting their basic features and differences.

Initial value problem (IVP)
For initial value ODE problem, all of the auxiliary conditions for the dependent variables are provided at the same value of the independent variable, which is commonly at time $t = 0$. These conditions are thus generally referred as the *initial conditions* and with the associated ODEs, they form the *ODE-IVP* system.

As an example, let us consider Equation (4.2b) with $n = 2$. Accordingly, we get a sample ODE-IVP structure as:

$$\frac{dy_1}{dx}=f_1(x,\ y_1,\ y_2)$$

$$\frac{dy_2}{dx}=f_2(x,\ y_1,\ y_2)$$

Initial conditions:

$$y_1(x=0)=y_{10}$$

$$y_2(x=0)=y_{20}$$

Suppose that here x is the time (t). When $0\le x\le 1$ (i.e., $x\in[0,\ 1]$), we have $x=t=0$ that indicates the starting time and $t=1$ is the end time.

Boundary value problem (BVP)
For boundary value ODE problem, all of the auxiliary conditions are provided at different values of the independent variable. These conditions are referred as the *boundary conditions* and with the associated ODEs, they form the *ODE-BVP* problem. Note that the boundary value problems are perhaps more concerned with *space*, while the initial value problems with *time*. Here, time and space both are independent variables.

As an example, let us consider Equation (4.2a):

$$p(x)\frac{d^2y}{dx^2}+q(x)\frac{dy}{dx}=r(x)$$

Boundary conditions:

$$y\ (x=a)=y_a$$

$$y\ (x=b)=y_b$$

Note that here x stands for the space. For $x \in [a, b]$, thus $x = a$ corresponds to the left boundary and $x = b$ to the right boundary.

Based on the above discussion, let us highlight the key differences between the ODE-IVP and ODE-BVP in Table 4.1.

TABLE 4.1 Basic differences between the IVP and BVP problems

IVP	BVP
1. The IVP system is concerned with time.	1. The BVP system is more concerned with space.
2. Auxiliary conditions for the dependent variables are provided at the same value of the independent variable (i.e., time $t = 0$).	2. Auxiliary conditions are provided at the different values of the independent variable.
3. Start at time $t = 0$ and continue solution step-by-step to nearby points.	3. Solution is determined everywhere (at each node) simultaneously.

4.4 Numerical Integration Methods: ODE-IVP Systems

Here we will mainly cover the following numerical integration methods that are widely employed to solve various types of the engineering and scientific problems:

- Euler method
- Heun method
- Taylor method
- Runge–Kutta family
- Runge–Kutta–Gill method
- Runge–Kutta–Fehlberg method
- Adams–Bashforth–Moulton (predictor–corrector) method

We will learn the formulation of these methods one-by-one, along with some of their stability analysis and application to example differential equation systems, including a couple of industrially relevant processes.

4.5 Euler Method

Let us consider an ODE,

$$\frac{dx}{dt} = f(x, t) \tag{4.4}$$

in which, x denotes the dependent variable, t (i.e., time) the independent variable and $f(x, t)$ a function (linear or non-linear). The initial condition (i.e., at time $t = 0$) for x we represent as: $x(0) = x_0$. With this, one can solve Equation (4.4) by employing the *Euler method*,[3] which has two versions,

3 Named after Swiss mathematician and engineer Leonhard Euler (1707–1783).

- explicit Euler
- implicit Euler

In the following, both these approaches are elaborated.

4.5.1 Explicit Euler or Forward Euler

A forward difference approximation yields:

$$\frac{dx}{dt} \approx \frac{x_{k+1} - x_k}{t_{k+1} - t_k} \tag{4.5}$$

Accordingly, Equation (4.4) is written as:

$$\frac{x_{k+1} - x_k}{\Delta t} = f(x_k, t_k) \tag{4.6}$$

Here, $\Delta t = h = t_{k+1} - t_k$. This time increment Δt $(= h)$ is referred as the *step size* or *integration interval.*

Rearranging Equation (4.6),

$$x_{k+1} = x_k + h f(x_k, t_k) \tag{4.7}$$

That is,

$$x_{k+1} = x_k + h\left(\frac{dx}{dt}\right)_{(x_k, t_k)} \tag{4.8}$$

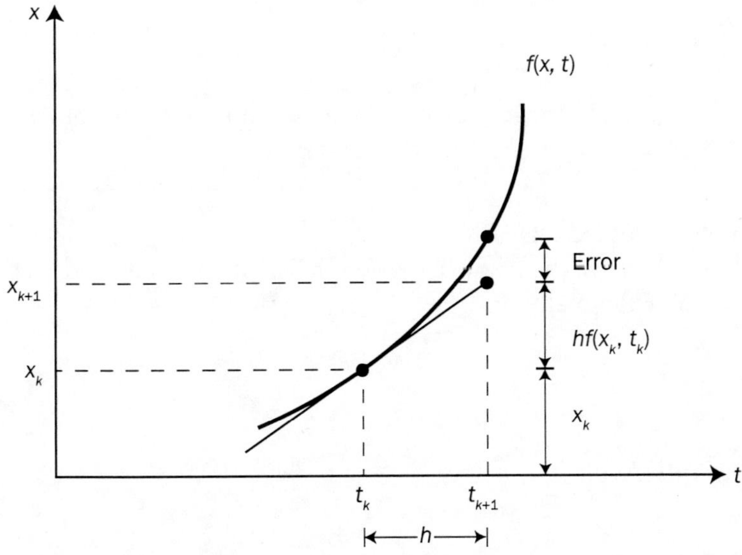

Figure 4.1 Illustrating explicit Euler method [slope = $f(x_k, t_k)$]

Equation (4.7) or (4.8) represents the *explicit Euler* formula. This simple integration method is graphically illustrated in Figure 4.1. It is obvious that if sufficiently small integration step size, h is chosen, the estimate of x_{k+1} will be reasonably close to its correct value. This explicit method will provide error-free predictions if the underlying function f is linear, because this approach uses straight line segments to approximate the solution.

Derivation: Explicit Euler Equation (4.7) from Taylor series

For the formal derivation of explicit Euler, let us use the Taylor series as:

$$x(t+h) = x(t) + \frac{h}{1!}x'(t) + \frac{h^2}{2!}x''(t) + \frac{h^3}{3!}x'''(t) + \ldots\ldots\ldots\ldots$$

That is,

$$x(t+h) = x(t) + \frac{h}{1!}x'(t) + R$$

The *remainder* or *local truncation error* (R) has the magnitude of $o(h^2)$. This notation $o(h^2)$ conveys that the error term is approximately a multiple of h^2.

Neglecting the second- and the higher-order terms (i.e., R),

$$x(t+h) = x(t) + \frac{h}{1!}x'(t)$$

This yields Equation (4.7) as:

$$x_{k+1} = x_k + h\,f(x_k, t_k)$$

since, $x'(t) = f(x_k, t_k)$. The derivation is complete here.[4]

Further, one can write from Taylor series:

$$x'_k = \frac{x_{k+1} - x_k}{h} + o(h)$$

Note that this is the actual format of explicit Euler and it leads to *first-order accuracy*. The order of truncation error has been reduced from $o(h^2)$ to $o(h)$.

Explicit Euler with coupled ODEs

Now we would like to solve the following system having two simultaneous coupled ODEs by using the explicit Euler method:

$$\frac{dx_1}{dt} = f_1(x_1, x_2, t) \tag{4.9a}$$

$$\frac{dx_2}{dt} = f_2(x_1, x_2, t) \tag{4.9b}$$

4 We should note that the Euler approach is basically the Taylor series method with terms containing only up to the first-order of h.

Applying the explicit Euler scheme,

$$x_{1\,k+1} = x_{1k} + h\left(\frac{dx_1}{dt}\right)_{(x_{1k},x_{2k},t_k)} \tag{4.10a}$$

$$x_{2\,k+1} = x_{2k} + h\left(\frac{dx_2}{dt}\right)_{(x_{1k},x_{2k},t_k)} \tag{4.10b}$$

Similarly, for any number of dependent variables (say, n), one can have:

$$X_{k+1} = X_k + h\,F(X_k,\ t_k) \tag{4.11}$$

where,

$$X_k = \begin{bmatrix} x_{1\,k} \\ x_{2\,k} \\ . \\ . \\ x_{n\,k} \end{bmatrix} \text{ and } F(X_k, t_k) = \begin{bmatrix} \left(\frac{dx_1}{dt}\right)_{(x_{1k},x_{2k},\cdots x_{nk},t_k)} \\ \left(\frac{dx_2}{dt}\right)_{(x_{1k},x_{2k},\cdots x_{nk},t_k)} \\ . \\ . \\ \left(\frac{dx_n}{dt}\right)_{(x_{1k},x_{2k},\cdots x_{nk},t_k)} \end{bmatrix}$$

This Euler integration approach is really simple and easily implementable to solve even highly non-linear multi-variable complex systems having a large number of ODEs.

Remarks

1. As illustrated in Figure 4.1, the explicit Euler method with smaller step size, h provides better approximation but at the expense of larger computational effort. So we should judiciously select h such that there is a balance maintained between the computational effort and accuracy of the solution. This point is discussed later with an example.
2. The step size, h affects the truncation and round-off errors in the solution. The small h leads to small truncation error and large round-off error. Whereas, for large step size, it is just reverse. Hence we must be careful enough in selecting h so that the *total error* (sum of truncation and round-off errors) is reasonably small.
3. The explicit Euler method is not always stable (i.e., conditionally stable). It depends on the clever choice of h as shown later with an example.

Example 4.2 Explicit Euler for a three-variable system

Let us take a formal example to apply the explicit Euler technique on a system modeled with three state variables as:

$$\frac{dy_1}{dt} = -y_1$$
$$\frac{dy_2}{dt} = -50y_2 \tag{4.12}$$
$$\frac{dy_3}{dt} = -2y_3$$

with an initial condition: $Y_0 = \begin{bmatrix} 1 & 1 & 1 \end{bmatrix}^T$.

Solution

The explicit Euler method yields:

$$\begin{bmatrix} y_{1k+1} \\ y_{2k+1} \\ y_{3k+1} \end{bmatrix} = \begin{bmatrix} y_{1k} \\ y_{2k} \\ y_{3k} \end{bmatrix} + h \begin{bmatrix} -y_{1k} \\ -50y_{2k} \\ -2y_{3k} \end{bmatrix}$$

Start with $k = 0$, at which, $y_{10} = y_{20} = y_{30} = 1$. With this and $h = 0.1$, a sample calculation is shown as:

$$\begin{bmatrix} y_{11} \\ y_{21} \\ y_{31} \end{bmatrix} = \begin{bmatrix} 0.9 \\ -4.0 \\ 0.8 \end{bmatrix}$$

Likewise, continue the calculation for $k = 1, 2, 3, \ldots$. and it is left for you.

4.5.2 Implicit Euler or Backward Euler

The implicit Euler approach uses the backward difference approximation. Accordingly, Equation (4.4) gives:

$$\frac{x_{k+1} - x_k}{h} = f(x_{k+1}, t_{k+1}) \tag{4.13}$$

Rearranging,

$$x_{k+1} = x_k + h\,f(x_{k+1}, t_{k+1}) \tag{4.14}$$

That is,

$$x_{k+1} = x_k + h\left(\frac{dx}{dt}\right)_{(x_{k+1},\, t_{k+1})} \tag{4.15}$$

These two Equations (4.14) and (4.15) represent the *implicit Euler* formula. Notice that the unknown x_{k+1} appears on both sides of the formula and thus, it is called an *implicit* method.

Implicit Euler with coupled ODEs

Extension of this implicit Euler formula to multi-variable system is quite easy. For example, for a system of n variables, Equation (4.11) modifies to:

$$X_{k+1} = X_k + h\, F(X_{k+1}, t_{k+1})$$

where,

$$F(X_{k+1}, t_{k+1}) = \begin{bmatrix} \left(\dfrac{dx_1}{dt}\right)_{(x_{1\,k+1},\, x_{2\,k+1}, \cdots x_{n\,k+1},\, t_{k+1})} \\ \left(\dfrac{dx_2}{dt}\right)_{(x_{1\,k+1},\, x_{2\,k+1}, \cdots x_{n\,k+1},\, t_{k+1})} \\ \cdot \\ \cdot \\ \left(\dfrac{dx_n}{dt}\right)_{(x_{1\,k+1},\, x_{2\,k+1}, \cdots x_{n\,k+1},\, t_{k+1})} \end{bmatrix}$$

Derivation: Implicit Euler Equation (4.14) from Taylor series

For backward difference,

$$x(t-h) = x(t) - \frac{h}{1!}x'(t) + \frac{h^2}{2!}x''(t) - \frac{h^3}{3!}x'''(t) + \dots\dots\dots\dots$$

That is,

$$x(t-h) = x(t) - \frac{h}{1!}x'(t) + R$$

It has the local truncation error (R) having magnitude of $o(h^2)$.

Neglecting the second- and the higher-order terms (i.e., R),

$$x(t-h) = x(t) - \frac{h}{1!}x'(t)$$

Rearranging yields Equation (4.14) and the derivation is complete.

Notice that like explicit Euler, its implicit counterpart is also first-order accurate.

A Short Note

This implicit Euler formula [i.e., Equation (4.15)] indicates that one needs to evaluate the derivative at the next step in time, t_{k+1}. But how is it possible without knowing x_{k+1}? Indeed, it is simple for a linear system. To elaborate this, let us take a simple example of a system of linear ODE:

$$\frac{dx}{dt} = 2x + 5 \tag{4.16}$$

Based on Equation (4.15), one can write:

$$x_{k+1} = x_k + h(2x_{k+1} + 5)$$

Simplifying,

$$x_{k+1} = \frac{x_k + 5h}{1 - 2h} \tag{4.17}$$

Now we can vary $k = 0, 1, 2, \ldots$ and get the results easily.

However, for a non-linear system, the implicit Euler method does not work so simply. To elaborate this point, consider a system of non-linear ODE:

$$\frac{dx}{dt} = 5\exp(-x^2) \tag{4.18}$$

From Equation (4.15):

$$x_{k+1} = x_k + 5h\exp(-x_{k+1}^2)$$

In order to solve it for x_{k+1}, a non-linear algebraic solution technique needs to be employed (shown later in Example 4.7).

Remarks

1. To solve a non-linear IVP, as indicated above, the implicit Euler method involves iterations, requiring an iterative convergence technique (e.g., Newton–Raphson method). Compared to the explicit Euler, thus it involves more complexity and computational load.
2. Obviously, the performance of the implicit Euler method is dependent on that of the iterative convergence technique.
3. Implicit Euler is always stable irrespective of the step size, h. It is elaborated further with an example at the later stage.

To address the issue concerning the effect of step size h on stability of the Euler method, let us take a simple linear system example in the following.

Example 4.3 Euler method

A liquid tank example

Let us consider a simple liquid tank system shown in Figure 4.2.

Here,

h = height of liquid in the tank

F_i = volumetric flow rate of inlet stream

F_0 = volumetric flow rate of outlet stream

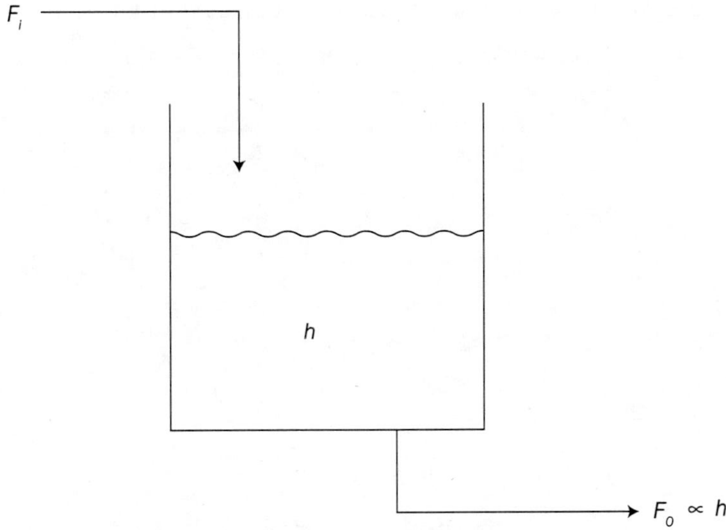

FIGURE 4.2 A simple liquid tank system

Performing mass balance based on Equation (A4.2),

$$A\frac{dh}{dt} = F_i - F_o \tag{4.19}$$

where, A is the cross sectional area. Note that this balance equation we get by assuming constant liquid density.

Considering the flow rate of the outlet stream F_0 in the liquid tank system proportional to the height of that liquid h (i.e., $F_0 \propto h$), Equation (4.19) yields:

$$A\frac{dh}{dt} = F_i - \beta\, h$$

Obviously, β is the proportionality constant. Let us now assume that F_i is fixed at its steady state value, F_{is} throughout the operation. Accordingly, the above model equation can be represented in terms of deviation variable as:

$$A\frac{dH}{dt} + \beta\, H = 0$$

Rearranging,

$$\tau_P\frac{dH}{dt} + H = 0$$

In which, the deviation variable, $H = h - h_s$, where h_s is the steady state value of h and the time constant, $\tau_P = \frac{A}{\beta}$. Accordingly,

$$\frac{dH}{dt} = -\frac{1}{\tau_P}H$$

For the liquid tank system, one can have the information as shown in Table 4.2.

TABLE 4.2 Information of liquid tank system

Given	Calculated
F_{is} = 8 m³/sec	$\tau_P = \frac{A}{\beta} = 4$ sec
β = 4 m²/sec	
A = 16 m²	$h_s = \frac{F_{is}}{\beta} = 2$ m

Let us now represent the liquid tank example in a general way with:

$$\text{System: } \frac{dx}{dt} = -\frac{1}{\tau_P}x \tag{4.20a}$$

where, $x = H$, $\tau_P = 4$ and initial condition, $x_0 = 5$.

With this, we would now like to compare the numerical solutions of the explicit and implicit Euler with reference to the analytical solutions by varying the step size,[5] Δt as 0.01, 1 and 10 sec.

Analytical solution

From Equation (4.20a),

$$x(t) = \exp\left(-\frac{t}{\tau_P}\right)x(0)$$

where, $x(0) = x_0 = 5$.

Numerical solution

Explicit Euler

$$x_{k+1} = x_k + \Delta t\left(-\frac{1}{\tau_P}x_k\right) = \left(1 - \frac{\Delta t}{\tau_P}\right)x_k \qquad k = 0, 1, 2, ... \tag{4.20b}$$

Implicit Euler

$$x_{k+1} = x_k + \Delta t\left(-\frac{1}{\tau_P}x_{k+1}\right)$$

This linear form of the equation yields:

$$x_{k+1} = \frac{1}{1 + \frac{\Delta t}{\tau_P}}x_k \qquad k = 0, 1, 2, ... \tag{4.20c}$$

5 To avoid any confusion, we use Δt (instead of h) for the liquid tank example to represent the step size since h is used here to denote the liquid height.

With this, we produce the comparative results in Table 4.3.

TABLE 4.3 Comparing numerical and analytical results

Δt	Analytical (x_A)	Numerical (x_N)		Absolute true error
		Explicit	Implicit	($\lvert x_N - x_A \rvert$)
0.01	4.9875	4.9875	4.9875	0 (explicit) 0 (implicit)
1	3.894	3.75	4	0.144 (explicit) 0.106 (implicit)
10	0.4104	–7.5	1.4286	7.9104 (explicit) 1.0182 (implicit)

Remarks

For this liquid level problem, we have the following findings and remarks:

1. At a reasonably large value of Δt (i.e., Δt = 10 sec), the explicit Euler provides a negative H of –7.5 m (that means, actual height, $h = H + h_s = -7.5 + 2 = -5.5$ m), which is physically unrealistic since height never be negative.

   This reveals that the explicit Euler method is very sensitive to the step size, whereas there is no such problem arises in case of implicit Euler.
2. Although there is no concern on stability of the implicit Euler, but its accuracy gets affected with step size. It is obvious from the tabulated results that the absolute true error increases with the increase of step size.
3. To get stable (realistic) solution from explicit Euler method, one needs to choose a reasonably small step size that, in turn, leads to large computations. On the other hand, although comparatively large step size can be used in implicit Euler, the involved iterative convergence technique (for non-linear systems) requires huge computational effort. So, it is not so straightforward to conclude that one is always better than other one on the stated issue.

In the following, we extend our discussion on numerical stability to have a proper guideline for selecting the step size, Δt in Euler methods. Prior to this, one would go through the convergence and stability analysis made for non-linear algebraic equation in the Appendix Section A3.1 (at the end of Chapter 3).

Numerical Stability of Explicit Euler: Liquid tank system (Example 4.3)

For explicit Euler, we obtained before for the liquid tank system [Equation (4.20b)],

$$x_{k+1} = \left(1 - \frac{\Delta t}{\tau_p}\right) x_k \qquad k = 0,\ 1,\ 2,\ ....$$

$$\begin{aligned}
\text{For } k=0 \quad & x_1 = \left(1-\frac{\Delta t}{\tau_p}\right)x_0 \\
k=1 \quad & x_2 = \left(1-\frac{\Delta t}{\tau_p}\right)x_1 = \left(1-\frac{\Delta t}{\tau_p}\right)^2 x_0 \\
k=2 \quad & x_3 = \left(1-\frac{\Delta t}{\tau_p}\right)x_2 = \left(1-\frac{\Delta t}{\tau_p}\right)^3 x_0 \\
\vdots \quad & \\
k=n \quad & x_{n+1} = \left(1-\frac{\Delta t}{\tau_p}\right)x_n = \left(1-\frac{\Delta t}{\tau_p}\right)^{n+1} x_0
\end{aligned} \tag{4.21}$$

This $\left(1-\frac{\Delta t}{\tau_p}\right)$ term is sometimes called the *growth or amplification factor*.

Now, we represent Equation (4.20b) as:

$$x_{k+1} = \left(1-\frac{\Delta t}{\tau_p}\right)x_k = g'\, x_k$$

for which, the stability condition is:

$$|g'| \le 1$$

Note that with this, the output x remains bounded.[6]

Accordingly,

$$\left|1-\frac{\Delta t}{\tau_p}\right| \le 1$$

which gives the *stability requirement* for the liquid tank system in terms of step size as:

$$0 \le \Delta t \le 2\tau_p$$

Since $\Delta t > 0$ (and $\tau_p > 0$), the stability condition is rewritten as:

$$0 < \Delta t \le 2\tau_p$$

Or simply,

$$\Delta t \le 2\tau_p$$

Now, to observe the different *nature of convergence* (or *solution*), let us divide the range of Δt. Accordingly, one can have:

---

6 But remember, the solution is convergent to zero when $|g'|<1$ (see Appendix A3.1).
Additional constraint $|g'| = 1$ corresponds to:
- neither convergence nor divergence (inconclusive)
- marginal stability condition (that also leads to keep the output bounded)

A. Stable monotonic solution when $0 < g' < 1$ that yields

$$0 < \Delta t < \tau_P$$

B. Stable oscillatory solution when $-1 < g' < 0$ that yields

$$\tau_P < \Delta t < 2\tau_P$$

Example 4.4 Stable monotonic solution of explicit Euler

A liquid tank example (revisited)

We have derived above the condition for stable monotonic solution as:

$$0 < \Delta t < \tau_P$$

Now to validate this for the liquid tank system, let us adopt $\Delta t = 2$ (recall that $\tau_p = 4$ and $x_0 = 5$). Using Equation (4.21), we produce a set of results in Table 4.4.

TABLE 4.4 Produced x_k versus k data

k	time t (sec)	x_k
0	0	5
1	2	2.5
2	4	1.25
3	6	0.625
4	8	0.3125
5	10	0.15625
6	12	0.078125
7	14	0.0390625
8	16	0.0195312
9	18	0.0097656
10	20	0.0048828

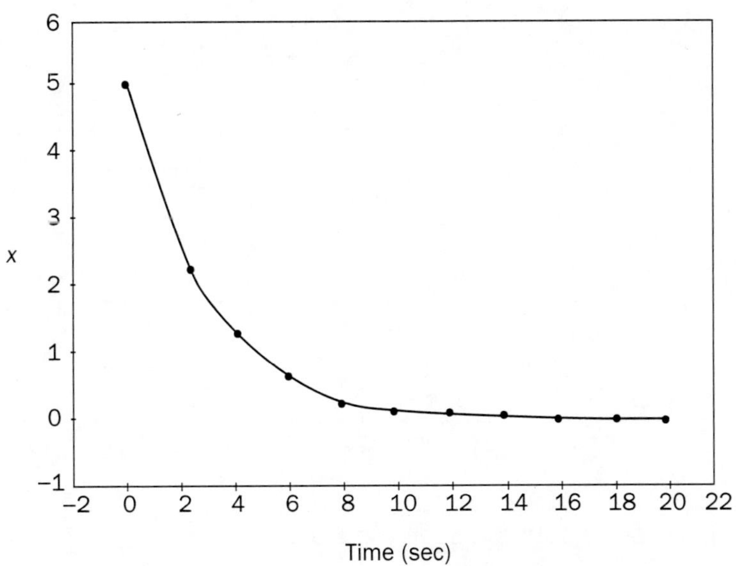

FIGURE 4.3 Monotonically decreasing liquid height, H

Plotting them in Figure 4.3, one can see how the liquid height, H (= $x = h - h_s$) decreases monotonically with time and converges to the steady state.

Example 4.5 Stable oscillatory solution of explicit Euler

A liquid tank example (revisited)

For stable oscillatory solution, we have derived the condition before as:

$$\tau_P < \Delta t < 2\tau_P$$

Within this range, let us adopt $\Delta t = 5$. With $\tau_p = 4$ and $x_0 = 5$, we generate the following set of data (Table 4.5) based on Equation (4.21):

Table 4.5 Produced x_k versus k data

k	time t (sec)	x_k
0	0	5
1	5	−1.25
2	10	0.3125
3	15	−0.078125
4	20	0.01953125
5	25	−0.0048828
6	30	0.00122070
7	35	−0.0003051758
8	40	0.00007629395

Oscillatory convergence is quite obvious from Figure 4.4 that is produced based on the tabulated data set. Recall that, as stated, here x is basically H (= $h - h_s$), where $h_s = 2$ m.

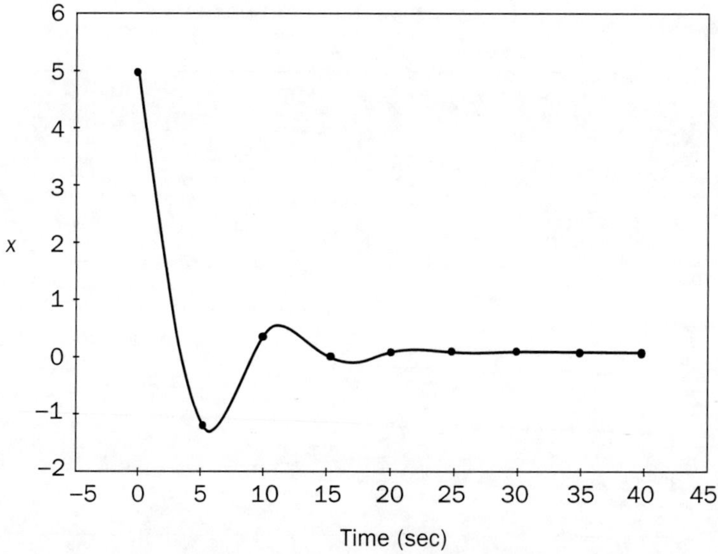

Figure 4.4 Stable oscillatory solution of explicit Euler

Remark

Note that x_1 of −1.25 (see Table 4.5) corresponds to a liquid height of 0.75 m (since liquid height $= x + h_s = -1.25 + 2 = 0.75$). However, if one adopts a step size Δt = 7 that is also within the range of $\tau_P < \Delta t < 2\tau_P$, then the liquid height becomes negative (that is, −1.75, since $x_1 = -3.75$), which is physically not realistic. In such a situation, one could start with a lower initial value for x (for example, x_0 = 2).

Example 4.6 Unstable solution of explicit Euler

A liquid tank example (revisited)

The explicit Euler method is unstable when $|g'| > 1$; which means, either $g' > 1$ or $g' < -1$. For the liquid tank system, we adopt a step size $\Delta t = 10$ ($> 2\tau_P$). With $\tau_P = 4$ and $x_0 = 5$, we generate a set of data in Table 4.6 based on Equation (4.21).

Table 4.6 Produced x_k versus k data

k	time t (sec)	x_k
0	0	5
1	10	−7.5
2	20	11.25
3	30	−16.875
4	40	25.3125
5	50	−37.9687

Divergence is quite obvious from Figure 4.5.

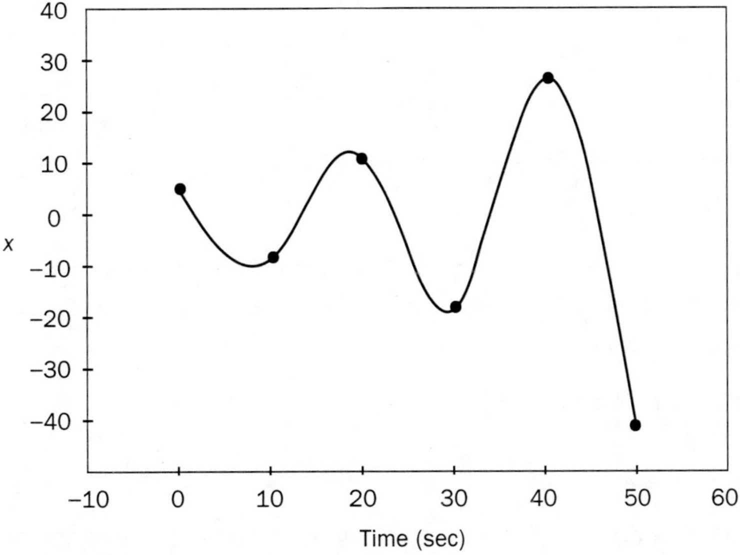

Figure 4.5 Unstable solution of explicit Euler

Numerical Stability of Implicit Euler: Liquid tank system (Example 4.3)

For implicit Euler, we previously derived,

$$x_{k+1} = \frac{1}{1+\frac{\Delta t}{\tau_p}} x_k = g' x_k \qquad k = 0,\ 1,\ 2,\ ....$$

Recall the stability condition: $|g'| \le 1$. Accordingly,

$$\left| \frac{1}{1+\frac{\Delta t}{\tau_p}} \right| \le 1$$

It is fairly true that both the step size Δt and time constant τ_p are positive quantity. So, the above stability condition is always satisfied. The implicit Euler method is thus *unconditionally stable.*

Example 4.7 Integrating a non-linear system by implicit Euler

Consider the following non-linear system equation:

$$\frac{dx}{dt} = 5\, e^{-x^2}$$

with $x_0 = 0$ and $h = \Delta t = 0.001$. Solve this problem and tabulate the results in the format of x_k versus k for the first 10 time steps.

Solution

For the given system,

$$\frac{dx}{dt} = 5 \exp(-x^2)$$

the implicit Euler formula [Equation (4.15)] yields:

$$x_{k+1} = x_k + 5h\ \exp(-x_{k+1}^2)$$

With known x_k, the above form of equation is implicit in x_{k+1}. For solving this non-linear form, one can reframe it as:

$$\eta(w) = w - x_k - 5h\ \exp(-w^2) \tag{4.22a}$$

Obviously, w is the next estimate of x, which means $w_k = x_{k+1}$.

Differentiating,

$$\eta'(w) = 1 + 10h\ w\ \exp(-w^2) \tag{4.22b}$$

Now to find the root of $\eta(w) = 0$ using the Newton–Raphson method, one can have:

$$w_k^{i+1} = w_k^i - \frac{\eta(w_k^i)}{\eta'(w_k^i)} \qquad \text{for } k = 0, 1, 2 \ldots \text{ and } i = 0, 1, 2 \ldots \tag{4.22c}$$

Notice that here the subscript k denotes the *time step* and the superscript i denotes the *iteration step*.

Initial guess

At each time step k, we need an initial guess for w $(= w_k^0)$. For this, let us use the given value (i.e., $x_0 = 0$) at the beginning ($k = 0$) as:

$$w_0^0 = x_0 = 0$$

and find w_k $(= x_1)$ at the end of iteration with Equation (4.22c). Then use this x_1 to initialize w at $k = 1$ and find w_k $(= x_2)$, and so on. Accordingly, for the subsequent time steps ($k \geq 1$), our guess is set as:

$$w_k^0 = x_k$$

With this background, we frame the following computer-assisted algorithm to solve the problem at each time step.

Computational steps

Step 1: Set initial guess for Equation (4.22c) as:

$$w_0^0 = x_0 = 0 \qquad \text{at } k = 0$$

$$w_k^0 = x_k \qquad \text{at } k = 1, 2, \ldots.$$

Step 2: Find $\eta(w)$ and $\eta'(w)$ from Equations (4.22a) and (4.22b), respectively.

Step 3: Update w by using Equation (4.22c).

Applying this algorithm, we produce the solutions for the first 10 time steps (tol = 10^{-6}) as shown in Table 4.7.

Table 4.7 x_k versus k for the first 10 time steps

k	time t	x_k
0	0	0.000000
1	0.001	5.000000e-3
2	0.002	9.99962502e-3
3	0.003	1.49986252e-2
4	0.004	1.99967511e-2
5	0.005	2.49937533e-2
6	0.006	2.99893830e-2
7	0.007	3.49833919e-2
8	0.008	3.99755322e-2
9	0.009	4.49655567e-2
10	0.010	4.99532192e-2

4.6 Heun Method

Now we would like to discuss a better numerical integration method, namely the *Heun method*,[7] than the explicit Euler for solving the same type of initial value problem. In formulating this Heun technique, the Euler method is used as the foundation and thus, the Heun method is sometimes referred as the *Improved Euler* method.

One can rewrite Equation (4.4) as:

$$\frac{dx(t)}{dt} = x'(t) = f[x(t),\ t] \tag{4.23}$$

Let us integrate it,

$$\int_{t_k}^{t_{k+1}} f[x(t),\ t]\,\mathrm{dt} = \int_{t_k}^{t_{k+1}} x'(t)\,\mathrm{dt} = x(t_{k+1}) - x(t_k) = x_{k+1} - x_k$$

Rearranging,

$$x_{k+1} = x_k + \int_{t_k}^{t_{k+1}} f[x(t),\ t]\,\mathrm{dt} \tag{4.24}$$

To approximate the definite integral term present in the right side, let us employ the *trapezoidal rule*[8] with a step size, $\Delta t = h = t_{k+1} - t_k$. Accordingly, one obtains:

$$x_{k+1} \approx x_k + \frac{h}{2}[f(x_k,\ t_k) + f(x_{k+1},\ t_{k+1})] \tag{4.25}$$

Using the explicit Euler formula [Equation (4.7)],

$$x_{k+1} = x_k + \frac{h}{2}[f(x_k,\ t_k) + f(x_k + h\,f(x_k,\ t_k),\ t_{k+1})] \tag{4.26}$$

This equation we can split, and present by the following structure:

$$\begin{cases} P_{k+1} = x_k + h\,f(x_k,\ t_k) \\ x_{k+1} = x_k + \dfrac{h}{2}[f(x_k,\ t_k) + f(P_{k+1},\ t_{k+1})] \end{cases} \tag{4.27}$$

where $t_{k+1} = t_k + h$. This equation set represents the *Heun method*. This algorithm can be derived in an alternative way, which is shown later (along with the derivation of second-order Runge–Kutta method).

Remarks

1. As indicated in Equation (4.27), the Euler method is described as a predictor algorithm, while the Heun method as a predictor–corrector algorithm.

---

7 Named after a German mathematician Karl Heun (1859–1929).

8 Detailed in Chapter 9.

2. Actually, the Heun method is developed by combining the explicit Euler (predictor) and trapezoidal rule (corrector). Notice that Equation (4.25) is obtained by averaging the two Euler methods (i.e., explicit and implicit).

Numerical Stability of Heun Method: Liquid tank system (Example 4.3)

For the liquid tank system, Equation (4.27) gives with $\Delta t = h$,

$$x_{k+1} = x_k + \frac{\Delta t}{2}\left[-\frac{1}{\tau_P}x_k - \frac{1}{\tau_P}P_{k+1}\right]$$

$$= x_k + \frac{\Delta t}{2}\left[-\frac{1}{\tau_P}x_k - \frac{1}{\tau_P}\left(x_k - \Delta t\frac{1}{\tau_P}x_k\right)\right] \qquad k = 0, 1, 2, ...$$

Simplifying,

$$x_{k+1} = \left(1 - \frac{\Delta t}{\tau_P} + \frac{1}{2}\frac{(\Delta t)^2}{\tau_P^2}\right)x_k$$

For stable solution, $|g'| \le 1$ and thus,

$$\left|1 - \frac{\Delta t}{\tau_P} + \frac{1}{2}\frac{(\Delta t)^2}{\tau_P^2}\right| \le 1$$

That is:

$$\left|\frac{(\Delta t - \tau_P)^2 + \tau_P^2}{2\tau_P^2}\right| \le 1$$

Notice that here g' is always positive and its minimum value is 1/2 at $\Delta t = \tau_P$. Accordingly, we have:

$$\left(1 - \frac{\Delta t}{\tau_P} + \frac{1}{2}\frac{(\Delta t)^2}{\tau_P^2}\right) \le 1$$

It yields the following form:

$$\frac{\Delta t}{\tau_P}\left(\frac{1}{2}\frac{\Delta t}{\tau_P} - 1\right) \le 0$$

Since $\dfrac{\Delta t}{\tau_P} > 0$, the stability condition is simply written as:

$$\Delta t \le 2\tau_P$$

Remark

It is obvious that although the Heun method is no more stable than the explicit Euler, but this predictor–corrector method has the advantage of higher global accuracy.

4.7 Taylor Method

The Taylor method[9] is one of the oldest numerical integration methods used to solve the ODEs. It is employed to formulate the various elegant numerical approaches; notable examples include the Euler and Newton–Raphson methods.

To understand this Taylor method, let us start with Equation (4.23):

$$\frac{dx(t)}{dt} = x'(t) = f[x(t),\, t] \tag{4.23}$$

The value of x at t_{k+1} (i.e., x_{k+1}) is approximated from the nth degree Taylor series of $x(t)$ at $t = t_{k+1}$. Considering $h = t_{k+1} - t_k$,

$$x\left(t_{k+1}\right) = x_{k+1} = x_k + \frac{d_1 h}{1!} + \frac{d_2 h^2}{2!} + \frac{d_3 h^3}{3!} + \ldots\ldots + \frac{d_n h^n}{n!} \tag{4.28}$$

In which,

$$d_j = x^{(j)}(t_k) = x_k^{(j)} \qquad \text{for } j = 1, 2, 3, \ldots. n$$

It implies $d_1 = x'_k$, $d_2 = x''_k$ and so on.

Remarks

1. The Taylor method can be used to test the accuracy of other numerical integration approaches.
2. The Taylor algorithm can be devised for higher level of accuracy. For example, if one keeps the first n+1 terms of the Taylor series expansion as shown in Equation (4.28), an nth order accurate method is obtained.
3. Our problem is given in the following form:

   $x'(t) = f[x(t),\ t]$

   Now to use the Taylor method, one must compute the higher derivatives by differentiating this function repeatedly. In this respect, at first we need to compute:

   $$x''(t) = f_x[x(t),\ t]\, x'(t) + f_t[x(t),\ t]$$

   $$= f_x[x(t),\ t]\, f[x(t),\ t] + f_t[x(t),\ t]$$

9 Named after the English mathematician Brook Taylor, FRS (1685–1731).

where,

$$f_x = \frac{\partial f}{\partial x}$$

$$f_t = \frac{\partial f}{\partial t}$$

Likewise, it is required to compute x''', $x^{(iv)}$ and so on as per the accuracy requirement. Obviously, they lead to very messy expressions and thus, the Taylor method has very limited use in practice.

4.8 Runge–Kutta (RK) Family

The Runge–Kutta (RK) methods[10] are a family of explicit *single step* numerical approaches, which are very popular in solving the ODE-IVP systems. This method is a better explicit integration approach than the standard explicit Euler method. We will realize this at a little later stage.

Importantly, the RK method is comparable with multistep algorithms in terms of accuracy but unlike them, the RK scheme does not have the starting problem involving several previous values of the solution and/or its derivative.

Let us first discuss some basic mathematics involved in the Runge–Kutta family. For this, we rewrite Equation (4.4),

$$\frac{dx}{dt} = f(t, x) \qquad x(t=0) = x_0 \tag{4.29}$$

We have the solution in a general way as:

$$x_{k+1} = x_k + h\phi(t_k, x_k, h) \tag{4.30}$$

The increment function, ϕ is defined as:

$$\phi = a_1k_1 + a_2k_2 + a_3k_3 + \cdots\cdots + a_nk_n \tag{4.31}$$

where, a refers to the weight, k the slope and n the order of the RK method.

The slopes are specified through a set of *recursive relations*:

$$k_1 = f(t_k, x_k) \tag{4.32a}$$

$$k_2 = f\left(t_k + p_1h,\ x_k + q_{11}k_1h\right) \tag{4.32b}$$

$$k_3 = f\left(t_k + p_2h,\ x_k + q_{21}k_1h + q_{22}k_2h\right) \tag{4.32c}$$

10 Named after two German mathematicians, Carl David Tolmé Runge (1856–1927) and Martin Wilhelm Kutta (1867–1944).

.

.

.

$$k_n = f\left(t_k + p_{n-1}h,\ x_k + q_{n-1\ 1}k_1 h + ... + q_{n-1\ n-1}k_{n-1}h\right) \tag{4.32d}$$

Again, p and q are weights, which along with a, need to be determined in some optimal manner that can enforce a desired accuracy. Recall that n refers to the order of the RK method. With this background, we derive the first- and second-order RK methods below.

4.8.1 First-Order Runge–Kutta (RK1)

For first-order (i.e., $n = 1$),

$$\phi = a_1 k_1$$

where,

$$k_1 = f(t_k,\ x_k) \tag{4.32a}$$

and $a_1 = 1$.

Accordingly, Equation (4.30) yields:

$$x_{k+1} = x_k + h\, f(t_k,\ x_k)$$

This is the representation of the *first-order Runge–Kutta* (RK1) method, which is exactly same with the explicit Euler technique, and it is thus, first-order accurate.

4.8.2 Second-Order Runge–Kutta (RK2)

The second-order Runge–Kutta method is an extension of the first-order Euler approach and it is also called the *midpoint Euler* method. Prior to formally deriving this RK2, let us first understand the concept involved in it. In this scheme, first the Euler method is used to estimate x at the midpoint of the integration interval (i.e., step size = $h/2$). If k_1 be the slope (dx/dt) at the initial point ($h = 0$), then we get from Equation (4.29):

$$k_1 = f(t_k,\ x_k) \tag{4.32a}$$

Thus, x at the midpoint will be $x_k + (h/2)k_1$.

Next, we evaluate the derivative (slope k_2) at the midpoint. It implies,

$$k_2 = f\left(t_k + \frac{h}{2},\ x_k + \frac{h}{2}k_1\right) \tag{4.33}$$

At the final stage, we estimate the value of x at the end of the step (step size = h) as:

$$x_{k+1} = x_k + hk_2 \tag{4.34}$$

This is the *second-order Runge–Kutta* (RK2) algorithm that consists of Equations (4.32a), (4.33) and (4.34).

Remarks

1. This is obviously a single-step, two-stage explicit method.
2. This integration scheme secures better accuracy than the explicit Euler method but the Euler approach is computationally less expensive since it involves one function evaluation, whereas RK2 requires two evaluations.

Derivation: Second-order Runge–Kutta (RK2)

For second-order method (i.e., $n = 2$), we get from Equation (4.31):

$$\phi = a_1 k_1 + a_2 k_2 \tag{4.35}$$

for which,

$$k_1 = f(t_k, x_k) \tag{4.32a}$$

$$k_2 = f\left(t_k + p_1 h,\ x_k + q_{11} k_1 h\right) \tag{4.32b}$$

Figure 4.6 shows the points A and B at which the respective slopes, k_1 and k_2, are determined.

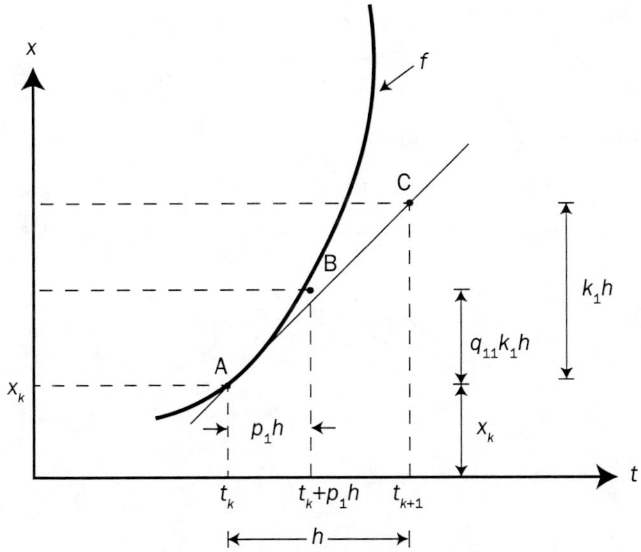

FIGURE 4.6 Illustration of the RK2 method

As shown in Equation (4.32a), determining k_1 is quite straightforward. Let us now evaluate k_2, which can be expanded about the point (t_k, x_k) with using a three-term Taylor series. We know,

$$f(t+h,\ x+k) = f(t,x) + h f_t + k f_x + o(h^2) \tag{4.36}$$

Recall that,

$$f_t = \frac{\partial f}{\partial t}$$

$$f_x = \frac{\partial f}{\partial x}$$

Using the three-term Taylor series [Equation (4.36)],

$$k_2 = f(t_k + p_1 h,\ \ x_k + q_{11} k_1 h)$$

$$= f(t_k, x_k) + p_1 h f_t(t_k, x_k) + q_{11} k_1 h f_x(t_k, x_k) + o(h^2) \tag{4.37}$$

From Equation (4.30),

$$x_{k+1} = x_k + h\phi(t_k, x_k, h)$$

$$= x_k + h(a_1 k_1 + a_2 k_2)$$

$$= x_k + h\ [a_1 f(t_k, x_k) + a_2\{f(t_k, x_k) + p_1 h f_t(t_k, x_k) + q_{11} k_1 h f_x(t_k, x_k)\}] + o(h^3)$$

Rearranging,

$$x_{k+1} = x_k + h\ [a_1 f(t_k, x_k) + a_2 f(t_k, x_k)] + h^2 [a_2 p_1 f_t(t_k, x_k) + a_2 q_{11} f(t_k, x_k) f_x(t_k, x_k)] + o(h^3) \tag{4.38}$$

A Taylor expansion of x_{k+1} about x_k gives:

$$x_{k+1} = x_k + h f(t_k, x_k) + \frac{h^2}{2!} f'(t_k, x_k) + \frac{h^3}{3!} f''(t_k, x_k) + \ldots\ldots \tag{4.39}$$

Since,

$$f'(t_k, x_k) = \frac{df}{dt}(t_k, x_k) = \left[\frac{\partial f}{\partial t} + \frac{\partial f}{\partial x}\frac{dx}{dt}\right]_k = \left[f_t + f_x f\right]_k \qquad \text{[Chain rule]}$$

Equation (4.39) yields:

$$x_{k+1} = x_k + h f(t_k, x_k) + \frac{h^2}{2!} f_t(t_k, x_k) + \frac{h^2}{2!} f_x(t_k, x_k) f(t_k, x_k) + o(h^3) \tag{4.40}$$

Comparing Equations (4.38) and (4.40),

$$a_1 + a_2 = 1 \tag{4.41a}$$

$$a_2 p_1 = \frac{1}{2} \tag{4.41b}$$

$$a_2 q_{11} = \frac{1}{2} \tag{4.41c}$$

Obviously, there are three equations [i.e., Equations (4.41a–c)] available for four unknowns (i.e., a_1, a_2, p_1 and q_{11}), and thus it is *underspecified.* As a result, there are infinitely many solutions and hence, infinitely many second-order RK approaches.

In the following, we will make the system completely specified by specifying one of the four constant parameters (say, a_1) and get different versions of the RK2.

Case 1: Midpoint (second-order Runge–Kutta) method
Choosing $a_1 = 0$, we get:

$$a_2 = 1$$

$$p_1 = q_{11} = \frac{1}{2}$$

Equation (4.32) gives:

$$k_1 = f(t_k, x_k)$$

$$k_2 = f\left(t_k + \frac{h}{2}, x_k + \frac{h}{2}k_1\right)$$

Combining Equations (4.30) and (4.35),

$$x_{k+1} = x_k + h\,(a_1 k_1 + a_2 k_2)$$

Substituting a_1, a_2, k_1 and k_2 yields the standard RK2 formula [i.e., Equation (4.34)] and this we get with arbitrarily choosing one parameter (i.e., $a_1 = 0$).

Remark
It is obvious from Equation (4.40) that this RK2 method is second-order accurate.

Case 2: Heun method

Choosing $a_1 = \frac{1}{2}$, we get:

$$a_2 = \frac{1}{2}$$

$$p_1 = q_{11} = 1$$

Equation (4.32) gives:

$$k_1 = f(t_k, x_k)$$

$$k_2 = f(t_k + h, x_k + hk_1)$$

Combining Equations (4.30) and (4.35),

$$x_{k+1} = x_k + \frac{h}{2}\,(k_1 + k_2)$$

It gives:

$$x_{k+1}= x_k+\frac{h}{2}\left[f(t_k, x_k)+\left\{f\left(t_k+h,\ x_k+h\,f(t_k, x_k)\right)\right\}\right]$$

which is the same as Equation (4.27) that represents the Heun formula.

Remark

Like RK2, Heun method is also second-order accurate.

At this point, we should note that one can extend this theory to *higher-order Runge–Kutta* algorithms (e.g., RK3 and RK4). There will be more number of free parameters, which need to be chosen somewhat arbitrarily (for RK2, only one parameter is selected arbitrarily).

Example 4.8 RK2 for a three-variable system (revisit Example 4.2)

A system of three state variables is modeled as:

$$\frac{dx}{dt}=-x$$
$$\frac{dy}{dt}=-50y \qquad (4.12)$$
$$\frac{dz}{dt}=-2z$$

with an initial condition of $x_0=y_0=z_0=1$ and $h=0.1$. Develop the computational steps involved in the RK2 method.

Solution

In the following, let us develop the computer-assisted RK2 algorithm with showing a sample calculation for the first time step (i.e., t = 0.1).

Computational steps

Step 1: Let us define:

$$x0=x_k$$
$$y0=y_k$$
$$z0=z_k$$

Obviously at $k=0$,

$$x0=x_0=1$$
$$y0=y_0=1$$
$$z0=z_0=1$$

Step 2: Compute the first set of slopes.

$$k_{11}=-x0=-1$$
$$k_{12}=-50y0=-50$$
$$k_{13}=-2z0=-2$$

Step 3: Compute *x*, *y* and *z* at the midpoint.

$$x1 = x_k + \frac{h}{2} \times k_{11} = 1 + 0.05 \times (-1) = 0.95$$

$$y1 = y_k + \frac{h}{2} \times k_{12} = 1 + 0.05 \times (-50) = -1.5$$

$$z1 = z_k + \frac{h}{2} \times k_{13} = 1 + 0.05 \times (-2) = 0.9$$

Step 4: Compute the second set of slopes.

$$k_{21} = -x1 = -0.95$$

$$k_{22} = -50\,y1 = 75$$

$$k_{23} = -2z1 = -1.8$$

Step 5: Update *x*, *y* and *z*.

$$x_{k+1} = x_k + k_{21} h = 1 - 0.95 \times 0.1 = 0.905$$

$$y_{k+1} = y_k + k_{22} h = 1 + 75 \times 0.1 = 8.5$$

$$z_{k+1} = z_k + k_{23} h = 1 - 1.8 \times 0.1 = 0.82$$

Step 6: Check the convergence.
Although it is optional, one can do the convergence check in terms of absolute error for all three variables as:

$$|x_{k+1} - x_k| \le \text{tol}$$

$$|y_{k+1} - y_k| \le \text{tol}$$

$$|z_{k+1} - z_k| \le \text{tol}$$

If these conditions are not satisfied together, go to Step 1 and continue the computation for next k = 1; else stop here.

A sample calculation is shown above to find x_1, y_1 and z_1 based on their known initial values (i.e., $x_0 = y_0 = z_0 = 1$) along with $h = 0.1$. To repeat this calculation, it is wise to develop a computer program and quickly get the results.

Remark

Obviously, the convergence check shown in Step 6 above is done based on those stopping criteria. This is followed only if the system [e.g., Equation set (4.12)] has the *steady state*[11] or

11 Steady state is defined as a state at which the dependent variables do not change with time.
Most of the systems (including unsteady state or transient processes) reach steady state or some kind of equilibrium. There are a few exceptions (e.g., fed-batch bioreactor with continuous feeding (Jana, 2018) and continuous growth of population in India!). For such type of unsteady state cases, one may not go for convergence check with the stopping criterion and rather, continue the computations for the specified *operational time*.

similar condition (e.g., chemical equilibrium, see Case Study 4.4). This issue will further be elaborated later with more examples.

4.8.3 Fourth-Order Runge–Kutta (RK4)

The concept involved in the fourth-order Runge–Kutta (RK4) method for estimating x_{k+1} is outlined below.

- At first, the first slope k_1 [i.e., $f(t_k, x_k)$] is determined and then it is used to generate a first estimate for x at the midpoint [i.e., $x_k + (h/2)k_1$].
- The first estimate of x is employed to determine the next slope k_2 at the midpoint.
- Then, a corrected midpoint slope k_3 is estimated by using k_2.
- The fourth slope k_4 is subsequently estimated by using k_3 at the end of integration interval h.
- Finally, a weighted average of these four slopes is used to calculate the final estimate of x that is x_{k+1}.

With this, the complete RK4 algorithm is constructed for *single variable* system with the following equations:

$$k_1 = f(t_k, x_k) \tag{4.32a}$$

$$k_2 = f\left(t_k + \frac{h}{2}, x_k + \frac{h}{2}k_1\right) \tag{4.33}$$

$$k_3 = f\left(t_k + \frac{h}{2}, x_k + \frac{h}{2}k_2\right) \tag{4.42}$$

$$k_4 = f(t_k + h, x_k + h\, k_3) \tag{4.43}$$

$$x_{k+1} = x_k + \frac{h}{6}[k_1 + 2k_2 + 2k_3 + k_4] + o(h^5) \tag{4.44}$$

This RK4 method is graphically illustrated in Figure 4.7.

For a multi-variable system, Equation (4.44) has the following form:

$$X_{k+1} = X_k + \frac{h}{6}[K_1 + 2K_2 + 2K_3 + K_4]$$

where,

$$X_k = \begin{bmatrix} x_{1,k} \\ x_{2,k} \\ \cdot \\ \cdot \\ x_{n,k} \end{bmatrix}$$

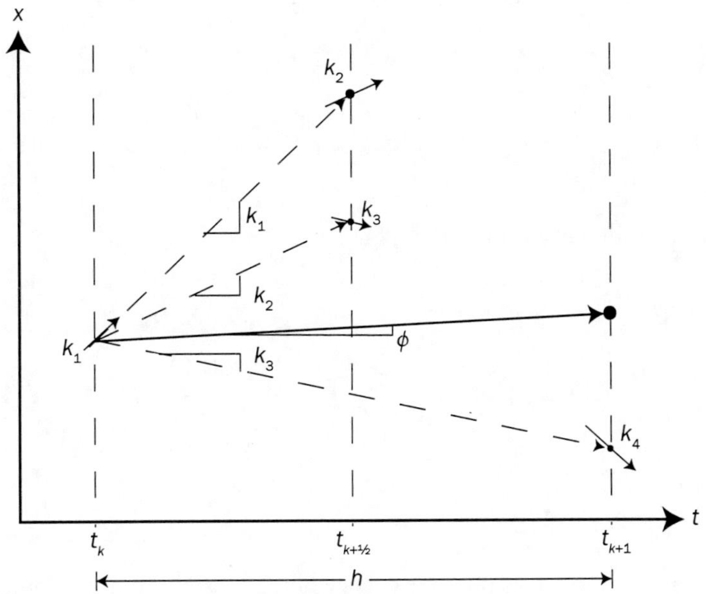

FIGURE **4.7** Illustrating RK4 method

One can easily find K_1, K_2, K_3 and K_4 based on the forms of Equations (4.32a), (4.33), (4.42) and (4.43), respectively.

Remarks

1. It is obvious that both the complexity and computational load increase as the order of the Runge–Kutta method increases.
2. RK4 requires four function evaluations compared to only one for the Euler per ODE at every time step. In this respect, the Euler scheme is about four times faster than the RK4.
3. Fourth-order RK secures better accuracy (fourth-order accurate) than the second-order RK (second-order accurate) and explicit Euler (first-order accurate).
4. It is quite easy to extend the Runge–Kutta methods to multi-variable systems as shown above for RK4.

Numerical Stability of RK4 Method: Liquid tank system (Example 4.3)

For the liquid tank system,

$$k_1 = -\frac{1}{\tau_p} x_k$$

$$k_2 = -\frac{1}{\tau_p}\left(x_k + \frac{\Delta t}{2} k_1\right) = -\frac{1}{\tau_p}\left(x_k - \frac{\Delta t}{2}\frac{1}{\tau_p} x_k\right) = -\frac{1}{\tau_p}\left(1 - \frac{\gamma}{2}\right) x_k \qquad \text{where } \gamma = \frac{\Delta t}{\tau_p}$$

$$k_3=-\frac{1}{\tau_p}\left(x_k+\frac{\Delta t}{2}k_2\right)=-\frac{1}{\tau_p}\left[x_k-\frac{\Delta t}{2}\frac{1}{\tau_p}x_k\left(1-\frac{\gamma}{2}\right)\right]=-\frac{1}{\tau_p}\left(1-\frac{\gamma}{2}+\frac{\gamma^2}{4}\right)x_k$$

$$k_4=-\frac{1}{\tau_p}(x_k+\Delta t\ k_3)=-\frac{1}{\tau_p}\left[x_k-\frac{\Delta t}{\tau_p}x_k\left(1-\frac{\gamma}{2}+\frac{\gamma^2}{4}\right)\right]=-\frac{1}{\tau_p}\left(1-\gamma+\frac{\gamma^2}{2}-\frac{\gamma^3}{4}\right)x_k$$

Substituting in the following Equation (4.44),

$$x_{k+1}=x_k+\frac{\Delta t}{6}[k_1+2k_2+2k_3+k_4]$$

$$=x_k-\frac{\Delta t\ x_k}{6\ \tau_p}\left[1+2\left(1-\frac{\gamma}{2}\right)+2\left(1-\frac{\gamma}{2}+\frac{\gamma^2}{4}\right)+\left(1-\gamma+\frac{\gamma^2}{2}-\frac{\gamma^3}{4}\right)\right]$$

$$=\left[1-\frac{\gamma}{6}\left(6-3\gamma+\gamma^2-\frac{\gamma^3}{4}\right)\right]x_k$$

Since the function in brackets is positive everywhere, the stability requirement is:

$$1-\frac{\gamma}{6}\left(6-3\gamma+\gamma^2-\frac{\gamma^3}{4}\right)\le 1$$

Now, using an iterative convergence method (e.g., Newton–Raphson), one can find the root of the following equation:

$$6-3\gamma+\gamma^2-\frac{\gamma^3}{4}=0$$

that has a single root (physically meaningful)[12] as $\gamma = 2.785$.

The stability condition thus becomes,

$$\gamma \le 2.785$$

which yields,

$$\Delta t \le 2.785\tau_p$$

Remark

The RK4 method allows a wider range of step size (i.e., $\Delta t \le 2.785\ \tau_P$) compared to the other explicit methods, namely the explicit Euler ($\Delta t \le 2\tau_p$) and Heun methods ($\Delta t \le 2\tau_p$).

Example 4.9 RK4 method

A liquid tank example (revisited)

Recall the liquid tank example [see Equation (4.20a)] with the following details:

System: $\frac{dx}{dt}=-\frac{1}{\tau_p}x$

where, $x = H$, $\tau_p = 4$ and initial guess $x_0 = 5$.

12 Other two conjugate imaginary roots are: $0.6075 \pm i\ 2.87203$.

For this, we would now like to test the stability of the RK4 method with reference to the derived stability limit, that is $\Delta t \le 2.785\,\tau_p$.

Case 1: $\Delta t = 5$ [that means, $\Delta t < 2.785\,\tau_p$]

Let us calculate x_1 from Equation (4.44). For this, $k = 0$ and accordingly,

$$k_1 = -\frac{1}{\tau_p} x_0 = -\frac{1}{4} \times 5 = -1.25$$

$$k_2 = -\frac{1}{\tau_p}\left(x_0 + \frac{\Delta t}{2} k_1\right) = -\frac{1}{4}\left(5 + \frac{5}{2} \times (-1.25)\right) = -0.46875$$

$$k_3 = -\frac{1}{\tau_p}\left(x_0 + \frac{\Delta t}{2} k_2\right) = -\frac{1}{4}\left(5 + \frac{5}{2} \times (-0.46875)\right) = -0.95703125$$

$$k_4 = -\frac{1}{\tau_p}(x_0 + \Delta t\, k_3) = -\frac{1}{4}(5 + 5 \times (-0.95703125)) = -0.05371094$$

Substituting in Equation (4.44),

$$x_1 = x_0 + \frac{\Delta t}{6}\left[k_1 + 2k_2 + 2k_3 + k_4\right] = 5 + \frac{5}{6}\left[-1.25 - 2(0.46875 + 0.95703125) - 0.05371094\right]$$
$$= 1.53727$$

By this way, one can further compute $x_2 = 0.47264$, $x_3 = 0.14531$ and so on, and they are documented in Table 4.8.

Table 4.8 Produced x with $\Delta t = 5$

k	x_k	x_{k+1}	Absolute error
0	5	1.53727	3.46273
1	1.53727	0.47264	1.06463
2	0.47264	0.14531	0.32733
3	0.14531	0.04467	0.10064
4	0.04467	0.01373	0.03094
5	0.01373	0.00422	0.00951
6	0.00422	0.00129	0.00293
7	0.00129	0.00039	0.00090
8	0.00039	0.00012	0.00027
9	0.00012	3.77378E-5	8.22622E-5
10	3.77378E-5	1.16026E-5	2.61352E-5
11	1.16026E-5	3.56729E-6	8.03531E-6
12	3.56729E-6	1.09678E-6	2.47051E-6
13	1.09678E-6	3.37210E-7	7.59570E-7

Obviously, the liquid height, H (= x) is converging toward zero [i.e., toward the actual solution h_s (=2 m)] for the step size adopted within stability limit.

Case 2: $\Delta t = 12$ [that means, $\Delta t > 2.785\ \tau_p$]
With this, we get the following set of results (Table 4.9):

TABLE 4.9 Produced x with $\Delta t = 12$

k	x_k	x_{k+1}	Absolute error
0	5	6.87500	1.87500
1	6.87500	9.45312	2.57812
2	9.45312	12.9980	3.54488
3	12.9980	17.8723	4.87430
4	17.8723	24.5744	6.70210
5	24.5744	33.7898	9.21540
6	33.7898	46.4610	12.6712
7	46.4610	63.8839	17.4229
8	63.8839	87.8403	23.9564
9	87.8403	120.780	32.9397
10	120.780	166.073	45.2930
11	166.073	228.350	62.2770
12	228.350	313.982	85.6320
13	313.982	431.725	117.743
14	431.725	593.622	161.897

These tabulated results clearly show the divergence of the RK4 method when the step size is outside of the stability limit.

4.9 Fourth-Order Runge–Kutta–Gill (RKG4) Method

The RKG4 method consists of the following set of equations (Gill, 1951):

$$k_1 = f(t_k,\ x_k) \tag{4.32a}$$

$$k_2 = f\left(t_k + \frac{h}{2},\ x_k + \frac{h}{2}k_1\right) \tag{4.33}$$

$$k_3 = f\left(t_k + \frac{h}{2},\ x_k + \left(\frac{\sqrt{2}}{2} - \frac{1}{2}\right)hk_1 + \left(1 - \frac{\sqrt{2}}{2}\right)h\ k_2\right) \tag{4.45}$$

$$k_4 = f\left(t_k + h,\ x_k - \frac{\sqrt{2}}{2}hk_2 + \left(1 + \frac{\sqrt{2}}{2}\right)h\ k_3\right) \tag{4.46}$$

$$x_{k+1} = x_k + \frac{h}{6}\left[k_1 + \left(2 - \sqrt{2}\right)k_2 + \left(2 + \sqrt{2}\right)k_3 + k_4\right] + o\left(h^5\right) \tag{4.47}$$

Remarks

1. As compared to the other fourth-order algorithms, this Runge–Kutta–Gill method minimizes the round-off errors, not local truncation errors (Lapidus and Seinfeld, 1971).
2. Like RK4, this RKG4 approach is fourth-order accurate and easily extendable to multi-variable systems. Moreover, both these methods involve four function evaluations.

Example 4.10 Runge–Kutta–Gill method

Consider the following system:

$$\frac{dx}{dt} = a\,x^2$$

where, $a = 5 \times 10^{-12}$ and $x(t=0) = 10^8$. If $t = 0$ corresponds to the year 1860, find $x(t = 2000)$ when $h = 20$ yr.

Solution

Using the RKG4 method outlined above, we get:
From Equation (4.32a)

$$k_1 = f(t_k, x_k) = 5 \times 10^{-12}\left(10^8\right)^2 = 50000$$

From Equation (4.33)

$$k_2 = f\left(t_k + \frac{h}{2},\ x_k + \frac{h}{2}k_1\right) = 5 \times 10^{-12}\left(10^8 + \frac{20}{2} \times 50000\right)^2 = 50501.25$$

From Equation (4.45)

$$k_3 = f\left(t_k + \frac{h}{2},\ x_k + \left(\frac{\sqrt{2}}{2} - \frac{1}{2}\right)hk_1 + \left(1 - \frac{\sqrt{2}}{2}\right)h\,k_2\right)$$

$$= 5 \times 10^{-12}\left(10^8 + 0.2071 \times 20 \times 50000 + 0.29289 \times 20 \times 50501.25\right)^2 = 50504.2$$

From Equation (4.46)

$$k_4 = f\left(t_k + h,\ x_k - \frac{\sqrt{2}}{2}hk_2 + \left(1 + \frac{\sqrt{2}}{2}\right)h\,k_3\right)$$

$$= 5 \times 10^{-12}\left(10^8 - 0.7071 \times 20 \times 50501.25 + 1.7071 \times 20 \times 50504.2\right)^2 = 51015.25$$

Finally, from Equation (4.47), we get:

$$x_{k+1} = x_k + \frac{h}{6}\left[k_1 + (2-\sqrt{2})k_2 + (2+\sqrt{2})k_3 + k_4\right]$$

$$= 10^8 + 20\begin{pmatrix} 0.16667 \times 50000 + 0.097631 \times 50501.25 + 0.569 \times 50504.2 + \\ 0.16667 \times 51015.25 \end{pmatrix}$$

$$= 101010072$$

With this methodology, the computer program produces the results as documented in Table 4.10. As shown, we get $x(t = 2000) = 107526648.41$.

Table 4.10 Numerical solutions using the RK Gill method

k	t_k	Numerical solution
0	1860	100000000.00
1	1880	101010071.59
2	1900	102040756.28
3	1920	103092691.58
4	1940	104166541.54
5	1960	105262998.18
6	1980	106382782.97
7	2000	107526648.41

4.10 Runge–Kutta–Fehlberg (RKF45) Method

For solving the ODE-IVP systems, there is another elegant method, namely *Runge–Kutta–Fehlberg 4th–5th order* (RKF45) method (Fehlberg, 1964; 1969). This advanced ODE integrator imposes some control over its own performance by adapting its step size (h) as required. Unlike other approaches studied so far under Runge–Kutta family, the RKF45 method updates the h value at every time step, thereby reducing the number of calculations and thus errors.

At each step, this RKF45 approach calculates two estimates, namely x_{k+1} and z_{k+1}. These two estimates are typically used in computing the proper step size. At this point, it becomes clear that why this RKF45 scheme is sometimes referred as the *adaptive Runge–Kutta* method (or the *embedded RK* method).

Basically, a numerical estimate of the error, $(z_{k+1} - x_{k+1})/h_k$, is computed at each time step, based on which, the RKF45 method updates its step size with the following logic:

- If the estimated error is lower than the desired tolerance limit ('tol'), the new step size ($= h_{k,\,new}$) to be used in the next step for generating x_{k+2} is increased to speed up the computations and reduce the computational effort.
- In contrary, if that error exceeds the tolerance limit, a lower value of h ($= h_{k,\,new}$) needs to be used and the calculation is repeated.

- If the error is quite close to 'tol', the two estimates are almost equal and the value of step size is kept unchanged.

To solve a system of Equation (4.29), let us note the Runge–Kutta–Fehlberg method with the following equations to calculate the six k-values:

$$k_1 = h\,f(t_k,\ x_k) \tag{4.48}$$

$$k_2 = h\,f\left(t_k + \frac{h}{4},\ x_k + \frac{k_1}{4}\right) \tag{4.49}$$

$$k_3 = h\,f\left(t_k + \frac{3}{8}h,\ x_k + \frac{3}{32}k_1 + \frac{9}{32}k_2\right) \tag{4.50}$$

$$k_4 = h\,f\left(t_k + \frac{12}{13}h,\ x_k + \frac{1932}{2197}k_1 - \frac{7200}{2197}k_2 + \frac{7296}{2197}k_3\right) \tag{4.51}$$

$$k_5 = h\,f\left(t_k + h,\ x_k + \frac{439}{216}k_1 - 8k_2 + \frac{3680}{513}k_3 - \frac{845}{4104}k_4\right) \tag{4.52}$$

$$k_6 = h\,f\left(t_k + \frac{h}{2},\ x_k - \frac{8}{27}k_1 + 2k_2 - \frac{3544}{2565}k_3 + \frac{1859}{4104}k_4 - \frac{11}{40}k_5\right) \tag{4.53}$$

With this, the two estimates one can obtain using the following two equations:

$$x_{k+1} = x_k + \left[\frac{25}{216}k_1 + \frac{1408}{2565}k_3 + \frac{2197}{4104}k_4 - \frac{1}{5}k_5\right] + o(h^5) \tag{4.54a}$$

$$z_{k+1} = x_k + \left[\frac{16}{135}k_1 + \frac{6656}{12825}k_3 + \frac{28561}{56430}k_4 - \frac{9}{50}k_5 + \frac{2}{55}k_6\right] + o(h^6) \tag{4.54b}$$

It should be noted that x_{k+1} [Equation (4.54a)] and z_{k+1} [Equation (4.54b)] are obtained using the Runge–Kutta method of order four (i.e., fourth order accurate) and order five (i.e., fifth order accurate), respectively.

The optimal step size, $h_{k,\,new}$ can be obtained from:

$$h_{k,\,new} = h_k\left(\frac{\text{tol} \times h_k}{2\,|z_{k+1} - x_{k+1}|}\right)^{0.25} \tag{4.55}$$

Rewriting,

$$h_{k,\,new} \approx 0.84 h_k\left(\frac{\text{tol} \times h_k}{|z_{k+1} - x_{k+1}|}\right)^{0.25} \tag{4.56}$$

Derivation of this Equation (4.56) is not shown here and it is beyond the scope of this book.

Remarks

1. The RKF45 method involves calculations that are tedious and time consuming. However, this approach provides faster response under the desired level of accuracy (≡ tol) because of its adaptive (step size) nature over the classical Runge–Kutta methods.
2. One can do the same job of adaptive step-size control by using two predictions of fourth-order and fifth-order Runge–Kutta method that typically involve total 10 function evaluations per step. To improve this situation, Fehlberg had derived fourth-order and fifth-order (RK) formulas with some common functions in them. By this way, the number of function evaluations gets reduced from 10 to 6, thereby reduces the computational effort and complexity.
3. Extension of this method to multi-variable systems is fairly obvious.

Example 4.11 RKF45 method

Consider the following problem:

$$\frac{dx}{dt} = 1 + a\,x^2$$

with $a = 1$, $x(0) = 0$, $\text{tol} = 2 \times 10^{-5}$ and $h = 0.2$.

(a) Solve the problem by using the RKF45 method and tabulate the solutions for the first 5 steps (i.e., up to $k = 5$), updating h in all steps with the use of Equation (4.56).

(b) Solve the above problem with updating h by the use of Equation (4.56) only if:

$$\left| z_{k+1} - x_{k+1} \right| > \text{tol}$$

else, $h_{k,\,new} = h_k$. Tabulate the results up to $t_k = 1$ when $a = 2$.

Solution

(a) Using the RKF45 method that is presented before, we get the numerical solutions as reported in Table 4.11. Further, these results are compared there with the analytical solution,

$$x_{kA} = \tan(t_k)$$

The results are shown up to $k = 5$. Note that the detailed step-wise calculation for Runge–Kutta–Gill approach is shown in Example 4.10. In the similar fashion, one can easily develop the computer-assisted methodology for the RKF45 method.

Table 4.11 Comparing numerical (RKF45) and analytical solutions

k	t_k	Numerical solution (RKF45), x_{kN}	Analytical solution, $x_{kA} = \tan(t_k)$	Error $\lvert x_{kN} - x_{kA} \rvert$
0	0.00	0.00	0.00	0.00
1	0.20	0.2027100	0.2027100	0.00
2	0.6442113	0.7511711	0.7511103	6.08e-5

Contd

Contd

k	t_k	Numerical solution (RKF45), x_{kN}	Analytical solution, $x_{kA} = \tan(t_k)$	Error $\lvert x_{kN}-x_{kA} \rvert$
3	1.0788670	1.8674296	1.8661280	0.0013016
4	1.2436389	2.9497895	2.9467938	0.0029957
5	1.3017717	3.6314093	3.6270213	0.0043880

(b) In this part, we consider $a = 2$ as given. Here, the step size is only updated if:

$$\left| z_{k+1} - x_{k+1} \right| > \text{tol}$$

else, $h_{k,\,new} = h_k$ is adopted. With this, we produce the numerical solutions up to about $t_k = 1.0$ and they are documented in Table 4.12.

Table 4.12 Numerical solutions of RKF45 method

k	t_k	Numerical solution
0	0.00	0.00
1	0.20	0.2055099
2	0.40	0.4489437
3	0.60	0.8025641
4	0.80	1.5045682
5	0.8769988	2.0614180
6	0.9539977	3.1394089
7	0.9759664	3.6674243
8	0.9979350	4.3983738
9	1.0199037	5.4796668

4.11 Adams–Bashforth–Moulton (Predictor–Corrector) Method

For solving the ODE-IVP systems, so far we covered the single-step numerical integration approaches (e.g., Euler, Taylor and Runge–Kutta methods). They are called so since those methods require the information from only one previous point for computing the successive points. For instance, to compute (t_{k+1}, x_{k+1}), the initial point (t_0, x_0) must be known.

Here, we would like to learn an advanced *multi-step predictor–corrector* method of Adams–Bashforth–Moulton.[13] The idea is to *predict* x_{k+1} employing the Adams–Bashforth approach and then to *correct* its value with using the Adams–Moulton approach. This predictor–corrector method is basically a four-step approach and at the starting, it requires total four initial points, namely (t_0, x_0), (t_1, x_1), (t_2, x_2) and (t_3, x_3), in advance in order to generate the next points [i.e., (t_{k+1}, x_{k+1}) for $k \geq 3$].

13 Named after two British mathematicians John Couch Adams, FRS (1819–1892) and Francis Bashforth (1819–1912), and an American astronomer Forest Ray Moulton (1872–1952).

Based on the fundamental theorem of calculus, this multi-step scheme is formulated with:

$$x_{k+1} = x_k + \int_{t_k}^{t_{k+1}} f[t,\ x(t)]\ dt \tag{4.57}$$

We need to integrate the function, $f(t, x)$ over a finite step using an interpolating polynomial approximation.[14]

Adams–Bashforth predictor

In this step, one needs to fit a cubic polynomial to the function $f(t, x)$ through the points (t_{k-3}, f_{k-3}), (t_{k-2}, f_{k-2}), (t_{k-1}, f_{k-1}) and (t_k, f_k). The function is then integrated over the interval t_k to t_{k+1}, which yields the Adams–Bashforth predictor as:

$$\textit{Predictor: } P_{k+1} = x_k + \frac{h}{24}(55f_k - 59f_{k-1} + 37f_{k-2} - 9f_{k-3}) + o(h^5) \tag{4.58}$$

Adams–Moulton corrector

In the similar fashion, the corrector part is formulated. A second cubic polynomial for $f(t, x)$ is devised based on the points (t_{k-2}, f_{k-2}), (t_{k-1}, f_{k-1}), (t_k, f_k) and the new point (t_{k+1}, f_{k+1}), in which, $f_{k+1} = f(t_{k+1}, P_{k+1})$. Integrating over t_k to t_{k+1} yields the Adams–Moulton corrector as:

$$\textit{Corrector: } x_{k+1} = x_k + \frac{h}{24}(9f_{k+1} + 19f_k - 5f_{k-1} + f_{k-2}) + o(h^5) \tag{4.59}$$

The main problem associated in the Adams–Bashforth–Moulton scheme is that it requires four initial points in advance. For an initial value problem, x_0 is typically specified and t_0 is obviously zero. In order to use this multi-step approach, thus three additional starting values, x_1, x_2 and x_3, need to be computed using another numerical integration approach (e.g., Runge–Kutta approach). With this, the computational steps involved in this four-step predictor–corrector method are outlined below:

Computational steps

Step 1: Employ the RK method to find x_1, x_2 and x_3 based on known x_0.

Step 2: Calculate P_{k+1} from the predictor Equation (4.58).

Step 3: Find $f_{k+1} = f(t_{k+1}, P_{k+1})$ and then compute x_{k+1} from the corrector Equation (4.59).

Step 4: Increment the t interval and go to Step 2 to repeat the computation until the convergence condition is satisfied.

Remarks

1. The multi-step Adams–Bashforth–Moulton method is fourth-order accurate.
2. This method is not self-starting. It basically requires another method (e.g., Runge–Kutta) at the beginning, which leads to make it computationally expensive.

14 Detailed in Chapter 8.

3. This predictor–corrector method requires only two new function evaluations in each step, whereas the RK4 method requires four function evaluations. In this respect, the predictor–corrector method is computationally less expensive.

Example 4.12 Adams–Bashforth–Moulton method

Consider the following system equation:

$$\frac{dx}{dt} = \sin x$$

Integrate it with the use of the predictor–corrector method with $x(0) = x_0 = 2$ and $h = 0.05$. Produce the results for a single time step (i.e., by finding x_4) with the employment of the RK4 approach to estimate x_1, x_2 and x_3 at the beginning.

Solution

Step 1: Finding x_1, x_2 and x_3 using the RK4 method.

For x_1:

$$k_{11} = \sin(2) = 0.909297$$

$$k_{12} = \sin\left(2+\frac{0.05}{2}k_{11}\right) = \sin\left(2+\frac{0.05}{2}\times 0.909297\right) = 0.8996$$

$$k_{13} = \sin\left(2+\frac{0.05}{2}k_{12}\right) = \sin\left(2+\frac{0.05}{2}\times 0.8996\right) = 0.89971$$

$$k_{14} = \sin(2+0.05k_{13}) = \sin(2+0.05\times 0.89971) = 0.889663$$

$$x_1 = x_0 + \frac{h}{6}[k_{11} + 2k_{12} + 2k_{13} + k_{14}]$$

$$= 2 + \frac{0.05}{6}[0.909297 + 2(0.8996 + 0.89971) + 0.889663]$$

$$= 2.045$$

Similarly, for x_2:

$$k_{21} = 0.88966$$

$$k_{22} = 0.8793$$

$$k_{23} = 0.8794$$

$$k_{24} = 0.8687$$

$$x_2 = x_1 + \frac{h}{6}[k_{21} + 2k_{22} + 2k_{23} + k_{24}]$$

$$= 2.089$$

Further, for x_3:

$k_{31} = 0.8687$

$k_{32} = 0.85776$

$k_{33} = 0.85789$

$k_{34} = 0.84667$

$$x_3 = x_2 + \frac{h}{6}[k_{31} + 2k_{32} + 2k_{33} + k_{34}]$$
$$= 2.132$$

Step 2: Calculate P_{k+1}.

Let us first determine:

$f_{k-3} = f_0 = \sin x_0 = \sin(2) = 0.9093$
$f_{k-2} = f_1 = \sin x_1 = 0.8896$
$f_{k-1} = f_2 = \sin x_2 = 0.8687$
$f_k = f_3 = \sin x_3 = 0.8466$

Then from Equation (4.58),
$P_{k+1} = P_4 = 2.174$

Step 3: Compute x_{k+1}.

We evaluate,

$$f_{k+1} = f(t_{k+1}, P_{k+1}) = f_4 = \sin(2.174) = 0.8235$$

Then from Equation (4.59), finally we get:

$$x_{k+1} = x_4 = 2.174.$$

Likewise, we can continue our calculation for x_5, x_6 and so on, and it is left as your homework.

4.12 Numerical Stability of Non-linear IVP System

Let us consider a non-linear system,

$$\frac{dx}{dt} = f(t, x) \tag{4.29}$$

Using the Taylor series for $f(t, x)$ around a solution point (t_s, x_s),

$$f(t, x) = f(t_s, x_s) + f_t(t_s, x_s)(t - t_s) + f_x(t_s, x_s)(x - x_s) + \ldots \tag{4.60}$$

Note that when (t, x) is reasonably close to (t_s, x_s), one can disregard the second- and higher-order terms. With this, Equation (4.60) gives:

$$\frac{dx}{dt} = [f(t_s, x_s) - f_t(t_s, x_s)t_s - f_x(t_s, x_s)x_s] + f_t(t_s, x_s)\, t + f_x(t_s, x_s)\, x \tag{4.61}$$

Rearranging,

$$\frac{dx}{dt} = \left.\frac{\partial f}{\partial x}\right|_{(t_s,\, x_s)} x + G(t) \tag{4.62}$$

in which, the term $G(t)$ has no influence on stability.

Now if,

$$\left.\frac{\partial f}{\partial x}\right|_{x=x_s} = -\frac{1}{\tau_p} \tag{4.63}$$

Equation (4.62) reduces to Equation (4.20a), which is a linear form of equation, for which the numerical stability is studied before. Consequently, at the local point (t_s, x_s), the same stability limits will hold. For example, the choice of h for stability is made for the explicit Euler as:

$$h \leq \frac{2}{\left(-\left.\frac{\partial f}{\partial x}\right|_{x=x_s}\right)}$$

As the solution of a non-linear system marches along t, $\frac{\partial f}{\partial x}$ will change, which leads to change the stability limits of an explicit method. This, in turn, makes a complex stability problem for non-linear IVP systems, and thus, it is recommended to judiciously use reasonably small h for the explicit methods or use the implicit techniques.

4.13 Numerical Stability of Multi-variable System

Stability Analysis: Multi-variable system

Let us consider a simple multi-variable linear system in the following form:

$$\frac{dX}{dt} = A\,X \qquad X(0) = X_0 \tag{4.64}$$

where, A is an $M \times M$ matrix (square matrix) and

$$X = \begin{bmatrix} x_1 \\ x_2 \\ \cdot \\ \cdot \\ x_M \end{bmatrix}$$

It is now straightforward to find the analytical solution as:

$$X(t) = e^{A\,t} X(0) \tag{4.65}$$

It is shown below that this solution is stable if all of the eigenvalues of A are less than zero.[15]

To analyze this problem, one can use the Routh stability criterion with a simple 2×2 square matrix A as:

$$A = \begin{bmatrix} a_{11} & a_{12} \\ a_{21} & a_{22} \end{bmatrix}$$

The characteristic polynomial for matrix A is:

$$\det(\lambda I - A) = 0 \tag{4.66}$$

where, λ denotes the eigenvalues of A.

From Equation (4.66),

$$\begin{vmatrix} \lambda - a_{11} & -a_{12} \\ -a_{21} & \lambda - a_{22} \end{vmatrix} = 0$$

That is,

$$(\lambda - a_{11})(\lambda - a_{22}) - a_{12}a_{21} = 0$$

$$\lambda^2 - (a_{11} + a_{22})\lambda + (a_{11}a_{22} - a_{12}a_{21}) = 0$$

Rewriting,

$$\lambda^2 - \mathrm{tr}(A)\lambda + \det(A) = 0 \tag{4.67}$$

where, tr (A) refers to the *trace* of matrix A. Obviously, the eigenvalues (λ) are the roots of the characteristic polynomial [Equation (4.67)].

According to the Routh stability criterion, all the coefficients associated in Equation (4.67) must be positive. Accordingly, the necessary condition includes:

tr $(A) < 0$

and

det $(A) > 0$

With this [i.e., $\mathrm{tr}(A) < 0$ and $\det(A) > 0$], it can easily be shown that the roots of Equation (4.67) will be negative.

In the next phase of Routh test, one needs to construct the *Routh array* for finding the sufficient condition of stability. Doing this, it can be shown that the analytical solution of

---

15 A single variable equation system:

$$\frac{dx}{dt} = ax$$

has the analytical solution:

$$x(t) = e^{at}x(0)$$

that is stable if $a < 0$.

Equation (4.64) is stable if all of the eigenvalues (λ) of matrix A are less than zero, or better to say, the eigenvalues have a negative real part (i.e., $\text{Re}(\lambda_i)<0$).[16] Note that this is a pre-requisite with which we are in a position to be talked about numerical stability.

Numerical stability

As indicated, we are still in the process of analyzing the numerical stability of a multi-variable system represented by Equation (4.64). For this system, we can have:

$$AV = V\Lambda \tag{4.68}$$

where, V is the matrix of eigenvectors and Λ the matrix of eigenvalues.

For example, for a 2 × 2 matrix A,

$$A\varsigma_1 = \lambda_1\varsigma_1$$

$$A\varsigma_2 = \lambda_2\varsigma_2$$

$$\text{where, } \varsigma_1 = \begin{bmatrix} v_{11} \\ v_{21} \end{bmatrix},\ \varsigma_2 = \begin{bmatrix} v_{12} \\ v_{22} \end{bmatrix},$$

$$V = [\varsigma_1 \quad \varsigma_2] = \begin{bmatrix} v_{11} & v_{12} \\ v_{21} & v_{22} \end{bmatrix}$$

$$\text{and } \Lambda = \begin{bmatrix} \lambda_1 & 0 \\ 0 & \lambda_2 \end{bmatrix}$$

Obviously, the eigenvector ς_i is associated with the eigenvalue λ_i.

From Equation (4.68),

$$AVV^{-1} = V\Lambda V^{-1}$$

or,

$$A = V\Lambda V^{-1}$$

Substituting in Equation (4.64),

$$\frac{dX}{dt} = V\Lambda V^{-1}X$$

Left multiplying both sides by V^{-1},

$$V^{-1}\frac{dX}{dt} = V^{-1}V\Lambda V^{-1}X$$

Since V is independent of time t,

$$\frac{d}{dt}(V^{-1}X) = \Lambda V^{-1}X \tag{4.69}$$

16 For the solution to be bounded (including under marginally stable condition), it is basically $\text{Re}(\lambda_i) \le 0$.

Now, let us define,

$$\widehat{X}=\text{V}^{-1}X \tag{4.70}$$

Accordingly, Equation (4.69) yields,

$$\frac{d\widehat{X}}{dt}=\Lambda\widehat{X} \tag{4.71}$$

This gives the following set of fully decoupled ODEs for an $M \times M$ system:

$$\frac{d\widehat{x}_1}{dt}=\lambda_1\widehat{x}_1$$

$$\frac{d\widehat{x}_2}{dt}=\lambda_2\widehat{x}_2$$

$$\vdots$$

$$\frac{d\widehat{x}_M}{dt}=\lambda_M\widehat{x}_M$$

With a suitably modified set of initial conditions (i.e., $\widehat{X}(0)$), one can easily find the solutions for $\widehat{X}$ from the above set of ODEs. Further using Equation (4.70), the solution for X can finally be obtained.

Recall that for numerical stability, the pre-requisite is that the eigenvalues must have non-positive real part, that is, $\text{Re}(\lambda_i)\leq 0$. With this, like a single ODE,[17] we can similarly show the existence of an analogous condition for numerical stability of multi-variable systems.

For explicit Euler,

$$h\leq\frac{2}{|\lambda_{\max}|}$$

where, $\lambda_{\max}$ denotes an eigenvalue having highest magnitude. Note that when the eigenvalue is complex,

$$\lambda=a+ib$$

then we have:

$$|\lambda|=\sqrt{a^2+b^2}$$

Similarly, for RK4,

$$h\leq\frac{2.785}{|\lambda_{\max}|}$$

17 For similarity, adopt the single variable system Equation (4.20a) with $\tau_p=\frac{1}{\mu}$ (where, the constant $\mu>0$). For this system, the stability requirement for the explicit Euler is shown before as: $h\leq 2\tau_p\left(\text{i.e., } h\leq\frac{2}{\mu}\right)$.

Remark

For a non-linear multi-variable case, the matrix A should be replaced by the Jacobian matrix J. In that case, the eigenvalues of J will play the role in determining the step size h.

4.14 Stiff Systems

Let us consider the example of a typical chemical reactor. In this, the fast reacting species reach the equilibrium in a very short period of time, while the slowly changing species are more or less fixed, which is *stiff*. This type of problem is quite common whenever fast and slow phenomena co-exist.

Along with chemical reactors, we often encounter the stiffness problem in the study of spring and damping systems, and in the analysis of control systems. Numerical stability of such stiff systems is a critical issue.

For an ODE-IVP system with:

$$\frac{dX}{dt} = A\,X \qquad X(0) = X_0 \tag{4.64}$$

The stiffness is determined by estimating the *stiffness ratio* (SR) defined as:

$$\mathrm{SR} = \frac{\left|\mathrm{Re}\ \lambda_i(A)\right|_{\max}}{\left|\mathrm{Re}\ \lambda_i(A)\right|_{\min}} \tag{4.72}$$

This means,

$$\mathrm{SR} = \frac{\max\left|\mathrm{Re}\ (\lambda_i)\right|}{\min\left|\mathrm{Re}\ (\lambda_i)\right|} \tag{4.73}$$

where, λ is the eigenvalue of matrix A. When this SR is large enough, the system is called the *stiff* system. In other words, an IVP system is said to be *stiff* if the Jacobian (local) contains at least one eigenvalue that has a little contribution over most of the domain of interest.

Stiff System: Some Facts

- A system is stiff if some components of its solution decay much faster than others.
- A stiff IVP system is numerically unstable unless the step size is reasonably small.
- For a stiff system, the step size is dictated by stability requirements, not by accuracy requirements.
- Stiff systems are characterized as those whose exact solution includes a term of the form $\exp(-\alpha t)$, where α is a reasonably large positive constant.
- For a linear system, the stiffness exists if all of its eigenvalues have non-positive real part and the stiffness ratio (SR) is large.
- For non-linear systems, the eigenvalues of the Jacobian matrix differ significantly in magnitude.

Most of these facts we will face in the following two examples.

Example 4.13 A simple stiff system

Consider a linear system having the following model:

$$\frac{dx_1}{dt} = -x_1 \tag{4.74a}$$

$$\frac{dx_2}{dt} = -\alpha\, x_2 \tag{4.74b}$$

where, $x_1(0)=1$, $x_2(0)=1$ and $t \in [0, 1]$. Analyze the stiffness of this system.

Solution

Rearranging Equation (4.74),

$$\begin{bmatrix} \dot{x}_1 \\ \dot{x}_2 \end{bmatrix} = \begin{bmatrix} -1 & 0 \\ 0 & -\alpha \end{bmatrix} \begin{bmatrix} x_1 \\ x_2 \end{bmatrix}$$

This can further be written as:

$$\dot{X} = A\,X \tag{4.75}$$

where,

$$X = \begin{bmatrix} x_1 \\ x_2 \end{bmatrix}$$

$$A = \begin{bmatrix} -1 & 0 \\ 0 & -\alpha \end{bmatrix}$$

From Equation (4.67),

$$\lambda^2 + (1+\alpha)\lambda + \alpha = 0$$

Solving,

$$\lambda_1 = -1$$

$$\lambda_2 = -\alpha$$

When this α is large (for example, 50000), then from Equation (4.73):

$$\text{SR} = \frac{\max|\text{Re}\ (\lambda_i)|}{\min|\text{Re}\ (\lambda_i)|} = \frac{|-50000|}{|-1|} = 50000$$

which is large and thus, the example system is stiff.

It is quite straightforward to find the analytical solution as:

$$x_1(t) = e^{-t} \tag{4.76a}$$

$$x_2(t) = e^{-\alpha t} \tag{4.76b}$$

It clearly indicates that for a large α value, x_2 approaches zero much faster than x_1. Using the explicit Euler method (with $h = 0.0005$), we observe this phenomena in Figures 4.8a–c with the progressive increase of α value.

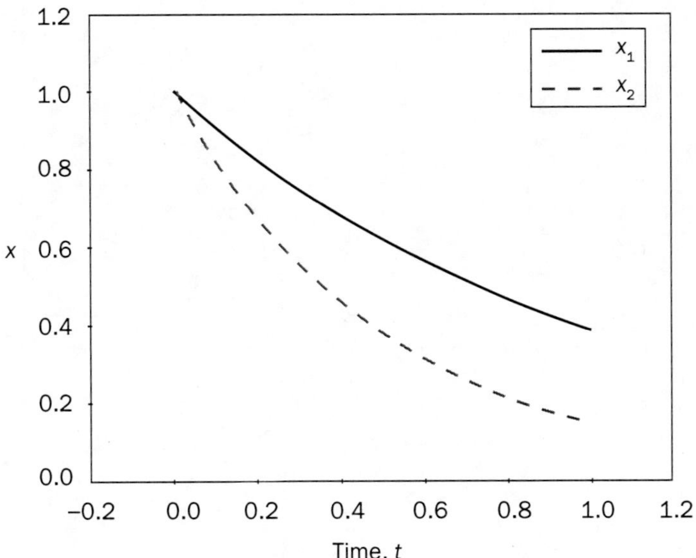

FIGURE 4.8a Numerical solution with $\alpha = 2$

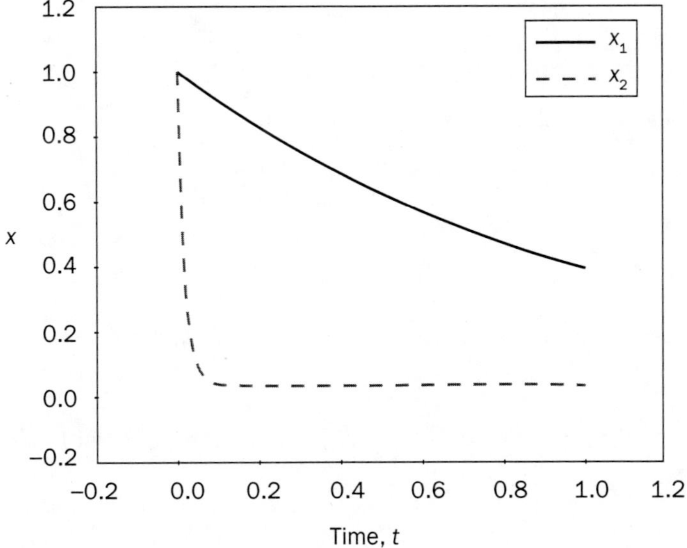

FIGURE 4.8b Numerical solution with $\alpha = 50$

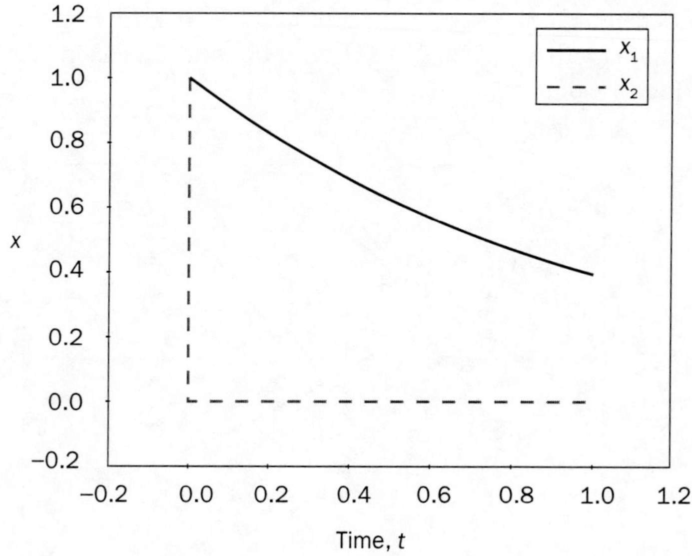

FIGURE 4.8c Numerical solution with $\alpha = 1000$

Remark

This example system is stiff (when α is large) because:

- It contains components that vary with different speeds
- The stiffness ratio is large

Example 4.14 A little complex stiff system

Consider the following stiff differential equation system:

$$\frac{dx_1}{dt} = x_2 \tag{4.77a}$$

$$\frac{dx_2}{dt} = -1000x_1 - 1001x_2 \tag{4.77b}$$

where, $x_1(0) = 1$ and $x_2(0) = 0$. Analyze the stiffness of this system.

Solution

Equation (4.77) can be represented in the form of Equation (4.75) with:

$$X = \begin{bmatrix} x_1 \\ x_2 \end{bmatrix}$$

$$A = \begin{bmatrix} 0 & 1 \\ -1000 & -1001 \end{bmatrix}$$

Using Equation (4.67),

$$\lambda^2-(0-1001)\lambda+(0+1000)=0$$

Solving, we get:

$$\lambda_1=-1$$

$$\lambda_2=-1000$$

Both these eigenvalues have negative real parts and the stiffness ratio is calculated as:

$$\text{SR}=\frac{|-1000|}{|-1|}=1000$$

Obviously, it is large and thus, this IVP system is stiff.

Let us next find the analytical solution of the given system, through which, we can analyze the stiffness better. For $\lambda_1=-1$, the eigenvector is obtained as:

$$\varsigma_1=\begin{bmatrix}1\\-1\end{bmatrix}$$

Similarly for $\lambda_2=-1000$,

$$\varsigma_2=\begin{bmatrix}-0.001\\1\end{bmatrix}$$

The analytical solution gets the following form,

$$\begin{bmatrix}x_1\\x_2\end{bmatrix}=c_1\begin{bmatrix}1\\-1\end{bmatrix}e^{-t}+c_2\begin{bmatrix}-0.001\\1\end{bmatrix}e^{-1000t}$$

Using the initial condition (at $t = 0$): $x_1(0) = 1$ and $x_2(0) = 0$,

$$\begin{bmatrix}1\\0\end{bmatrix}=c_1\begin{bmatrix}1\\-1\end{bmatrix}+c_2\begin{bmatrix}-0.001\\1\end{bmatrix}$$

It gives,

$$c_1=c_2=\frac{1000}{999}$$

The final analytical solution of this problem is:

$$x_1=\frac{1000}{999}e^{-t}-\frac{1}{999}e^{-1000t}$$

$$x_2=-\frac{1000}{999}e^{-t}+\frac{1000}{999}e^{-1000t}$$

Most interesting feature depicted in Figure 4.9 is the co-existence of fast and slow components. The numerical solution by the use of explicit Euler (with $h = 0.0005$) is shown in Figures 4.9a and 4.9b for a shorter period of time, and in Figure 4.9c for a longer time.

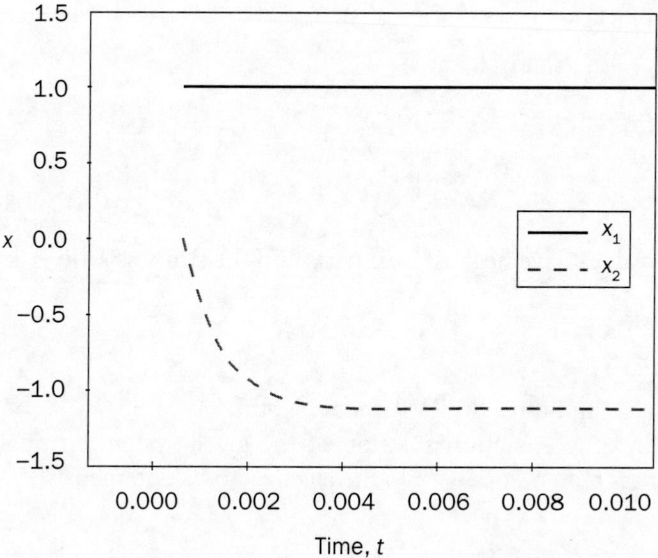

FIGURE 4.9a Numerical solution for a very short time (= 0.01)

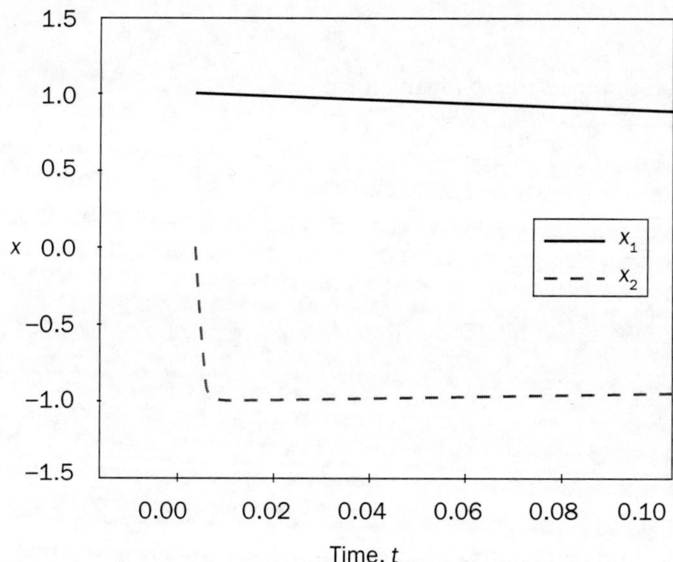

FIGURE 4.9b Numerical solution for a short time (= 0.1)

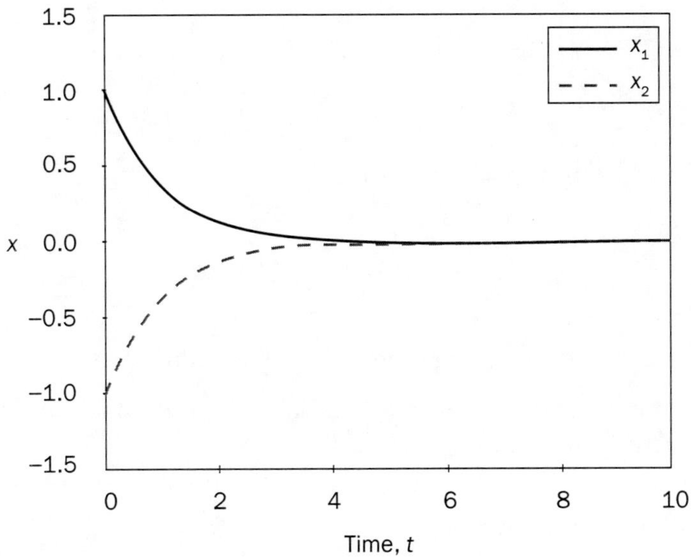

FIGURE 4.9c Numerical solution for a long time (= 10)

In Figure 4.9c (long time solution), it seems that the initial conditions for this problem include:

$$x_1(0) = \frac{1000}{999}$$

$$x_2(0) = -\frac{1000}{999}$$

Due to the large separation between the two eigenvalues ($\lambda_1 = -1$, $\lambda_2 = -1000$), one cannot even visualize the short-time behavior in Figure 4.9c.

It is a fact that the fast changes are typically caused by the eigenvalue with the largest magnitude (i.e., $\lambda_2 = -1000$). Thus, it leads to set the limit for numerical stability at the beginning. Accordingly, one needs to follow the step size limit. As shown before, for explicit Euler (see Subsection 4.13):

$$h \le \frac{2}{|-1000|} = 0.002$$

It should be noted that at longer time t, the largest eigenvalue in magnitude (i.e., 1000) becomes *insignificant* as soon as the fast exponential (i.e., e^{-1000t}) gets decayed. Then the solution is naturally dominated by the smallest eigenvalue with:

$$x_1 = \frac{1000}{999} e^{-t}$$

$$x_2 = -\frac{1000}{999} e^{-t}$$

For this, one can adopt the explicit Euler with,

$$h \le \frac{2}{|-1|} = 2$$

Remark

It becomes obvious that for the entire time period, if one adopts

- $h \le 0.002$, the solution involves significant computational effort
- $h \le 2$ (actually, $0.002 < h \le 2$), the solution is unstable

Handling of stiff differential systems

- Keeping the numerical stability issue in mind, it is always safe to use the implicit numerical technique, which is unconditionally stable.
- Another approach is to efficiently use the explicit method with adapting its step size. Once the fast changes decay, one can use large step size (e.g., initially use $h \le 0.002$ and then $h \le 2$ as shown in the last Example 4.14) and comfortably maintain stability.

4.15 Solved Case Studies

Case study 4.1 Heating tank system

Figure 4.10 demonstrates a heating tank system, in which a stirrer rotates continuously to maintain uniformity in temperature. As shown, a coil is immersed in the tank liquid, through which a heating medium (i.e., steam) is passed. The saturated steam basically releases latent heat and thereafter, leaves the coil as a saturated liquid (condensate).

Here,

F = volumetric liquid flow rate
T = temperature
T_i = inlet liquid temperature
m_{st} = mass flow rate of steam
V = liquid volume in the tank

Deriving model

We assume that there is no phase change involved in the tank liquid. With this, since the inlet and outlet liquid streams have the same flow rate (= F), the volume of liquid content in the tank (= V) remains constant. So the temperature (= T) is our variable of interest for this heating tank system.

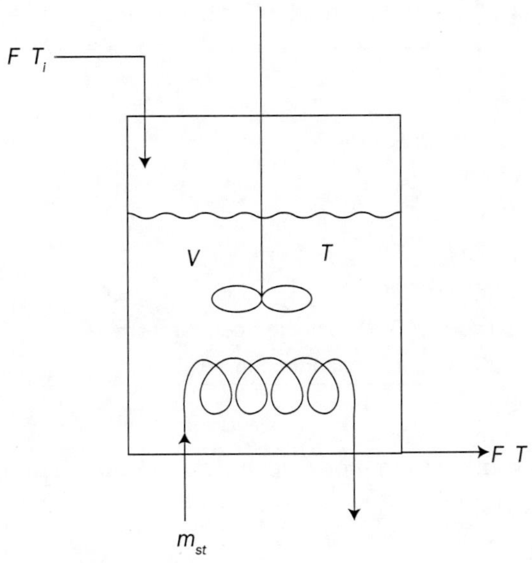

FIGURE 4.10 A heating tank system

Recall the energy balance equation:

Rate of accumulation of energy = Rate of input of energy − Rate of output of energy + Rate of generation of energy (A4.3)

It yields:

$$\frac{d}{dt}(V\rho C_p T) = F\rho C_p T_i - F\rho C_p T + m_{st}\lambda \tag{4.78}$$

where,

ρ = liquid density
λ = latent heat of steam
C_p = specific heat capacity of liquid

Simplifying,

$$\frac{dT}{dt} = \frac{F}{V}(T_i - T) + \frac{m_{st}\lambda}{V\rho C_p} \tag{4.79}$$

This is the final model equation of the heating tank system that we need to solve now as an ODE-IVP problem.

Problem statement

Solve Equation (4.79) using the RK4 method based on the given information in Table 4.13.

Table 4.13 Given information for the heating tank system

$F = 10$ m^3/hr
$V = 10$ m^3
$T_i = 25$ °C = 298 K
$m_{st} = 340.2$ kg/hr
$\lambda = 2250$ kJ/kg
$\rho = 1260$ kg/m^3
$C_p = 2.43$ kJ/kg.K
$T_0 = 30$ °C = 303 K (guess temperature)
$h = 0.005$ hr (integration time interval)

Solution

To simulate the heating tank system model, let us first develop the computational steps with the application of RK4 method.

Step 1: Let us consider,

$$T0 = T_k$$

At $k = 0$, obviously $T0 = T_0 =$ 30 °C = 303 K.

Step 2: Find initial slope k_1.

$$k_1 = \frac{F}{V}(T_i - T0) + \frac{m_{st}\lambda}{V\rho C_p}$$

Step 3: Determine first estimate for T at the midpoint.

$$T1 = T_k + \frac{h}{2}k_1$$

Step 4: Find slope k_2 at the midpoint.

$$k_2 = \frac{F}{V}(T_i - T1) + \frac{m_{st}\lambda}{V\rho C_p}$$

Step 5: Determine next estimate for T.

$$T2 = T_k + \frac{h}{2}k_2$$

Step 6: Find corrected midpoint slope k_3.

$$k_3 = \frac{F}{V}(T_i - T2) + \frac{m_{st}\lambda}{V\rho C_p}$$

Step 7: Determine third estimate for T.

$$T3 = T_k + hk_3$$

Step 8: Find slope k_4 at the end of integration interval.

$$k_4 = \frac{F}{V}(T_i - T3) + \frac{m_{st}\lambda}{V\rho C_p}$$

Step 9: Update T.

$$T_{k+1} = T_k + \frac{h}{6}[k_1 + 2k_2 + 2k_3 + k_4]$$

Repeat these steps with $k = 1, 2, 3, \ldots$ until you reach the steady state of the heating tank system.

Based on these computational steps, we produce the results. For the first 10 time steps, the T values are tabulated below (Table 4.14). Moreover, in Figure 4.11, we have shown the entire temperature profile. The final temperature (at steady state) is obtained as 323 K (= 50 °C).

Table 4.14 Numerical solutions in the first 10 time steps

Step (k)	T_k
0	303.0000
1	303.0998
2	303.1990
3	303.2978
4	303.3960
5	303.4938
6	303.5911
7	303.6879
8	303.7842
9	303.8800
10	303.9754

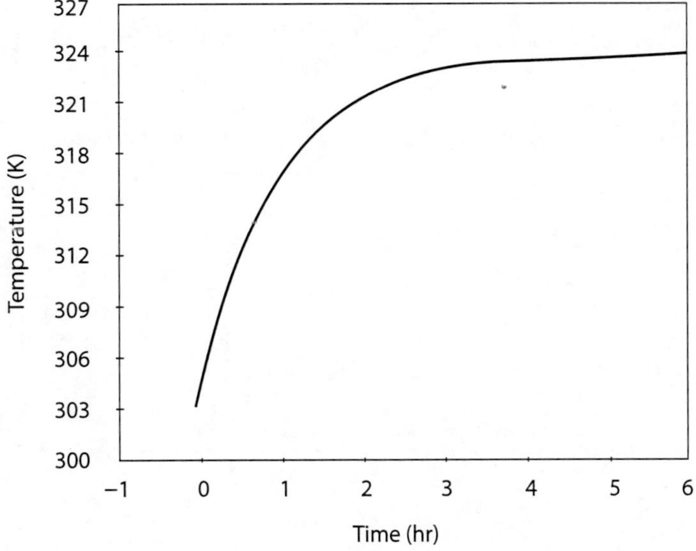

Figure 4.11 Temperature profile of the heating tank system

Case study 4.2 Moving bed reactor

A moving bed reactor (Fogler, 2005) is modeled as:

$$\frac{dx}{dw} = \frac{kaC_{Ao}^2(1-x)^2}{F_{Ao}} \tag{4.80}$$

$$\frac{da}{dw} = -\frac{k_d a^{0.8}}{\upsilon_s} \tag{4.81}$$

where,

x = conversion of reactant A
w = solid catalyst weight
k = reaction rate constant
C_{A0} = initial concentration of A
F_{A0} = inlet flow rate of A
υ_s = mass flow rate of catalyst moving through the bed

The rate of catalyst decay is basically expressed as:

$$-\frac{da}{dt} = k_d a^n$$

where,

a = catalyst activity
t = contact time between the moving catalyst and reactant A $\left(=\frac{w}{\upsilon_s}\right)$
k_d = decay constant
n = decay reaction order

Problem statement

Solve Equations (4.80) and (4.81) using the RK4 method over $w \in [0, 10]$. Relevant information are given in Table 4.15.

TABLE 4.15 Relevant information of the moving bed reactor

$k = 1000$
$C_{A0} = 0.1$
$F_{A0} = 50$
$k_d = 0.5$
$\upsilon_s = 5$
$\Delta w = 1$
Initial condition: $x(0) = 0$, $a(0) = 1$

Solution

The computer-assisted algorithm for the RK4 method is developed before (see for example, Case Study 4.1). Using that, we simulate the reactor model Equations (4.80) and (4.81). In Table 4.16, the simulation results for the first 10 steps (i.e., the catalyst weight up to 10) are shown.

Table 4.16 Numerical solutions for the first 10 steps (k = 0, 1, 2,..., 10)

w	x_k	a_k
0.0	0.00	1.00
1.0	0.1598550	0.9039208
2.0	0.2658301	0.8153728
3.0	0.3407601	0.7339041
4.0	0.3961685	0.6590817
5.0	0.4385016	0.5904902
6.0	0.4716480	0.5277321
7.0	0.4980943	0.4704273
8.0	0.5195054	0.4182122
9.0	0.5370387	0.3707401
10.0	0.5515251	0.3276803

The simulation results are further plotted in Figure 4.12 to observe the conversion and catalyst activity profile at a glance.

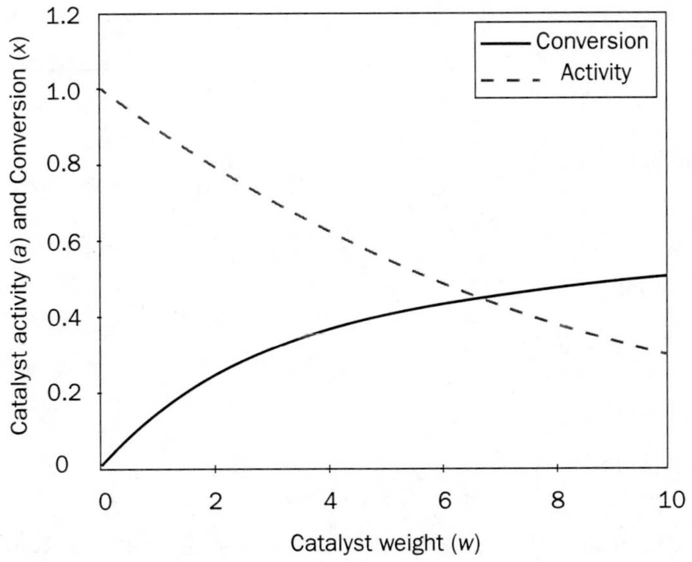

Figure 4.12 Catalyst activity and conversion versus catalyst weight in moving bed reactor

Case study 4.3 Membrane reactor

In a catalytic membrane reactor, the following reversible reaction takes place:

$$A \Leftrightarrow B + C$$

The reactor that operates at 8.2 atm and 227 °C receives pure gaseous A at a rate of 9 mol/min. The complete model (Fogler, 2005) of this reactor is documented in Table 4.17.

TABLE 4.17 Complete membrane reactor model

$$\frac{dF_A}{dV} = r_A \tag{4.82}$$

$$\frac{dF_B}{dV} = -r_A - k_c C_{To} \frac{F_B}{F_T} \tag{4.83}$$

$$\frac{dF_C}{dV} = -r_A \tag{4.84}$$

$$-r_A = kC_{To}\left[\frac{F_A}{F_T} - \left(\frac{C_{To}}{K_c}\right)\left(\frac{F_B F_C}{F_T^2}\right)\right] \tag{4.85}$$

$$F_T = F_A + F_B + F_C \tag{4.86}$$

$$C_{T0} = \frac{P}{RT} \tag{4.87}$$

Here,

F_A = flow rate of species A
F_B = flow rate of species B
F_C = flow rate of species C
V = reactor volume
K_c = equilibrium constant
k = reaction rate constant
k_c = transport coefficient
$-r_A$ = rate of disappearance of A
P = operating pressure (reactant A enters at this same pressure)
T = operating temperature (reactant A enters at this same temperature)

Problem statement

Integrate this ODE-IVP system, consisting of Equations (4.82)–(4.87), by using the explicit Euler method with $\Delta V = 1.0$ L. The supporting information is documented in Table 4.18.

TABLE 4.18 Supporting information of the membrane reactor

$k_C = 0.2\ \text{min}^{-1}$

$K_c = 0.05$ mol/L

$k = 0.7\ \text{min}^{-1}$

$$C_{T0} = \frac{P}{RT} = \frac{8.2\ \text{atm}}{0.082\ \text{atm.L/mol.K} \times 500\ \text{K}} = 0.2\ \text{mol/L}$$

Initial conditions: $F_{A0} = 9$ mol/min

$F_{B0} = F_{C0} = 0$ mol/min

Final reactor volume $V_f = 500$ L

Solution

Using the explicit Euler method, we produce the results in Table 4.19 for the first 10 volume steps. Further, we show the complete flow profiles in Figure 4.13 for all three species, *A*, *B* and *C* that took part in the reaction.

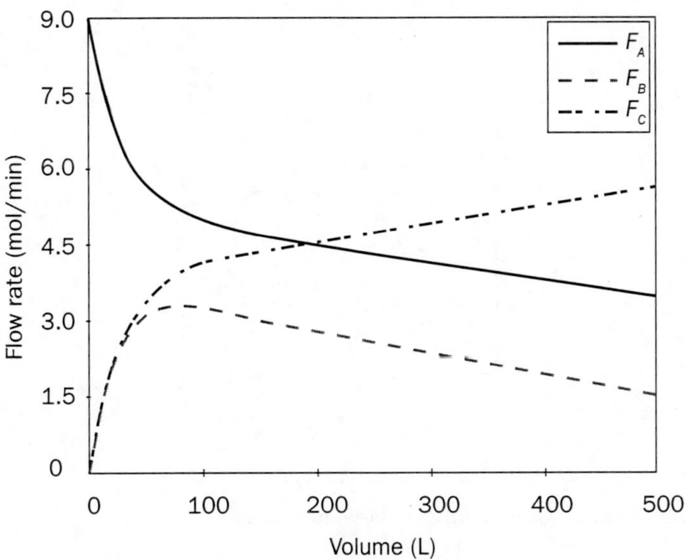

FIGURE 4.13 Component flow rate profile

Table 4.19 Numerical results for the first 10 volume steps

V (L)	F_A (mol/min)	F_B (mol/min)	F_C (mol/min)
0	9	0	0
1	8.86	0.14	0.14
2	8.72442	0.27497	0.27558
3	8.59322	0.40498	0.40678
4	8.46635	0.53013	0.53365
5	8.34372	0.65053	0.65628
6	8.22525	0.76631	0.77475
7	8.11082	0.8776	0.88918
8	8.00034	0.98452	0.99966
9	7.89369	1.08723	1.10631
10	7.79076	1.18585	1.20924

Case study 4.4 Isothermal batch reactor

A constant volume isothermal batch reactor is schematically shown in Figure 4.14.

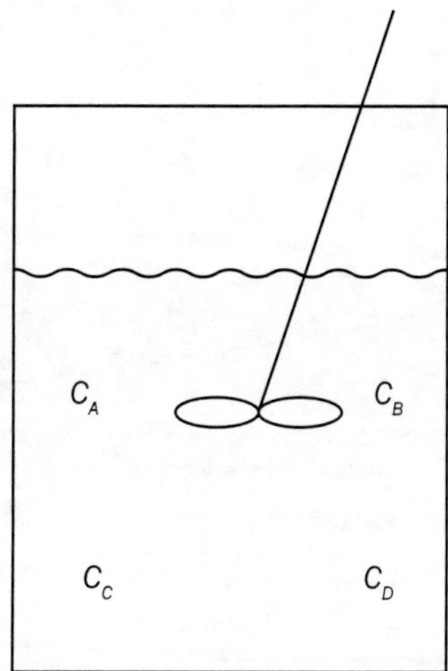

Figure 4.14 An isothermal batch reactor

Here the following reactions (elementary) take place:

$$A \xrightarrow{k_1} B \xrightarrow{k_2} C$$

$$B \xrightarrow{k_3} D$$

Accordingly, the mass balance equations for all four species, namely A, B, C and D, can be derived as:

$$\frac{dC_A}{dt} = -k_1 C_A \tag{4.88a}$$

$$\frac{dC_B}{dt} = k_1 C_A - k_2 C_B - k_3 C_B \tag{4.88b}$$

$$\frac{dC_C}{dt} = k_2 C_B \tag{4.88c}$$

$$\frac{dC_D}{dt} = k_3 C_B \tag{4.88d}$$

where,

C_i = concentration of species i ($i = A, B, C, D$)
k = reaction rate constant

Problem statement

Solve the isothermal batch reactor model [i.e., Equation (4.88)] using the RK2 method based on the given information in Table 4.20.

TABLE **4.20** Isothermal batch reactor specification

$k_1 = 2$
$k_2 = 0.5$
$k_3 = 0.3$
$C_{A0} = 1$
$C_{B0} = 0.05$
$h = 0.005$

Note that the fresh feed consists of species A and B only (i.e., $C_{C0} = C_{D0} = 0$).

Solution

For RK2, the computational steps are developed earlier in Example 4.8. Following the same methodology, we repeat those steps for the multi-variable isothermal batch reactor below.

Step 1: Let us consider,

$$C_A0 = C_{Ak}$$

$$C_B0 = C_{Bk}$$

$$C_C0 = C_{Ck}$$

$$C_D0 = C_{Dk}$$

At $k = 0$,

$$C_A0 = C_{A0} = 1$$

$$C_B0 = C_{B0} = 0.05$$

$$C_C0 = C_D0 = 0$$

Step 2: Find first set of slopes.

$$k_{11} = -2C_A0$$

$$k_{12} = 2C_A0 - 0.5C_B0 - 0.3C_B0$$

$$k_{13} = 0.5C_B0$$

$$k_{14} = 0.3C_B0$$

Step 3: Find C_i at the midpoint.

$$C_A1 = C_{Ak} + \frac{h}{2}k_{11}$$

$$C_B1 = C_{Bk} + \frac{h}{2}k_{12}$$

$$C_C1 = C_{Ck} + \frac{h}{2}k_{13}$$

$$C_D1 = C_{Dk} + \frac{h}{2}k_{14}$$

Step 4: Find second set of slopes.

$$k_{21} = -2C_A1$$

$$k_{22} = 2C_A1 - 0.5C_B1 - 0.3C_B1$$

$$k_{23} = 0.5C_B1$$

$$k_{24} = 0.3C_B1$$

Step 5: Update C_i.

$$C_{A\ k+1} = C_{Ak} + k_{21}h$$

$$C_{B\ k+1} = C_{Bk} + k_{22}h$$

$$C_{C\ k+1} = C_{Ck} + k_{23}h$$

$$C_{D\ k+1} = C_{Dk} + k_{24}h$$

Step 6: Check the convergence.
If the following criterion,
$\left|C_{i\ k+1} - C_{i\ k}\right| \leq \text{tol}$
is satisfied, go to next Step 7, else go to Step 1.

Step 7: Stop.

Following these computational steps, we produce the results. For the first 10 time steps, the C_i values are tabulated in Table 4.21. Moreover, in Figure 4.15, we have shown the entire concentration profile for all four components, namely *A*, *B*, *C* and *D*. At the end, the four concentrations are obtained as:

$C_A = 7.0064923\text{e-}44$
$C_B = 3.5032462\text{e-}43$
$C_C = 0.6562446$
$C_D = 0.3937478$

Notice that species *A* and *B* are almost completely converted to *C* and *D*.

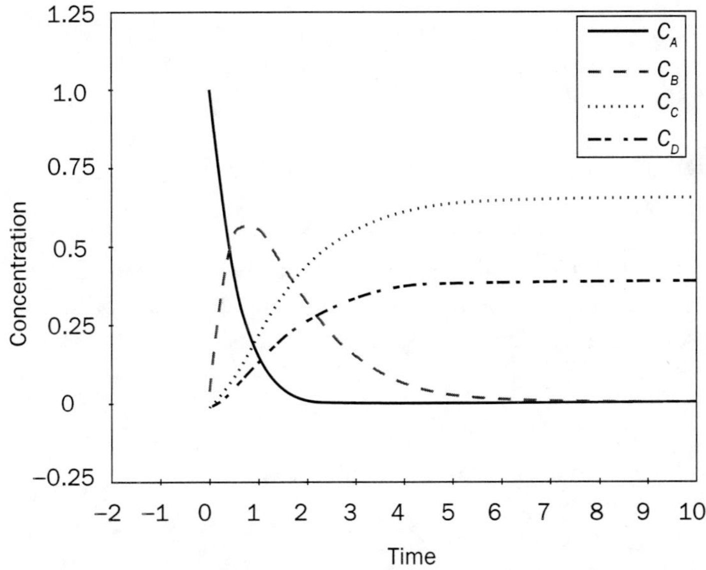

FIGURE **4.15** Concentration profile of isothermal batch reactor for all four components (*A*, *B*, *C* and *D*)

TABLE 4.21 Numerical solutions for the first 10 time steps

Step (k)	C_{AK}	C_{BK}	C_{CK}	C_{DK}
0	1.000000	5.0000001e-02	0.0000000e+00	0.0000000e+00
1	0.990050	5.9730399e-02	1.3725000e-04	8.2350009e-05
2	0.9801990	6.9323152e-02	2.9865297e-04	1.7919179e-04
3	0.9704461	7.8779794e-02	4.8386672e-04	2.9032005e-04
4	0.9607901	8.8101834e-02	6.9255289e-04	4.1553174e-04
5	0.9512302	9.7290777e-02	9.2437683e-04	5.5462611e-04
6	0.9417655	0.1063481	1.1790077e-03	7.0740463e-04
7	0.9323949	0.1152753	1.4561183e-03	8.7367097e-04
8	0.9231176	0.1240738	1.7553851e-03	1.0532311e-03
9	0.9139326	0.1327451	2.0764882e-03	1.2458930e-03
10	0.9048389	0.1412905	2.4191113e-03	1.4514668e-03

Case study 4.5 Batch bioreactor

In a batch bioreactor shown in Figure 4.16, the biomass typically grows by consuming the substrate. Thus, we consider that there are two components present, namely biomass and substrate. With this, the bioreactor model is developed below.

Deriving model

Reaction Rate

It is experimentally observed that the quantity of biomass (thus its concentration) increases exponentially over time. Under ideal conditions, this behavior one can explain by the fact that all biomass cells have the same probability to multiply. Accordingly, the rate of cell generation is proportional to the biomass itself:

$$r_x = \mu\, x \tag{4.89}$$

where,

r_x = rate of cell generation (g of cells generated/L.hr)
x = cell concentration (g cell/L)
μ = kinetic growth constant or specific growth rate (1/hr)

Biomass Balance

For the constant volume batch bioreactor,

$$\frac{dx}{dt} = r_x$$

That gives,

$$\frac{dx}{dt} = \mu\, x \tag{4.90}$$

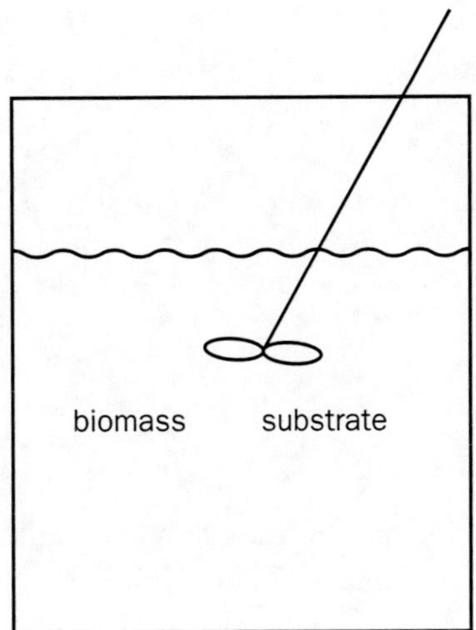

FIGURE 4.16 A batch bioreactor

The Monod model is widely used to explain the specific growth rate as:

$$\mu=\mu_{\max}\frac{s}{k_s+s} \tag{4.91}$$

where,

$\mu_{\max}$ = maximum specific growth rate (1/hr)
s = substrate concentration (g substrate/L)
k_s = saturation constant[18] (g/L)

Accordingly, Equation (4.90) yields:

$$\frac{dx}{dt}=\mu_{\max}\frac{s}{k_s+s}x$$

Substrate Balance

It yields:

$$\frac{ds}{dt}=-r_s \tag{4.92}$$

where, r_s is the rate of substrate consumption (g of substrate consumed/L.hr).

Let us introduce a term called *yield coefficient* that is defined as:

$$Y_{xs}=\frac{\text{mass of cells generated}}{\text{mass of substrate consumed}}$$

18 It is actually called *half-saturation constant* since when $s = k_s$, the specific growth rate becomes $\frac{\mu_{\max}}{2}$.

That gives:

$$Y_{xs} = \frac{r_x}{r_s} \tag{4.93}$$

Equation (4.92) becomes:

$$\frac{ds}{dt} = -\frac{r_x}{Y_{xs}}$$

Inserting Equation (4.89),

$$\frac{ds}{dt} = -\frac{1}{Y_{xs}}\mu\, x \tag{4.94}$$

So the complete batch bioreactor model leads to a coupled differential algebraic equation (DAE) system as documented in Table 4.22.

TABLE 4.22 Complete model of a batch bioreactor

$\frac{dx}{dt} = \mu\, x$	(4.90)
$\frac{ds}{dt} = -\frac{1}{Y_{xs}}\mu\, x$	(4.94)
$\mu = \mu_{max}\frac{s}{k_s + s}$	(4.91)

Problem statement

Solve the batch bioreactor model (Table 4.22) using the explicit Euler method based on the given information in Table 4.23.

TABLE 4.23 Given information for the batch bioreactor

μ_{max} = 0.1 hr^{-1}
k_s = 0.1 g/L
Y_{xs} = 0.1
x_0 = 0.3 g/L
s_0 = 100 g/L
h = 0.001 hr

Solution

The model equations of the batch bioreactor are solved and the simulation data produced for the first 10 time steps are provided in Table 4.24.

TABLE 4.24 Numerical solutions for the first 10 time steps

Step (k)	x_k (g/L)	s_k (g/L)
0	0.3000000	100.0000
1	0.3000300	99.99970
2	0.3000600	99.99940
3	0.3000900	99.99911
4	0.3001199	99.99881
5	0.3001499	99.99851
6	0.3001799	99.99821
7	0.3002099	99.99792
8	0.3002399	99.99762
9	0.3002698	99.99732
10	0.3002999	99.99702

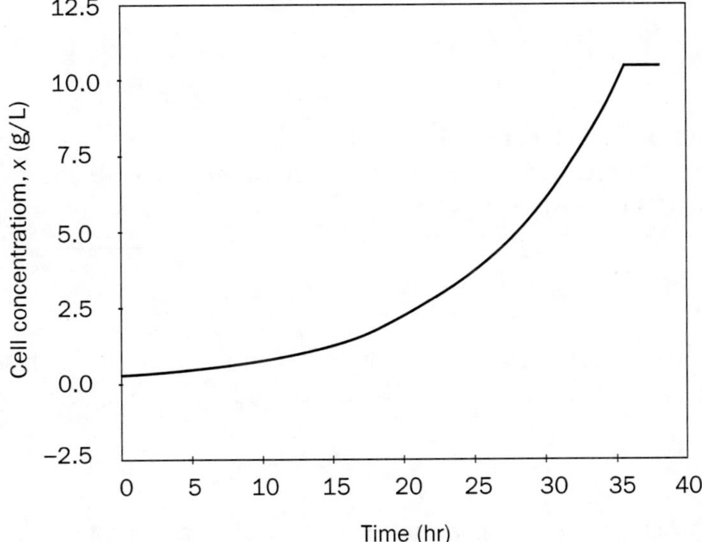

FIGURE 4.17 Cell concentration profile

Further, we produce the dynamic profile for cell concentration in Figure 4.17 and substrate concentration in Figure 4.18.

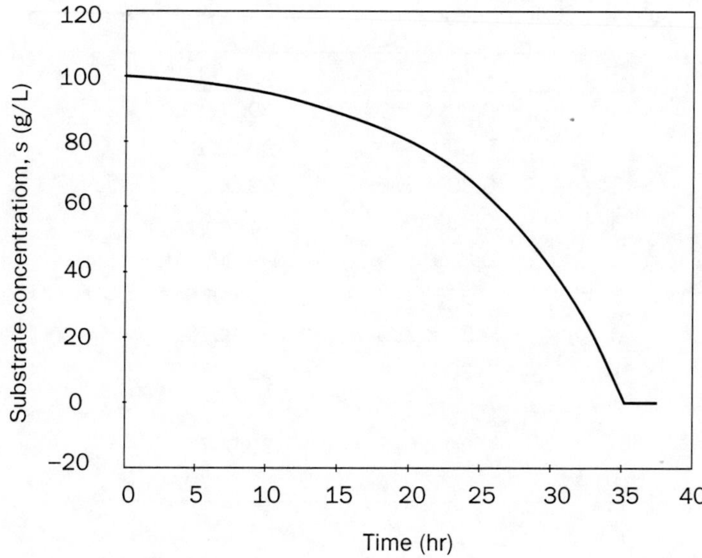

FIGURE 4.18 Substrate concentration profile

Remark

It becomes obvious from Figures 4.17 and 4.18 that the cells grow ($x\uparrow$) with consuming substrate ($s\downarrow$) in the bioreactor. When there is no substrate (i.e., $s = 0$), the cell growth stops and it corresponds to $x = 10.3$ g/L.

Case study 4.6 Batch adsorption process

In an adsorption process, the gas or liquid molecules (or atoms or ions) diffuse to the surface of a solid. The adsorbed solutes are called *adsorbate* and the solid material is the *adsorbent.*

Deriving model

Adsorption Isotherm

There is adsorption isotherm (at constant temperature) by means of solute loading on the adsorbent versus concentration (if the fluid is liquid) or partial pressure (if the fluid is gas) of the adsorbate in the fluid. This is developed based on the experimental data sets since, unlike vapor–liquid or liquid–liquid equilibria, there is no established theory to readily estimate the fluid–solid adsorption equilibria.

Freundlich isotherm that has the following form (for gas adsorption), is widely used:

$$q = k\,p^{\frac{1}{n}} \tag{4.95}$$

Obviously, this is a non-linear empirical expression, in which,

q = equilibrium loading (= mass adsorbed/mass of adsorbent)
k = empirical constant (temperature dependent)
p = partial pressure of solute in the gas
n = temperature dependent constant (that typically lies in the range of 1–5)

When $n = 1$, Equation (4.95) leads to a linear isotherm:

$$q = k\,p$$

Rearranging,

$$p = \frac{1}{k} q$$

which is the *Henry's law* equation.

An alternative form of Freundlich isotherm (for liquid adsorption) is:

$$C^{*} = \left(\frac{q}{k}\right)^{n} \tag{4.96}$$

where, C^{*} is the equilibrium concentration (mass of solute/volume).

In the following, we continue the model development for liquid adsorption.

Rate of Adsorption

The rate of adsorption of solute is modeled as:

$$-\frac{dC}{dt} = K_L\, a\left(C - C^{*}\right) \tag{4.97}$$

where,

C = concentration of solute in the bulk liquid (mass of solute/volume)
K_L = external liquid phase mass transfer coefficient
a = external surface area of the adsorbent per unit volume of liquid

Component (solute) Balance

Assuming constant liquid volume (reasonable for dilute feeds), the solute balance yields:

$$C_F V = C V + q m \tag{4.98}$$

where,

C_F = feed concentration (mass of solute/volume)
V = liquid volume
m = mass of adsorbent

The complete batch adsorption column model is documented in Table 4.25.

Table 4.25 Complete model of a batch adsorption column

$$-\frac{dC}{dt} = K_L\, a\left(C - C^*\right) \quad (4.97)$$

$$C^* = \left(\frac{q}{k}\right)^n \quad (4.96)$$

$$q = \frac{(C_F - C)}{m/V} \quad \text{[from Equation (4.98)]}$$

Problem statement

There is an agitated vessel operated in the batch mode to adsorb phenol (adsorbate) by the use of activated carbon (adsorbent). Relevant information (Seader and Henley, 2001) are given in Table 4.26.

Table 4.26 Relevant process information

$C_F = 0.01$ mol phenol/L = 0.01 kmol/m^3
$K_L = 5 \times 10^{-5}$ m/sec
$m/V = 8$ kg/m^3
$a = 40$ m^{-1}
$h = 1$ sec

Freundlich isotherm is given as:

$$C^* = \left(\frac{q}{0.0106}\right)^{4.3}$$

where, q is in kmol/kg and C^* in kmol/m^3. Compute the time required by the use of the implicit Euler method to reduce the phenol concentration to 0.0005 mol/L (= 0.0005 kmol/m^3).

Solution

From Equation (4.98),

$$q = \frac{(C_F - C)}{m/V} = \frac{0.01 - C}{8} \text{(kmol/kg)}$$

From the given Freundlich isotherm:

$$C^* = \left(\frac{q}{0.0106}\right)^{4.3} = \left(\frac{0.01 - C}{0.0848}\right)^{4.3}$$

Substituting this in Equation (4.97),

$$-\frac{dC}{dt}=\left(5\times10^{-5}\right)\times40\left[C-\left(\frac{0.01-C}{0.0848}\right)^{4.3}\right]$$

This gives:

$$\frac{dC}{dt}=-0.002\left[C-\left(\frac{0.01-C}{0.0848}\right)^{4.3}\right]\text{(kmol/m}^3\text{.sec)} \tag{4.99}$$

Obviously, the batch adsorption model reduces to an ODE-IVP system having Equation (4.99) with the following initial condition:

at time $t = 0$, $C = C_F = 0.01$ kmol/m^3

Now, we need to find the time t by solving Equation (4.99) so that C becomes 0.0005 kmol/m^3.

Using the implicit Euler method, Equation (4.99) yields:

$$C_{k+1}=C_k-0.002h\left[C_{k+1}-\left(\frac{0.01-C_{k+1}}{0.0848}\right)^{4.3}\right] \tag{4.100}$$

For finding the root of this non-linear form of algebraic equation, one can use the N–R method in each time step k. For this,

$$\eta(w)=w-C_k+0.002h\left[w-\left(\frac{0.01-w}{0.0848}\right)^{4.3}\right]=0 \tag{4.101}$$

Notice that here, $w_k=C_{k+1}$. Now, let us revisit the computational steps below.[19]

Computational steps

Step 1: Set initial guess for Equation (4.22c) as:

$$w_0^0=C_0=0.01 \qquad at\ k=0$$
$$w_k^0=C_k \qquad at\ k=1,2,....$$

Step 2: Find $\eta(w)$ and $\eta'(w)$.

Equation (4.101) is used to estimate $\eta(w)$. Differentiating that equation, we have:

$$\eta'(w)=1+0.002h\left[1+50.7075\left(\frac{0.01-w}{0.0848}\right)^{3.3}\right]$$

Step 3: Update w.

Use Equation (4.22c) for this purpose.

19 Discussed before in Example 4.7.

The representative batch adsorption process is an unsteady state system and thus, repeat the calculation until you get $w = 0.0005$.

Based on this computer-assisted algorithm, we produce the solute concentrations for the first 10 time steps in Table 4.27.

TABLE 4.27 C_k versus t for the first 10 time steps

k	time t (sec)	C_k (kmol/m^3)
0	0	0.01000000
1	1	0.009980039696
2	2	0.009960119456
3	3	0.009940238977
4	4	0.009920398180
5	5	0.009900596985
6	6	0.009880835314
7	7	0.009861113086
8	8	0.009841430225
9	9	0.009821786650
10	10	0.009802182285

Complete solute concentration profile is depicted in Figure 4.19. The adsorption process reaches about 0.0005 kmol/m^3 of solute concentration by 1559 sec.

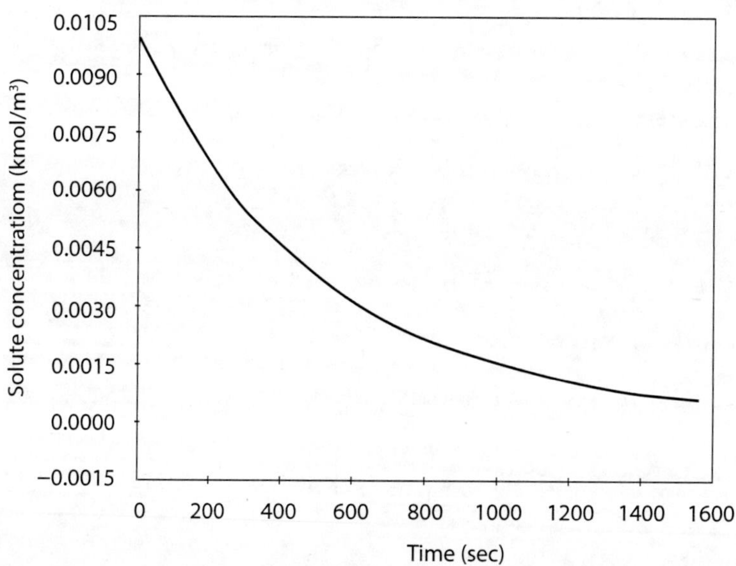

FIGURE 4.19 Solute concentration profile

Case study 4.7 Binary distillation column

A distillation column, shown in Figure 4.20, fractionates a binary mixture of components p and q. It consists of total 10 ideal trays (i.e., 100% efficient), a total condenser at the top and a partial reboiler at the bottom. The feed stream, having flow rate F and composition z (mole fraction), is a saturated liquid that enters at Tray 5. This column operates at atmospheric pressure and yields two product streams, namely distillate (flow rate = D and composition = x_D) and bottoms (flow rate= B and composition = x_B). The reflux (flow rate = R and composition = x_D) and boil-up vapor (flow rate = V_B and composition = y_B) are recycled back above the topmost tray (i.e., Tray 10) and below the bottommost tray (i.e., Tray 1), respectively. The liquid holdup in both the condenser-reflux drum (i.e., m_D) and column base-reboiler system (i.e., m_B) can be properly maintained by the feedback controller. With this, we assume that both these m_D and m_B remain constant.

Deriving model

Tray-wise model equations for the sample binary distillation column are developed by splitting the column into 7 envelops as presented below.

- ***Condenser–reflux drum system (envelop 1)***

Total mole balance

$$\frac{dm_D}{dt} = V_{10} - R - D \tag{4.102}$$

Component mole balance

$$\frac{dx_D}{dt} = \frac{1}{m_D}\left[V_{10}y_{10} - (D+R)x_D\right] \tag{4.103}$$

Recall that here m_D is maintained constant by a feedback controller.

Here, L, V, x and y denote the liquid flow rate, vapor flow rate, liquid phase composition and vapor phase composition, respectively.

- ***Topmost tray (Tray 10) (envelop 2)***

Total mole balance

$$\frac{dm_{10}}{dt} = R + V_9 - L_{10} - V_{10} \tag{4.104}$$

Component mole balance

$$\frac{d(m_{10}x_{10})}{dt} = Rx_D + V_9 y_9 - L_{10}x_{10} - V_{10}y_{10} \tag{4.105}$$

- ***nth tray*** ($n = 2$ to 4 and 6 to 9) ***(envelop 3 and 5)***

Total mole balance

$$\frac{dm_n}{dt} = L_{n+1} + V_{n-1} - L_n - V_n \tag{4.106}$$

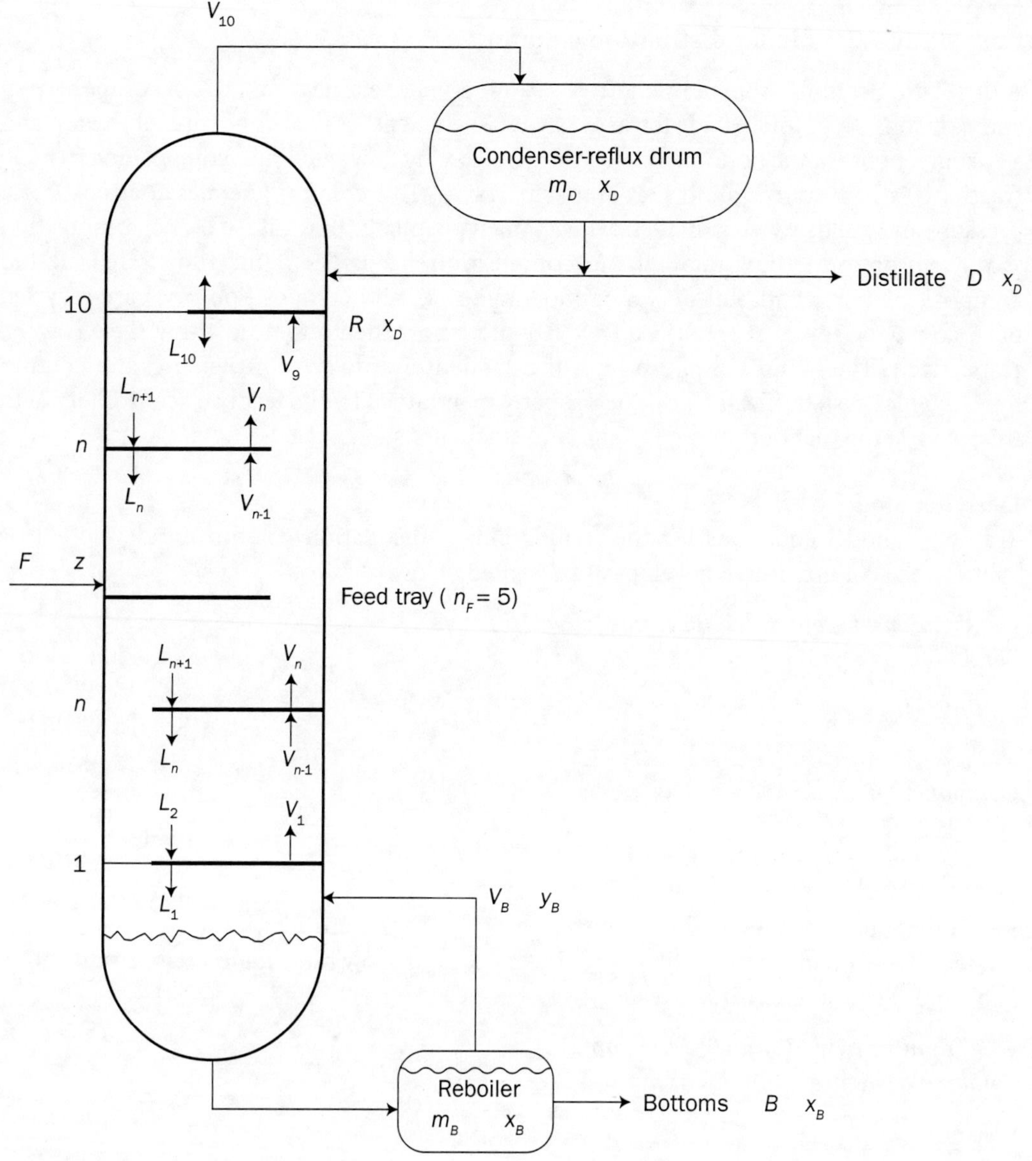

FIGURE 4.20 Schematic of a binary distillation column

Component mole balance

$$\frac{d(m_n x_n)}{dt} = L_{n+1}x_{n+1} + V_{n-1}y_{n-1} - L_n x_n - V_n y_n \tag{4.107}$$

- ***Feed tray (Tray 5) (envelop 4)***

Total mole balance

$$\frac{dm_5}{dt} = L_6 + V_4 + F - L_5 - V_5 \tag{4.108}$$

Component mole balance

$$\frac{d(m_5x_5)}{dt} = L_6x_6 + V_4y_4 + Fz - L_5x_5 - V_5y_5 \quad (4.109)$$

As stated, feed is a saturated liquid.

- ***Bottommost tray (Tray 1) (envelop 6)***

Total mole balance

$$\frac{dm_1}{dt} = L_2 + V_B - L_1 - V_1 \quad (4.110)$$

Component mole balance

$$\frac{d(m_1x_1)}{dt} = L_2x_2 + V_By_B - L_1x_1 - V_1y_1 \quad (4.111)$$

- ***Column base-reboiler system (envelop 7)***

Total mole balance

$$\frac{dm_B}{dt} = L_1 - B - V_B \quad (4.112)$$

Component mole balance

$$\frac{dx_B}{dt} = \frac{1}{m_B}\left[L_1x_1 - Bx_B - V_By_B\right] \quad (4.113)$$

Recall that like m_D, here m_B is kept constant by using a feedback controller.

Computing Vapor Flow Rate (V)

Let us assume for simplicity that all the vapor flow rates are same with the vapor boil-up rate (V_B). Accordingly, we have:

$$V_1 = V_2 = \dots\dots\dots = V_{10} = V_B \quad (4.114)$$

Computing Liquid Flow Rate

Distillate rate

For constant m_D, Equation (4.102) yields:

$$D = V_{10} - R = V_B - R \quad (4.115)$$

Bottoms rate

From Equation (4.112) with constant m_B,

$$B = L_1 - V_B \quad (4.116)$$

For all other trays

Use the following Francis weir formula (Luyben, 1990),

$$L_n = L_{n0} + \frac{m_n - m_{n0}}{\beta} \quad (4.117)$$

where,

$$L_{n0}=\begin{cases} R & \text{for Tray 6 to 10} \\ R+F & \text{for Tray 1 to 5} \end{cases} \tag{4.118}$$

m_{n0} = reference value of m_n

β = hydraulic time constant (= 3–6 sec per tray) (Luyben, 1990).

Computing Vapor Phase Composition (y)

The following phase equilibrium equation can be used to find y:

$$y=\frac{\alpha\, x}{1+(\alpha-1)\, x} \tag{4.119}$$

where, α is the relative volatility.

Problem statement

Solve the binary distillation model derived above by the use of the explicit Euler method based on the given operating condition and column information in Table 4.28. For this, you are also asked to develop the computational steps involved in the simulation.

Table 4.28 Operating condition and column specification

m_D = 100 lbmol
m_B = 600 lbmol
V_B = 400 lbmol/hr
R = 380 lbmol/hr
F = 500 lbmol/hr
z = 0.4 (mole fraction of more volatile component p)
α = 3.5
β = 3.6 sec (= 0.001 hr)
Integration time interval (h) = 0.0001 hr
Total number of trays (excluding condenser and reboiler) = 10
Feed tray = 5
m_{n0} = 10 lbmol (for Tray 1–10) [for Equation (4.117)]
Initial conditions (at *time*, t = 0)
$x_1(0)=x_2(0)=\ldots\ldots=x_{10}(0)=x_B(0)=x_D(0)=0.4$ (same with feed composition)
$m_1(0)=m_2(0)=\ldots\ldots=m_{10}(0)=6$ lbmol

Solution

To solve the model Equations (4.102)–(4.119), one needs to develop a computer program for the simulation of those coupled differential algebraic equations (DAEs). For this, let us formulate the computational steps in sequence.

Computational steps

Step 1: Specify the followings (see Table 4.28):

- Feed (flow rate, F and composition, z)
- Liquid holdup (for reflux drum, m_D and column base, m_B)
- Reference liquid holdup (m_{n0} for Tray 1–10)
- Relative volatility, α
- Hydraulic time constant, β
- Integration time interval, h

Step 2: To solve this ODE-IVP system, initialize the following variables (see Table 4.28):

- Liquid holdup (m) for each tray, excluding reflux drum and column base
- Liquid composition (x) for each tray, including reflux drum and column base

Step 3: Either manipulate (employing controller) or specify the following variables:

- Reflux rate, R
- Vapor boil-up rate, V_B

Here, they both are specified (see Table 4.28).

Step 4: Compute vapor flow rates (V) through all trays using Equation (4.114) with known V_B (see Table 4.28).

Step 5: Compute liquid flow rates (L) for Tray 1–10 using Equations (4.117) and (4.118).

Step 6: Compute distillate (D) and bottom flow rate (B) from Equations (4.115) and (4.116), respectively.

Step 7: Compute vapor phase composition (y_B for reboiler and y for Tray 1–10) from Equation (4.119).

Step 8: Solve ordinary differential equations [Equations (4.102)–(4.113)] and compute liquid holdup (m for Tray 1–10) and liquid phase composition (for all trays (i.e., x), x_D and x_B) for the next time step.

Go back to Step 3 to continue the computation until the steady state is attained.

Using the computational steps highlighted above for the binary distillation column, we solve the coupled DAE system. The explicit Euler method is employed and the given information in Table 4.28 is used. With this, we attain the steady state within about 5 hours. The obtained liquid phase compositions (mole fraction of component p) at fifth hour are listed in Table 4.29.

TABLE 4.29 Obtained liquid phase compositions at fifth hour

$x_B = 0.37085$
$x_1 = 0.51068$
$x_2 = 0.56140$
$x_3 = 0.57615$
$x_4 = 0.58016$
$x_5 = 0.58122$
$x_6 = 0.82032$
$x_7 = 0.93803$
$x_8 = 0.98052$
$x_9 = 0.99409$
$x_{10} = 0.99824$
$x_D = 0.99950$

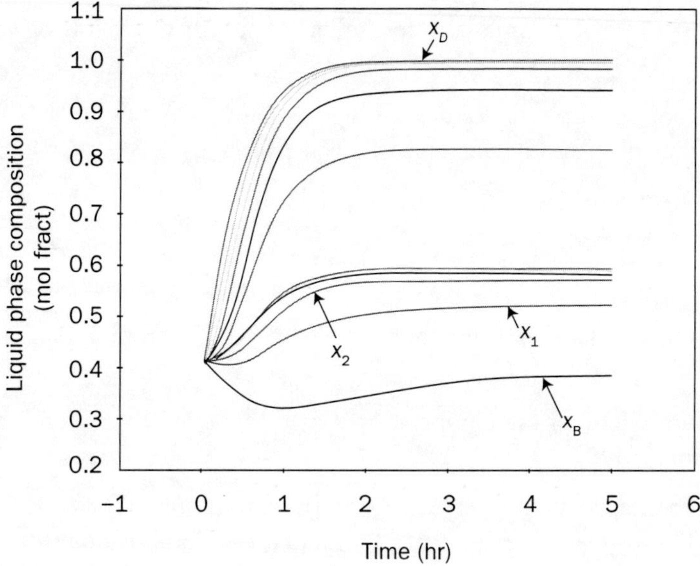

FIGURE 4.21 Distillation start-up profile in terms of liquid composition

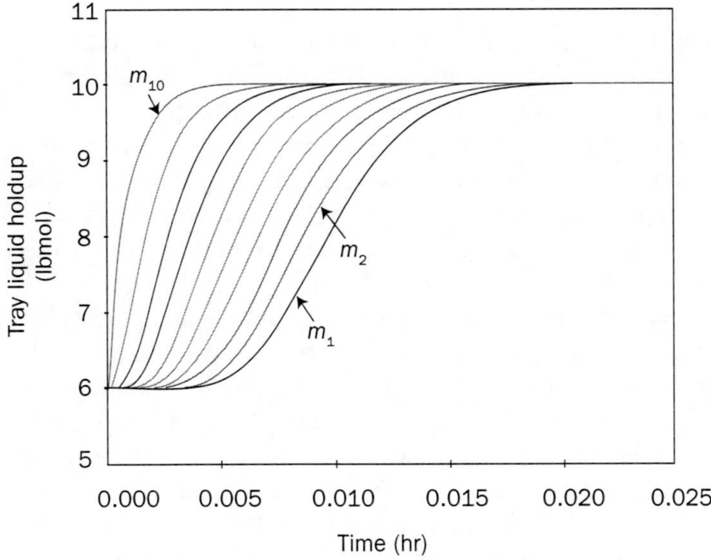

Figure 4.22 Distillation start-up profile in terms of tray liquid holdup (m_D and m_B are not shown here since they are fixed at 100 and 600 lbmol, respectively)

The liquid phase composition profiles for all trays (including reflux drum and reboiler) with time are depicted in Figure 4.21. It is obvious that the top section of the distillation column reaches steady state faster than its bottom section.

In the next Figure 4.22, the liquid holdup profiles are shown. It is evident that all the liquid holdups converge to their reference value (i.e., 10 lbmol).

Remark

Comparing Figures 4.21 and 4.22, we see that the composition dynamics is much slower than the holdup dynamics. Also, it becomes clear that the dynamics of column top section is faster than that of bottom section.

4.16 Summary and Concluding Remarks

In this chapter, we learned various numerical integration methods to solve the initial value problems. They notably include the Euler method, Runge–Kutta family and predictor–corrector approach for single and multi-variable systems of first-order and higher-order ODEs. Along with their theoretical derivation, the numerical stability analysis is also made for many of them for the proper choice of step size. A few case studies are finally modelled and the resulting differential algebraic equations are solved with employing the suitable numerical methods.

General Recommendations

For solving ODE-IVP systems, it is always safe to use the implicit Euler method without having any risk of instability. However, this method is although simple for linear IVP systems, it additionally requires an iterative convergence approach (e.g., Newton–Raphson) for non-linear systems, involving some added complexity and computational effort. Since there is as such no guideline for implicit Euler in selecting the h (step size) value, one can arbitrarily choose any large h with the aim of reduced computational effort and invite inaccuracy in the numerical solutions. We must be cautious about these issues before adopting this implicit Euler scheme.

On the other hand, explicit Euler is perhaps the simplest approach (with only 1 function evaluation in each step), because of which, we frequently prefer to use it, particularly for solving the large multi-variable IVP systems. However, for this approach, one must choose h within a restricted stability limit (e.g., $h \leq 2\tau_p$ for the liquid tank system (Example 4.3)). To have better flexibility on h that usually affects the speed of response, one can select RK4 method (i.e., $h \leq 2.785\tau_p$) but at the expense of complexity and increased number of function evaluations (total 4). However, do not forget that the RK4 provides better accuracy (fourth-order accurate) than the Euler method (first-order accurate).

To enhance the speed of response of the RK4 method, it is further recommended to employ the RKF45 approach although it requires more function evaluations (total 6) than the RK4 approach (total 4).

Appendix 4A

A4.1 Conservation Principle

The conservation principle states that:

$$\text{Accumulation} = \text{Input} - \text{Output} + \text{Generation} \tag{A4.1}$$

For the three fundamental quantities, namely mass, energy and momentum, this balance equation can be presented as follows.

Conservation of mass

$$\underset{\text{of mass}}{\text{Rate of accumulation}} = \underset{\text{of mass}}{\text{Rate of input}} - \underset{\text{of mass}}{\text{Rate of output}} + \underset{\text{of mass}}{\text{Rate of generation}} \tag{A4.2}$$

This form of equation can be used for both total and component mass balance.

Conservation of energy

$$\underset{\text{of energy}}{\text{Rate of accumulation}} = \underset{\text{of energy}}{\text{Rate of input}} - \underset{\text{of energy}}{\text{Rate of output}} + \underset{\text{of energy}}{\text{Rate of generation}} \tag{A4.3}$$

It should be noted that the first law of thermodynamics is a version of the conservation principle of energy for thermodynamic processes.

Conservation of momentum

Rate of accumulation of momentum = Rate of input of momentum – Rate of output of momentum + Rate of forces acting on volume element (A4.4)

The momentum balance equation is a generalization of Newton's law of motion.

EXERCISE

Review Questions

4.1 Why do we need to transform a higher-order ODE into a set of first-order ODEs?

4.2 Implicit Euler involves large computational effort, whereas explicit Euler has certain stability limit. Will they have similar computational load if their stated problems are taken care of?

4.3 Integrating Equation (4.4), we get:

$$x(t) = x(t_0) + \int_{t_0}^{t} f(x,\ t)\ dt$$

i) Is this an implicit or explicit scheme?

ii) Write its both implicit and explicit[20] forms.

4.4 List the major drawbacks of Taylor method as an IVP solver.

4.5 How does the Heun method differ from the RK2?

4.6 Explain the development of Figure 4.7 for the RK4 method.

4.7 List the relative merits and demerits of the RK4 and RKF45.

4.8 Why is the corrector part required in the Adams–Bashforth–Moulton approach?

4.9 Develop the solution algorithm for a coupled ODE-algebraic equation system.

4.10 Fill in the blanks in Table 4.30.

TABLE 4.30 Comparison of numerical integration approaches

Approach	Number of function evaluation(s)	Order of accuracy	Stability limit
Explicit Euler	1	1	2
RK2	2	2	?
Heun	?	?	2
RK4	4	4	2.785
RKG4	?	?	?
Adams–Bashforth–Moulton	?	4	?

20 Note that the following explicit form is called *Picard method of successive approximation*:

$$x_{k+1} = x_0 + \int_{t_0}^{t} f(x_k,\ t)dt.$$

Practice Problems

4.11 Solve the following differential equation system with guessing the initial condition:

$$y' = \cos t + \alpha(y - \sin t)$$

for $\alpha = -0.2$.

4.12 An ODE – IVP system is given as:

$$\frac{dx}{dt} = t^3 - 3\sqrt{x}$$

with $x(0) = 0$ and $h = 0.1$. Find $x(0.3)$ using the Heun and RK4 method.

4.13 Consider the following problem:

$$y''' - ty'' + 5yy' = 3$$

with $y(0) = 0$, $y'(0) = y''(0) = 1$. Determine $y(0.1)$ with the use of Taylor series method.

4.14 For the following system:

$$\frac{dy}{dt} = t^2 - \cos y$$

find $y(2)$ using a suitable method when $y(0) = 1$ and $h = 0.2$.

4.15 For the following system:

$$\frac{dy}{dt} = t + y$$

find $y(1)$ using the predictor–corrector method when $y(0) = 0.5$ and $h = 0.2$. For this, use the RK4 to determine the three starting values: $y(0.2)$, $y(0.4)$ and $y(0.6)$.

4.16 Solve the following problem:

$$x' = 2x^2 + \ln t$$

with $x(1) = 0.5$ and $h = 0.1$. Use the RK2 method to find $x(2)$.

4.17 Consider the following system equation:

$$\frac{dx}{dt} = \frac{1}{3}(t - x)$$

Compare the RK2 solutions with different step sizes (h = 1, 1/2, 1/4, 1/8) over [0, 3] with $x(0) = 1$.

4.18 Using the implicit Euler technique, integrate the following equation:

$$\frac{dx}{dt} = \sin x$$

with $x(0) = 1$ and $h = 0.05$. Compute x_1 and x_2.

4.19 Solve the following ODE-IVP system:

$$\frac{dx}{dt} = \frac{1}{x^2} + e^{-x}$$

using the implicit Euler and $h = 0.01$ with guessing an initial condition.

4.20 Use the Runge–Kutta–Gill method to integrate:

$$\frac{dx}{dt} = e^{-2xt}$$

with $x(0) = 0$ and $h = 0.01$.

4.21 Perform the numerical stability analysis of the explicit Euler for:

$$y' = -2.5y$$

with $y(0) = 1$ and

i) $h = 0.5$

ii) $h = 1$

4.22 Consider a first-order process represented by:

$$\frac{dx}{dt} = \frac{5x}{\lambda}$$

with $x(0) = x_0$ and $\lambda < 0$. Determine the stability condition for this system with the trapezoidal rule [Equation (4.25)] in terms of h.

4.23 Repeat the above Problem 4.22 to find the stability condition with the Ralston method in terms of h.

Ralston method: Substitute $a_1 = \frac{1}{3}$, $a_2 = \frac{2}{3}$ and $P_1 = q_{11} = \frac{3}{4}$ in Equation (4.38) and get:

$$x_{k+1} = x_k + \frac{h}{3}[k_1 + 2k_2]$$

where,

$$k_1 = f(t_k, x_k)$$

$$k_2 = f\left(t_k + \frac{3h}{4}, x_k + \frac{3h}{4}k_1\right)$$

4.24 Repeat Problem 4.22 to find the stability condition of the third-order Runge–Kutta method (RK3) in terms of h.

Third-order Runge–Kutta: $x_{k+1} = x_k + \frac{h}{6}[k_1 + 4k_2 + k_3]$

where,

$$k_1 = f(t_k, x_k)$$

$$k_2 = f\left(t_k + \frac{h}{2}, x_k + \frac{h}{2}k_1\right)$$

$$k_3 = f(t_k + h, x_k - hk_1 + 2hk_2)$$

4.25 For the removal of phenol in a semi-continuous adsorption column, the following form of model is developed (Seader and Henley, 2001):

$$\frac{dq}{dt} = 0.00963\left[0.01 - \left(\frac{q}{0.0106}\right)^{4.3}\right]$$

where, q denotes the amount adsorbed/unit mass of adsorbent. Solve this equation using the RKF45 method with $q = 0$ at $t = 0$ and $h = 1$.

4.26 The following empirical equation is used to predict the world's population:

$$\frac{dx}{dt} = 4.257 \times 10^{-12} x^{2.0101}$$

with $x_0 = 10^9$ and $h = 5$ yr. With $t_0 = 0$ that represents the base year, 1840 AD, predict the population in 2020 using the Runge–Kutta–Gill method.

4.27 Consider the following IVP system:

$$\frac{dx}{dt} = 1+9x^2$$

with $x(0) = 0$, $tol = 1 \times 10^{-5}$ and $h = 0.1$. Solve the problem using the RKF45 method and tabulate the solutions for the first 5 steps (i.e., up to $k = 5$). Update h at all kth steps.

4.28 Consider a liquid tank system shown in Figure 4.23. Develop the model [consult Equation (4.19)] and simulate it using the RK4 method with the process information given in Table 4.31.

TABLE **4.31** Given information for the liquid tank

Given
$F_i = 8$ m³/sec
β (proportionality constant) = 4 m$^{2.5}$/sec
$A = 16$ m²

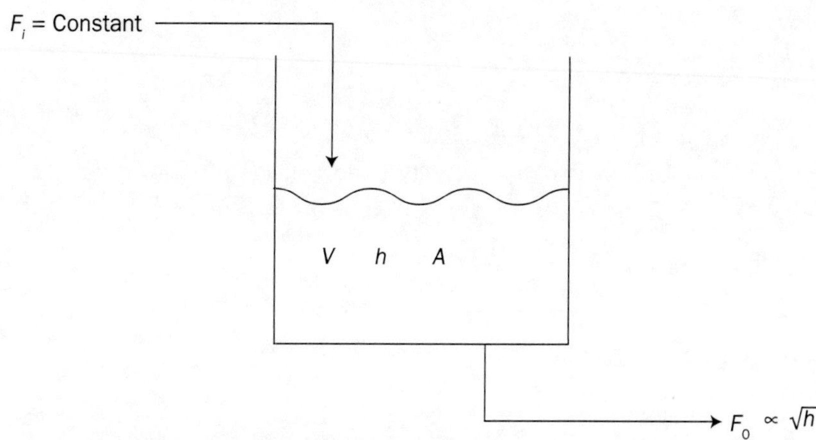

FIGURE **4.23** A liquid tank

4.29 Figure 4.24 shows a batch heating tank system, in which, the liquid is heated up by saturated steam. The steam (at T_{st}) flows through the coil and releases Q amount of heat in the tank. Assuming constant liquid volume, V in the tank (i.e., no phase change), derive the tank model as:

$$V\rho\, C_p \frac{dT}{dt} = Q = UA_t(T_{st} - T)$$

where,

T = tank temperature
U = overall heat transfer coefficient
A_t = heat transfer area
ρ = liquid density
C_p = liquid heat capacity

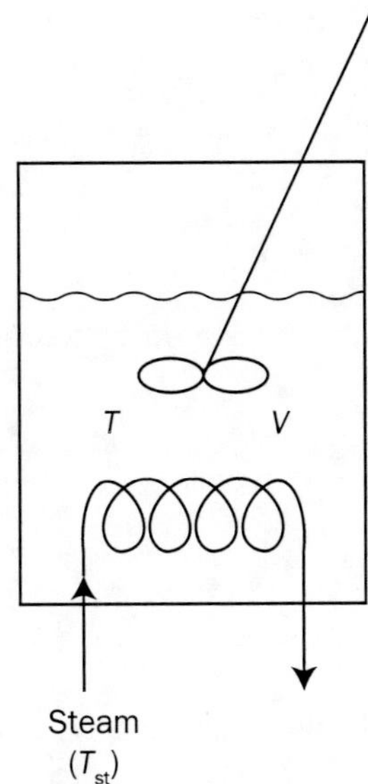

FIGURE **4.24** A batch heating tank system

(a) Develop the computer-assisted simulation steps in sequence by the use of the RKG4 method.

(b) Tabulate the numerical solutions up to $t = 0.01$ with the following known information:

$$\frac{UA_t}{V\rho C_p} = 10 \text{ min}^{-1}$$

$$T(0) = 25°C$$

$$T_{st} = 100°C$$

$$h = 0.001 \text{ min}$$

(c) Compare the numerical solutions obtained in part (b) with the analytical one:

$$\ln\frac{T_{st} - T_0}{T_{st} - T} = \frac{UA_t}{V\rho C_p}t$$

4.30 Simulate the model of the continuous heating tank system (see Case Study 4.1) by the use of the:

i) RKF45 method

ii) Runge–Kutta–Gill method

Compare the numerical solutions and make comment on your findings.

4.31 Consider a first-order process represented by:

$$\frac{dx}{dt} = -5x\eta$$

i) Solve this equation using the fourth-order Runge–Kutta method.
Given: $x_0 = 3.0$, $\eta = 0.15$ and $h = 1.0$.

ii) Find the stability condition for explicit and implicit Euler, and RK2 method.

4.32 Consider the following system,

$$u' = -\lambda\, u; \qquad u(t = 0) = u_0 \text{ and } \lambda > 0$$

Find the numerical stability condition in terms of h $(= \Delta t)$ and λ for the following methods:

i) explicit Euler
ii) Heun
iii) second-order Runge–Kutta
iv) third-order Runge–Kutta
v) fourth-order Runge–Kutta
vi) Adams–Bashforth–Moulton

4.33 Solve the following systems using the predictor–corrector (Adams–Bashforth–Moulton) approach:

i) $$\frac{dx}{dt} = -xe^{t}$$

with $x(0) = 2$.

ii) $$\frac{dx}{dt} = -x + 3t$$

with $x(0) = -1.5$.

4.34 Compare the analytical and numerical (explicit and implicit Euler) solutions of the model of a batch reactor, in which, a second-order reaction occurs. An initial reactant concentration of $C_0 = 2$ mol/L and a reaction rate constant of $k = 1.5$ L/mol.sec are given for the following model:

$$\frac{dC}{dt} = -kC^{2}$$

Tabulate the results for three step sizes of 0.001, 0.01 and 1 sec.

4.35 Consider an isothermal tubular reactor in which benzene hydrogenation takes place on a supported Ni/kieselguhr catalyst. The one-dimensional steady state material balance for the reactor (Davis, 1984) yields:

$$\frac{dy}{dx} = -\rho_B \theta P_{H2} k_0 K_0 T \exp\left(-\frac{Q+E}{RT}\right) y$$

where, y denotes the dimensionless concentration of benzene $(= C_B / C_{B0})$ and x the dimensionless axial coordinate $(= z/L)$, C_{B0} the feed concentration of benzene (mol/cm^3), z the axial reactor coordinate (cm) and L the reactor length. In addition, the following information is provided:

ρ_B = density of the reactor bed, 1.2 gm cat/cm^3

θ = contact time, 0.226 sec

P_{H_2} = hydrogen partial pressure, 685 torr

k_0 = 4.22 mol/gm cat.sec.torr

K_0 = 2.63×10^{-6} cm^3/mol.K

T = absolute temperature, 423 K

$-(Q+E) = 2700$ cal/mol

R = universal gas constant, 1.987 cal/mol.K

List the analytical solutions comparing with the numerical solutions for the Euler and RK4 methods ($\Delta x = 0.05$). Note that for the example reactor, $y = 1$ at $x = 0$.

4.36 In an ideal plug flow reactor (PFR), the following reaction takes place:

$A \rightarrow \text{product}$

The steady state model is derived as:

$$\upsilon \frac{dC_A}{dx} + (-r_A) = 0$$

where, the reaction rate is given by the Langmuir–Hinshelwood model as:

$$-r_A = \frac{k\, C_A}{\sqrt{1 + k_r C_A^2}}$$

Given:

$k = 2.5 \text{ sec}^{-1}$

$k_r = 0.8 \text{ m}^6/\text{mol}^2$

Initial concentration, $C_{A0} = 1 \text{ mol/m}^3$

Axial velocity, $\upsilon = 0.4 \text{ m/sec}$

Simulate the model using the Heun method and produce the concentration, C_A profile along the length of 1 m long PFR.

4.37 Simplify the above PFR model with:

$-r_A = k\, C_A$

Given:

$k = 1.2 \text{ sec}^{-1}$

Initial concentration, $C_{A0} = 1 \text{ mol/m}^3$

Axial velocity, $\upsilon = 1 \text{ m/sec}$

Simulate the model using the RKG4 method and produce the concentration, C_A profile along the length of 5 m long PFR.

4.38 An object having an absolute temperature of T is getting cooled in a medium that is at a constant absolute temperature of T_0. This heat transfer can be described by the Stefan's law of radiation:

$$\frac{dT}{dt} = k(T^4 - T_0^4)$$

where,

T_0 = ambient temperature (= 298 K)

k = constant (= – 0.2)

Solve this problem by the use of the RK4 with T(0) = 373 K and $\Delta t = h = 0.5$, and find $T(10)$.

4.39 Solve the above problem with the RKF45 method when the heat transfer is described by the Newton's law of cooling/warming:

$$\frac{dT}{dt} = k(T - T_0)$$

4.40 An ODE-IVP system is given as:

$$\frac{dx}{dt} = -25x$$

with $x(0) = 1$. Compare the numerical solutions of:

i) explicit Euler with $h = 0.4$

ii) explicit Euler with $h = 0.1$

iii) RK4 with $h = 0.1$

iv) Adams–Bashforth–Moulton with $h = 0.1$

4.41 Consider a system represented by:

$$\frac{d^2y}{dx^2}=\frac{1}{3+x}\frac{dy}{dx}-(3+x)y$$

Transform this ODE into a set of two first-order ODEs. Using the following initial values:

$$y(-1)=0$$
$$y'(-1)=1$$

find $y(1)$ when $\Delta x=0.1$.

4.42 The van der Pol equation has the following form:

$$\frac{d^2x}{dt^2}+\mu(x^2-1)\frac{dx}{dt}+x=0$$

where the parameter $\mu=1.1$. Solve this equation using the RK2 method. Do you have any specific observation?

4.43 The equation of motion is given as:

$$\frac{d^2x}{dt^2}+64x=0$$

Initial conditions include:

$$x(0)=\frac{2}{3}$$
$$x'(0)=-\frac{4}{3}$$

Transform the system equation into a set of two ODEs and then solve the resulting IVP problem using the explicit Euler method.

4.44 An ODE-IVP system is given as:

$$\frac{d^2y}{dt^2}=5y+10t(1-t)$$

Initial conditions include:

$$y(0)=0$$
$$y'(0)=3.75$$

Find $y(9)$ when $h=3$.

4.45 Consider a forced spring-mass system shown in Figure 4.25. It is modeled by making the momentum balance based on the Newton's second law of motion as:

$$m\frac{d^2x}{dt^2}=F_d+F_s-mg-F_e$$

Substituting all terms and rearranging,

$$\frac{d^2x}{dt^2}+\frac{c}{m}\frac{dx}{dt}+\frac{k}{m}x=-g-\frac{F}{m}\cos\omega t$$

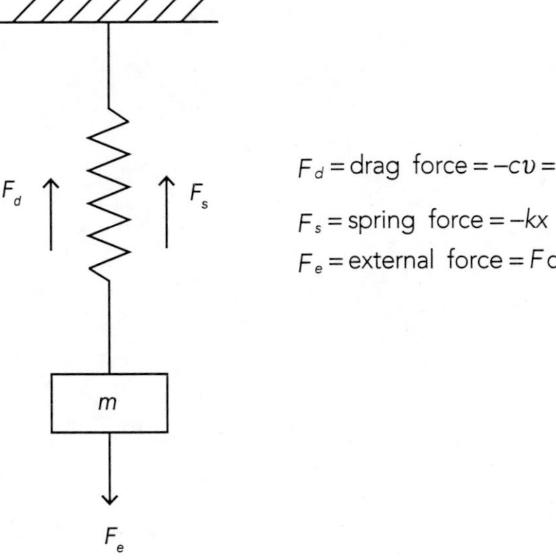

Figure 4.25 A forced spring-mass system

where,

x = length of extension or compression
c = drag coefficient = 0.01
m = mass hanged over there = 1.1
k = spring constant = 1.1
g = acceleration due to gravity = 9.81
F = amplitude of the external force, F_e = 1
ω = external frequency = 0.2

The initial conditions include:

$x(0) = 1$
$x'(0) = 0$

Solve this problem using a suitable numerical method and plot the results in terms of $x(t)$ vs. t.

4.46 Consider a pendulum having mass m and length l shown in Figure 4.26. Making momentum balance (Newton's second law of motion):

$$ma = -mg\sin\theta + F_d \qquad \text{(tangential direction)}$$

In which,

θ = angular displacement of the pendulum (i.e., the angle that the pendulum makes with the vertical)

a = tangential acceleration due to velocity change

$$= l\frac{d^2\theta}{dt^2}$$

F_d = drag force

$$= -cv = -cl\frac{d\theta}{dt}$$

[angular velocity (v) is converted to linear velocity $\left(l\dfrac{d\theta}{dt}\right)$]

c = drag coefficient

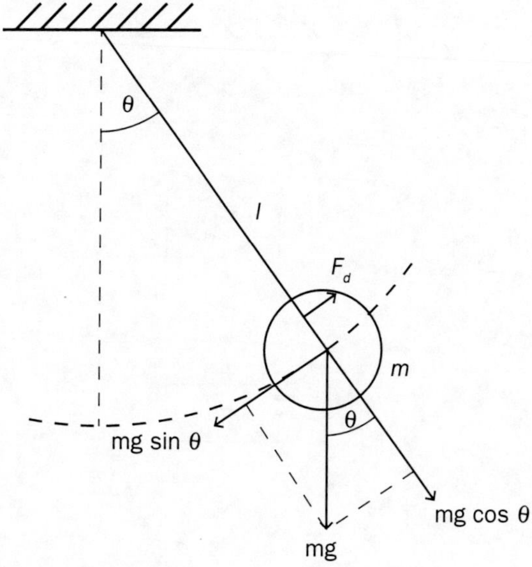

FIGURE 4.26 A pendulum with air drag

Substituting and rearranging, we get:

$$\frac{d^2\theta}{dt^2}+\frac{c}{m}\frac{d\theta}{dt}+\frac{g}{l}\sin\theta=0$$

The initial conditions include:

$$\theta(0)=\frac{\pi}{2}$$

$$\dot{\theta}(0)=0$$

Given information:

$c = 0.9$ kg/sec

$m = 1$ kg

$l = 1$ m

i) Convert the given governing equation into a system of first-order ODEs.
ii) Solve this problem using the RK4 method and plot the results in terms of $\theta(t)$ vs. t for $0 \le t \le 50$ sec.

4.47 Using an IVP solver of your choice, solve the following equations:

$$\frac{dx}{dt}=y$$

$$\frac{dy}{dt}=t+x-x^2$$

with $x(0) = -0.5$ and $y(0) = 1.0$.

4.48 The Predator–Prey model has the following form:

$$\frac{dx_1}{dt} = 1.2x_1 - 0.6x_1x_2$$

$$\frac{dx_2}{dt} = -0.8x_2 + 0.3x_1x_2$$

with $x_1(0) = 1.5$ and $x_2(0) = 1$. Compare the numerical results between the explicit Euler and RK2 method for $h = 0.05$.

4.49 Consider a coupled system:

$$\frac{dy_1}{dt} = y_2$$

$$\frac{dy_2}{dt} = -\frac{1}{y_1} y_2^2$$

with $y_1(0) = \sqrt{2}$ and $y_2(0) = 0.35$. Find $y_1(t = 2)$ when $h = 0.1$.

4.50 Solve the following IVP system by the use of the RK4 method:

$$\frac{d\theta_1}{dt} = \theta_2$$

$$\frac{d\theta_2}{dt} = -\frac{1}{1+\theta_1}\theta_2^2$$

Initial conditions:

$$\theta_1(t = 0) = 0$$

$$\theta_2(t = 0) = 1.5$$

Find $\theta_1(t = 1)$ when $h = 0.1$.

4.51 Consider the following IVP system (Burden and Faires, 1985):

$$\frac{dx_1}{dt} = 9x_1 + 24x_2 + 5\cos t - \frac{1}{3}\sin t$$

$$\frac{dx_2}{dt} = -24x_1 - 51x_2 - 9\cos t + \frac{1}{3}\sin t$$

with $x_1(0) = \frac{4}{3}$ and $x_2(0) = \frac{2}{3}$. Solve the problem using the RK4 method with $h = 0.02$.

4.52 Solve the following system using the RK3 method:

$$\frac{dx_1}{dt} - 5x_1 + 2x_2 = 5e^t$$

$$\frac{dx_2}{dt} - 2x_1 - x_2 = 8e^t$$

with $x_1(0) = -1$ and $x_2(0) = 1$.

4.53 The van der Pol model used to describe an electronic circuit (in the days of vacuum tubes) has the following form:

$$\frac{d^2x}{dt^2}+\mu(x^2-1)\frac{dx}{dt}+x=0$$

with

$$x(0)=x'(0)=1$$
$$\mu=1$$

Analyze this stiff problem with solving the IVP system from t = 0 to 25.

4.54 There is a coupled system:

$$\frac{dy_1}{dt}=-100y_1$$
$$\frac{dy_2}{dt}=2y_1-y_2$$

with $Y_0=[2 \quad 1]^T$. Using the explicit Euler method with h = 0.05, solve this stiff system and comment on the results.

4.55 Analyze the stiffness of a system described by:

$$\frac{dy_1}{dt}=-5y_1+3y_2$$
$$\frac{dy_2}{dt}=100y_1-301y_2$$

with $Y_0=[50 \quad 80]^T$. Use the explicit and implicit Euler method with h = 0.01. Suggest the remedy for explicit Euler.

4.56 Consider a coupled system:

$$\frac{dx_1}{dt}=-80.6\,x_1+119.4x_2$$
$$\frac{dx_2}{dt}=79.6\,x_1-120.4x_2$$

with $X_0=[0 \quad 1]^T$. Solve this stiff system and comment on the results.

4.57 Consider the following system (Kharab and Guenther, 2011):

$$\frac{dx_1}{dt}=131\,x_1-931x_2$$
$$\frac{dx_2}{dt}=133\,x_1-933x_2$$

with $X_0=[6 \quad -6]^T$. Using the explicit and implicit Euler method, solve it and analyze this stiff problem. Compare the numerical solutions with analytical ones.

4.58 Figure 4.27 shows the schematic of two interacting tanks in series. The outlet flow rates are described by (Bequette, 1998):

$$F_1\,\alpha\,\sqrt{h_1-h_2}$$
$$F_2\,\alpha\,\sqrt{h_2}$$

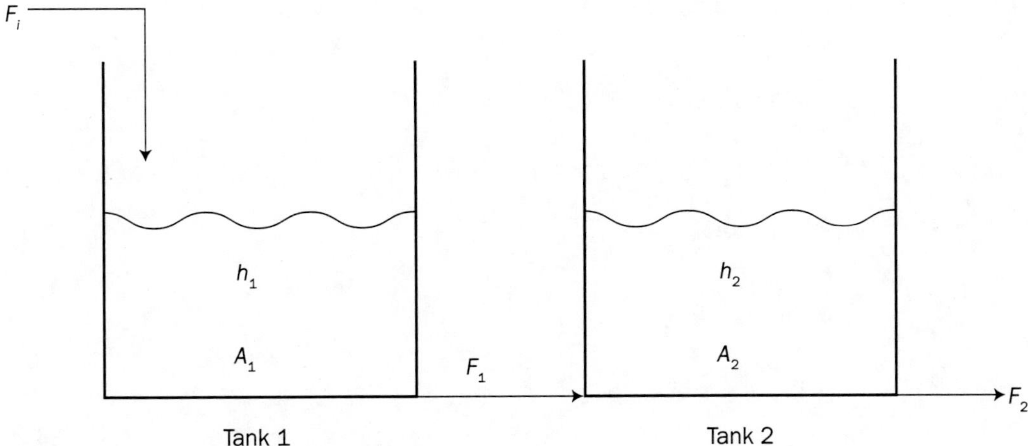

FIGURE 4.27 Two interacting tanks in series

With this, the mass balance yields:

Tank 1

$$\frac{dh_1}{dt} = \frac{F_i}{A_1} - \frac{\beta_1}{A_1}\sqrt{h_1 - h_2}$$

Tank 2

$$\frac{dh_2}{dt} = \frac{\beta_1}{A_2}\sqrt{h_1 - h_2} - \frac{\beta_2}{A_2}\sqrt{h_2}$$

Given information:

Inlet flow rate, $F_i = 5\ \text{m}^3/\text{min}$

Cross sectional area, $A_1 = 5\ \text{m}^2$

$A_2 = 10\ \text{m}^2$

Proportionality factor, $\beta_1 = 2.5\,\frac{\text{m}^{2.5}}{\text{min}}$

$\beta_2 = \frac{5}{\sqrt{6}}\,\frac{\text{m}^{2.5}}{\text{min}}$

Solve this ODE-IVP system with $h_1(0) = h_2(0) = 8$ m with $\Delta t = 0.1$ min. Produce the profile of liquid height in two interacting tanks.

4.59 Analyze the stiffness of a system described by:

$$\frac{dC_1}{dt} = -0.013C_1 - 1000C_1C_3$$

$$\frac{dC_2}{dt} = -2500C_2C_3$$

$$\frac{dC_3}{dt} = -0.013C_1 - 1000C_1C_3 - 2500C_2C_3$$

with $C_0 = [1 \quad 1 \quad 0]^T$. Using the explicit Euler method with $h = 0.005$, solve the system from $t = 0$ to 40.

4.60 In a continuous stirred tank reactor (CSTR), the following reaction takes place:

$$A \rightarrow B \rightarrow C$$

The model equations are given as:

$$\frac{dC_A}{dt} = \frac{F}{V}(C_{Ai} - C_A) - k_1 C_A$$

$$\frac{dC_B}{dt} = -\frac{F}{V}C_B + k_1 C_A - k_2 C_B$$

Given:

$$C_{Ai} = 2$$

$$\frac{F}{V} = 1$$

$$k_1 = 1$$

$$k_2 = 100$$

$$C_A(0) = 1$$

$$C_B(0) = 0.001$$

Analyze this stiff problem.

4.61 An isothermal plug flow reactor (Jothi and Varma, 1981) is modeled in the following form of dimensionless equations:

$$\frac{dx_1}{dl} = 0.66(1 - x_1)(1 - x_2)$$

$$\frac{dx_2}{dl} = 2\ (1 - x_1)(1 - x_2)$$

$$\frac{dx_3}{dl} = x_3(x_1 - 1)$$

with $X_0 = [0 \quad 0 \quad 1]^T$. Here x_1, x_2 and x_3 are the conversions of three reacting species (CO, NO and O_2), and l is the dimensionless reactor length. Using the RK2 method, integrate these coupled equations and produce a plot of x versus l.

4.62 The Lorenz equations include:

$$\frac{dx_1}{dt} = -\sigma(x_1 - x_2)$$

$$\frac{dx_2}{dt} = r\, x_1 - x_2 - x_1 x_3$$

$$\frac{dx_3}{dt} = x_1 x_2 - b x_3$$

where the parameters, $\sigma = 10$, $r = 28$ and $b = 8/3$. Initial conditions include: $x_1(0) = x_2(0) = 5$ and $x_3(0) = 4$. Use the Heun method, and produce numerical solutions and x versus t plot for $h = 0.01$.

4.63 Simulate the model of the batch bioreactor (see Case Study 4.5) by the use of:

i) RK4 method

ii) RKF45 method

Compare the numerical solutions.

4.64 Solve the dynamic model of a continuous flow bioreactor given in Problem 3.61 (Chapter 3) by using the RK4 method and following initial conditions:

$x_0 = 5$ g/L

$s_0 = 4$ g/L

$p_0 = 20$ g/L

4.65 Washing machine causes indoor air pollution by releasing volatile organic compounds (VOCs). It is schematically shown in Figure 4.28. In this machine, hypochlorites are typically used as the bleaching agent. It supplies free chlorine that further reacts with the organics of municipal water and forms chloroform (i.e., VOC), which gets released into the indoor through the circulated air stream.

The mathematical model (Shepherd and Coral, 1996) is developed as:

$$V_1\frac{dC_w}{dt} = V_1 r - V_1 K_L a\left(C_w - \frac{C_g}{H}\right)$$

$$V_2\frac{dC_g}{dt} = V_1 K_L a\left(C_w - \frac{C_g}{H}\right) - Q_g C_g,$$

with the rate of VOC generation:

$$r = kC = kC_0 e^{-kt}$$

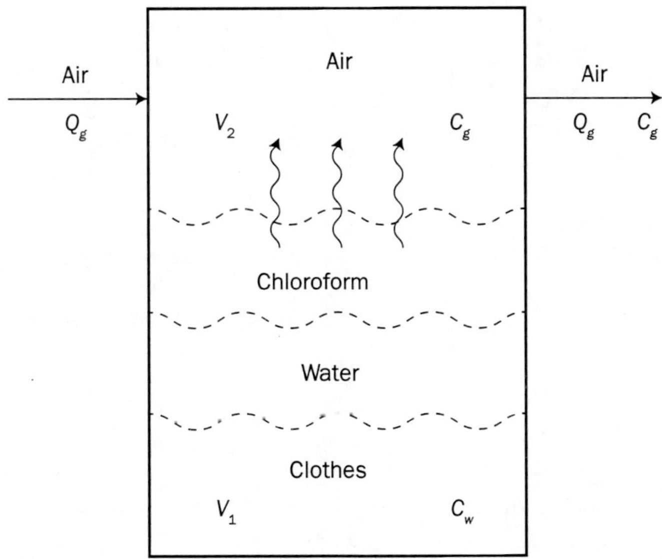

FIGURE **4.28** Schematic of a washing machine

Relevant information is given in Table 4.32.

Table 4.32 Given information

Water and clothes volume	$V_1 = 75$ L
Air space volume (above water level)	$V_2 = 50$ L
Initial concentration of residual chlorine	$C_0 = 300$ mg/L
Reaction rate constant for chloroform formation	$k = 0.1$ min^{-1}
Overall volumetric mass transfer coefficient	$K_La = 0.02$ min^{-1}
Henry's constant	$H = 0.2$
Air circulation rate	$Q_g = 50$ L/min
Initial conditions (at $t = 0$)	
Chloroform (or VOC) concentration in water	$C_{w0} = 0.01$ mg/L
Chloroform concentration in the headspace	$C_{g0} = 0.0$ mg/L

Simulate the coupled ordinary differential equations system using the RK3 method to find C_w and C_g at time, $t = 10$ min.

4.66 In a batch reactor, the following reaction takes place:

$$A \xrightarrow{k_1} B \xrightarrow{k_2} C$$

The mole balance for the three species yields:

Species A

$$\frac{dC_A}{dt} = -k_1 C_A$$

Species B

$$\frac{dC_B}{dt} = k_1 C_A - k_2 C_B$$

Species C

$$\frac{dC_C}{dt} = k_2 C_B$$

Given:

rate constants, $k_1 = 1.2$ sec^{-1}
$k_2 = 1$ sec^{-1}

initial conditions, $C_A(0) = 1$ mol/m^3
$C_B(0) = C_C(0) = 0$ mol/m^3

Simulate this ODE-IVP system using the RK4 method and produce the concentration profile (C versus t) for all three species.

4.67 Solve the following IVP system by the use of the RK4 method:

$$\frac{d\theta_1}{dt} = \theta_2$$

$$\frac{d\theta_2}{dt} = -\frac{1}{1+\theta_1}\theta_2^2$$

$$\frac{d\theta_3}{dt} = \theta_4$$

$$\frac{d\theta_4}{dt} = \left(\frac{\theta_2}{1+\theta_1}\right)^2 \theta_3 - \frac{2\theta_2}{1+\theta_1}\theta_4$$

Initial conditions

$$\theta_1(t=0) = 0$$

$$\theta_2(t=0) = 1.5125$$

$$\theta_3(t=0) = 0$$

$$\theta_4(t=0) = 1$$

Find $\theta_1(t=1)$ when $h = 0.1$.

4.68 In a semi-batch reactor, the synthesis of hexyl monoester of maleic acid takes place (Chang and Hseih, 1995):

maleic anhydride + hexanol → hexyl monoester of maleic acid + water
(A) (B) (C) (D)

The dynamic model is given as:

Total mass balance

$$\frac{dV}{dt} = U_1$$

Component A balance

$$\frac{dC_A}{dt} = -\frac{C_A U_1}{V} - kC_A C_B$$

Component B balance

$$\frac{dC_B}{dt} = \frac{(C_{BL} - C_B)U_1}{V} - kC_A C_B$$

Energy balance

$$\frac{dT}{dt} = \left(\frac{-\Delta H_r}{\rho\, C_p}\right) kC_A C_B - (T - 327)\frac{U_1}{V} - U_2(T - 327)$$

The operating parameters are given in Table 4.33.

Table 4.33 Parameters and operating conditions

$C_{BL} = 9.7$ kmol/m^3

$k = 1.37 \times 10^{12} \exp\left(-\frac{12628}{T}\right)$ m^3/kmol.sec

$(-\Delta H_r) = 33.5 \times 10^3$ kJ/kmol

$\rho C_p = 1980$ kJ/m^3.K

$U_1 = 0.0013$ m^3/sec

$U_2 = 0.000126$ sec^{-1}

Initial conditions

$C_A(0) = 10.1$ kmol/m^3

$C_B(0) = 0.0$ kmol/m^3

$T(0) = 328$ K

$V(0) = 2.2$ m^3

Solve the coupled ODEs using the Runge–Kutta–Gill method for the final volume (V) of 4.7 m^3.

4.69 The following model (Kurtz et al., 2000) is developed for a competitive mixed-culture bioreactor combined with an adhesion column that separates the cell populations continuously:

$$V\frac{dx}{dt} = -(F_i + F_w)x + V\mu_1 x$$

$$V\frac{dy}{dt} = -(F_i + F_w + F_s)y + V\mu_2 y$$

$$V\frac{ds}{dt} = F_i(s_i - s) + F_w(s_w - s) - \frac{\mu_1}{Y_1}V\,x - \frac{\mu_2}{Y_2}V\,y$$

with (Monod kinetics)

$$\mu_1 = \mu_{1\max}\frac{s}{k_1 + s}$$

$$\mu_2 = \mu_{2\max}\frac{s}{k_2 + s}$$

Simulate this dynamic model using the RKG4 method and the information given in Table 4.34.

Table 4.34 Given information for the bioreactor

Maximum specific growth rates	$\mu_{1\max} = 0.75$ hr^{-1}
	$\mu_{2\max} = 0.525$ hr^{-1}
Saturation constants	$k_1 = 6.4 \times 10^{-2}$ g/L
	$k_2 = 7.76 \times 10^{-3}$ g/L
Yield coefficients	$Y_1 = Y_2 = 0.44$
Reactor volume	$V = 350$ ml
Substrate concentration in feed stream	$s_i = 0.5$ g/L

Contd

Contd

Substrate concentration in washing stream	$s_w = 0.001$ g/L
Input flow rate of feed stream	$F_i = 56.8$ ml/hr
Input flow rate of washing stream	$F_w = 40$ ml/hr
Input flow rate of recycle stream	$F_s = 42.9$ ml/hr
Initial conditions	
Concentration of slower growing cells	$x_0 = 0.04$ g/L
Concentration of faster growing cells	$y_0 = 0.04$ g/L
Concentration of substrate	$s_0 = 0.05$ g/L

4.70 In a batch reactor, the following reactions take place:

$$A + B \xrightarrow{k_1} C$$

$$A + C \xrightarrow{k_2} D$$

The mole balance for the four species yields:

Species A

$$\frac{dC_A}{dt} = -k_1 C_A C_B - k_2 C_A C_C$$

Species B

$$\frac{dC_B}{dt} = -k_1 C_A C_B$$

Species C

$$\frac{dC_C}{dt} = k_1 C_A C_B - k_2 C_A C_C$$

Species D

$$\frac{dC_D}{dt} = k_2 C_A C_C$$

Given:

rate constants, $k_1 = 1.2$ m^3/mol.sec
$k_2 = 1$ m^3/mol.sec

initial conditions, $C_A(0) = C_B(0) = 1$ mol/m^3
$C_C(0) = C_D(0) = 0$ mol/m^3

Simulate this IVP problem using the RK2 method and produce the concentration profile (C versus t) for all four species.

4.71 In a continuous stirred tank reactor (CSTR) shown in Figure 4.29, an isothermal free-radical polymerization of methyl methacrylate occurs in presence of toluene as a solvent and azo-bis-isobutyronitrile as an initiator (Maner and Doyle III, 1997). The dynamic model has the following form:

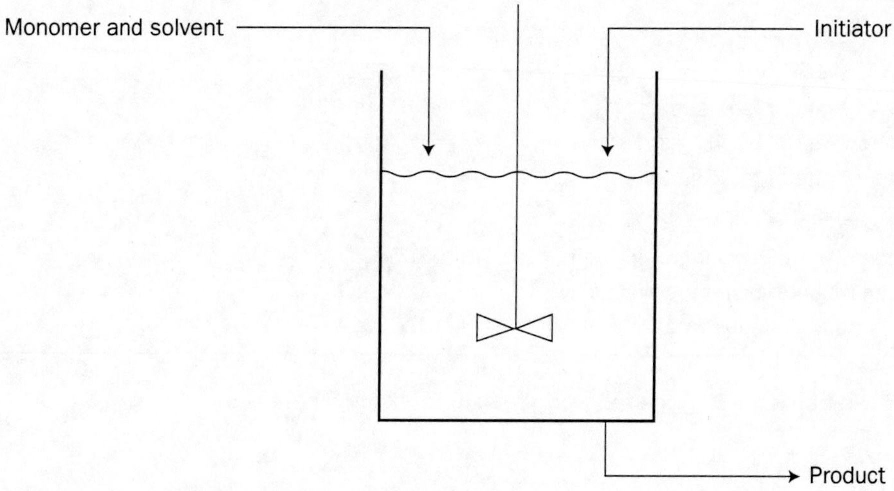

FIGURE 4.29 A polymerization reactor (CSTR)

$$\frac{dC_m}{dt} = \frac{F}{V}(C_{mi} - C_m) - (k_p + k_{fm})C_m p_0$$
$$\frac{dC_I}{dt} = \frac{F_I C_{Ii} - FC_I}{V} - k_I C_I$$
$$\frac{dD_0}{dt} = -\frac{FD_0}{V} + k_{fm} C_m p_0 + (0.5k_c + k_d)p_0^2$$
$$\frac{dD_1}{dt} = -\frac{FD_1}{V} + M_m(k_p + k_{fm})C_m p_0$$

where,

$$p_0 = \left(\frac{2f^* k_I C_I}{k_c + k_d}\right)^{0.5}$$

Solve this CSTR model using the RK4 method and the parameter values reported in Table 4.35.

TABLE 4.35 Model parameters and initial conditions

$F = 1.0$ m³/hr
$F_I = 0.016783$ m³/hr
$V = 0.1$ m³
$C_{Ii} = 8.0$ kmol/m³
$M_m = 100.12$ kg/kmol
$C_{mi} = 6.0$ kmol/m³
$k_c = 1.3281 \times 10^{10}$ m³/kmol.hr
$k_d = 1.0930 \times 10^{11}$ m³/kmol.hr

Contd

Contd

$k_I = 1.0225 \times 10^{-1}$ hr^{-1}

$k_p = 2.4952 \times 10^{6}$ m^3/kmol.hr

$k_{fm} = 2.4522 \times 10^{3}$ m^3/kmol.hr

$f^* = 0.58$

Initial conditions

$C_{m0} = 5.5$ kmol/m^3

$C_{I0} = 0.13$ kmol/m^3

$D_{00} = 0.002$ kmol/m^3

$D_{10} = 49.38$ kg/m^3

4.72 The model equations involved in ethane cracking include:

$$\frac{dC_1}{dt} = -k_1C_1 - k_2C_1C_2 - k_4C_1C_6$$

$$\frac{dC_2}{dt} = 2k_1C_1 - k_2C_1C_2$$

$$\frac{dC_3}{dt} = k_2C_1C_2$$

$$\frac{dC_4}{dt} = k_2C_1C_2 - k_3C_4 + k_4C_1C_6 - k_5C_4^2$$

$$\frac{dC_5}{dt} = k_3C_4$$

$$\frac{dC_6}{dt} = k_3C_4 - k_4C_1C_6$$

$$\frac{dC_7}{dt} = k_4C_1C_6$$

$$\frac{dC_8}{dt} = \frac{1}{2}k_5C_4^2$$

with

$$k_1 = 12\exp\left[\frac{85000}{R}\left(\frac{1}{1250} - \frac{1}{T}\right)\right]\text{sec}^{-1}$$

$$k_2 = 8.5\times10^6\exp\left[\frac{13000}{R}\left(\frac{1}{1250} - \frac{1}{T}\right)\right]\text{dm}^3/\text{mol.sec}$$

$$k_3 = 3.2\times10^6\exp\left[\frac{40000}{R}\left(\frac{1}{1250} - \frac{1}{T}\right)\right]\text{sec}^{-1}$$

$$k_4 = 2.5\times10^9 \exp\left[\frac{9500}{R}\left(\frac{1}{1250}-\frac{1}{T}\right)\right] \text{dm}^3/\text{mol.sec}$$

$$k_5 = 4.0\times10^9 \ \ \text{dm}^3/\text{mol.sec}$$

Initial conditions

$C_{10} = 0.1$ mol/dm^3

$C_{i0} = 0.0$ mol/dm^3 $\qquad i = 2, 3, \ldots, 8$

Solving the model for $T = 1000$ K, produce the concentration profile for $t = 12$ sec.

References

Bequette, B. W. (1998). *Process Dynamics, Modeling, Analysis, and Simulation*, 1st ed., Upper Saddle River, New Jersey: Prentice-Hall.

Burden, R. L. and Faires, J. D. (1985). *Numerical Analysis*, 3rd ed., Boston, MA: Prindle, Weber and Schmidt.

Chang, J. -S. and Hseih, W. -Y. (1995). Optimization and control of semibatch reactors. *Ind. Eng. Chem. Res.*, **34**, 545–556.

Davis, M. E. (1984). *Numerical Methods and Modeling for Chemical Engineers*, 2nd ed., New York: John Wiley & Sons.

Fehlberg, E. (1964). *Z. Angewandte Mathem. Mech.*, **44**, 17.

Fehlberg, E. (1969). Low-order classical Runge-Kutta formulas with stepsize control and their application to some heat transfer problems. *NASA Technical Report*, NASA TR R-315.

Fogler, H. S. (2005). *Elements of Chemical Reaction Engineering*, 3rd ed., New Delhi: Prentice-Hall.

Gill, S. (1951). A process for the step-by-step integration of differential equations in an automatic digital computing machine. *Mathematical Proceedings of the Cambridge Philosophical Society*, **47**, 96–108.

Jana, A. K. (2018). *Chemical Process Modeling and Computer Simulation*, 3rd ed., New Delhi: PHI Learning.

Jothi, N. and Varma, A. (1981). Simultaneous reactions of CO, NO and O_2 in a tubular reactor. *AIChE J.*, **27**, 848–851.

Kharab, A. and Guenther, R. B. (2019). *An Introduction to Numerical Methods: A MATLAB Approach*, 4th ed., New York: CRC Press.

Kurtz, M. J., Henson, M. A., and Hjorts, M. A. (2000). Nonlinear control of competitive mixed-culture bioreactors via specific cell adhesion. *The Canadian Journal of Chemical Engineering*, **78**, 237–247.

Lapidus, L. and Seinfeld, J. H. (1971). *Numerical Solution of Ordinary Differential Equations*, New York: Academic Press.

Luyben, W. L. (1990). *Process Modeling, Simulation, and Control for Chemical Engineers*, 2nd ed., Singapore: McGraw-Hill Book Company.

Maner, B. R. and Doyle III, F. J. (1997). Polymerization reactor control using autoregressive-plus Voltera-based MPC. *AIChE J.*, **43**, 1763–1784.

Seader, J. D. and Henley, E. J. (2001). *Separation Process Principles*, 2nd ed., New Delhi: John Wiley & Sons Inc.

Shepherd, J. L. and Coral, R. L. (1996). Chloroform in indoor air and wastewater: the role of residential washing machines. *J. Air Waste Manage. Assoc.*, **46**, 631–642.

5

Ordinary Differential Equation: Boundary Value Problem (ODE-BVP)

"Science is a differential equation. Religion is a boundary condition." – Alan Turing

Key Learning Objectives

- Knowing various types of boundary conditions
- Learning numerical methods to solve boundary value problems
- Applying numerical methods on linear and non-linear systems
- Modeling case studies for their numerical simulation
- Analyzing simulated process behaviors at various conditions

5.1 Introduction

We have seen in the last chapter that for an initial value problem (IVP), the initial conditions need to be specified at the starting (i.e., at time $t = 0$). On the other hand, for a boundary value problem (BVP), one needs to specify the boundary conditions that hold for all times. Basically, the conditions that one imposes on the boundary of the domain are termed as the *boundary conditions* (BCs) and the resulting problems are called the *boundary value problems* (BVPs). For instance, for a second-order differential equation, one needs to specify two boundary conditions.

There are typically three common types of boundary conditions and they are highlighted in the following section.

A. Dirichlet BC

The Dirichlet boundary condition[1] is perhaps the most common type of constraint that is concerned with the value of the dependent variable, y on the boundary. For example,

$$y(x_0)=\beta \tag{5.1}$$

1 Named after the German mathematician Peter Gustav Lejeune Dirichlet (1805–1859).

in which, β is a constant. Note that this BC is specific to the left boundary (i.e., at x_0) and one can have a similar expression for the right boundary (i.e., at x_M) as $y(x_M)$ = constant.

In physical systems, this BC is typically used in fixing the state variables, namely temperature, composition and so on, at a boundary.

B. Neumann BC

The Neumann boundary condition[2] is concerned with the derivative of y on the boundary. For example,

$$\left.\frac{dy}{dx}\right|_{x_M} = \beta \tag{5.2}$$

Similar boundary condition one can have for the left boundary (i.e., at x_0).

In physical systems, this BC is used in fixing the mass or energy flux. Note that for no-flux condition (when, say the boundary is impermeable or perfectly insulated), $\beta = 0$.

C. Robin BC

The Robin boundary condition[3] involves a combination of the above two (i.e., y and its derivative on the boundary). Thus, it is also called as *mixed BC*. For example,

$$\beta_1 \left.\frac{dy}{dx}\right|_{x_M} + \beta_2 y(x_M) = \beta_3 \tag{5.3}$$

Similar boundary condition one can have for the left boundary (i.e., at x_0).

This BC corresponds to a heat or mass transport correlation for the flux at the boundary. Danckwerts BC is a popular example of this category.

5.2 Numerical Methods: ODE-BVP Systems

There are a wide varieties of numerical methods employed to solve the boundary value problems. They typically include:

- Finite difference method (FDM)
- Initial value method (Shooting method)
- Finite element method (FEM)
- Orthogonal collocation

Here, we will learn the FDM and Shooting method with their applications to a couple of relevant scientific and engineering problems.

2 Named after the German mathematician Carl Gottfried Neumann (1832–1925).

3 Named after the French mathematician Victor Gustave Robin (1855–1897).

5.3 Finite Difference Method

Any approximation of a derivative made in terms of values at a discrete set of points is referred to as *finite difference approximation*. The approach used for this approximation is called the *finite difference method* (FDM) and it was first proposed by A. Thom in 1920s under the name of *the method of squares*. The target was to solve the non-linear hydrodynamic equations. This FDM is the simplest approach and it is widely used in various applications.

5.3.1 Discretization in Space

As stated, the boundary value problems are more concerned with *space*. In finite difference method, thus we first proceed to discretize the spatial domain. For this, let us divide the domain of x ($0 \le x \le 1$ or $x \in [0,\ 1]$) into $M + 1$ equispaced or equidistant *grid points* or *nodes*, namely $x_0,\ x_1,\ x_2,\ .....,\ x_M$, such that

$$x_m = x_0 + m\,k$$

for $m = 0,\ 1,\ 2, ...,\ M$, where the *spatial increment* (also called *mesh size*):

$$k = \Delta x = \frac{x_M - x_0}{M}$$

This is illustrated in Figure 5.1. It should be noted that $m = 1,\ 2, ...,\ M-1$ correspond to the *interior* or *internal nodes*, whereas $m = 0$ and M indicate the *boundary nodes*.

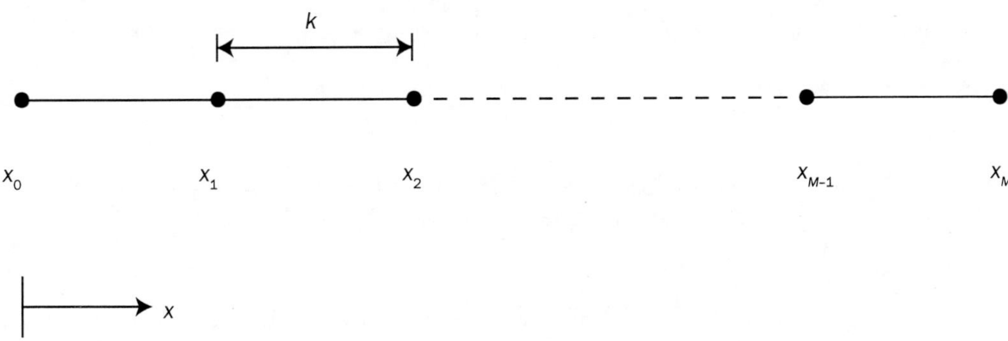

Figure 5.1 Discretization of spatial domain

5.3.2 Deriving Finite Difference Equations

There are several systems, which are typically modeled by the ordinary differential equations (ODEs). The ODEs are also there to represent the boundary conditions (i.e., Neumann and Robin BC) associated with those systems. To discretize these ODEs, we all perhaps prefer the finite difference approximation formulas since they are simple and easy to program. In the following, we will briefly discuss the three different finite difference approximations.

A. Forward Difference (FD)

Taylor series expansion of $y(x)$ around node x_{m+1} yields:

$$y_{m+1} = y(x_m + \Delta x) = y_m + \left.\frac{dy}{dx}\right|_{x_m} \Delta x + \frac{1}{2\,!}\left.\frac{d^2 y}{dx^2}\right|_{x_m} (\Delta x)^2 + o[(\Delta x)^3] \tag{5.4}$$

Neglecting the second- and higher-order terms and considering $\Delta x = k$,

$$y_{m+1} = y_m + \left.\frac{dy}{dx}\right|_{x_m} k + o(k^2)$$

Rearranging,

$$\left.\frac{dy}{dx}\right|_{x_m} = \frac{y_{m+1} - y_m}{k} + o(k) \tag{5.5}$$

This is the first-order *forward difference* (FD) approximation and it is sometimes called as a *two-point* FD scheme. Obviously, this formula is first-order accurate. Note that the same approximation is made before for explicit Euler with time derivative (i.e., for ODE-IVP systems).

B. Backward Difference (BD)

Similarly, the Taylor series expansion of $y(x)$ around node x_{m-1} gives:

$$y_{m-1} = y(x_m - \Delta x) = y_m - \left.\frac{dy}{dx}\right|_{x_m} \Delta x + \frac{1}{2\,!}\left.\frac{d^2 y}{dx^2}\right|_{x_m} (\Delta x)^2 + o[(\Delta x)^3] \tag{5.6}$$

Neglecting the second- and higher-order terms, and rearranging with $\Delta x = k$, we get:

$$\left.\frac{dy}{dx}\right|_{x_m} = \frac{y_m - y_{m-1}}{k} + o(k) \tag{5.7}$$

This is known as the first-order *backward difference* (BD) approximation, which leads to first-order accuracy. Like FD, it is also referred to as a *two-point* BD formula.

C. Central Difference (CD)

It is seen above that both the FD and BD methods are first-order accurate. To have a better accuracy, let us proceed to develop a formula by subtracting Equation (5.6) from (5.4):

$$y_{m+1} - y_{m-1} = 2\left.\frac{dy}{dx}\right|_{x_m} k + o(k^3) \tag{5.8}$$

Rearranging,

$$\left.\frac{dy}{dx}\right|_{x_m} = \frac{y_{m+1} - y_{m-1}}{2k} + o(k^2) \tag{5.9}$$

This is the *central* or *centered finite difference* approximation for the *first derivative*, which is second-order accurate.

Further, adding Equations (5.4) and (5.6), we get:

$$y_{m+1}+y_{m-1}=2y_m+\left.\frac{d^2y}{dx^2}\right|_{x_m}k^2+o(k^4)$$

Rearranging,

$$\left.\frac{d^2y}{dx^2}\right|_{x_m}=\frac{y_{m+1}-2y_m+y_{m-1}}{k^2}+o(k^2) \tag{5.10}$$

This is the *centered finite difference* approximation for the *second derivative*, which is also second-order accurate.

A few more finite difference formulas are given in Table A5.1 (see Appendix 5A).

5.3.3 Solution Methodology with FDM

Discretizing the spatial domain, the ODE-BVP system is subsequently transformed into a set of algebraic equations through the finite difference approximation discussed above. Then solving the resulting algebraic equations simultaneously, one can get the final solution of the given problem. This methodology is framed through the following steps in sequence:

Step 1: Discretize the spatial domain into total $M + 1$ nodes as shown in Figure 5.1.

Step 2: Discretize the system ODE(s) (excluding BCs) by using the finite difference approximation and get total $M-1$ algebraic equations for $M-1$ interior nodes (i.e., $m=1,\ 2,\dots\dots,\ M-1$).

Step 3: For boundary nodes (i.e., $m = 0$ and M) with

- Dirichlet BC: one can directly use the specified boundary values and proceed to solve the $M-1$ algebraic equations obtained in Step 2 for total $M-1$ unknowns, namely $y_1,\ y_2,\ y_3,\dots\dots,y_{M-1}$.
- Other BC: one can discretize the boundary condition (that may result in fictitious node, depending on the type of FDM used) and proceed to solve the resulting algebraic equation along with $M-1$ algebraic equations obtained in Step 2. This point is further elaborated later.

Step 4: Finally, employ a suitable numerical technique to solve the produced set of algebraic equations.

Remarks

1. It becomes obvious that the finite difference approximations of the differential equation systems lead to form the systems of linear algebraic equations (for linear differential equations) or non-linear algebraic equations (for non-linear differential equations).

2. For the systems of linear algebraic equations, we can choose a suitable method among the several approaches discussed in Chapter 2 (e.g., backward substitution or Gauss–Seidel method). For the non-linear case, use one of the iterative methods discussed in Chapter 3 (e.g., Newton–Raphson or Muller method).

5.4 ODE-BVP WITH DIRICHLET BC

Here, we would like to solve the boundary value problems having the simple Dirichlet boundary conditions. For this, the FDM is to be used to discretize the system model represented by the ordinary differential equation. Then the resulting set of algebraic equations is to be solved. Two examples are used below to discuss the numerical solution of this type of problem.

Example 5.1 Dirichlet BC (a general system)

Consider a system represented by a general form of a second-order ODE:

$$p(x)\frac{d^2y}{dx^2}+q(x)\frac{dy}{dx}=r(x) \tag{4.2a}$$

over the domain $x\in[a,b]$ subject to:

$$\text{Dirichlet BC}\ \begin{cases} y(x=a)=y_a \\ y(x=b)=y_b \end{cases}$$

Discuss the solution methodology of this boundary value problem.

Solution

The example system governed by Equation (4.2a) includes one first-order and one second-order derivative. To discretize them, we would like to apply the centered finite difference approximations [i.e., Equations (5.9) and (5.10), both of which are second-order accurate][4] derived before:

$$\left.\frac{dy}{dx}\right|_{x_m}=\frac{y_{m+1}-y_{m-1}}{2k}+o(k^2) \tag{5.9}$$

$$\left.\frac{d^2y}{dx^2}\right|_{x_m}=\frac{y_{m+1}-2y_m+y_{m-1}}{k^2}+o(k^2) \tag{5.10}$$

where, $m=1,\ 2,\ 3,\ .....,\ M-1$. As seen in Figure 5.1, the given boundary conditions are specific to m = 0 and M, respectively, as:

---

4 For the same independent variable, it is recommended to use the finite difference formulas with same order of accuracy.

$$y_0 = y_a$$
$$y_M = y_b$$

With this, one can have total $M-1$ equations and the same number of unknowns, namely, $y_1, y_2, y_3, \ldots\ldots, y_{M-1}$. Now, let us discuss how to find these unknowns with transforming the ODE-BVP system to an algebraic equation system.

Inserting Equations (5.9) and (5.10) into (4.2a),

$$p_m \frac{y_{m+1} - 2y_m + y_{m-1}}{k^2} + q_m \frac{y_{m+1} - y_{m-1}}{2k} = r_m$$

which gives,

$$2p_m y_{m+1} - 4p_m y_m + 2p_m y_{m-1} + kq_m y_{m+1} - k\, q_m y_{m-1} = 2k^2 r_m$$

Rearranging,

$$(2p_m - kq_m) y_{m-1} - 4p_m y_m + (2p_m + k\, q_m) y_{m+1} = 2k^2 r_m \tag{5.11}$$

This is the final discretized form of Equation (4.2a).

With known boundary conditions, $y_0 (= y_a)$ and $y_M (= y_b)$, Equation (5.11) yields for:

- $m = 1$

$$-4p_1 y_1 + (2p_1 + k\, q_1) y_2 = 2k^2 r_1 - (2p_1 - k\, q_1) y_a$$

- $m = 2$

$$(2p_2 - kq_2) y_1 - 4p_2 y_2 + (2p_2 + k\, q_2) y_3 = 2k^2 r_2$$

$$\vdots$$

- $m = M - 1$

$$(2p_{M-1} - kq_{M-1}) y_{M-2} - 4p_{M-1} y_{M-1} = 2k^2 r_{M-1} - (2p_{M-1} + k\, q_{M-1}) y_b$$

Notice that all these are algebraic forms of equations. Representing them in matrix form,

$$\begin{bmatrix} -4p_1 & (2p_1 + kq_1) & 0 & 0\ldots & 0 & \ldots & 0 \\ (2p_2 - kq_2) & -4p_2 & (2p_2 + kq_2) & 0\ldots & 0 & \ldots & 0 \\ \ldots & \ldots & \ldots & \ldots & \ldots & \ldots & \ldots \\ \ldots & \ldots & \ldots & \ldots & \ldots & \ldots & \ldots \\ \ldots & \ldots & \ldots & 0\ldots & 0 & (2p_{M-1} - kq_{M-1}) & -4p_{M-1} \end{bmatrix} \begin{bmatrix} y_1 \\ y_2 \\ \cdot \\ \cdot \\ y_{M-1} \end{bmatrix}$$

$$= \begin{bmatrix} 2k^2 r_1 - (2p_1 - kq_1) y_a \\ 2k^2 r_2 \\ \cdot \\ \cdot \\ 2k^2 r_{M-1} - (2p_{M-1} + kq_{M-1}) y_b \end{bmatrix} \tag{5.12}$$

Obviously, this leads to a linear system in the form of,

$$AY = B$$

where,

$$A = \begin{bmatrix} -4p_1 & (2p_1+kq_1) & 0 & 0\ldots & 0 & \ldots & 0 \\ (2p_2-kq_2) & -4p_2 & (2p_2+kq_2) & 0\ldots & 0 & \ldots & 0 \\ \ldots & \ldots & \ldots & \ldots & \ldots & \ldots & \ldots \\ \ldots & \ldots & \ldots & \ldots & \ldots & \ldots & \ldots \\ \ldots & \ldots & \ldots & 0\ldots & 0 & (2p_{M-1}-kq_{M-1}) & -4p_{M-1} \end{bmatrix}$$

$$Y = \begin{bmatrix} y_1 \\ y_2 \\ . \\ . \\ y_{M-1} \end{bmatrix}$$

$$B = \begin{bmatrix} 2k^2r_1 - (2p_1 - kq_1)\, y_a \\ 2k^2r_2 \\ . \\ . \\ 2k^2r_{M-1} - (2p_{M-1} + kq_{M-1})\, y_b \end{bmatrix}$$

Notice that A is a tridiagonal matrix of constants. To solve the resulting linear system for Y, one can use one of the methods (e.g., Thomas algorithm) detailed in Chapter 2.

Remark

The given problem reduces to solving the tridiagonal system, $AY = B$ whose solution will provide approximate values for the solution of the ODE-BVP at discrete points on the interval $[a, b]$. We will discuss a few other options later for better accuracy.

Example 5.2 Dirichlet BC (a specific system)

Consider another problem with

$$-\frac{d^2u}{dx^2} = 2 \tag{5.13}$$

on $x \in [0, 1]$ subject to:

$$\text{Dirichlet BC} \begin{cases} u(0) = 0 \\ \\ u(1) = 0 \end{cases}$$

Solve this boundary value problem, and compare the numerical and analytical solutions.

Solution

Numerical solution

Let us first discretize the spatial domain into say, seven equidistant nodes as shown in Figure 5.2. So, there are total six equally spaced intervals and thus, the mesh size $k = \frac{1}{6}$.

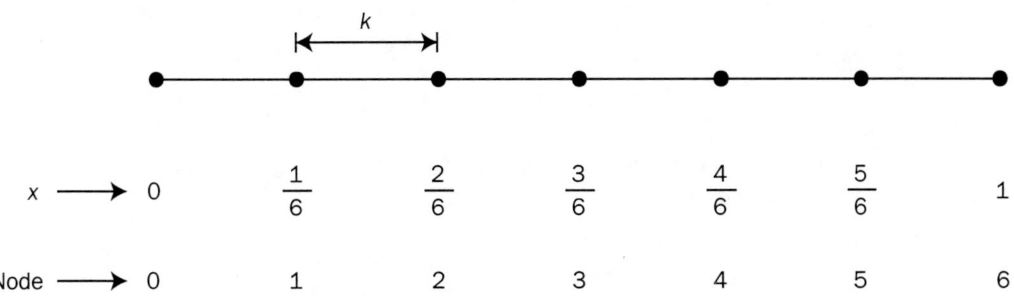

FIGURE 5.2 Discretization of spatial domain into seven equidistant nodes

In the next step, use the centered finite difference approximation in Equation (5.13),

$$-\frac{u_{m+1} - 2u_m + u_{m-1}}{k^2} = 2$$

It gives,

$$-u_{m-1} + 2u_m - u_{m+1} = 2k^2 \tag{5.14}$$

This is the final discretized form of the system Equation (5.13).

There are two given Dirichlet boundary conditions: $u_0 = 0$ and $u_6 = 0$, which are specific to $m = 0$ and $m = 6$, respectively. To find u_m with $m = 1, 2, ..., 5$, we are going to use Equation (5.14) that yields for:

- $m = 1$ $\quad 2u_1 - u_2 = 2k^2$ $\quad$ [since $u_0 = 0$ (first BC)]
- $m = 2$ $\quad -u_1 + 2u_2 - u_3 = 2k^2$
- $m = 3$ $\quad -u_2 + 2u_3 - u_4 = 2k^2$
- $m = 4$ $\quad -u_3 + 2u_4 - u_5 = 2k^2$
- $m = 5$ $\quad -u_4 + 2u_5 = 2k^2$ $\quad$ [since $u_6 = 0$ (second BC)]

In matrix form,

$$\begin{bmatrix} 2 & -1 & 0 & 0 & 0 \\ -1 & 2 & -1 & 0 & 0 \\ 0 & -1 & 2 & -1 & 0 \\ 0 & 0 & -1 & 2 & -1 \\ 0 & 0 & 0 & -1 & 2 \end{bmatrix} \begin{bmatrix} u_1 \\ u_2 \\ u_3 \\ u_4 \\ u_5 \end{bmatrix} = k^2 \begin{bmatrix} 2 \\ 2 \\ 2 \\ 2 \\ 2 \end{bmatrix} \tag{5.15}$$

This is again a linear form of $AU = B$, solving which, one obtains:

$u_1 = 0.138889$

$u_2 = 0.222222$

$u_3 = 0.25$

$u_4 = 0.222222$

$u_5 = 0.138889$

Analytical solution

To validate the numerical solution obtained above, let us solve Equation (5.13) analytically. Using the two boundary conditions,

$u(x=0)=0$

$u(x=1)=0$

one obtains the following analytical solution:

$$-u = x^2 - x \tag{5.16}$$

Figure 5.3 and Table 5.1 compare the analytical solution with numerical one. It becomes obvious that there is almost no mismatch existed between them.

Table 5.1 Comparing numerical and analytical solutions

x	Analytical solution	Numerical solution
0	0	0
0.16667	0.138891	0.138889
0.33333	0.222221	0.222222
0.5	0.25	0.25
0.66667	0.222221	0.222222
0.83333	0.138891	0.138889
1	0	0

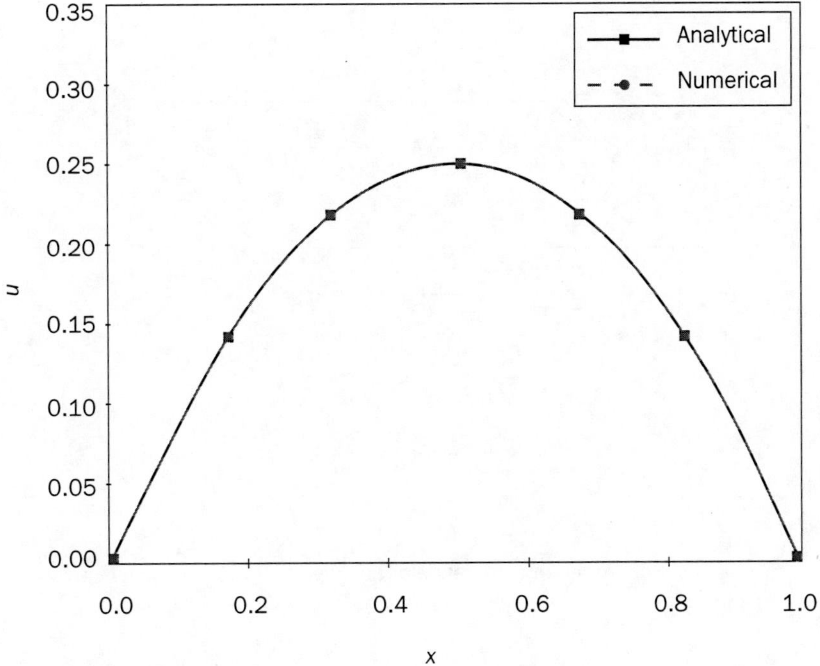

FIGURE 5.3 Comparing numerical and analytical solutions

5.5 ODE-BVP WITH NEUMANN BC

In our next assignment, we would like to solve the boundary value problem having the Neumann boundary condition, along with the Dirichlet BC. For this, the FDM is to be used to discretize both the system equation and the Neumann BC. To discuss the numerical solution of such type of problem, a case study is adopted below.

Case Study 5.1 Neumann BC

A cooling fin: steady state

The schematic representation of a cooling fin is demonstrated in Figure 5.4. For developing a steady state heat transfer model, we consider the following assumptions:

Assumptions

1. There is no internal heat generation.
2. Material properties are independent of temperature (i.e., constant).
3. Heat conduction through the material occurs in x-direction only.
4. Heat is lost to the surroundings by convection.

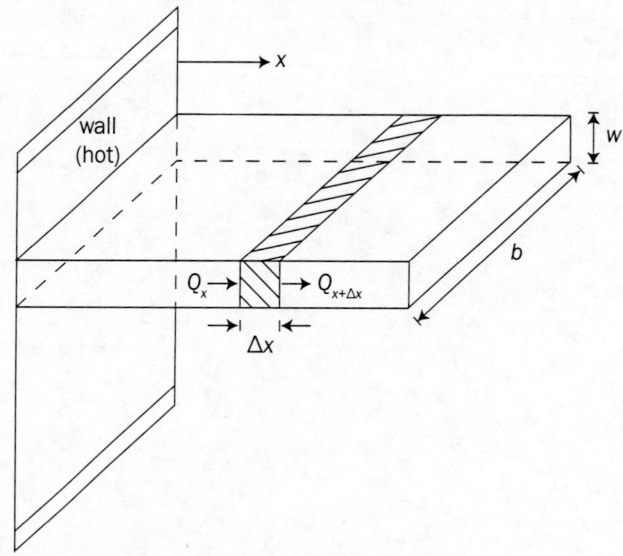

FIGURE 5.4 Schematic of a rectangular cooling fin

Here, we use the following notations:

A = area of heat conduction = bw

b = fin breadth

w = fin thickness

h = convective heat transfer coefficient

K = thermal conductivity

l = fin length

p = fin perimeter = $2(b + w)$

Q = heat flux (i.e., heat transfer rate per unit area)

T = local temperature of the fin

T_0 = ambient temperature

T_w = wall or base temperature

Deriving model

We substitute all individual terms (input, output and heat loss) in Equation (A4.3),

$$0 = AQ_x - AQ_{x+\Delta x} - hp\Delta x(T - T_0)$$

Since it is a steady state process, thus rate of accumulation of energy = 0.

Considering $\Delta x \to 0$ and rearranging,

$$A\frac{dQ}{dx} = -hp(T - T_0) \tag{5.17}$$

Using the Fourier law,

$$Q = -K\frac{dT}{dx} \tag{5.18}$$

one obtains from Equation (5.17),

$$\frac{d^2T}{dx^2} = \frac{hp}{KA}(T - T_0) \tag{5.19}$$

This is the final modeling equation of heat transfer for a cooling fin.

Problem statement

The cooling fin system represented by Equation (5.19) is subject to the following boundary conditions:

At the base $(x = 0)$

Dirichlet BC: $T(x = 0) = T_w$ (5.20a)

At the tip $(x = l)$

$$\text{Neumann BC: } \left.\frac{dT}{dx}\right|_{x=l} = 0 \tag{5.20b}$$

This Neumann boundary condition reveals that the heat flux, Q at the tip is negligible. Additional information is given as:

$$\frac{hp}{KA} = 1$$

$$l = 4$$

As recommended before, always prefer to use the finite difference approximations, for a process model and the associated boundary condition having differential term, with same order of accuracy. With this, solve the cooling fin problem using the CD method, and compare the numerical and analytical results.

Solution

Numerical solution

For the cooling fin system, let us introduce the following dimensionless variables:

$$\theta = \frac{T - T_0}{T_w - T_0} \tag{5.21}$$

$$\zeta = \frac{x}{l} \tag{5.22}$$

$$H = \sqrt{\frac{l^2 hp}{KA}} \tag{5.23}$$

Differentiating Equation (5.21),

$$\frac{d\theta}{dx} = \frac{1}{T_w - T_0}\frac{dT}{dx} \tag{5.24}$$

and thus,

$$\frac{d^2T}{dx^2} = (T_w - T_0)\frac{d^2\theta}{dx^2}$$

Using Equation (5.22),

$$\frac{d^2T}{dx^2} = (T_w - T_0)\frac{1}{l^2}\frac{d^2\theta}{d\zeta^2} \tag{5.25}$$

Substituting this in Equation (5.19),

$$\frac{T_w - T_0}{l^2}\frac{d^2\theta}{d\zeta^2} = \frac{hp}{KA}(T - T_0)$$

which gives,

$$\frac{d^2\theta}{d\zeta^2} = \frac{l^2 hp}{KA}\theta$$

Finally,

$$\frac{d^2\theta}{d\zeta^2} = H^2\theta \tag{5.26}$$

Accordingly, the boundary conditions get transformed into:

Dirichlet BC: $\theta(\zeta = 0) = 1$ since at $x = 0:\ \zeta = 0$ and $T = T_w$ (5.27a)

Neumann BC: $\left.\frac{d\theta}{d\zeta}\right|_{\zeta=1} = 0$ (5.27b)

So the cooling fin ODE-BVP system gets the following form:

ODE-BVP system	
$\frac{d^2\theta}{d\zeta^2} = H^2\theta$	(5.26)
Dirichlet BC: $\theta(\zeta = 0) = 1$	(5.27a)
Neumann BC: $\left.\frac{d\theta}{d\zeta}\right|_{\zeta=1} = 0$	(5.27b)

For numerical solution of this ODE-BVP, we use the central difference approximation (as asked) for the second-order derivative in Equation (5.26):

$$\frac{\theta_{m+1}-2\theta_m+\theta_{m-1}}{k^2}-H^2\theta_m=0$$

Simplifying,

$$\theta_{m-1}-(2+H^2k^2)\theta_m+\theta_{m+1}=0 \tag{5.28}$$

Let us discretize the dimensionless spatial domain of ζ ($0\le\zeta\le1$) into four equally spaced intervals (without counting *fictitious node*[5]) as shown in Figure 5.5. Accordingly, Equation (5.28) can be used for $m=1, 2, 3, 4$. Here,

$$k=\frac{1-0}{4}=0.25 \qquad \text{and} \qquad H=\sqrt{\frac{l^2hp}{KA}}=4$$

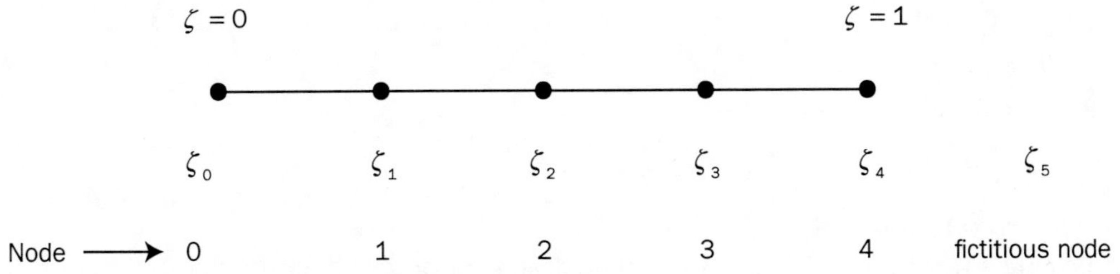

$\zeta = 0$ $\zeta = 1$

ζ_0 ζ_1 ζ_2 ζ_3 ζ_4 ζ_5

Node ⟶ 0 1 2 3 4 fictitious node

FIGURE 5.5 Discretization of the dimensionless spatial domain into five equidistant nodes (excluding fictitious node)

Based on Equation (5.28), we can have for:

- $m = 1$ $\quad -3\theta_1+\theta_2=-1$ $\quad$ [since $\theta_0=1$ (first BC)] (5.29a)
- $m = 2$ $\quad \theta_1-3\theta_2+\theta_3=0$ (5.29b)
- $m = 3$ $\quad \theta_2-3\theta_3+\theta_4=0$ (5.29c)
- $m = 4$ $\quad \theta_3-3\theta_4+\theta_5=0$ (5.29d)

Obviously, there are four equations [i.e., Equations (5.29a–d)] and five unknowns (i.e., θ_1, θ_2, θ_3, θ_4, θ_5). Among them, as shown in Figure 5.5, θ_5 is located at the fictitious boundary outside the domain (i.e., at ζ_5).

In order to make the system completely specified, we need to either add one more equation or reduce the number of unknowns by 1 (so that the number of unknowns and available equations become equal). Here, we pick up the first option since Neumann boundary Equation (5.27b) is there. Discretizing it with the central difference formula:

$$\left.\frac{d\theta}{d\zeta}\right|_{\zeta_4}=\frac{\theta_5-\theta_3}{2k}=0$$

5 Called so since it is located outside of the domain.

This gives,

$$\theta_5-\theta_3=0 \tag{5.30}$$

Including this, we have at this moment total five equations and five unknowns (i.e., a completely specified system).

Representing these 5 equations in matrix form:

$$\begin{bmatrix} -3 & 1 & 0 & 0 & 0 \\ 1 & -3 & 1 & 0 & 0 \\ 0 & 1 & -3 & 1 & 0 \\ 0 & 0 & 1 & -3 & 1 \\ 0 & 0 & -1 & 0 & 1 \end{bmatrix} \begin{bmatrix} \theta_1 \\ \theta_2 \\ \theta_3 \\ \theta_4 \\ \theta_5 \end{bmatrix} = \begin{bmatrix} -1 \\ 0 \\ 0 \\ 0 \\ 0 \end{bmatrix} \tag{5.31}$$

Solving this algebraic equation system (linear), one obtains:

$\theta_1 = 0.3830$
$\theta_2 = 0.1489$
$\theta_3 = 0.0638$
$\theta_4 = 0.0425$
$\theta_5 = 0.0638$

Analytical solution

It is quite straightforward to find the analytical solution of Equation (5.26) based on the given BCs in Equation (5.27) as:

$$\theta=0.0003355\, e^{4\zeta}+0.9996645\, e^{-4\zeta} \tag{5.32}$$

Using this equation, we produce the analytical solutions in Table 5.2. In the same table, along with Figure 5.6, the numerical and analytical results are compared.

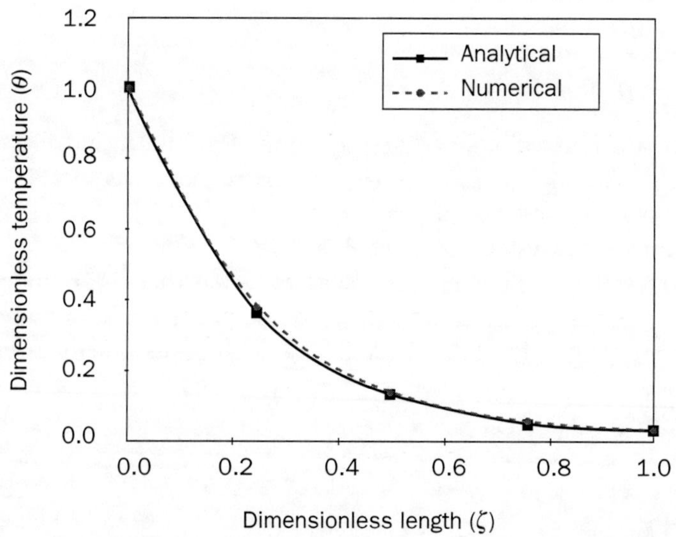

FIGURE 5.6 Comparative temperature profile (θ versus ζ) of the cooling fin

TABLE 5.2 Comparing numerical and analytical solutions for the cooling fin

ζ	Analytical solution	Numerical solution
0	1	1
0.25	0.3687	0.3830
0.5	0.1378	0.1489
0.75	0.0565	0.0638
1.0	0.0366	0.0425

Remarks

1. In solving the cooling fin problem above governed by Equation (5.26) with the use of the central difference method, we have obtained an unknown variable, θ_5 that corresponds to the fictitious node. To deal with such kind of BVP systems, there are a couple of ways as highlighted below.
   i) Having Neumann BC (or Robin BC), replace the unknown variable at fictitious node (i.e., θ_5 in the cooling fin) by that at a node within the domain (i.e., θ_3 in the cooling fin). This strategy is followed (as in the cooling fin problem) by the use of the CD technique for the BC as well and it is popularly known as the *ghost-point* method.
   ii) One can use the backward difference (BD) formula at the right boundary node[6] (i.e., at ζ_4 for the cooling fin) and consequently, there is no chance of having the fictitious node. But remember, this forces us to use the two different methods [BD (for BC) and CD (for system equation)] having different order of accuracy. To overcome this problem, one can adopt the following option.
   iii) If Neumann BC (or Robin BC) is there, one can use an alternative method, namely *three-point* finite difference approximation (second-order accurate), for its first-order derivative. This approach also avoids the generation of any fictitious boundary node and it is discussed later with example (see Subsection 5.5.2).
2. In order to have improved matching between the analytical and numerical solutions, one could do *mesh refinement*. This approach is described below prior to its application to the cooling fin problem.

In the following, we will know how to do the mesh refinement and three-point finite difference approximation. Further, we will illustrate them through the example systems to verify whether they lead to achieve any improvement in numerical solution.

5.5.1 Mesh Refinement

It is fairly true that the accuracy of the numerical solution is dependent on the number of nodes generated by discretizing the domain. To improve the accuracy level, thus one could

6 Similarly, forward difference approximation is employed at the left boundary node if Neumann or Robin BC is specified. However, this approximation is not required for the cooling fin problem since Dirichlet BC is given there for the left boundary.

prefer to do mesh refinement. For this, start with a few nodes and keep on reducing the discretization step (i.e., increasing the number of nodes or grid points) until no remarkable change in the solution is noticed. Doing so, we should keep the increased computational load in mind involved in mesh refinement.

As stated, we should start solving an ODE-BVP system with relatively few nodes (i.e., a coarse discretization). Then we extend the solution methodology to a new grid where we put an additional point between each of the old grid points. Accordingly,

$$\begin{pmatrix}\text{Number of new}\\ \text{grid points}\end{pmatrix} = 2\times\begin{pmatrix}\text{Number of old}\\ \text{grid points}\end{pmatrix} - 1 \tag{5.33}$$

It is illustrated in Figure 5.7. As indicated, continue this refinement exercise until the difference in the numerical solution between the two grids at their common grid points is within the desired tolerance. To visualize the benefit of mesh refinement, we apply this strategy on the cooling fin system in the following assignment.

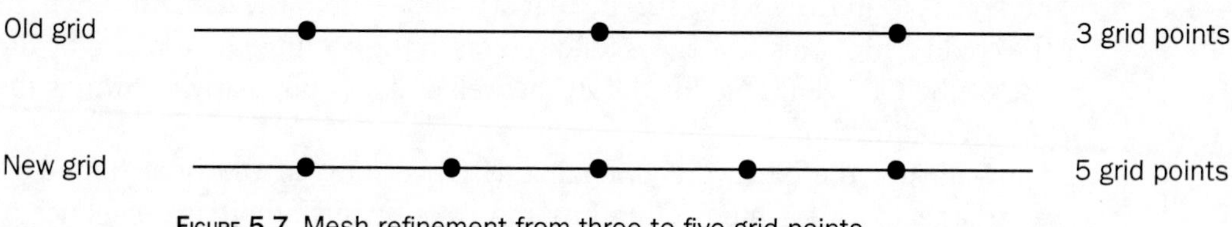

FIGURE 5.7 Mesh refinement from three to five grid points

Case Study 5.2 Neumann BC

A cooling fin (revisited): mesh refinement

Let us continue the cooling fin system with:

$$\frac{d^2\theta}{d\zeta^2} = H^2\theta \tag{5.26}$$

$$\text{Dirichlet BC: } \theta(\zeta=0)=1 \tag{5.27a}$$

$$\text{Neumann BC: } \left.\frac{d\theta}{d\zeta}\right|_{\zeta=1} = 0 \tag{5.27b}$$

where $H = 4$. Now our goal is to observe the improvement in accuracy level, if any, through the mesh refinement.

Solution

We have solved this cooling fin problem previously (Case Study 5.1) with five nodes (i.e., $M = 4$). For mesh refinement, we use Equation (5.33) and obtain nine nodes (i.e., $M = 8$). As a result, we have the mesh size, $k = \frac{1}{8} = 0.125$.

Accordingly, Equation (5.28) yields:

$$\theta_{m-1} - 2.25\theta_m + \theta_{m+1} = 0 \tag{5.34}$$

for $m = 1, 2, \ldots\ldots, 8$. Using this Equation (5.34), we get for:

- $m = 1$ $\quad -2.25\theta_1 + \theta_2 = -1$ $\quad$ [since $\theta_0 = 1$ (first BC)] (5.35a)
- $m = 2$ $\quad \theta_1 - 2.25\theta_2 + \theta_3 = 0$ (5.35b)
- $m = 3$ $\quad \theta_2 - 2.25\theta_3 + \theta_4 = 0$ (5.35c)
- $m = 4$ $\quad \theta_3 - 2.25\theta_4 + \theta_5 = 0$ (5.35d)
- $m = 5$ $\quad \theta_4 - 2.25\theta_5 + \theta_6 = 0$ (5.35e)
- $m = 6$ $\quad \theta_5 - 2.25\theta_6 + \theta_7 = 0$ (5.35f)
- $m = 7$ $\quad \theta_6 - 2.25\theta_7 + \theta_8 = 0$ (5.35g)
- $m = 8$ $\quad 2\theta_7 - 2.25\theta_8 = 0$ $\quad$ [since $\theta_9 = \theta_7$ (second BC)] (5.35h)

Solving this set of algebraic equations simultaneously, we obtain the numerical solutions that are documented in Table 5.3. In the same table, we compare the numerical results with $M = 4$ and $M = 8$ with reference to the analytical solutions.

Table 5.3 Comparing numerical solutions with and without mesh refinement with reference to analytical solutions

ζ	Analytical solution	Numerical solution ($M = 4$)	Numerical solution ($M = 8$)
0	1	1	1
0.125	0.6069	–	0.6100
0.25	0.3687	0.3830	0.3725
0.375	0.22456	–	0.2281
0.5	0.1378	0.1489	0.1407
0.625	0.08614	–	0.0885
0.75	0.0565	0.0638	0.0584
0.875	0.0413	–	0.0429
1.0	0.0366	0.0425	0.0381

Remark

It becomes obvious that the accuracy gets improved with mesh refinement (i.e., reduced truncation error). But remember, this is achieved at the expense of increased computations (i.e., increased round-off error). Thus, care must be exercised in selecting the value of mesh size, k such that the total error is minimum as discussed before with Figure 1.3.

5.5.2 Three-Point Finite Difference Approximation

Using Taylor series, one can rewrite Equation (5.4) as:

$$y_{m+1}=y(x_m+k)=y_m+\frac{k}{1\,!}y'_m+\frac{k^2}{2\,!}y''_m+o(k^3) \tag{5.36}$$

Similarly,

$$y_{m+2}=y(x_m+2k)=y_m+\frac{2k}{1\,!}y'_m+\frac{(2k)^2}{2\,!}y''_m+o(k^3) \tag{5.37}$$

Using these two equations,

$$y_{m+2}-4y_{m+1}=-3y_m-2ky'_m+o(k^3)$$

Rearranging, one obtains:

$$y'_m=\frac{-3y_m+4y_{m+1}-y_{m+2}}{2k}+o(k^2) \tag{5.38}$$

This is the final form for first-order derivative at x_m and it is sometimes called as the *three-point FD* formula. As shown, it is second-order accurate.

In the similar fashion, one can show that

$$y_{m-2}-4y_{m-1}=-3y_m+2ky'_m+o(k^3)$$

Rearranging,

$$y'_m=\frac{y_{m-2}-4y_{m-1}+3y_m}{2k}+o(k^2) \tag{5.39}$$

This is the final form of *three-point BD* scheme for first-order derivative that leads to second-order accuracy. The three-point formula for second-order derivative is documented in Table A5.1 (see Appendix 5A).

In the following, the *three-point finite difference* method is illustrated by the cooling fin system.

Case Study 5.3 Neumann BC

A cooling fin (revisited): three-point finite difference

The cooling fin system is demonstrated by:

$$\frac{d^2\theta}{d\zeta^2}=H^2\theta \tag{5.26}$$

Dirichlet BC: $\theta(\zeta=0)=1$ (5.27a)

Neumann BC: $\left.\frac{d\theta}{d\zeta}\right|_{\zeta=1}=0$ (5.27b) with $H = 4$ and $M = 4$. Here we fix our goal to observe whether there is any improvement to be achieved through the three-point finite difference scheme.

Solution

Using the three-point BD scheme [Equation (5.39)], the Neumann BC yields:

$$\left.\frac{d\theta}{d\zeta}\right|_{\zeta_4} = \frac{\theta_2 - 4\theta_3 + 3\theta_4}{2k} = 0$$

Rearranging,

$$\theta_2 - 4\theta_3 + 3\theta_4 = 0 \tag{5.40}$$

Obviously, this is true for $m = 4$ (right boundary node). The Dirichlet BC [Equation (5.27a)] is valid for $m = 0$ (left boundary node) and for rest $m = 1, 2, 3$ (interior nodes), we have developed before Equation (5.28) that can be rewritten as:

$$\theta_{m-1} - 3\theta_m + \theta_{m+1} = 0 \tag{5.41}$$

It basically yields Equations (5.29a–c). Notice that there is *no* fictitious node in both Equations (5.40) and (5.41) with $m = 1, 2, 3$. Moreover, both these equations have same second-order of accuracy. Solving them simultaneously, we obtain the numerical solutions reported in Table 5.4.

TABLE **5.4** Comparing numerical and analytical solutions

ζ	Analytical solution	Numerical solution with fictitious node[1]	Numerical solution without fictitious node[2]
0	1	1	1
0.25	0.3687	0.3830	0.38235
0.5	0.1378	0.1489	0.14706
0.75	0.0565	0.0638	0.05882
1.0	0.0366	0.0425	0.02941

[1] CD is used to discretize the Neumann BC [Equation (5.27b)].

[2] Three-point BD scheme is used to discretize the same Neumann BC.

Remark

Calculating absolute true error from Table 5.4, one can quantify that there is eventually some improvement achieved in numerical results of the cooling fin system when for the Neumann BC, the three-point finite difference approximation is used (i.e., there is no fictitious node) instead of the CD approximation (i.e., there is fictitious node existed).

5.6 ODE-BVP with Robin BC

Our next study is to solve the ODE-BVP system having the Robin boundary condition that is constituted by combining the Dirichlet and Neumann BC. It is discussed below with the use of FDM to discretize the ordinary differential equations.

Example 5.3 Robin BC

An ODE-BVP system is described by:

$$p(x)\frac{d^2y}{dx^2}+q(x)\frac{dy}{dx}=r(x) \tag{4.2a}$$

over the domain $x\in[a,b]$ subject to:

$$\text{Robin BC: } \begin{cases} a_0 y(a)+b_0 y'(a)=c_0 & (5.42a) \\ a_1 y(b)+b_1 y'(b)=c_1 & (5.42b) \end{cases}$$

Discuss the solution methodology of this boundary value problem.

Solution

Using the central difference method, we have obtained previously (see Example 5.1) the discretized form of Equation (4.2a) as:

$$(2p_m-kq_m)y_{m-1}-4p_m y_m+(2p_m+k\,q_m)y_{m+1}=2k^2 r_m \tag{5.11}$$

In Example 5.1, since y at $m=0$ (i.e., y_a) and y at $m=M$ (i.e., y_b) are directly available from the given Dirichlet BC, there we use Equation (5.11) to find the unknown y for m equals 1 to $M-1$. However, in this example with Robin BC, we need to consider m starting from 0 to M. Accordingly, for:

- $m=0$, use first BC [i.e., Equation (5.42a)]
- $m=1,\ 2,\ \ldots\ldots.,\ M-1$, use Equation (5.11)
- $m=M$, use second BC [i.e., Equation (5.42b)]

Let us discretize Equation (5.42) and make the problem ready for solution. To avoid the generation of fictitious node, we would like to employ the three-point finite difference approximation (i.e., three-point FD for left boundary and three-point BD for right boundary). Accordingly, for $m=0$ (i.e., $x=a$), the three-point FD method [Equation (5.38)] yields:

$$y'_0=\frac{-3y_0+4y_1-y_2}{2k}$$

Inserting in Equation (5.42a),

$$a_0 y_0+b_0\frac{-3y_0+4y_1-y_2}{2k}=c_0$$

This gives,

$$\left(a_0-\frac{3b_0}{2k}\right)y_0+\frac{2b_0}{k}y_1-\frac{b_0}{2k}y_2=c_0 \tag{5.42c}$$

Similarly for $m=M$ (i.e., $x=b$), the three-point BD method [Equation (5.39)] yields:

$$y'_M = \frac{y_{M-2} - 4y_{M-1} + 3y_M}{2k}$$

Plugging in Equation (5.42b),

$$a_1 y_M + b_1 \frac{y_{M-2} - 4y_{M-1} + 3y_M}{2k} = c_1$$

Rearranging, we have:

$$\frac{b_1}{2k} y_{M-2} - \frac{2b_1}{k} y_{M-1} + \left(a_1 + \frac{3b_1}{2k}\right) y_M = c_1 \tag{5.42d}$$

Now, let us consolidate Equation (5.42c) for $m = 0$, Equation (5.11) for $m = 1, 2, \ldots\ldots, M-1$ and Equation (5.42d) for $m = M$ in matrix form as:[7]

$$\begin{array}{l} \text{For } m=0 \\ m=1 \\ m=2 \\ \vdots \\ m=M-1 \\ m=M \end{array}
\begin{bmatrix}
\left(a_0 - \frac{3b_0}{2k}\right) & \left(\frac{2b_0}{k}\right) & \left(\frac{-b_0}{2k}\right) & 0 & \cdots & 0 & 0 & 0 \\
(2p_1 - kq_1) & (-4p_1) & (2p_1 + kq_1) & 0 & \cdots & 0 & 0 & 0 \\
0 & (2p_2 - kq_2) & (-4p_2) & (2p_2 + kq_2) & \cdots & 0 & 0 & 0 \\
\vdots & \vdots & \vdots & \vdots & \cdots & \vdots & \vdots & \vdots \\
0 & 0 & 0 & 0 & \cdots & (2p_{M-1} - kq_{M-1}) & (-4p_{M-1}) & (2p_{M-1} + kq_{M-1}) \\
0 & 0 & 0 & 0 & \cdots & \left(\frac{b_1}{2k}\right) & \left(\frac{-2b_1}{k}\right) & \left(a_1 + \frac{3b_1}{2k}\right)
\end{bmatrix}_{(M+1)\times(M+1)}
\begin{bmatrix} y_0 \\ y_1 \\ y_2 \\ \vdots \\ y_{M-1} \\ y_M \end{bmatrix}_{(M+1)\times 1}
=
\begin{bmatrix} c_0 \\ 2k^2 r_1 \\ 2k^2 r_2 \\ \vdots \\ 2k^2 r_{M-1} \\ c_1 \end{bmatrix}_{(M+1)\times 1}$$

Finally, one needs to solve this linear algebraic system, $AY = B$ to find Y.

5.7 Non-linear Boundary Value Problem

In the next phase, we are willing to solve a non-linear ODE-BVP system having specified boundaries. The issue related to non-linearity of an ODE is discussed in Chapter 4.

Let us find the numerical solution of a simple non-linear BVP system.

7 Note that all these three equations are formulated with the same order of accuracy (i.e., second-order).

Example 5.4 Non-linear ODE-BVP system

Solve the following problem:

$$\frac{d^2y}{dx^2}+2y\frac{dy}{dx}=0 \tag{5.43}$$

over the domain $x\in[0,4]$ subject to:

$$\text{Dirichlet BC: } \begin{cases} y(x=0)=1 & (5.44a) \\ y(x=4)=0.2 & (5.44b) \end{cases}$$

Compare the numerical solution with the analytical one, which has the following form:

$$y=\frac{1}{1+x} \tag{5.45}$$

Solution

Using the CD method,

$$\frac{y_{m+1}-2y_m+y_{m-1}}{k^2}+2y_m\frac{y_{m+1}-y_{m-1}}{2k}=0 \tag{5.46}$$

Rearranging,

$$y_{m+1}-2y_m+y_{m-1}+ky_my_{m+1}-ky_my_{m-1}=0 \tag{5.47}$$

Adopting $M = 4$, we have $k = 1$. Accordingly, for:

- $m = 1$ $\quad y_1y_2+y_2-3y_1+1=0$ $\quad$ [since $y_0=1$ (first BC)]

Let us consider,

$$y_1y_2+y_2-3y_1+1=g_1 \tag{5.48a}$$

- $m = 2$ $\quad y_3-2y_2+y_1+y_2y_3-y_1y_2=g_2$ $\quad$ (5.48b)
- $m = 3$ $\quad -y_2y_3-1.8y_3+y_2+0.2=g_3$ $\quad$ [since $y_4=0.2$ (second BC)] (5.48c)

To solve this non-linear set of Equations (5.48a–c), we would like to use the N–R method that is reframed as:

$$Y^{i+1}=Y^i-(J^i)^{-1}G^i \tag{5.49}$$

Here the Jacobian matrix,

$$J=\begin{bmatrix} \frac{\partial g_1}{\partial y_1} & \frac{\partial g_1}{\partial y_2} & \frac{\partial g_1}{\partial y_3} \\ \frac{\partial g_2}{\partial y_1} & \frac{\partial g_2}{\partial y_2} & \frac{\partial g_2}{\partial y_3} \\ \frac{\partial g_3}{\partial y_1} & \frac{\partial g_3}{\partial y_2} & \frac{\partial g_3}{\partial y_3} \end{bmatrix}$$

This gives,

$$J = \begin{bmatrix} y_2 - 3 & y_1 + 1 & 0 \\ 1 - y_2 & -2 + y_3 - y_1 & 1 + y_2 \\ 0 & -y_3 + 1 & -y_2 - 1.8 \end{bmatrix}$$

With this, we use Equation (5.49) and get the numerical solutions as reported in Table 5.5. A comparison is also made there between the numerical and analytical solutions. Notice that there is no difference existed between the numerical and analytical results for this simple non-linear system.

Table 5.5 Comparing numerical and analytical solutions

x	Analytical solution	Numerical solution
0	1	1
1	0.5	0.5
2	0.33333	0.33333
3	0.25	0.25
4	0.2	0.2

Note that the numerical solution of a few more non-linear boundary value problems is shown later (e.g., see Case Study 5.4–5.7).

5.8 Shooting Method

The Shooting method is used to solve the second- or higher-order boundary value problems (BVPs) by reducing them into the systems of initial value problems (IVPs). For this, we transform a second- or higher-order ODE into a set of first-order ODEs as described in Section 4.1. However, to solve those transformed ODE-IVP equations, one cannot directly use an IVP solver. This is because the given set of boundary conditions for the BVP is not same with the set of initial conditions of the transformed IVP (boundary conditions are not given at the same value of x). Thus, we need to find the unknown initial condition and as soon as this condition is correctly obtained (satisfying the desired tolerance with reference to the concerned BC), we stop our computation and get the final solution of the ODE-BVP system.

To go into detail of this Shooting method, let us consider the general form of a non-linear ODE-BVP system:

$$\frac{d^2y}{dx^2} = f\left(x, y, \frac{dy}{dx}\right) \tag{5.50}$$

on $x \in [0, a]$ subject to:

$$y(x=0)=\alpha \tag{5.51a}$$

$$y(x=a)=\beta \tag{5.51b}$$

To solve this problem, we develop the computational steps below.

Computational steps

Step 1: Convert Equation (5.50) into a set of two ODEs (with considering $y = y_1$) as:

$$\frac{dy_1}{dx} = y_2 \tag{5.52a}$$

$$\frac{dy_2}{dx} = f(x, y_1, y_2) \tag{5.52b}$$

Initial conditions include:

$$y_1(x=0)=y_{10}=\alpha \qquad \text{(first BC)} \tag{5.53a}$$

$$y_2(x=0)=y_{20} \qquad \text{(assumed)} \tag{5.53b}$$

Notice that the missing initial condition for y_2 (i.e., y'_1) is assumed here.

Step 2: Integrate (i.e., *shoot*) the transformed ODE-IVP system framed above in Step 1 [Equations (5.52) and (5.53)] from 0 to a by using a numerical integration approach (e.g., Euler method) and finally obtain $y_1(x=a)$ for $\Delta x = k$.

Step 3: Check the convergence. If the estimated y_1 at $x = a$ is close to the value given in second BC (i.e., β), then stop the computation. Else, use an iterative convergence method (e.g., Newton–Raphson [N–R]) to generate a new guess value for y_2 at $x = 0$, then go to Step 2 and repeat the computation.

Remarks

1. One can simply use a linear interpolation[8] in Step 3 (when convergence condition is not satisfied) to generate a new guess value[9] for y_2 at $x = 0$.
2. This method is analogous to the procedure of a firing object at a stationary target and thus, it is referred to as the *Shooting method*. This fact is somewhat reflected in Figure 5.8 (see Figure 5.12 for better clarity).
3. This Shooting method is often preferred to use for non-linear ODE-BVP systems over the finite difference method.

8 This technique is detailed in Chapter 8.

9 This is actually the third guess value (elaborated in the subsequent Example 5.5).

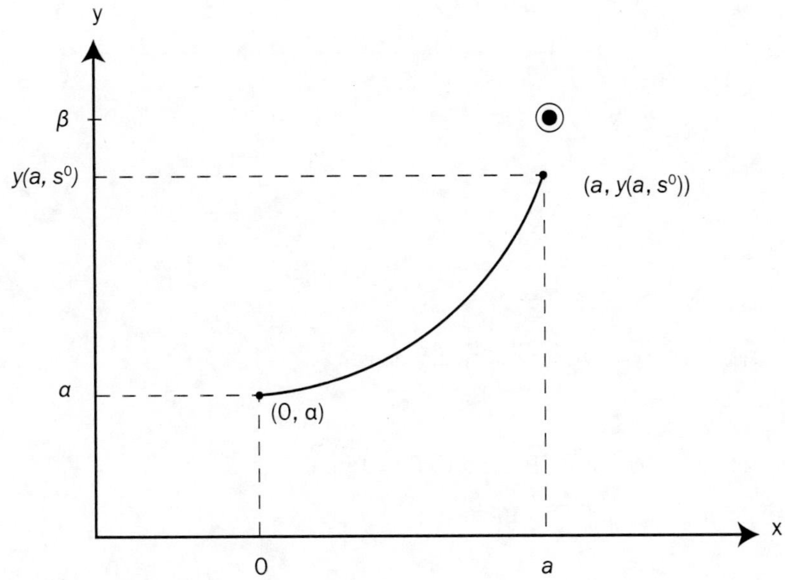

FIGURE 5.8 Illustrating the Shooting method

5.8.1 Applying Shooting Method to Linear ODE-BVP System

Prior to solving a non-linear system represented by Equation (5.50), let us first understand some basics of the Shooting method through its application to a simple linear example system discussed below.

Example 5.5 Shooting method with linear interpolation

We take a simple linear ODE-BVP system:

$$\frac{d^2y}{dx^2} = 5y + 10x(1-x) \tag{5.54}$$

on $x \in [0,\ 9]$ subject to:

$$y(0) = 0 \tag{5.55a}$$

$$y(9) = 0 \tag{5.55b}$$

Solve this problem for $M = 3$ (i.e., $k = 3$) with an accuracy of 10^{-6} (= tol).

Solution

To solve this linear problem, let us follow the computational steps developed before for the Shooting method. The discretization of spatial domain is made in Figure 5.9.

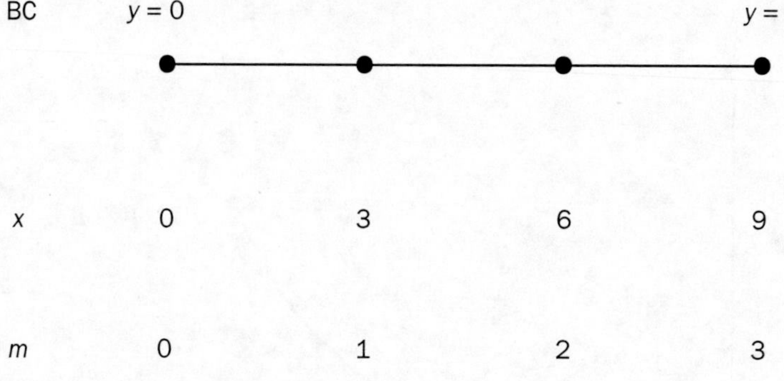

FIGURE 5.9 Discretization of spatial domain

Step 1: Transform the given BVP to an IVP system.

Transformed set of ODEs includes:

$$\frac{dy_1}{dx} = y_2 \tag{5.56a}$$

$$\frac{dy_2}{dx} = 5y_1 + 10x(1-x) \tag{5.56b}$$

Initial conditions:

$y_1(x=0) = 0$ (first BC)

$y_2(x=0) = 4$ (assumed)

Obviously, the missing initial condition for y_2 is assumed here.

Step 2: Integrate the transformed IVP system.

For the two first-order simultaneous ODEs [Equation (5.56)], let us apply the simple forward Euler method:

$$y_{1\ m+1} = y_{1m} + k\ y_{2m} \tag{5.57a}$$

$$y_{2\ m+1} = y_{2m} + k[5y_{1m} + 10x(1-x)] \tag{5.57b}$$

for $m = 0,\ 1,\ 2$. Accordingly,

- $m = 0$ (i.e., $x = 0$)

  $y_{11} = y_{10} + 3y_{20} = 0 + 12 = 12$

  $y_{21} = y_{20} + 3[5y_{10} + 10x(1-x)] = 4 + 3[0+0] = 4$

So, $y_1(3) = 12$ and $y_2(3) = 4$.

- $m = 1$ (i.e., $x = 3$)

  $$y_{12} = y_{11} + 3y_{21} = 12 + 3 \times 4 = 24$$

  $$y_{22} = y_{21} + 3[5y_{11} + 10x(1-x)] = 4 + 3[5 \times 12 + 30 \times (1-3)] = 4$$

  It means, $y_1(6) = 24$ and $y_2(6) = 4$.

- $m = 2$ (i.e., $x = 6$)

  $$y_{13} = y_{12} + 3y_{22} = 24 + 3 \times 4 = 36$$

  Thus, $y_1(9) = 36$.

Step 3: Convergence check.
This is done by comparing the estimated and given values of y at $x = 9$ as:

$$|y_1(9) - y(9)| = |36 - 0| = 36 > \text{tol} \ (= 10^{-6})$$

Note that at $x = 9$, y is calculated as 36 but the actual y is 0 [see second BC in Equation (5.55b)]. Since they are not sufficiently close, let us further arbitrarily guess $y_2(0) = -2$ and repeat the same computational Steps 1 to 3 followed above. Basically we are repeating this to generate two sets of solution that are required for linear interpolation.

Accordingly, the initial conditions are modified to:

$$y_1(x = 0) = 0 \qquad \text{(first BC)}$$

$$y_2(x = 0) = -2 \qquad \text{(assumed)}$$

From Equation (5.57), we have:

- $m = 0$ (i.e., $x = 0$)

  $$y_{11} = y_{10} + 3y_{20} = -6$$

  $$y_{21} = y_{20} + 3[5y_{10} + 10x(1-x)] = -2$$

So here, $y_1(3) = -6$ and $y_2(3) = -2$.

- $m = 1$ (i.e., $x = 3$)

  $$y_{12} = y_{11} + 3y_{21} = -12$$

  $$y_{22} = y_{21} + 3[5y_{11} + 10x(1-x)] = -272$$

Thus, $y_1(6) = -12$ and $y_2(6) = -272$.

- $m = 2$ (i.e., $x = 6$)

  $$y_{13} = y_{12} + 3y_{22} = -828$$

This means, $y_1(9) = -828$.

Doing the convergence check,

$$|y_1(9) - y(9)| = 828 > \text{tol} \ (=10^{-6})$$

we see that the tolerance condition is not satisfied. Thus, in the next stage, we are going to initialize y_2 through the linear interpolation as discussed below.

Linear interpolation

Let us tabulate the results (Table 5.6) obtained based on the two guess values of y_2.

TABLE 5.6 Numerical results

Assumed $y_2(0)$	Estimated $y(9)$
4	36
→ q = ?	0 (actual) = p
–2	–828

We know that the actual $y(9) = 0$, for which, we are trying to find the correct $y_2(0)$. Suppose the corresponding $y_2(0)$ is q and to find this, let us use the linear interpolation technique illustrated in Figure 5.10. It gives:

$$\frac{q-q_0}{p-p_0}=\frac{q_1-q_0}{p_1-p_0}$$

Rearranging,

$$\begin{aligned} q &= q_0+\frac{q_1-q_0}{p_1-p_0}(p-p_0) \\ &= 4+\frac{-2-4}{-828-36}(0-36)=3.75 \end{aligned}$$

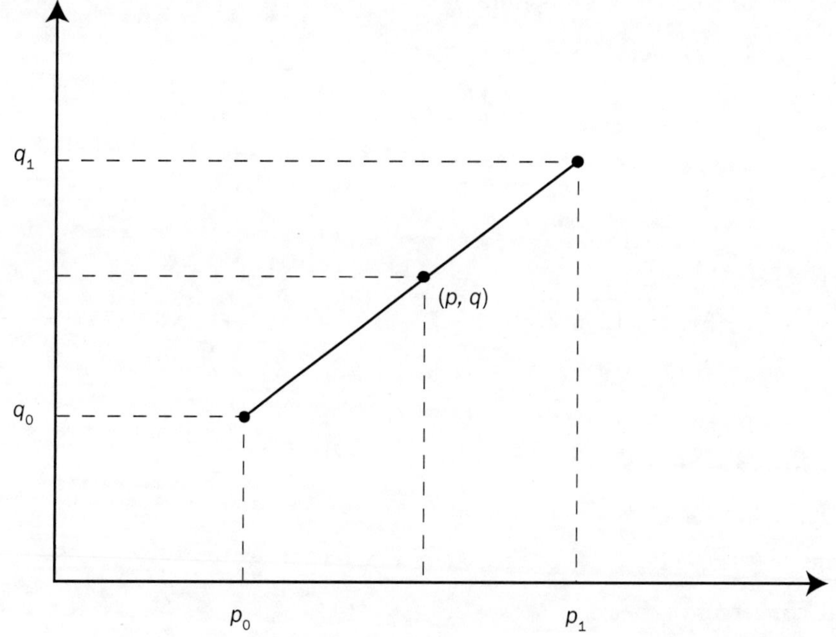

FIGURE 5.10 Illustrating linear interpolation

With this new initial guess (= 3.75), the transformed IVP is rewritten below as:

$$\frac{dy_1}{dx} = y_2 \tag{5.56a}$$

$$\frac{dy_2}{dx} = 5y_1 + 10x(1-x) \tag{5.56b}$$

Initial conditions:

$y_1(x=0)=0$ (first BC)

$y_2(x=0)=3.75$ (estimated)

Let us now verify whether the estimated $y_2(0)$ of 3.75 leads to $y(9) = 0$. Applying the forward Euler method further, we get the results shown in Table 5.7.

TABLE 5.7 Numerical results

m	x	$y_m = y_{1m}$	y_{2m}
0	0	0	3.75
1	3	11.25	3.75
2	6	22.5	−7.5
3	9	0	not required

Our verification is successful since we got $y(x = 9)$ exactly equal to 0. Since the desired tolerance (= 10^{-6}) is satisfied, we stop our computation here and note down the final solution of the ODE-BVP system obtained by the Shooting method as:

$y(x=0)=0$

$y(x=3)=11.25$

$y(x=6)=22.5$

$y(x=9)=0$

The first and last solutions are basically the given boundary conditions. Notice that along with the solutions (y), Table 5.7 also reports the respective slopes (y_2).

Remarks

1. The solution of the ODE-BVP system varies with the initial value of the transformed ODE-IVP system guessed in the Shooting method. In this regard, let us precisely present the results obtained for the above example in Table 5.8 that shows how the numerical solutions vary with the initial value of y_2. As shown, there are three guess values of y_2 used, namely 4, −2 and 3.75 (actually 3.75 is estimated through linear interpolation). The results are also compared in Figure 5.11.

Table 5.8 Effect of initial guess of y_2

x	Numerical solutions (y) when		
	$y_{20} = 4$	$y_{20} = -2$	$y_{20} = 3.75$
0	0	0	0
3	12	−6	11.25
6	24	−12	22.5
9	36	−828	0

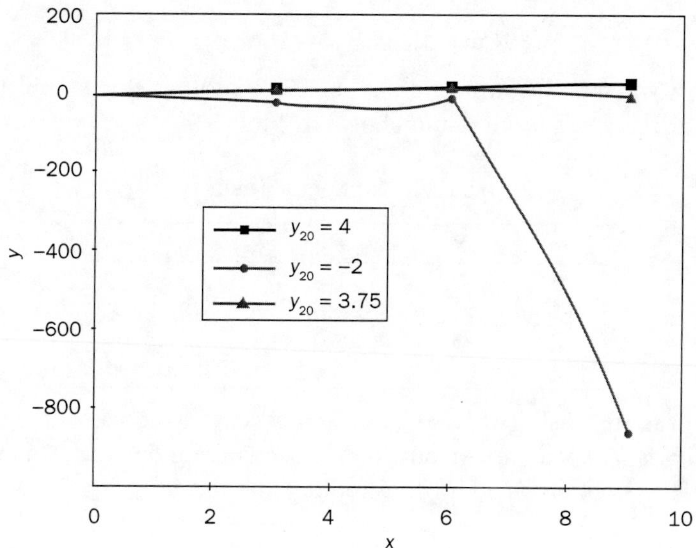

Figure 5.11 Comparative y profiles against different guess values of y_{20}

2. In the Shooting method, one can use the linear interpolation technique to estimate the third guess value for y_2 with adopting the first and second guess values arbitrarily. If it fails to satisfy the tolerance condition, it is required to repeat the procedure with changing the initial guess values. However, the same job we can do more efficiently and rigorously by using an iterative convergence method and it is elaborated below.

5.8.2 Reformulating Shooting Method for Non-linear ODE-BVP System [Equation (5.50)]

Let us further elaborate the solution methodology of the non-linear BVP system represented by Equation (5.50) on $x \in [0, 1]$ by the use of the Shooting method. The boundary conditions are rewritten as:

$$y(0)=\alpha \tag{5.58a}$$

$$y(1)=\beta \tag{5.58b}$$

Notice that here Equation (5.58b) is obtained by putting $a = 1$ in Equation (5.51b).

Recall the transformed IVP system,

$$\frac{dy_1}{dx} = y_2 \tag{5.52a}$$

$$\frac{dy_2}{dx} = f(x, y_1, y_2) \tag{5.52b}$$

for which, $y_1(0)$ is specified as α. One should assume a value for $y_2(0)$ as, say:

$$y_2(0) = s$$

Here s is a real number.

As stated, then apply an IVP solver (e.g., Euler or RK4 method) to integrate Equation (5.52), and get $y_1(x)$ and $y_2(x)$. It is true that their profiles will depend on the value of s, which we attempt to guess such that

$$\lim_{i\to\infty} y_1(1, s^i) = y_1(1) = \beta \tag{5.59}$$

For this, it is recommended to initialize s as:

$$s^0 = y_2(0) \approx \frac{y_1(1) - y_1(0)}{1-0} = \beta - \alpha \tag{5.60}$$

This s^0 is actually the first guess value of s. With this, one should first verify whether $y_1(1, s^0)$ is sufficiently close to β. If not, correct the approximation by choosing elevations s^1, s^2 and so on until $y_1(1, s^i)$ comes close to β with satisfying the desired tolerance criterion. This is demonstrated in Figure 5.12.

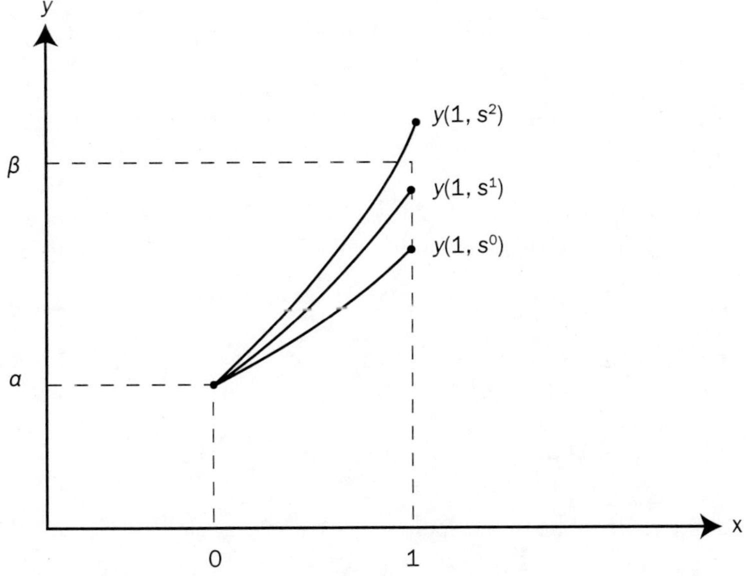

FIGURE 5.12 Illustrating the Shooting method for non-linear system

Now question is: how to choose s^i for $i \geq 1$? Indeed, we should choose it such that:

$$g(s) = y_1(1, s) - \beta = 0 \tag{5.61}$$

Actually, it should be $g(s) \leq \text{tol}$ (convergence criterion). Obviously, this is a non-linear algebraic equation. To solve it, one can use an iterative convergence technique, like Secant or N–R method (detailed in Chapter 3), and it is further discussed below.

A. Secant method

This numerical convergence method yields with two guess values (i.e., s^0 and s^1):

$$s^{i+1} = s^i - \frac{s^i - s^{i-1}}{g^i - g^{i-1}} g^i \tag{5.62a}$$

It gives,

$$s^{i+1} = s^i - \frac{s^i - s^{i-1}}{y_1(1, s^i) - y_1(1, s^{i-1})} [y_1(1, s^i) - \beta] \tag{5.62b}$$

Obviously, $y_1(1, s^i)$ is the last element in the array y_1. For this Secant method, one can assume the first guess s^0 $(= \beta - \alpha)$ based on Equation (5.60) and the second guess arbitrarily as s^1. With this, compute s^2 using Equation (5.62b) in order to continue the iterations.

B. Newton–Raphson method

This approach yields:

$$s^{i+1} = s^i - \frac{g(s^i)}{\left[g(s^i)\right]'} \tag{5.63a}$$

Note that as usual $g(s^i) = g^i$. The above equation gives:

$$s^{i+1} = s^i - \frac{y_1(1, s^i) - \beta}{\frac{dy_1}{ds}(1, s^i)} \tag{5.63b}$$

This N–R method involves only one guess (i.e., s^0), which we can directly get from Equation (5.60) and this is a great advantage over the Secant method. However, there is a difficulty associated with the N–R approach in finding $\frac{dy_1}{ds}(1, s^i)$ since an explicit representation for $y_1(1, s)$ is not known; we only have the values $y_1(1, s^0)$, $y_1(1, s^1)$, ……., $y_1(1, s^i)$. Let us discuss now how to resolve this issue.

Finding $\left[g(s^i)\right]'$ for N–R Method

Here we would like to know how to calculate g', which is actually $\frac{dy_1}{ds}$. For this, we assume:

$$\frac{\partial y_1}{\partial s}(x,\, s) = y_3(x,\, s) \tag{5.64}$$

$$\frac{\partial y_2}{\partial s}(x,\, s) = y_4(x,\, s) \tag{5.65}$$

We can then write,

$$\frac{dy_3}{dx} = \frac{d}{dx}\left(\frac{\partial y_1}{\partial s}\right) = \frac{\partial}{\partial s}\left(\frac{dy_1}{dx}\right) = \frac{\partial y_2}{\partial s} = y_4 \tag{5.66}$$

Similarly,

$$\frac{dy_4}{dx} = \frac{d}{dx}\left(\frac{\partial y_2}{\partial s}\right) = \frac{\partial}{\partial s}\left(\frac{dy_2}{dx}\right) = \frac{\partial f}{\partial s}(x,\, y_1,\, y_2)$$

$$= \frac{\partial f}{\partial x}\frac{\partial x}{\partial s} + \frac{\partial f}{\partial y_1}\frac{\partial y_1}{\partial s} + \frac{\partial f}{\partial y_2}\frac{\partial y_2}{\partial s}$$

Since x and s are independent, $\frac{\partial x}{\partial s} = 0$. Accordingly, we get:

$$\frac{dy_4}{dx} = y_3(x,\, s)\frac{\partial f}{\partial y_1}(x,\, y_1,\, y_2) + y_4(x,\, s)\frac{\partial f}{\partial y_2}(x,\, y_1,\, y_2) \tag{5.67}$$

With this, we have:

$$g' = \frac{\partial y_1}{\partial s}(1,\, s^i) = y_3(1,\, s^i)$$

Accordingly, Equation (5.63b) gets the following final form:

$$s^{i+1} = s^i - \frac{y_1(1,\, s^i) - \beta}{y_3(1,\, s^i)} \tag{5.68}$$

Now we would like to discuss how to obtain this y_3 for the N–R method [Equation (5.68)]. For this, we have to walk a little more as directed below.

Finding initial conditions for Equations (5.66) and (5.67)

At $x = 0$, we can rewrite Equation (5.58a) as:

$y_1(0,\, s) = \alpha$ (given BC)

Thus,

$$\frac{\partial y_1}{\partial s}(0,\, s) = \frac{d\alpha}{ds} = 0$$

It means,

$$y_3(x = 0) = 0$$

Again at $x = 0$, we assumed:

$y_2(0, s) = s$ (assumed)

Differentiating,

$$\frac{\partial y_2}{\partial s}(0, s) = \frac{ds}{ds} = 1$$

and thus,

$$y_4(x = 0) = 1$$

Representing transformed IVP for the Shooting method (with N–R approach)

In summary, we get the following coupled equations for the transformed ODE-IVP system:

$$\frac{dy_1}{dx} = y_2$$

$$\frac{dy_2}{dx} = f(x, y_1, y_2)$$

$$\frac{dy_3}{dx} = y_4$$

$$\frac{dy_4}{dx} = y_3\frac{\partial f}{\partial y_1} + y_4\frac{\partial f}{\partial y_2}$$

Initial conditions:

$$y_1(x=0)=\alpha$$

$$y_2(x=0)=s$$

$$y_3(x=0)=0$$

$$y_4(x=0)=1$$

Note that by simulating the above set of ODEs, we get y_3 required for the N–R formula [i.e., Equation (5.68)] to update s. Once we obtain the correct value of s with satisfying the given tolerance condition, we save the corresponding values of y (= y_1) at various nodes as our final solution of the given ODE-BVP system governed by Equation (5.50).

This is the general formulation of the Shooting method with an iterative convergence approach (i.e., Secant or N–R). Let us now apply this method to a specific non-linear BVP system below.

Case Study 5.4 Non-linear ODE-BVP: Shooting method (with Euler + Secant)

Steady state one-dimensional heat conduction in a slab

An elementary volume in a slab for one-dimensional heat conduction is schematically depicted in Figure 5.13. As shown, the temperature of one side of the slab is maintained at T_0, while the other side is at T_1.

At the first stage of this study, we would like to develop the model of this heat transfer problem with the following assumptions.

Assumptions

1. Steady state.
2. Volumetric rate of heat generation remains uniform throughout the wall (i.e., constant).
3. Heat conduction occurs only in x-direction that is normal to the wall.

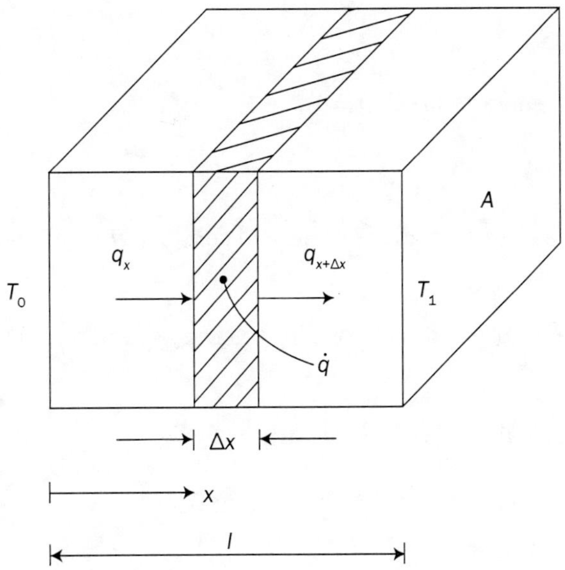

FIGURE 5.13 An elementary volume in a slab

Deriving model

Based on Equation (A4.3), we have the following heat balance equation at steady state condition:

$$\begin{pmatrix}\text{Rate of heat input}\\ \text{to the element at } x\end{pmatrix}+\begin{pmatrix}\text{Rate of heat generation}\\ \text{in the element}\end{pmatrix}=\begin{pmatrix}\text{Rate of heat output from}\\ \text{the element at } x+\Delta x\end{pmatrix}$$

Recall that the accumulation term is zero at steady state.

For the metal slab, we get:

$$q_x + \dot{q}A\Delta x = q_{x+\Delta x}$$

In which,

A = heat transfer area
q = heat transfer rate
$\dot{q}$ = rate of heat generation per unit volume

Using the Fourier law [modified form of Equation (5.18)],

$$q = -KA\frac{dT}{dx}$$

we obtain,

$$-KA\left(\frac{dT}{dx}\right)_x + KA\left(\frac{dT}{dx}\right)_{x+\Delta x} + \dot{q}A\Delta x = 0$$

As usual, K denotes the thermal conductivity and it is a function of temperature T. Dividing by $A\Delta x$ and taking limit $\Delta x \to 0$, we have:

$$\frac{d}{dx}\left[K(T)\frac{dT}{dx}\right] + \dot{q} = 0 \tag{5.69}$$

If there is no heat generation (i.e., $\dot{q} = 0$),

$$\frac{d}{dx}\left[K(T)\frac{dT}{dx}\right] = 0 \tag{5.70}$$

Now, we consider a linear variation of thermal conductivity with temperature as:

$$K = K_0 + K_1(T - T_0) \tag{5.71}$$

which makes the heat transfer model non-linear. Here, K_0 and K_1 are the constants.

Problem statement

Let us use the following dimensionless quantities:

$$\theta = \frac{T - T_0}{T_1 - T_0} \quad \text{(dimensionless temperature)} \tag{5.72}$$

$$\zeta = \frac{x}{l} \quad \text{(dimensionless position)} \tag{5.22}$$

$$a = \frac{K_1}{K_0}(T_1 - T_0) \quad \text{(dimensionless constant)} \tag{5.73}$$

Here, l denotes the slab length (shown in Figure 5.13).

Accordingly, Equation (5.70) yields:[10]

$$(1 + a\theta)\frac{d^2\theta}{d\zeta^2} + a\left(\frac{d\theta}{d\zeta}\right)^2 = 0 \qquad 0 \le \zeta \le 1 \tag{5.74}$$

Boundary conditions:

$$\theta(\zeta = 0) = 0 \tag{5.75a}$$

$$\theta(\zeta = 1) = 1 \tag{5.75b}$$

Solve this non-linear ODE-BVP system, which consists of Equations (5.74) and (5.75), by the use of the Shooting method with $a = 1$ and $M = 10$. For this, apply the forward Euler technique to integrate the transformed IVP system and for generating the guess values of θ' (i.e., s), use the Secant method (tol = 10^{-4}).

10 Derivation of Equation (5.74) is quite straightforward and thus, it is not shown here.

Solution

To solve this non-linear boundary value problem, let us reframe the computational algorithm developed before for Shooting method in step-wise fashion.

Step 1: *Transform the BVP to IVP system.*
Considering $\theta = \theta_1$,

$$\frac{d\theta_1}{d\zeta} = \theta_2 \tag{5.76a}$$

$$\frac{d\theta_2}{d\zeta} = -\frac{1}{1+\theta_1}\theta_2^2 \tag{5.76b}$$

Initial conditions:

$\theta_1(\zeta = 0) = 0$ (first BC)

$\theta_2(\zeta = 0) = \frac{1-0}{1-0} = 1$ [assumed based on Equation (5.60)]

Obviously, here the first guess, $s^0 = 1$.

Step 2: *Integrate the transformed IVP system.*
Using the forward Euler method with step size = 0.1 (since $M = 10$), we get (details not shown):

$\theta_1(1, s^0) = 0.7348$

Step 3: *Convergence check.*
Let us quantify how close we are to the actual value with finding:

$\left|\theta_1(1, s^0) - 1\right| = 0.2652 > \text{tol } (=10^{-4})$

It is obvious that the assumed value of $\theta_2(0)$ (= 1) fails to meet the convergence criterion. Thus, we need to update it by using an iterative convergence technique, namely Secant method.

To start the Secant method, we need the second guess for θ_2 (i.e., s^1) chosen as:

$s^1 = 1.2$

Although it is quite unlikely, still let us check whether this second guess, s^1 can satisfy the tolerance condition. With this, we solve the IVP system [Equation (5.76)] with $\theta_1(0) = 0$ and $\theta_2(0) = 1.2$, and obtain:

$\theta_1\left(1, s^1\right) = 0.8441$

for which,

$\left|\theta_1(1, s^1) - 1\right| = 0.1559 > \text{tol}$

Obviously, our both guess values, s^0 and s^1, are far from the actual s, which should lead to yield $\theta(\zeta=1)=1$. Thus, in the next phase, we need to update the initial guess of θ_2 with formally using the Secant method until the convergence criterion is satisfied. It is discussed below.

Secant Method

From Equation (5.62b) with:

- $i = 1$

$$s^2 = s^1 - \frac{s^1 - s^0}{\theta_1(1, s^1) - \theta_1(1, s^0)} [\theta_1(1, s^1) - 1]$$

$$= 1.2 - \frac{1.2 - 1}{0.8441 - 0.7348}(-0.1559)$$

$$= 1.4853$$

Again solving the IVP system [Equation (5.76)] with $\theta_1(0)=0$ and $\theta_2(0)=1.4853$,

$$\theta_1(1, s^2) = 0.9870$$

for which,

$$|\theta_1(1, s^2) - 1| = 0.013 > \text{tol}$$

- $i = 2$

$$s^3 = s^2 - \frac{s^2 - s^1}{\theta_1(1, s^2) - \theta_1(1, s^1)} [\theta_1(1, s^2) - 1]$$

$$= 1.4853 - \frac{1.4853 - 1.2}{0.9870 - 0.8441}(-0.013)$$

$$= 1.5112$$

With this, we get:

$$\theta_1(1, s^3) = 0.9993$$

and

$$|\theta_1(1, s^3) - 1| = 7 \times 10^{-4} > \text{tol}$$

- $i = 3$

$$s^4 = 1.5112 - \frac{1.5112 - 1.4853}{0.9993 - 0.9870}(-0.0007)$$

$$= 1.5127$$

With this, again we get:

$$\theta_1(1, s^4)=0.99998$$

$$\left|\theta_1(1, s^4)-1\right|=2\times10^{-5}<\text{tol}$$

So, s = 1.5127 satisfies our convergence criterion (i.e., tol = 10^{-4}). With this, we produce the final solutions by applying the forward Euler method for the example steady state heat conduction ODE-BVP system in Table 5.9.

Table 5.9 Numerical solutions for θ (i.e., θ_m with m = 0, 1, 2, ... 10)

$\theta_0 = 0.00$	$\theta_1 = 0.1513$	$\theta_2 = 0.2796$
$\theta_3 = 0.3937$	$\theta_4 = 0.4976$	$\theta_5 = 0.5938$
$\theta_6 = 0.6838$	$\theta_7 = 0.7687$	$\theta_8 = 0.8493$
$\theta_9 = 0.9262$	$\theta_{10} = 0.99998$	

Case Study 5.5 Non-linear ODE-BVP: Shooting method (with Euler + N–R)

Steady state one-dimensional heat conduction in a slab (revisited)

The non-linear ODE-BVP system is formulated before with:

$$(1+a\theta)\frac{d^2\theta}{d\zeta^2}+a\left(\frac{d\theta}{d\zeta}\right)^2=0 \qquad 0\le\zeta\le1 \tag{5.74}$$

Boundary conditions:

$$\theta(\zeta=0)=0 \tag{5.75a}$$

$$\theta(\zeta=1)=1 \tag{5.75b}$$

Now our goal is to solve this problem by the use of the Shooting method (with forward Euler + N–R) for a = 1, M = 10 and a tolerance limit of 10^{-4}.

Solution

Let us follow the same computational steps as in Case Study 5.4.

Step 1: Transform the BVP to IVP system.

The transformation is already done in the last Case Study 5.4 that yields the following equations:

$$\frac{d\theta_1}{d\zeta}=\theta_2 \tag{5.76a}$$

$$\frac{d\theta_2}{d\zeta}=-\frac{1}{1+\theta_1}\theta_2^2 \tag{5.76b}$$

These two coupled ODEs are to be solved using the initial conditions [i.e., $\theta_1(0)=0$ and $\theta_2(0)=s$] by the use of the forward Euler method. To reach at the correct value of s, the N–R iterative convergence method is to be employed. For this, we need to reformulate the IVP system [i.e., Equation (5.76)] representing the metal slab based on the procedure outlined before with associating the N–R algorithm.

Let us modify Equations (5.64) and (5.65) to:

$$\frac{\partial \theta_1}{\partial s}(\zeta, s)=\theta_3(\zeta, s) \tag{5.77}$$

$$\frac{\partial \theta_2}{\partial s}(\zeta, s)=\theta_4(\zeta, s) \tag{5.78}$$

With this, we get from Equation (5.66),

$$\frac{d\theta_3}{d\zeta}=\frac{d}{d\zeta}\left(\frac{\partial \theta_1}{\partial s}\right)=\frac{\partial}{\partial s}\left(\frac{d\theta_1}{d\zeta}\right)=\frac{\partial \theta_2}{\partial s}=\theta_4 \tag{5.79}$$

Similarly,

$$\begin{aligned}\frac{d\theta_4}{d\zeta} &= \frac{d}{d\zeta}\left(\frac{\partial \theta_2}{\partial s}\right)=\frac{\partial}{\partial s}\left(\frac{d\theta_2}{d\zeta}\right)\\ &= \frac{\partial}{\partial s}\left(-\frac{1}{1+\theta_1}\theta_2^2\right)\\ &= \frac{\partial}{\partial \theta_1}\left(-\frac{\theta_2^2}{1+\theta_1}\right)\frac{\partial \theta_1}{\partial s}+\frac{\partial}{\partial \theta_2}\left(-\frac{\theta_2^2}{1+\theta_1}\right)\frac{\partial \theta_2}{\partial s}\\ &= \left(\frac{\theta_2}{1+\theta_1}\right)^2\frac{\partial \theta_1}{\partial s}-\frac{2\theta_2}{1+\theta_1}\frac{\partial \theta_2}{\partial s}\end{aligned}$$

This gives,

$$\frac{d\theta_4}{d\zeta}=\left(\frac{\theta_2}{1+\theta_1}\right)^2\theta_3-\frac{2\theta_2}{1+\theta_1}\theta_4 \tag{5.80}$$

Moreover, we have the following form for the N–R algorithm [based on Equation (5.68)]:

$$s^{i+1}=s^i-\frac{\theta_1(1, s^i)-1}{\theta_3(1, s^i)} \tag{5.81}$$

Next, let us discuss how to obtain this θ_3 for Equation (5.81).

Initial conditions for Equations (5.79) and (5.80)

At $\zeta=0$,

	$\theta_1(0, s)=0$	$\theta_2(0, s)=s$
Thus,	$\dfrac{\partial\theta_1}{\partial s}(0, s)=0$	$\dfrac{\partial\theta_2}{\partial s}(0, s)=\dfrac{ds}{ds}=1$
It gives,	$\theta_3(\zeta=0)=0$	$\theta_4(\zeta=0)=1$

Transformed IVP for the Shooting method (with N–R approach)
We get the following coupled IVP system equations:

$$\frac{d\theta_1}{d\zeta}=\theta_2$$

$$\frac{d\theta_2}{d\zeta}=-\frac{1}{1+\theta_1}\theta_2^2$$

$$\frac{d\theta_3}{d\zeta}=\theta_4$$

$$\frac{d\theta_4}{d\zeta}=\left(\frac{\theta_2}{1+\theta_1}\right)^2\theta_3-\frac{2\theta_2}{1+\theta_1}\theta_4$$

Initial conditions:

$\theta_1(\zeta=0)=0$

$\theta_2(\zeta=0)=s$

$\theta_3(\zeta=0)=0$

$\theta_4(\zeta=0)=1$

Step 2: *Integrate the transformed IVP system.*

Using the forward Euler method with step size = 0.1 (since M = 10) and

$$s^0=\frac{1-0}{1-0}=1 \qquad \text{[assumed based on Equation (5.60)]}$$

we get,

$\theta_1(1, s^0)=0.7348$

$\theta_3(1, s^0)=0.5679$

Step 3: *Convergence check.*

$$\left|\theta_1(1, s^0)-1\right|=0.2652>\text{tol } (=10^{-4})$$

It is obvious that the assumed value of $\theta_2(0)$ (=1) fails to meet the convergence criterion. Thus, we need to update it by using the N–R method and it is as follows.

N–R Method

From Equation (5.81) with:

- $i = 0$

$$s^1 = s^0 - \frac{\theta_1(1, s^0) - 1}{\theta_3(1, s^0)}$$

$$= 1 - \frac{0.7348 - 1}{0.5679}$$

$$= 1.4670$$

Again solving the IVP system with updated $\theta_2(0) = 1.4670$, along with $\theta_1(0) = 0$, $\theta_3(0) = 0$ and $\theta_4(0) = 1$, we have:

$$\theta_1(1, s^1) = 0.9782$$

$$\theta_3(1, s^1) = 0.4794$$

and $|\theta_1(1, s^1) - 1| = 0.0218 >$ tol

- $i = 1$

$$s^2 = s^1 - \frac{\theta_1(1, s^1) - 1}{\theta_3(1, s^1)}$$

$$= 1.4670 - \frac{0.9782 - 1}{0.4794}$$

$$= 1.5125$$

With this, we subsequently get:

$$\theta_1(1, s^2) = 0.9999$$

$$\theta_3(1, s^2) = 0.4721$$

and $|\theta_1(1, s^2) - 1| = 10^{-4} =$ tol

Notice that $s = 1.5125$ satisfies the convergence criterion (i.e., tol = 10^{-4}). Thus, we stop the iteration here and note down the final solutions in Table 5.10.

Table 5.10 Numerical solutions for θ (i.e., θ_m with m = 0, 1, 2, ... 10)

$\theta_0 = 0.00$	$\theta_1 = 0.1512$	$\theta_2 = 0.2796$
$\theta_3 = 0.3937$	$\theta_4 = 0.4976$	$\theta_5 = 0.5937$
$\theta_6 = 0.6837$	$\theta_7 = 0.7686$	$\theta_8 = 0.8492$
$\theta_9 = 0.9261$	$\theta_{10} = 0.9999$	

Dimensionless temperature (θ) profile

In Figure 5.14, the profiles of θ against different s^i values are compared. Note that all these values [i.e., $s^0 = 1$ (assumed), and $s^1 = 1.4670$ and $s^2 = 1.5125$ (both estimated)] are specific to the Shooting method (with Euler + N–R) discussed above for the metal slab. In that figure, one can see that how the dimensionless temperature, θ is progressively approaching the end point that corresponds to $\zeta=1$ at which $\theta(\zeta=1)=1$.

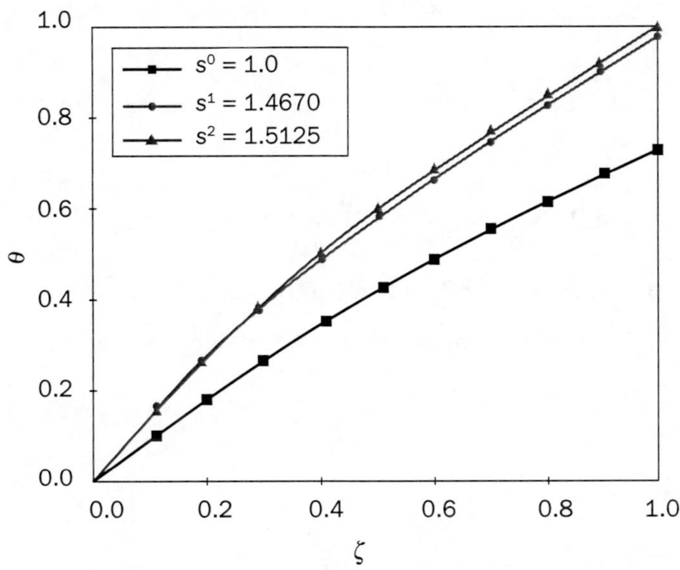

Figure 5.14 Profiles of θ against different s^i values

5.9 Coupled Boundary Value Problems

Our next assignment is to solve a BVP system with coupled ODEs. For this, let us consider a simple system of two general non-linear equations:

$$\frac{d^2y_1}{dx^2} = f_1(y_1, y_2) \tag{5.82}$$

$$\frac{d^2y_2}{dx^2} = f_2(y_1, y_2) \tag{5.83}$$

subject to boundary conditions on y_1 and y_2.

In order to solve this two-variable system, let us formulate the solution methodology through the following steps.

Discretize the Spatial Domain

Like single variable BVP system, at first we proceed to discretize the spatial domain of the double variable system into M equally spaced intervals. As shown in Figure 5.15, at each node x_m, there is a corresponding y_{1m} and y_{2m}. With this, there are total $2M + 2$ unknowns, namely $y_{10}, y_{11}, \ldots, y_{1M}, y_{20}, y_{21}, \ldots, y_{2M}$.

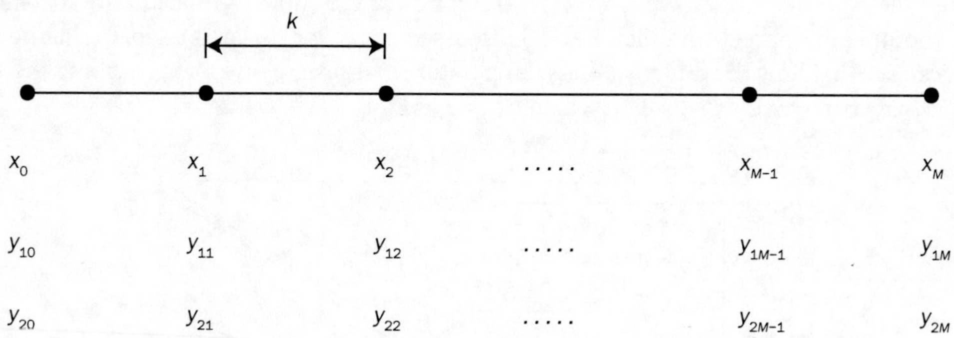

FIGURE 5.15 Discretization of spatial domain

Converting Local Variables (y_1 and y_2) to a Global Variable (z)

Appending those $2M + 2$ unknowns in a single vector, one obtains:

$$Z = \begin{bmatrix} y_{10} \\ y_{11} \\ \cdot \\ \cdot \\ y_{1M} \\ \hline y_{20} \\ y_{21} \\ \cdot \\ \cdot \\ y_{2M} \end{bmatrix} = \begin{bmatrix} z_0 \\ z_1 \\ \cdot \\ \cdot \\ z_M \\ \hline z_{M+1} \\ z_{M+2} \\ \cdot \\ \cdot \\ z_{2M+1} \end{bmatrix}$$

Accordingly, we can write:

$$z_m = \begin{cases} y_{1m} & \text{for } m \in [0,\ M] \\ \\ y_{2\ m-(M+1)} & \text{for } m \in [M+1,\ 2M+1] \end{cases} \tag{5.84}$$

As stated, here z is a global variable.

Discretize the ODEs in terms of Global Variable

Using the centered finite difference approximation, we get from Equation (5.82):

$$\frac{y_{1\ m+1}-2y_{1\ m}+y_{1\ m-1}}{k^2}=f_1(y_{1\ m},y_{2\ m}) \tag{5.85}$$

In terms of z,

$$\frac{z_{m+1}-2z_m+z_{m-1}}{k^2}-f_1(z_m,\ z_{m+M+1})=0 \tag{5.86}$$

Rearranging,

$$z_{m-1}-2z_m+z_{m+1}-k^2f_1(z_m,\ z_{m+M+1})=g_m\ \text{(say)} \tag{5.87}$$

Similarly, from Equation (5.83):

$$\frac{y_{2\ m+1}-2y_{2\ m}+y_{2\ m-1}}{k^2}=f_2(y_{1\ m},y_{2\ m}) \tag{5.88}$$

In terms of z,

$$\frac{z_{m+M+2}-2z_{m+M+1}+z_{m+M}}{k^2}-f_2(z_m,\ z_{m+M+1})=0 \tag{5.89}$$

Rearranging,

$$z_{m+M}-2z_{m+M+1}+z_{m+M+2}-k^2f_2(z_m,\ z_{m+M+1})=g_{m+M+1}\ \text{(say)} \tag{5.90}$$

Solve Discretized Equations

At the final stage, we need to solve the resulting algebraic Equations (5.87) and (5.90). Notice that they both are non-linear in nature and thus, let us use the Newton–Raphson method, which has the following form:

$$Z^{i+1}=Z^i-(\mathrm{J}^i)^{-1}G^i \tag{5.91}$$

Along with Equations (5.87) and (5.90), one must consider the given boundary conditions for finding J, and then compute Z vector as the solution of the coupled BVP.

Note

This type of coupled BVPs [i.e., Equations (5.82) and (5.83)] can alternatively be solved by representing Z with interlacing the two variable sets as:

$$Z=\begin{bmatrix} y_{10} \\ y_{20} \\ y_{11} \\ y_{21} \\ \cdot \\ \cdot \\ y_{1M} \\ y_{2M} \end{bmatrix}=\begin{bmatrix} z_0 \\ z_1 \\ z_2 \\ z_3 \\ \cdot \\ \cdot \\ z_{2M} \\ z_{2M+1} \end{bmatrix}$$

Then use the concept of global variable with the same logic as shown a little before and solve the resulting algebraic equations.

Case Study 5.6 Coupled BVP

Coupled heat and mass transfer problem

A non-isothermal reaction–diffusion system at steady state is schematically demonstrated in Figure 5.16. As shown, at $x = l$, the reactant concentration (C) is C_0 and at $x = 0$, $C = 0$. In both the left and right boundary of the reactor, the temperature is fixed at T_0.

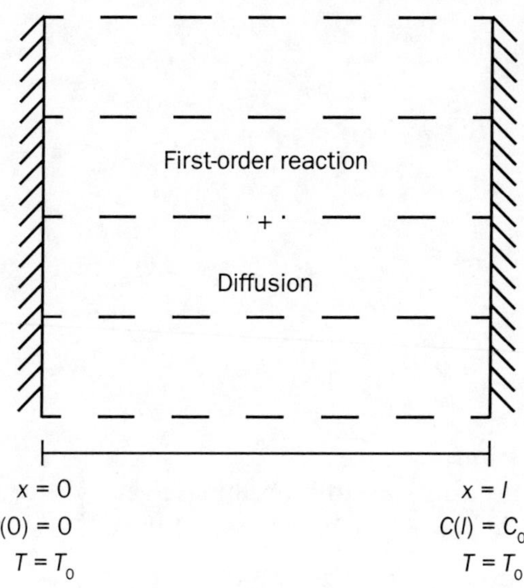

FIGURE 5.16 A non-isothermal reaction–diffusion system between two reservoirs (an empty reservoir on the left and a full reservoir on the right)

For the rectangular coordinate system, we know the following mass and energy balance equations (see Appendix A5.2):

Component mass balance (constant density, ρ and mass diffusivity, D)

$$\underbrace{\frac{\partial C}{\partial t}}_{\text{accumulation}}+\underbrace{v_x\frac{\partial C}{\partial x}+v_y\frac{\partial C}{\partial y}+v_z\frac{\partial C}{\partial z}}_{\text{convection}}=\underbrace{D\left(\frac{\partial^2 C}{\partial x^2}+\frac{\partial^2 C}{\partial y^2}+\frac{\partial^2 C}{\partial z^2}\right)}_{\text{diffusion}}+\underbrace{r}_{\text{reaction (generation)}} \qquad \text{[from Equation (A5.1)]}$$

Energy balance (constant ρ and thermal diffusivity, α)

$$\underbrace{\frac{\partial T}{\partial t}}_{\text{accumulation}}+\underbrace{\upsilon_x\frac{\partial T}{\partial x}+\upsilon_y\frac{\partial T}{\partial y}+\upsilon_z\frac{\partial T}{\partial z}}_{\text{convection}}=\underbrace{\alpha\left(\frac{\partial^2 T}{\partial x^2}+\frac{\partial^2 T}{\partial y^2}+\frac{\partial^2 T}{\partial z^2}\right)}_{\text{diffusion}}+\underbrace{H}_{\text{heat (source)}} \quad \text{[from Equation (A5.3)]}$$

Based on these two governing equations, we would like to formulate the model for the example non-isothermal reaction–diffusion system, making further simplifications through the following set of assumptions.

Assumptions

1. Steady state, i.e.,

$$\frac{\partial C}{\partial t}=\frac{\partial T}{\partial t}=0$$

2. One dimensional flow (i.e., only x-direction is considered). Thus, all differential terms with respect to y and z get cancelled.
3. There is no convective flow and consequently, the convective terms (mass and heat) are neglected, i.e.,

$$\upsilon_x\frac{\partial C}{\partial x}=\upsilon_x\frac{\partial T}{\partial x}=0$$

4. Constant physical properties (ρ, D, K and C_p (heat capacity)).
5. First-order endothermic reaction, i.e.,

$$-r=kC \quad (5.92)$$

Here, $(-r)$ denotes the rate of disappearance of reactant species. Thus, $-(-r)$ represents the rate of generation of reactant species, which is used later in the mass balance Equation (5.94).

With this, we have:

$$H=\frac{(\Delta H_r)(-r)}{\rho\, C_p}=\frac{(\Delta H_r)(kC)}{\rho\, C_p} \quad (5.93)$$

In which, (ΔH_r) denotes the heat of reaction [energy per mole (e.g., J/mol)]. Note that when heat is depleted in a system with endothermic reaction, the H term should be negative (i.e., heat sink). It is considered later in the energy balance Equation (5.95).

Deriving model

Based on these assumptions, the general mass and energy balance Equations (A5.1) and (A5.3) take the following respective forms:

$$D\frac{d^2C}{dx^2}-(-r)=0 \tag{5.94}$$

$$\alpha\frac{d^2T}{dx^2}-H=0 \tag{5.95}$$

Inserting Equations (5.92) and (5.93) along with the Arrhenius expression,

$$k=k_0\exp\left(-\frac{E}{RT}\right) \tag{5.96}$$

we get from Equations (5.94) and (5.95), respectively as:

$$D\frac{d^2C}{dx^2}-Ck_0\exp\left(-\frac{E}{RT}\right)=0 \tag{5.97}$$

$$K\frac{d^2T}{dx^2}-(\Delta H_r)Ck_0\exp\left(-\frac{E}{RT}\right)=0 \tag{5.98}$$

Here,

k_0 = pre-exponential factor or frequency factor
E = activation energy
R = universal gas constant

Converting model to its dimensionless form

The developed model of the coupled reaction–diffusion system consists of Equations (5.97) and (5.98). Now we would like to transform them into their dimensionless forms. For this, let us consider the following dimensionless terms:

$$\bar{C}=\frac{C}{C_0} \qquad \bar{T}=\frac{T}{T_0}$$

$$\bar{x}=\frac{x}{l} \qquad \beta=\frac{E}{RT_0}.$$

Component Mass Balance

Let us convert the component mass balance Equation (5.97) to a form with the above dimensionless terms. Substituting them in Equation (5.97),

$$\frac{DC_0}{l^2}\frac{d^2\bar{C}}{d\bar{x}^2}=\bar{C}C_0k_0\exp\left(-\frac{\beta}{\bar{T}}\right)$$

Rearranging,

$$\frac{d^2\bar{C}}{d\bar{x}^2} = \frac{k_0 l^2}{DC_0}\bar{C}C_0 \exp\left(-\frac{\beta}{\bar{T}}\right)$$

$$= \frac{k_0 l^2}{D}\bar{C} \exp\left(-\frac{\beta}{\bar{T}}\right)$$

Defining the Thiele modulus,

$$\phi = \sqrt{\frac{k_0 l^2}{D}} \tag{5.99}$$

we get,

$$\frac{d^2\bar{C}}{d\bar{x}^2} = \phi^2 \bar{C} \exp\left(-\frac{\beta}{\bar{T}}\right) \tag{5.100}$$

This represents the component mass balance equation in dimensionless form. The boundary conditions specific to this model equation include:

$$\bar{C}(\bar{x} = 0) = 0 \tag{5.101a}$$

$$\bar{C}(\bar{x} = 1) = 1 \tag{5.101b}$$

Energy Balance

Similarly, from Equation (5.98),

$$\frac{KT_0}{l^2}\frac{d^2\bar{T}}{d\bar{x}^2} = (\Delta H_r)\,\bar{C}C_0 k_0 \exp\left(-\frac{\beta}{\bar{T}}\right)$$

Rearranging,

$$\frac{d^2\bar{T}}{d\bar{x}^2} = \left(\frac{k_0 C_0 (\Delta H_r) l^2}{KT_0}\right)\bar{C} \exp\left(-\frac{\beta}{\bar{T}}\right)$$

Denoting,

$$\gamma = \left(\frac{k_0 C_0 (\Delta H_r) l^2}{KT_0}\right)$$

we get,

$$\frac{d^2\bar{T}}{d\bar{x}^2} = \gamma\,\bar{C} \exp\left(-\frac{\beta}{\bar{T}}\right) \tag{5.102}$$

This represents the energy balance equation in dimensionless form. The corresponding boundary conditions include:

$$\bar{T}(\bar{x} = 0) = 1 \tag{5.103a}$$

$$\bar{T}(\bar{x} = 1) = 1 \tag{5.103b}$$

Problem statement

The reaction–diffusion problem is modeled above and it yields the following non-linear ODE-BVP system:

$$\frac{d^2\bar{C}}{d\bar{x}^2} = \phi^2\bar{C}\exp\left(-\frac{\beta}{\bar{T}}\right) \tag{5.100}$$

$$\frac{d^2\bar{T}}{d\bar{x}^2} = \gamma\,\bar{C}\exp\left(-\frac{\beta}{\bar{T}}\right) \tag{5.102}$$

over the domain $\bar{x} \in [0,\ 1]$ subject to:

$\bar{C}(\bar{x}=0)=0$ (BC1) (5.101a)

$\bar{C}(\bar{x}=1)=1$ (BC2) (5.101b)

$\bar{T}(\bar{x}=0)=1$ (BC3) (5.103a)

$\bar{T}(\bar{x}=1)=1$ (BC4) (5.103b)

Solve this problem with:

$M = 4$
$\beta = 0.01$
$\phi^2 = 1$
$\gamma = 15$

Solution

For solving these two coupled BVP equations, we would like to apply the solution methodology developed before through finite difference approximation with the use of global variable.

Discretize the spatial domain

At first, let us discretize $\bar{x}$ into 4 equally spaced intervals since M is given as 4 (i.e., $k = \Delta\bar{x} = 0.25$). With this,

$z_0=\bar{C}_0=0$ $z_1=\bar{C}_1$ $z_2=\bar{C}_2$ $z_3=\bar{C}_3$ $z_4=\bar{C}_4=1$

$z_5=\bar{T}_0=1$ $z_6=\bar{T}_1$ $z_7=\bar{T}_2$ $z_8=\bar{T}_3$ $z_9=\bar{T}_4=1$

Notice that along with clubbing the two dependent variables ($\bar{C}$ and $\bar{T}$) into a global variable (z), we have also used a common index, m for them. It leads to yield $z_0,\ z_1,\ z_2,\ \cdots,\ z_9$, and this m can be called as the *global index.*[11]

11 This concept will further be extended in next Chapter 6.

Discretize the ODEs in terms of global variable

Component Mass Balance

Applying the centered finite difference approximation on Equation (5.100),

$$\frac{\bar{C}_{m+1}-2\bar{C}_m+\bar{C}_{m-1}}{k^2}-\phi^2\bar{C}_m\exp\left(-\frac{\beta}{\bar{T}_m}\right)=0 \tag{5.104}$$

In terms of z [from Equation (5.86)],

$$\frac{z_{m+1}-2z_m+z_{m-1}}{k^2}-\phi^2 z_m\exp\left(-\frac{\beta}{z_{m+5}}\right)=0$$

Rearranging,

$$z_{m-1}-2z_m-k^2\phi^2 z_m\exp\left(-\frac{\beta}{z_{m+5}}\right)+z_{m+1}=g_m \tag{5.105}$$

Notice that this is an equivalent form of the general Equation (5.87). Here, this resulting non-linear algebraic equation is to be solved for m = 1, 2, 3. If one wishes to employ the Newton–Raphson method, the Jacobian matrix needs to be constructed. Accordingly, we do the following formulations, for:

- $m = 1$ $\qquad z_0-2z_1-k^2\phi^2 z_1\exp\left(-\frac{\beta}{z_6}\right)+z_2=g_1$

  With $z_0=\bar{C}_0=0$ (BC1),

  $$\frac{\partial g_1}{\partial z_1}=-2-k^2\phi^2\exp\left(-\frac{\beta}{z_6}\right)=g_{11} \qquad \frac{\partial g_1}{\partial z_2}=1=g_{12}$$

  $$\frac{\partial g_1}{\partial z_6}=-k^2\phi^2 z_1\frac{\beta}{z_6^2}\exp\left(-\frac{\beta}{z_6}\right)=g_{16}$$

- $m = 2$ $\qquad z_1-2z_2-k^2\phi^2 z_2\exp\left(-\frac{\beta}{z_7}\right)+z_3=g_2$

  $$\frac{\partial g_2}{\partial z_1}=1=g_{21} \qquad \frac{\partial g_2}{\partial z_2}=-2-k^2\phi^2\exp\left(-\frac{\beta}{z_7}\right)=g_{22}$$

  $$\frac{\partial g_2}{\partial z_3}=1=g_{23} \qquad \frac{\partial g_2}{\partial z_7}=-k^2\phi^2 z_2\frac{\beta}{z_7^2}\exp\left(-\frac{\beta}{z_7}\right)=g_{27}$$

- $m = 3$ $\qquad z_2-2z_3-k^2\phi^2 z_3\exp\left(-\frac{\beta}{z_8}\right)+z_4=g_3$

  With $z_4=\bar{C}_4=1$ (BC2),

  $$\frac{\partial g_3}{\partial z_2}=1=g_{32} \qquad \frac{\partial g_3}{\partial z_3}=-2-k^2\phi^2\exp\left(-\frac{\beta}{z_8}\right)=g_{33}$$

  $$\frac{\partial g_3}{\partial z_8}=-k^2\phi^2 z_3\frac{\beta}{z_8^2}\exp\left(-\frac{\beta}{z_8}\right)=g_{38}$$

Energy Balance

Similarly, applying the centered finite difference approximation on Equation (5.102),

$$\frac{\overline{T}_{m+1}-2\overline{T}_m+\overline{T}_{m-1}}{k^2}-\gamma\overline{C}_m\exp\left(-\frac{\beta}{\overline{T}_m}\right)=0 \tag{5.106}$$

In terms of z [from Equation (5.89)],

$$\frac{z_{m+6}-2z_{m+5}+z_{m+4}}{k^2}-\gamma z_m\exp\left(-\frac{\beta}{z_{m+5}}\right)=0$$

Rearranging, one obtains the following non-linear algebraic form of equation:

$$z_{m+4}-2z_{m+5}-k^2\gamma z_m\exp\left(-\frac{\beta}{z_{m+5}}\right)+z_{m+6}=g_{m+5} \tag{5.107}$$

For m = 1, 2, 3, we further do the following formulations.

- $m = 1$ $\quad z_5-2z_6-k^2\gamma z_1\exp\left(-\frac{\beta}{z_6}\right)+z_7=g_6$

  With $z_5=\overline{T}_0=1$ (BC3),

  $$\frac{\partial g_6}{\partial z_1}=-k^2\gamma\exp\left(-\frac{\beta}{z_6}\right)=g_{61} \qquad \frac{\partial g_6}{\partial z_6}=-2-k^2\gamma z_1\frac{\beta}{z_6^2}\exp\left(-\frac{\beta}{z_6}\right)=g_{66}$$

  $$\frac{\partial g_6}{\partial z_7}=1=g_{67}$$

- $m = 2$ $\quad z_6-2z_7-k^2\gamma z_2\exp\left(-\frac{\beta}{z_7}\right)+z_8=g_7$

  $$\frac{\partial g_7}{\partial z_2}=-k^2\gamma\exp\left(-\frac{\beta}{z_7}\right)=g_{72} \qquad \frac{\partial g_7}{\partial z_6}=1=g_{76}$$

  $$\frac{\partial g_7}{\partial z_7}=-2-k^2\gamma z_2\frac{\beta}{z_7^2}\exp\left(-\frac{\beta}{z_7}\right)=g_{77} \qquad \frac{\partial g_7}{\partial z_8}=1=g_{78}$$

- $m = 3$ $\quad z_7-2z_8-k^2\gamma z_3\exp\left(-\frac{\beta}{z_8}\right)+z_9=g_8$

  With $z_9=\overline{T}_4=1$ (BC4),

  $$\frac{\partial g_8}{\partial z_3}=-k^2\gamma\exp\left(-\frac{\beta}{z_8}\right)=g_{83} \qquad \frac{\partial g_8}{\partial z_7}=1=g_{87}$$

  $$\frac{\partial g_8}{\partial z_8}=-2-k^2\gamma z_3\frac{\beta}{z_8^2}\exp\left(-\frac{\beta}{z_8}\right)=g_{88}$$

Solve discretized equations

To solve the non-linear algebraic Equations (5.105) and (5.107), as stated, we would like to use the N–R convergence method. For this, all the concerned elements of the Jacobian matrix are formulated above. Then using Equation (5.91), one can represent the resulting system in matrix form as:

$$\begin{bmatrix} z_1^{i+1} \\ z_2^{i+1} \\ z_3^{i+1} \\ z_6^{i+1} \\ z_7^{i+1} \\ z_8^{i+1} \end{bmatrix} = \begin{bmatrix} z_1^{i} \\ z_2^{i} \\ z_3^{i} \\ z_6^{i} \\ z_7^{i} \\ z_8^{i} \end{bmatrix} - \begin{bmatrix} g_{11}^i & g_{12}^i & 0 & g_{16}^i & 0 & 0 \\ g_{21}^i & g_{22}^i & g_{23}^i & 0 & g_{27}^i & 0 \\ 0 & g_{32}^i & g_{33}^i & 0 & 0 & g_{38}^i \\ g_{61}^i & 0 & 0 & g_{66}^i & g_{67}^i & 0 \\ 0 & g_{72}^i & 0 & g_{76}^i & g_{77}^i & g_{78}^i \\ 0 & 0 & g_{83}^i & 0 & g_{87}^i & g_{88}^i \end{bmatrix}^{-1} \begin{bmatrix} g_1^{i} \\ g_2^{i} \\ g_3^{i} \\ g_6^{i} \\ g_7^{i} \\ g_8^{i} \end{bmatrix}$$

Solving this, we finally get the numerical solutions reported in Table 5.11. The results are also plotted in Figure 5.17, which depicts the dimensionless concentration and temperature profile.

TABLE 5.11 Numerical solutions of the coupled BVP

m	Dimensionless length ($\bar{x}_m$)	Dimensionless concentration ($\bar{C}_m$)	Dimensionless temperature ($\bar{T}_m$)
0	0	0.0 (given BC1)	1.0 (given BC3)
1	0.25	0.2163	0.4942
2	0.50	0.4458	0.1872
3	0.75	0.7017	0.2763
4	1.0	1.0 (given BC2)	1.0 (given BC4)

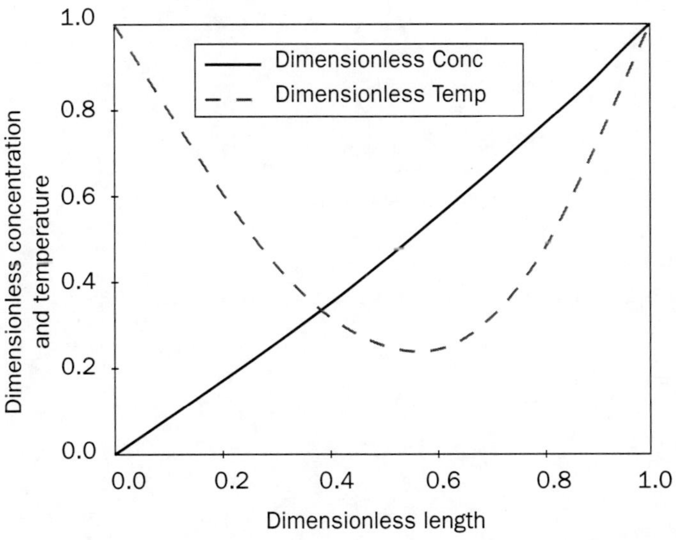

FIGURE 5.17 Concentration and temperature profile

5.10 Solved Case Studies

Case Study 5.7 A simple gravity pendulum

A simple gravity pendulum is demonstrated in Figure 5.18. As shown, a bob having a certain weight at the end of a rod or cord is suspended from a pivot. When there is no air resistance and the bob is displaced sideways from its resting (equilibrium) position, it swings back and forth with a constant amplitude. For developing the model of such a gravity pendulum[12] to predict the angular displacement, the following assumptions are considered.

Assumptions

1. There is no friction (frictionless pivot) and air resistance.
2. The rod is massless and inextensible.
3. Motion occurs only in two dimensions (i.e., the bob (a point mass) does trace an arc, not an ellipse).

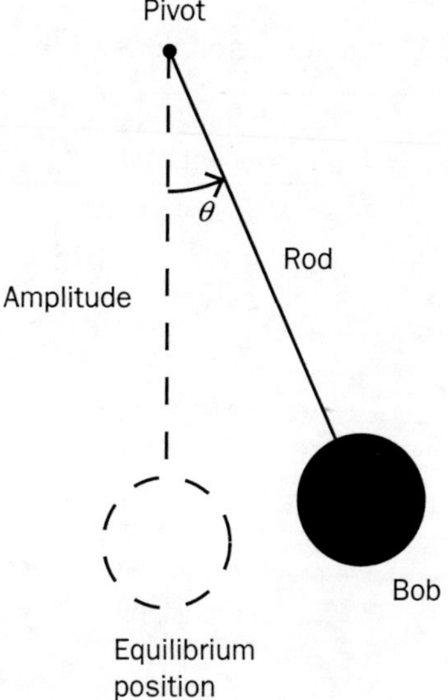

Figure 5.18 A simple gravity pendulum

Model equation

With these assumptions, a simple pendulum model is developed as:

$$\frac{d^2\theta}{dt^2}+\beta\sin\theta=0 \tag{5.108}$$

where,

$$\beta=\frac{g}{l}$$

12 This is used in quartz clock and in scientific instruments (e.g., accelerometer and seismometer).

in which,

g = acceleration due to gravity
l = pendulum length
θ = angular displacement (in radian)

Problem statement

For the simple pendulum modeled by Equation (5.108), we have:

Boundary conditions:

$$\theta(t=0)=0.7$$

$$\theta(t=2\pi)=0.5$$

Solve this non-linear ODE-BVP system[13] with $\beta = 0.7$ and $M = 6$ $\left(\text{i.e., } k=\frac{2\pi}{6}=1.0472\right)$.

Solution

In the gravity pendulum model [i.e., Equation (5.108)], the non-linearity comes from $\sin\theta$. One can linearize it as:

$$\sin\theta=\theta-\frac{\theta^3}{3\,!}+\frac{\theta^5}{5\,!}-.....$$

Approximating,

$$\sin\theta=\theta+o(\theta^3) \tag{5.109}$$

This is sometimes used when θ is reasonably small (in radian). However, using this approximation in the pendulum model Equation (5.108) leads to lose the key feature related to the pendulum oscillation. Thus, we would like to avoid this linearization approach.

Using the CD method, Equation (5.108) gives:

$$\frac{\theta_{m+1}-2\theta_m+\theta_{m-1}}{k^2}+\beta\sin\theta_m=f(\theta_m) \tag{5.110}$$

Rearranging,

$$\theta_{m-1}-2\theta_m+\theta_{m+1}+k^2\beta\sin\theta_m=k^2f(\theta_m) \tag{5.111}$$

This is the final discretized form of the pendulum model. Now this non-linear algebraic equation is to be used for all interior nodes (i.e., m = 1 to 5). If one wishes to employ the N–R method for numerical solution, it is required to formulate the Jacobian matrix. For this, we do the following formulation, for:

- $m = 1$ $\quad\theta_0-2\theta_1+\theta_2+k^2\beta\sin\theta_1=k^2f(\theta_1)=k^2f_1$

  With $\theta_0 = 0.7$ (BC1),

  $$\frac{\partial f_1}{\partial\theta_1}=\frac{1}{k^2}(-2+k^2\beta\cos\theta_1) \qquad \frac{\partial f_1}{\partial\theta_2}=\frac{1}{k^2}$$

13 Notice that in this boundary value problem, time (instead of space) is the independent variable.

- $m = 2$ $\qquad \theta_1 - 2\theta_2 + \theta_3 + k^2\beta\sin\theta_2 = k^2 f_2$

$$\frac{\partial f_2}{\partial\theta_1} = \frac{1}{k^2} \qquad \frac{\partial f_2}{\partial\theta_2} = \frac{1}{k^2}(-2 + k^2\beta\cos\theta_2) \qquad \frac{\partial f_2}{\partial\theta_3} = \frac{1}{k^2}$$

$\vdots$

- $m = M - 1$ $\qquad \theta_{M-2} - 2\theta_{M-1} + \theta_M + k^2\beta\sin\theta_{M-1} = k^2 f_{M-1}$

With $\theta_M = \theta_6 = 0.5$ (BC2),

$$\frac{\partial f_{M-1}}{\partial\theta_{M-2}} = \frac{1}{k^2} \qquad \frac{\partial f_{M-1}}{\partial\theta_{M-1}} = \frac{1}{k^2}(-2 + k^2\beta\cos\theta_{M-1})$$

Next we construct the Jacobian matrix for $M = 6$ as:

$$J = \frac{1}{k^2}\begin{bmatrix} -2+k^2\beta\cos\theta_1 & 1 & 0 & 0 & 0 \\ 1 & -2+k^2\beta\cos\theta_2 & 1 & 0 & 0 \\ \cdot & \cdot & \cdot & \cdot & \cdot \\ \cdot & \cdot & \cdot & \cdot & \cdot \\ 0 & 0 & 0 & 1 & -2+k^2\beta\cos\theta_5 \end{bmatrix}$$

The N–R method [see sample Equation (5.49)] yields the numerical solutions given in Table 5.12.

Table 5.12 Numerical solutions of the simple gravity pendulum

m	t_m	θ_m
0	0	0.7 (BC1)
1	1.0472	0.3333
2	2.0944	−0.2844
3	3.1416	−0.6868
4	4.1888	−0.6025
5	5.2360	−0.0831
6	6.2832	0.5 (BC2)

With this, we further produce the pendulum dynamics in Figure 5.19.

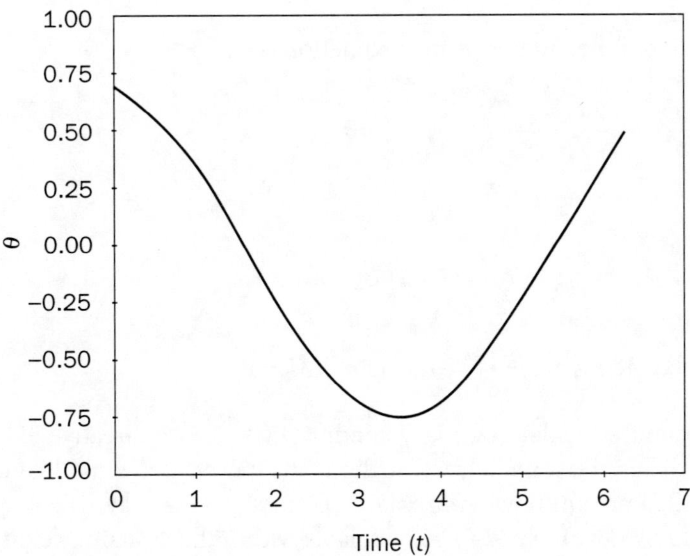

FIGURE 5.19 Angular displacement profile of the pendulum

Case Study 5.8 Reaction–diffusion in a catalyst pellet

For a porous spherical catalyst pellet, the mole balance in dimensionless form yields:

$$\frac{d^2\bar{C}}{d\bar{r}^2}+\frac{2}{\bar{r}}\frac{d\bar{C}}{d\bar{r}}-\phi^2\bar{C}=0 \tag{5.112}$$

This is obtained for the first-order irreversible reaction and at steady state isothermal condition. In this model equation,

$\bar{C}$ = dimensionless reactant concentration

$\bar{r}$ = dimensionless radius

ϕ = Thiele modulus

Boundary conditions include:

BC1: $\bar{C}$ is finite at the center of the pellet, i.e.,

$$\text{at } \bar{r}=0, \quad \frac{d\bar{C}}{d\bar{r}}=0$$

BC2: at $\bar{r}=1, \ \bar{C}=1$

Solve this linear ODE-BVP system with M = 10 for the three different values of Thiele modulus (i.e., $\phi=0.9, 2, 10$). Use the centered finite difference approximation to discretize all ODEs and compare the numerical results.

Solution

Using the centered finite difference approximation,

$$\frac{\overline{C}_{m+1}-2\overline{C}_{m}+\overline{C}_{m-1}}{k^2}+\frac{2}{\overline{r}}\frac{\overline{C}_{m+1}-\overline{C}_{m-1}}{2k}-\phi^2\,\overline{C}_m=0 \tag{5.113}$$

Rearranging,

$$(\overline{r}-k)\,\overline{C}_{m-1}-2\overline{r}\,\overline{C}_m-\overline{r}\,k^2\phi^2\overline{C}_m+(\overline{r}+k)\,\overline{C}_{m+1}=0 \tag{5.114}$$

Here, we have the mesh size, $k=\frac{1}{10}=0.1$ (since $M = 10$).

From the second Dirichlet boundary condition (BC2), we directly get $\overline{C}_{10}=1$. For the rest 10 unknown variables, namely $\overline{C}_m$ with $m=0, 1, ..., 9$, we would like to use Equation (5.114) that yields an unknown variable at fictitious node when $m = 0$ (i.e., $\overline{C}_{-1}$). To replace this unknown variable ($\overline{C}_{-1}$) by a variable within the domain, we discretize the first Neumann boundary condition (BC1) by the use of the CD approximation (ghost-point method), which yields:

$$\overline{C}_{-1}=\overline{C}_1$$

Using this, we solve the linear algebraic Equation (5.114) for $m=0, 1, ..., 9$ and produce the numerical results in Table 5.13 for three different values of Thiele modulus (i.e., $\phi=0.9, 2, 10$). The results are also compared in Figure 5.20.

Table 5.13. Numerical solutions for the catalyst pellet

$\overline{r}$	$\overline{C}$ ($\phi=0.9$)	$\overline{C}$ ($\phi=2$)	$\overline{C}$ ($\phi=10$)
0	0.878011	0.556103	0.001478
0.1	0.878011	0.556103	0.001478
0.2	0.881567	0.567225	0.002217
0.3	0.887512	0.586058	0.003942
0.4	0.895877	0.613056	0.007761
0.5	0.906701	0.648873	0.01626
0.6	0.920037	0.69438	0.035477
0.7	0.935951	0.750693	0.079611
0.8	0.954519	0.819201	0.182373
0.9	0.975834	0.901613	0.424407
1	1	1	1

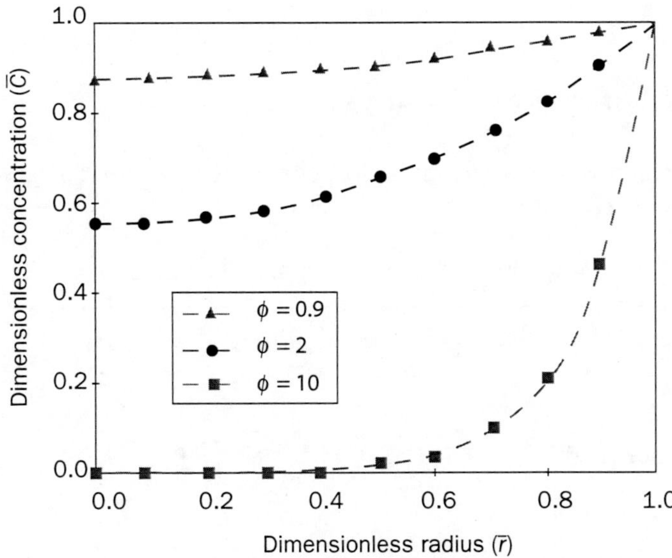

FIGURE 5.20 Comparing results for the catalyst pellet

5.11 Summary and Concluding Remarks

In this chapter, we have learned various numerical approaches and their applications in solving the systems described by ordinary differential equations with boundary conditions. There are three common types of boundary conditions we have come to know. Understanding them, we adopt a few systems having those boundary conditions individually or in combination with any two of them. We use the finite difference method as well as initial value method (Shooting method) to solve the boundary value problems, including coupled BVPs. A few case studies are also modeled and numerically solved for detailed analysis.

General Recommendations

For linear BVP systems, one can simply use the finite difference method (preferably CD that is second-order accurate) with a reasonably fine mesh size for the system model subjected to Dirichlet BC. For Neumann or Robin BC, one can also employ the CD method to discretize that BC and maintain the same order of accuracy.

Alternatively, we can use the CD method (second-order accurate) for the system ODE and three-point scheme (second-order accurate) for the Neumann or Robin BC (i.e., three-point FD for left boundary and three-point BD for right boundary) to secure the same order of accuracy with having no fictitious node. Additionally, it is suggested to do mesh refinement with the target of improving the accuracy level further.

For non-linear BVP systems, it is recommended to prefer the Shooting method that involves an IVP solver (preferably Runge–Kutta fourth-order (RK4) technique) and an iterative convergence method (e.g., Secant or Newton–Raphson). Judiciously select the mesh size for improved accuracy with taking into account both the truncation and round-off error.

Appendix 5A

A5.1 Finite Difference Formulas

The widely used finite difference formulas for numerical differentiation are documented in Table A5.1 for your ready use.

Table A5.1 Finite difference formulas

A. For First-Order Derivative	
Two-point formula	
Forward difference (FD)	$y'_m = \dfrac{y_{m+1} - y_m}{k} + o(k)$
Backward difference (BD)	$y'_m = \dfrac{y_m - y_{m-1}}{k} + o(k)$
Central difference (CD)	$y'_m = \dfrac{y_{m+1} - y_{m-1}}{2k} + o(k^2)$
Three-point formula	
Forward difference (FD)	$y'_m = \dfrac{-3y_m + 4y_{m+1} - y_{m+2}}{2k} + o(k^2)$
Backward difference (BD)	$y'_m = \dfrac{y_{m-2} - 4y_{m-1} + 3y_m}{2k} + o(k^2)$
B. For Second-Order Derivative	
Three-point formula	
Forward difference (FD)	$y''_m = \dfrac{y_{m+2} - 2y_{m+1} + y_m}{k^2} + o(k)$
Backward difference (BD)	$y''_m = \dfrac{y_{m-2} - 2y_{m-1} + y_m}{k^2} + o(k)$
Central difference (CD)	$y''_m = \dfrac{y_{m+1} - 2y_m + y_{m-1}}{k^2} + o(k^2)$

A Note on Derivation of Finite Difference Formula

Except the three-point FD and BD formulas for second-order derivative, the derivation of all other finite difference formulas documented in Table A5.1 is shown in the main text of this chapter. Now, let us move to first derive the following three-point FD formula:

$$y''_m = \frac{y_{m+2} - 2y_{m+1} + y_m}{k^2} + o(k)$$

Derivation

Let us start with:

$$y''_m = \left.\frac{d^2y}{dx^2}\right|_{x_m} = \frac{d}{dx}\left(\left.\frac{dy}{dx}\right|_{x_m}\right) = \frac{d}{dx}\left(\frac{y_{m+1} - y_m}{k} + o(k)\right)$$

$$= \frac{1}{k}\left(\frac{dy_{m+1}}{dx} - \frac{dy_m}{dx}\right) + o(k)$$

Note that the first-order FD formula is used above. Further using it:

$$y''_m = \frac{1}{k}\left(\frac{y_{m+2} - y_{m+1}}{k} - \frac{y_{m+1} - y_m}{k}\right) + o(k)$$

It gives:

$$y''_m = \frac{y_{m+2} - 2y_{m+1} + y_m}{k^2} + o(k)$$

The derivation is complete here. In the similar fashion, one can formulate the three-point BD scheme for the second-order derivative.

Merits and Demerits of Finite Difference Methods

- Quite easy to understand and apply to a wide variety of problems
- Point-wise discretization is made to obtain approximate solutions in those points
- Their derivations involve the simple Taylor series expansion
- Easy to implement for regular domains (i.e., rectangular, cylindrical and spherical coordinates)
- Often difficult to apply on complicated geometries
- Stronger regularity requirements (existence of high-order derivatives) over the finite element methods

A5.2 Reaction–Diffusion System

There are various reactive systems (e.g., reactors and reactive separators), for which, we typically consider the occurrence of the reaction and diffusion simultaneously. Indeed, in a reactive system, the reaction term acts as a source (or sink) of heat as well as mass. With this, here we would like to know the governing equations involving reaction–diffusion in rectangular, cylindrical and spherical coordinate systems.

A5.2.1 Rectangular Coordinate

The mass and energy balance equations for the reaction–diffusion problems in rectangular coordinate system are documented in Table A5.2 in general form. The derivation of these balance equations is quite straightforward and thus, the formulation of Equation (A5.1) is only shown below as a sample derivation.

Table A5.2 Mass and energy balance equations in rectangular coordinate

Mass balance for species A (constant ρ and D)

$$\frac{\partial C_A}{\partial t}+\upsilon_x\frac{\partial C_A}{\partial x}+\upsilon_y\frac{\partial C_A}{\partial y}+\upsilon_z\frac{\partial C_A}{\partial z}=D\left(\frac{\partial^2 C_A}{\partial x^2}+\frac{\partial^2 C_A}{\partial y^2}+\frac{\partial^2 C_A}{\partial z^2}\right)+R_A \tag{A5.1}$$

Total mass balance (constant ρ)

$$\frac{\partial \upsilon_x}{\partial x}+\frac{\partial \upsilon_y}{\partial y}+\frac{\partial \upsilon_z}{\partial z}=0 \tag{A5.2}$$

Energy balance (constant ρ and α)

$$\frac{\partial T}{\partial t}+\upsilon_x\frac{\partial T}{\partial x}+\upsilon_y\frac{\partial T}{\partial y}+\upsilon_z\frac{\partial T}{\partial z}=\alpha\left(\frac{\partial^2 T}{\partial x^2}+\frac{\partial^2 T}{\partial y^2}+\frac{\partial^2 T}{\partial z^2}\right)+H \tag{A5.3}$$

Here,

C_A = concentration of species A

υ = convective velocity

R_A = reaction term[14]

ρ = density

D = mass diffusivity

α = thermal diffusivity

H = heat source term[15]

Deriving Equation (A5.1)

Consider a three-dimensional (3D) differential control volume shown in Figure A5.1.

14 It is positive for the generation of species A (i.e., source) and negative for the disappearance of species A (i.e., sink).

15 It is positive for the exothermic heat generation (i.e., source) and negative for the endothermic heat depletion (i.e., sink).

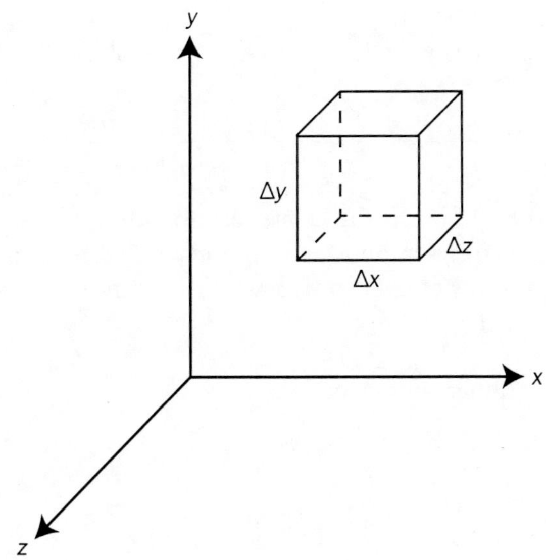

FIGURE A5.1 Differential control volume

The conservation principle states that [based on Equation (A4.1)]:

Rate of accumulation = Rate of input – Rate of output + Rate of generation (A5.4)

For *conservation of mass* (species A balance),

Accumulation of A:		$\Delta x\ \Delta y\ \Delta z\ [(C_A)_{t+\Delta t} - (C_A)_t]$
Input of A (diffusion):	x direction	$(J_{A,x})_x \Delta y\ \Delta z\ \Delta t$
	y direction	$(J_{A,y})_y \Delta x\ \Delta z\ \Delta t$
	z direction	$(J_{A,z})_z \Delta x\ \Delta y\ \Delta t$
Input of A (flow):	x direction	$(\upsilon_x C_A)_x \Delta y\ \Delta z\ \Delta t$
	y direction	$(\upsilon_y C_A)_y \Delta x\ \Delta z\ \Delta t$
	z direction	$(\upsilon_z C_A)_z \Delta x\ \Delta y\ \Delta t$
Output of A (diffusion):	x direction	$(J_{A,x})_{x+\Delta x} \Delta y\ \Delta z\ \Delta t$
	y direction	$(J_{A,y})_{y+\Delta y} \Delta x\ \Delta z\ \Delta t$
	z direction	$(J_{A,z})_{z+\Delta z} \Delta x\ \Delta y\ \Delta t$
Output of A (flow):	x direction	$(\upsilon_x C_A)_{x+\Delta x} \Delta y\ \Delta z\ \Delta t$
	y direction	$(\upsilon_y C_A)_{y+\Delta y} \Delta x\ \Delta z\ \Delta t$
	z direction	$(\upsilon_z C_A)_{z+\Delta z} \Delta x\ \Delta y\ \Delta t$
Generation of A (reaction):		$R_A \Delta x\ \Delta y\ \Delta z\ \Delta t$

Here, J_A denotes the diffusion flux.

Substituting all these terms in Equation (A5.4) and taking Δx, Δy, Δz, $\Delta t \to 0$,

$$\frac{\partial C_A}{\partial t} = -\frac{\partial(\upsilon_x C_A)}{\partial x} - \frac{\partial(\upsilon_y C_A)}{\partial y} - \frac{\partial(\upsilon_z C_A)}{\partial z} - \frac{\partial J_{A,x}}{\partial x} - \frac{\partial J_{A,y}}{\partial y} - \frac{\partial J_{A,z}}{\partial z} + R_A \tag{A5.5}$$

This is the general model for unsteady state reaction–diffusion in a rectangular coordinate system. With suitable considerations/assumptions, one can further use this mass balance equation in different forms of it. Let us discuss two common cases below.

Case 1: Constant D

The Fick's first law of diffusion is given by:

$$J_{A,x} = -D\frac{\partial C_A}{\partial x}$$

Assuming constant D, Equation (A5.5) yields:

$$\frac{\partial C_A}{\partial t} = -\frac{\partial(\upsilon_x C_A)}{\partial x} - \frac{\partial(\upsilon_y C_A)}{\partial y} - \frac{\partial(\upsilon_z C_A)}{\partial z} + D\frac{\partial^2 C_A}{\partial x^2} + D\frac{\partial^2 C_A}{\partial y^2} + D\frac{\partial^2 C_A}{\partial z^2} + R_A \tag{A5.6}$$

In vector form,

$$\frac{\partial C_A}{\partial t} = -\nabla.\ (\boldsymbol{\upsilon}\ C_A) + D\nabla^2 C_A + R_A$$

Case 2: Constant ρ and D

For this, let us revisit the differential control volume shown in Figure A5.1. For total mass balance, we have:

Accumulation:		$\Delta x\ \Delta y\ \Delta z\ [(\rho)_{t+\Delta t} - (\rho)_t]$
Input (flow):	x direction	$(\upsilon_x \rho)_x\ \Delta y\ \Delta z\ \Delta t$
Output (flow):	x direction	$(\upsilon_x \rho)_{x+\Delta x}\ \Delta y\ \Delta z\ \Delta t$

Similarly, write the input and output (flow) terms for other two directions (y and z directions).

Using Equation (A5.4) and then assuming constant ρ, one can get the following total mass balance equation:

$$\frac{\partial \upsilon_x}{\partial x} + \frac{\partial \upsilon_y}{\partial y} + \frac{\partial \upsilon_z}{\partial z} = 0 \tag{A5.2}$$

Recall that this is the total mass balance equation.

Substituting Equation (A5.2) in Equation (A5.6), one gets Equation (A5.1) that is basically obtained for constant ρ and D. Notice that along with Equation (A5.1), we did the derivation of Equation (A5.2) as well in the above.

A5.2.2 Cylindrical Coordinate

Like the rectangular coordinate, one can develop the balance equations for the cylindrical coordinate system $(r,\ \theta,\ z)$ listed in Table A5.3.

Table A5.3 Mass and energy balance equations in cylindrical coordinate

Mass balance for species A (constant ρ and D)

$$\frac{\partial C_A}{\partial t}+\upsilon_r\frac{\partial C_A}{\partial r}+\upsilon_\theta\frac{1}{r}\frac{\partial C_A}{\partial \theta}+\upsilon_z\frac{\partial C_A}{\partial z}=D\left(\frac{1}{r}\frac{\partial}{\partial r}\left(r\frac{\partial C_A}{\partial r}\right)+\frac{1}{r^2}\frac{\partial^2 C_A}{\partial \theta^2}+\frac{\partial^2 C_A}{\partial z^2}\right)+R_A \qquad \text{(A5.7)}$$

Energy balance (constant ρ and K (thermal conductivity))

$$\frac{\partial T}{\partial t}+\upsilon_r\frac{\partial T}{\partial r}+\upsilon_\theta\frac{1}{r}\frac{\partial T}{\partial \theta}+\upsilon_z\frac{\partial T}{\partial z}=\alpha\left(\frac{1}{r}\frac{\partial}{\partial r}\left(r\frac{\partial T}{\partial r}\right)+\frac{1}{r^2}\frac{\partial^2 T}{\partial \theta^2}+\frac{\partial^2 T}{\partial z^2}\right)+H \qquad \text{(A5.8)}$$

A5.2.3 Spherical Coordinate

In the similar fashion, the mass and energy balance equations can be developed in spherical coordinate system (r, θ, ϕ) and they are documented in Table A5.4.

Table A5.4. Mass and energy balance equations in spherical coordinate

Mass balance for species A (constant ρ and D)

$$\frac{\partial C_A}{\partial t}+\upsilon_r\frac{\partial C_A}{\partial r}+\upsilon_\theta\frac{1}{r}\frac{\partial C_A}{\partial \theta}+\upsilon_\phi\frac{1}{r\sin\theta}\frac{\partial C_A}{\partial \phi}$$
$$=D\left(\frac{1}{r^2}\frac{\partial}{\partial r}\left(r^2\frac{\partial C_A}{\partial r}\right)+\frac{1}{r^2\sin\theta}\frac{\partial}{\partial \theta}\left(\sin\theta\frac{\partial C_A}{\partial \theta}\right)+\frac{1}{r^2\sin^2\theta}\frac{\partial^2 C_A}{\partial \phi^2}\right)+R_A \qquad \text{(A5.9)}$$

Energy balance (constant ρ and K)

$$\frac{\partial T}{\partial t}+\upsilon_r\frac{\partial T}{\partial r}+\upsilon_\theta\frac{1}{r}\frac{\partial T}{\partial \theta}+\upsilon_\phi\frac{1}{r\sin\theta}\frac{\partial T}{\partial \phi}$$
$$=\alpha\left(\frac{1}{r^2}\frac{\partial}{\partial r}\left(r^2\frac{\partial T}{\partial r}\right)+\frac{1}{r^2\sin\theta}\frac{\partial}{\partial \theta}\left(\sin\theta\frac{\partial T}{\partial \theta}\right)+\frac{1}{r^2\sin^2\theta}\frac{\partial^2 T}{\partial \phi^2}\right)+H \qquad \text{(A5.10)}$$

Exercise

Review Questions

5.1 How initial value problems differ from boundary value problems?

5.2 Derive the three-point BD formula for the second-order derivative that is given in Table A5.1.

5.3 List the basic differences between FD, BD and CD methods.

5.4 How can the generation of fictitious node be avoided? Is there any gain?

5.5 What are the key differences between finite difference and the Shooting method?

5.6 Is there any gain by mesh refinement? What is the major drawback of this?

5.7 For non-linear BVP systems, is there any instability issue that arises when we use the finite difference method?

5.8 Is there any chance of having instability or divergence problem in the Shooting method? Discuss it.

5.9 In the Shooting method, what are the merits and demerits of using:

i) an iterative convergence technique (e.g., N–R) over linear interpolation?

ii) the N–R technique over the Secant method?

5.10 Recommend a suitable numerical solution strategy for solving the BVP system keeping the important issues (i.e., accuracy, total error and computational effort) in mind.

5.11 Why do we convert the BVP system model to its dimensionless form in many applications before solving it numerically? Does it happen for other differential systems (i.e., IVP or PDE)?

Practice Problems

5.12 Heat transfer in an infinite slab is characterized by the following ODE-BVP when internal heat generation is remained constant:

$$\frac{d^2y}{dx^2} = -1$$

on $x \in [0, 1]$ subject to:

$$y(0) = 0$$

$$y(1) = 0$$

Solve this system with $M = 10$.

5.13 An ODE-BVP system has the following form:

$$\frac{d^2u}{dx^2} = 5u$$

over the domain $x \in [0, 2]$ subject to:

at $x = 0$: $\quad \frac{du}{dx} = 0$

at $x = 2$: $\quad u = 5$

i) Solve this system by the use of CD method with $M = 10$

ii) Compare the numerical and analytical solutions

5.14 A boundary value problem is formulated as:

$$\frac{d^2u}{dx^2} = 1 + x$$

on $x \in [0, 1]$ subject to:

$$u(0) = 0$$

$$u(1) = 1$$

Considering $k = 0.2$, tabulate the numerical solutions with the use of central difference method.

5.15 Steady state one-dimensional heat conduction in a slab is modeled in Case Study 5.4. Further considering constant thermal conductivity, K and negligible heat generation ($\dot{q} = 0$), Equation (5.69) yields:

$$K\frac{d^2T}{dx^2} = 0$$

The rectangular slab having a length of 5 m maintains its left side at 500 °C and right side at 200 °C. Solve the problem to find the temperature distribution along the length.

5.16 The governing differential equation in terms of the dimensionless reactant concentration ($\bar{C} = C/C_o$) with respect to the dimensionless distance ($\bar{x} = x/l$) is given as:

$$\frac{d^2\bar{C}}{d\bar{x}^2} - 2\phi = 0$$

subjected to:

at $\bar{x} = 0$: $\quad \bar{C} = 1$

at $\bar{x} = 1$: $\quad \frac{d\bar{C}}{d\bar{x}} = 0$

Here,

$$\phi = \frac{kl^2}{2C_o D}$$

with

k = rate constant (zero-order reaction) = 10^{-3}
l = half-length of cylindrical pellet = 0.085
C_o = reactant concentration at the external surface of the catalyst pellet = 2×10^{-4}
D = mass diffusivity = 3.6×10^{-2}

Solve this problem using the finite difference method of your choice.

5.17 Solve the following reaction–diffusion problem:

$$\frac{d^2\bar{C}}{d\bar{x}^2} = \phi^2\bar{C}$$

on $\bar{x} \in [0, 1]$ subject to:

$$\bar{C}(0) = 0$$
$$\bar{C}(1) = 1$$

Here, the Thiele modulus, $\phi = 1.2$. Compare the numerical results with the analytical solution given as:

$$\bar{C} = \frac{\sinh\phi\,\bar{x}}{\sinh\phi}$$

5.18 The non-linear gravity pendulum model is numerically solved in Case Study 5.7. Compare those results with the solutions of the linearized model of that pendulum:

$$\frac{d^2\theta}{dt^2} + \beta\theta = 0$$

Use the same information given there.

5.19 Solve the above gravity pendulum model:

$$\frac{d^2\theta}{dt^2} + \beta\theta = 0$$

subjected to:

$$\theta(\pi/4) = -\pi/2$$

$$\dot{\theta}(\pi/2) = \pi/2$$

and $\beta = 0.7$.

5.20 Solve the following BVP system with the use of a numerical method of your choice:

$$\frac{d^2u}{dx^2} = -1 + u$$

over the domain $x \in [0,\ 1]$ subject to the following boundary conditions:

at $x = 0$: $\quad \frac{du}{dx} = 0$

at $x = 1$: $\quad u = 3$

with $k = 0.2$.

5.21 Reaction and diffusion in a pore is described by:

$$\frac{d^2C}{dx^2} - \frac{k}{D}C = 0$$

on $x \in [0,\ 1]$ subject to:

at $x = 0$ mm: $\quad C = 1$ mol/m^3

at $x = 1$ mm: $\quad \frac{dC}{dx} = 0$

Given:

rate constant, $k = 1.2 \times 10^{-3}$ sec^{-1}
effective diffusivity, $D = 1.2 \times 10^{-9}$ m^2/sec
$M = 50$

Solve the problem using the ghost-point method and produce the concentration profile (C versus x).

5.22 Repeat the Problem 5.21 with the use of:

– CD method for the system ODE.
– three-point BD method for the BC (Neumann).

Compare the solutions with those obtained in the last problem by the use of the ghost-point approach.

5.23 The distribution of neutrons in a slab is governed by:

$$\frac{d^2y}{dx^2} - \frac{1}{\xi^2}y = 0$$

over the domain $x \in [-2,\ 2]$ subject to:

$$y(-2) = 0$$

$$y(2) = 0$$

Solve it using a FDM with $M = 6$.

5.24 Solve the following non-linear ODE-BVP system:

$$\frac{d^2y}{dx^2} = -5y\frac{dy}{dx}$$

on $x \in [0,\ 2]$ subject to:

$$y(0) = 0$$
$$y(2) = 1$$

5.25 For the following boundary value problem,

$$\frac{d^2c}{dx^2} = 6c^2$$

$c(0) = 1$ and $c(0.5) = \frac{4}{9}$.

i) Find the solution using the Shooting method (with forward Euler + N–R) with tol = 10^{-5}.
ii) Compare the numerical solution with analytical one:

$$c(x) = \frac{1}{(1+x)^2}$$

5.26 A linear system represented by:

$$\frac{d^2y}{dx^2} = P(x)\frac{dy}{dx} + Q(x)y + R(x)$$

on $x \in [a,\ b]$ subject to:

$$y(a) = y_a$$
$$y(b) = y_b$$

Formulate the solution methodology by the use of the central finite difference approximation.

5.27 Consider the following ODE-BVP system:

$$\frac{d^2y}{dx^2} - 5\frac{dy}{dx} - 10y^2 = 0 \qquad 0 \le x \le 1$$

Boundary conditions:

at $x = 0$: $\quad \frac{dy}{dx} = 5(y-1)$

at $x = 1$: $\quad \frac{dy}{dx} = 0$

Solve this problem using the Shooting method (with forward Euler + linear interpolation).

5.28 Solve the above Problem 5.27 using the Shooting method (with RK2 + N–R).

5.29 A second-order boundary value problem is described by:

$$\frac{d^2y}{dx^2} + 4\frac{dy}{dx} + 1.75y = 0$$

Boundary conditions include:

$$y(x = 0) = 2$$
$$y\ (x = 1) = 0.5$$

Solve this linear system using the Shooting method (with RK3 + linear interpolation).

5.30 Solve the following ODE-BVP system:

$$y'' + 10y = \cos x$$

Boundary conditions:

$$y(\pi / 2) = -1.7$$
$$y'(0) = 5$$

5.31 Solve the following non-linear system using the Shooting method:

$$\frac{d^2y}{dx^2} + \frac{5}{y}\left(\frac{dy}{dx}\right)^3 = 0$$

on $x \in [1, 3]$ subject to the boundary conditions:

$$y(1) = \sqrt{2}$$
$$y(3) = 2$$

with $M = 20$. Use the explicit Euler along with the Secant as well as N–R method for comparison.

5.32 Use the Shooting method (with forward Euler + N–R) to solve the following boundary value problem:

$$x\frac{d^2y}{dx^2} - \frac{dy}{dx} + 2xy = x$$

Boundary conditions:

$$y(x = 1) = 1$$
$$y(x = 3) = 4$$

5.33 Consider the cooling fin system (see Case Study 5.1),

$$\frac{d^2T}{dx^2} = \frac{hp}{KA}(T - T_0)$$

subject to:

$$T(x = 0) = 200$$
$$T(x = l = 1) = 100$$

Solve this ODE-BVP using the Shooting method (with RK4 + linear interpolation) with

T_0 = ambient temperature = 0

$$\frac{hp}{KA} = 0.1$$

Compare the numerical and analytical solutions.

5.34 Reproduce the numerical results of Case Study 5.8 for Thiele modulus, $\phi = 1$ and compare with analytical solutions.

5.35 Consider a metal rod (Figure 5.21), in which, heat is transferred in axial direction by conduction. Heat is lost from the rod (at T) to the surroundings (at T_∞) by convection. Here radiation is assumed negligible.

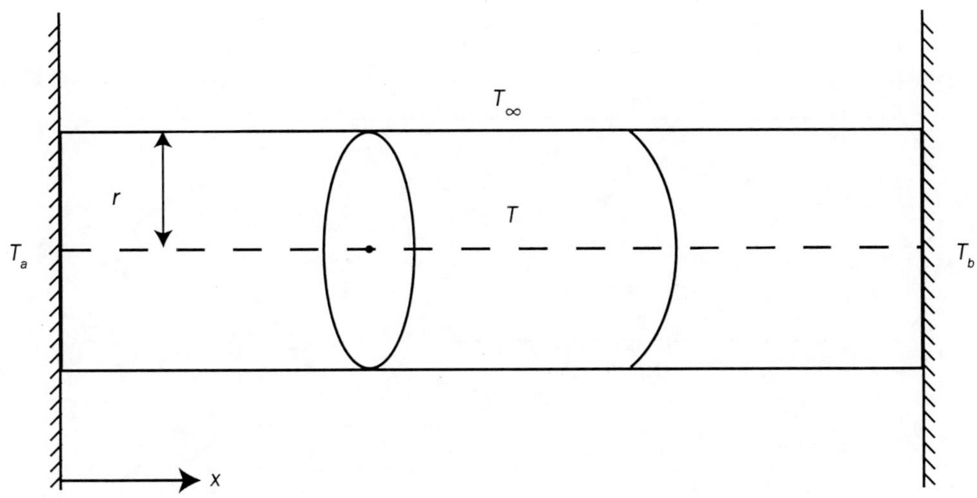

FIGURE 5.21 A metal rod (no radiative heat loss)

i) Develop the heat transfer model as:

$$\frac{d^2T}{dx^2} + \frac{2h}{rK}(T_\infty - T) = 0$$

ii) Solve this ODE-BVP system by the Shooting method for:

BCs: $T(x = 0 \text{ m}) = T(0) = T_a = 400 \text{ K}$

$T(x = l = 5 \text{ m}) = T(5) = T_b = 500 \text{ K}$

with

Surroundings temperature,	$T_\infty = 300$ K
Convective heat transfer coefficient,	$h = 1$ W/m².K
Radius of metal rod,	$r = 0.1$ m
Length of metal rod,	$l = 5$ m
Thermal conductivity,	$K = 100$ W/m.K

5.36 Consider heat conduction in a metal rod (Figure 5.22) with radiative heat loss to the surroundings. The heat transfer model has the following form:

$$K\frac{d^2T}{dx^2} - \varepsilon\,\sigma\,A(T^4 - T_\infty^4)$$

on $x \in [0, 5]$ subject to:

$T(x = 0 \text{ m}) = T_0 = 700 \text{ K}$

$T(x = l = 5 \text{ m}) = T_\infty = 300 \text{K}$

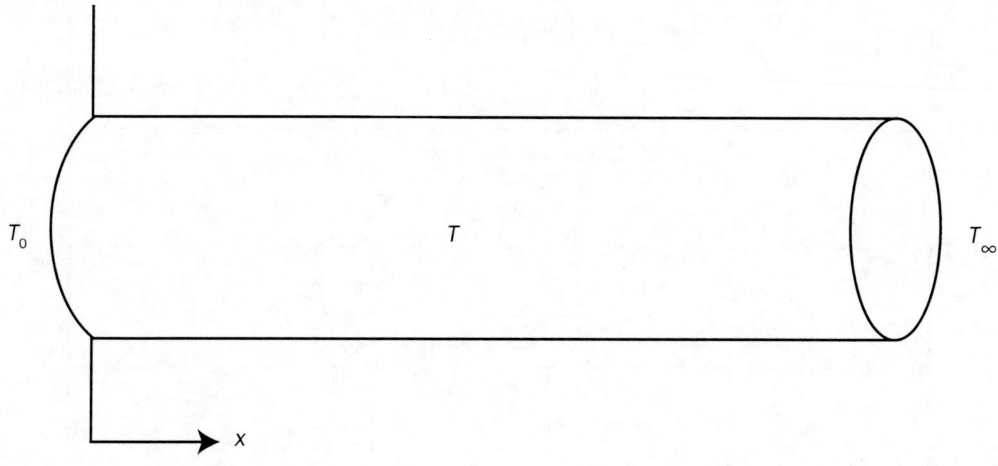

FIGURE 5.22 A metal rod

Given information:

ε = emissivity = 0.1

σ = Stefan–Boltzmann constant = 5.669×10^{-8} W/m^2.K^4

A = external area of the rod per unit volume = 5 m^{-1}

K = thermal conductivity = 100 W/m.K

Solve this problem employing the central difference method with $M = 50$.

5.37 To make the heat transfer model for the metal rod (Figure 5.21) more realistic, let us consider the heat loss occurred by both convection and radiation. Here, the heat is transferred in axial direction by conduction. Accordingly, the model gets the following form:

$$\frac{d^2T}{dx^2}+\frac{2h}{rK}(T_\infty - T)+\frac{\varepsilon\,\sigma\,A}{K}(T_\infty^4 - T^4)=0$$

Notice that this non-linear model is obtained by combining the models discussed in the last two problems on metal rod.

Solve this problem by using the Shooting method for:

BCs: $T(0) = 400$ K

$T(5) = 500$ K

Given:

$T_\infty = 300$ K

$\varepsilon = 0.1$

$\sigma = 5.669 \times 10^{-8}$ W/m^2.K^4

$A = 5$ m^{-1}

$h = 1$ W/m^2.K

$K = 100$ W/m.K

$r = 0.1$ m

$l = 5$ m

5.38 Solve the following ODE-BVP system:

$$\frac{d^2u}{dx^2}+\frac{du}{dx}-2u-e^{2u}=0$$

over the domain $x \in [0,\ 2]$ subject to:

$$u(0)=0$$
$$u(2)=0$$

Use a suitable FDM with $M=10$.

5.39 Solve the following ODE-BVP system:

$$\frac{d^2u}{dx^2}=3\frac{du}{dx}-u+xe^x-x$$

on $x \in [0,\ 3]$ subject to:

$$u(0)=0$$
$$u(3)=-4$$

Use the CD method with $M=10$.

5.40 Solve the following boundary value problem:

$$\frac{d^2y}{dx^2}+(1+x)\frac{dy}{dx}-1.8y=(1.2-x^2)e^{-x}$$

on $x \in [0,\ 1]$ subject to:

$$y(0)=0$$
$$y(1)=0$$

Use the FD method with $M=5$.

5.41 Solve the following boundary value problem (Kharab and Guenther, 2019):

$$\frac{d^2y}{dx^2}-\frac{3}{x}\frac{dy}{dx}+\frac{3y}{x^2}=2x^2e^x$$

on $x \in [1,\ 2]$ subject to:

$$y(1)=0$$
$$y(2)=29.556$$

Use the Shooting method with $M=10$.

5.42 A BVP system is described by:

$$\frac{d^2y}{dx^2}=-\frac{2x}{1+x^2}\frac{dy}{dx}+y+\frac{2}{1+x^2}-\log(1+x^2)$$

on $x \in [0,\ 1]$ subject to the following boundary conditions:

$$y(0)=0$$
$$y(1)=\log 2$$

Solve this ODE-BVP by using the FDM and compare the results with the analytical solution [that is, $y=\log(1+x^2)$].

5.43 Find the numerical solution of (Jain et al., 1993):

$$\frac{d^2y}{dx^2}+8y\sin^2(\pi x)=0$$

on $x\in[0, 1]$ subject to:

$$y(0)=1$$
$$y(1)=1$$

with $k = 0.2$.

5.44 Solve the following boundary value problem (Jain et al., 1993):

$$\frac{d^2y}{dx^2}+\frac{5x^2+3}{x(1+x^2)}\frac{dy}{dx}+\frac{4y}{3(1+x^2)}+\frac{1}{1+x^2}=0$$

on $x\in[-2, 2]$ subject to the following boundary conditions:

$$y(-2)=0.5$$
$$y(2)=0.5$$

Use a suitable finite difference method.

5.45 Reaction–diffusion model for a spherical catalyst pellet is given as:

$$D\frac{d^2C}{dr^2}+\frac{2D}{r}\frac{dC}{dr}-kC=0$$

subject to:

at $r = R$: $C=1$ mol/m^3

at $r = 0$: $\frac{dC}{dr}=0$

Given information:

radius of pellet, $R = 2.0$ cm
rate constant, $k = 0.001$ sec^{-1}
effective diffusivity, $D=1.2\times 10^{-9}$ m^2/sec

Solve this problem (using the CD for system ODE and three-point FD for BC) and produce the concentration profile.

5.46 Repeat the above problem concerning reaction–diffusion in a spherical catalyst particle with:

$$\frac{d^2C}{dr^2}+\frac{2}{r}\frac{dC}{dr}-4C=0$$

on $r\in[0, 1]$ subject to:

$$C(1)=1$$
$$\left.\frac{dC}{dr}\right|_{r=0}=0$$

Compute the concentration profile as a function of r using finite differences with five equally spaced intervals.

5.47 A spherical catalyst pellet is modeled in dimensionless form as (see Case Study 5.8):

$$\frac{d^2\bar{C}}{d\bar{r}^2}+\frac{2}{\bar{r}}\frac{d\bar{C}}{d\bar{r}}-\phi^2\bar{C}=0$$

where,

$$\bar{C}=\frac{C_A}{C_{A0}}=\text{dimensionless concentration}$$

$$\bar{r}=\frac{r}{R}=\text{dimensionless radius}$$

$$\phi^2=\frac{kR^2}{D}=\text{Thiele modulus}$$

D = diffusivity

Boundary conditions include:

at $r = 0$: $\frac{dC_A}{dr}=0$

at $r = R$: $C_A=C_{A0}=5\times10^{-5}$ mol/cm^3

Given parameters:

R = 4 mm

D = 10^{-5} cm^2/sec

k = 5 sec^{-1}

Solve this problem with M = 25.

5.48 The following model is used to describe the diffusion in a biological floc:

$$\frac{d^2y}{dx^2}-\frac{\varpi^2 y}{1+\sigma y}=0$$

subject to:

at x = 0: $y = 1$

at x = 1: $y' = 0$

Solve this problem for M = 5, $\varpi=1.5$ and $\sigma=1.1$.

5.49 The following model represents a spherical catalyst pellet, in which, both the chemical reaction and diffusion occur:

$$D\frac{d^2C_A}{dr^2}+\frac{2D}{r}\frac{dC_A}{dr}-r_A=0$$

Here, D is the effective diffusivity. Reaction kinetics is governed by the Langmuir–Hinshelwood model as:

$$-r_A=\frac{kC_A}{\sqrt{1+K_rC_A^2}}$$

where,

k = 80 sec^{-1}

$$K_r=1.5\times10^9\left(\frac{\text{cm}^3}{\text{mol}}\right)^2$$

Boundary conditions:

at $r = 0$: $\frac{dC_A}{dr} = 0$

at $r = R$: $C_A = 5\times10^{-5}$ mol/cm^3

Given parameters:

$R = 4$ mm

$D = 10^{-5}$ cm^2/sec

Solve this BVP with $M = 100$.

5.50 The following model (Schlichting, 1987) is used to describe the boundary layer flow over a stationary flat plate as:

$$2\frac{d^3y}{dx^3} + y\frac{d^2y}{dx^2} = 0$$

subject to:

$$y(0) = 0$$
$$y'(0) = 0$$
$$y'(5) = 1$$

Here

x = dimensionless coordinate that involves the location of a point

y = dimensionless stream function

y' = dimensionless velocity

Transform the model into a set of three first-order ODEs and develop the solution methodology using a suitable method.

5.51 Microwave heating of a slab (one-dimensional) is modeled by:

$$\frac{d^2T}{dx^2} + \eta e^T = 0$$

on $x \in [0, 1]$ subject to:

at $x = 0$: $T = 120$ (finite) and $\frac{dT}{dx} = 0$

at $x = 1$: $T = 170$

Solve this highly non-linear second-order ODE for $\eta = 1$.

5.52 Solve the following non-linear system:

$$\frac{d^2u}{dx^2} + 5u(1 - u^2) = 0$$

on $x \in [-1, 1]$ subject to:

$$u(-1) = -1$$
$$u(1) = 1$$

Solve this BVP with using central difference approximation for $M = 4$.

5.53 Repeat the above problem to find the solutions of the ODE-BVP by using the Shooting method with:

i) forward Euler + linear interpolation
ii) RK2 + Secant
iii) RK4 + N–R

5.54 A tubular reactor is modeled as:

$$D\frac{d^2C_A}{dx^2} - \upsilon\frac{dC_A}{dx} - kC_A^2 = 0$$

The boundary conditions include:

at the inlet ($x = 0$): $\upsilon C_{Ai} = \upsilon C_A - D\frac{dC_A}{dx}$

at the exit ($x = l$): $\frac{dC_A}{dx} = 0$

Given information:

length, $l = 10$ m
axial velocity, $\upsilon = 0.5$ m/sec
rate constant, $k = 2$ m^3/mol.sec
diffusivity, $D = 10^{-4}$ m^2/sec
inlet concentration of reactant A, $C_{Ai} = 1$ mol/m^3

Solve the problem using the CD method for $M = 10$.

5.55 In the method of lacquer application (called *curtain coating*), the following dimensionless model is used to characterize a thin sheet of viscous liquid emerging from a slot at the base of a converging channel as (Salariya, 1980):

$$\frac{d^2u}{dx^2} - \frac{1}{u}\left(\frac{du}{dx}\right)^2 - u\frac{du}{dx} = -1$$

subject to:

at $x = 0$: $u = 0.325$

at $x = 50$: $\frac{du}{dx} = \frac{1}{\sqrt{2x}}$

Here,

u = dimensionless velocity of the sheet
x = dimensionless distance from the slot

Solve the problem using the CD method and Shooting method (with RK2 + N–R).

5.56 The following second-order chemical reaction,

$$2A \rightarrow P$$

takes place in an isothermal tubular reactor. A mass balance on the reactor yields:

$$\frac{d^2c}{dz^2} + Pe\frac{dc}{dz} = Pe\,\phi^2\,c^2$$

subject to the following Danckwerts boundary conditions:

at $z = 0$: $\frac{dc}{dz} = 0$

at $z = 1$: $c_0 = c + \frac{1}{Pe}\frac{dc}{dz}$

Here,

c = dimensionless concentration of reactant A

z = dimensionless reactor length $\left(=1-\frac{x}{l}\right)$

l = reactor length

$\phi=\sqrt{\frac{k\,l^2}{D}}$ = Thiele modulus

k = reaction rate constant, volume/time.mol

D = diffusivity

Pe = Peclet number $=\frac{l\upsilon}{D}$

υ = fluid velocity

Develop the solution methodology for this non-linear BVP.

5.57 In an isothermal tubular reactor with axial mixing, a second-order irreversible chemical reaction takes place. This reactor is modeled as:

$$\frac{1}{Pe}\frac{d^2c}{dz^2}-\frac{dc}{dz}-Dac^2=0$$

subject to:

at $z = 0$: $\quad \frac{dc}{dz}=Pe\,(c-1)$

at $z = 1$: $\quad \frac{dc}{dz}=0$

Here,

c = dimensionless reactant concentration

z = dimensionless axial position

Da = Damkohler number (= 2)

Pe = Peclet number (= 6)

Solve this problem using the Shooting method (with Euler + Secant) for M = 5.

5.58 The heat transfer model (non-dimensionalized) of a tapered conical fin (Figure 5.23) has the following form (Chapra, 2012):

$$\frac{d^2u}{dz^2}+\frac{2}{z}\left(\frac{du}{dz}-\gamma\,u\right)=0$$

where,

u = dimensionless temperature ($0 \le u \le 1$)

z = dimensionless axial distance ($0 \le z \le 1$)

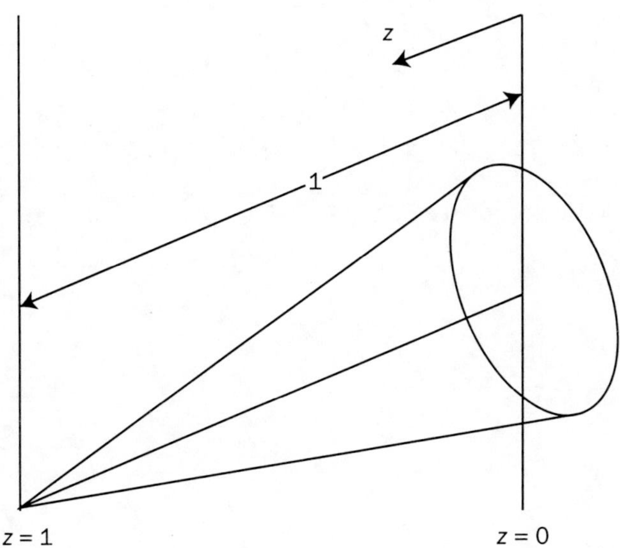

Figure 5.23 A tapered conical cooling fin

The non-dimensional parameter:

$$\gamma = \frac{hl}{K}\sqrt{1+\frac{4}{2m^2}}$$

In which,

h = convective heat transfer coefficient
K = thermal conductivity
l = length or height of the cone
m = slope of the cone wall

The following boundary conditions hold:

at $z = 0$: $u(0) = 0$
at $z = 1$: $u(1) = 1$

Solve this problem using a finite difference method for $\gamma = 25$.

5.59 Consider an insulated conductor for different current flow in two halves modeled by:

$$\frac{d^2T_1}{dx^2} + \frac{S_1}{K} = 0$$

$$\frac{d^2T_2}{dx^2} + \frac{S_2}{K} = 0$$

on $x \in [0,\ 2]$ subject to:

at $x = 0$: $T = T_1 = 25$ °C
at $x = 2$: $T = T_2 = 25$ °C
at $x = 1$: $T_1 = T_2$
at $x = 1$: $\frac{dT_1}{dx} = \frac{dT_2}{dx}$

Given information:

$S_1 = 2 \times 10^4$ W/m^3

$S_2 = 10^4$ W/m^3

$K = 100$ W/m.°C

Solve the problem using a FDM and produce the temperature profile. State necessary assumption, if any.

5.60 Solve the following coupled equations:

$$D_1 \frac{d^2C_1}{dx^2} = k_1C_1 - k_2C_2$$

$$D_2 \frac{d^2C_2}{dx^2} = -k_1C_1 + k_2C_2$$

on $0 \le x \le 1$ subject to:

$$C_1(x=0) = 0$$

$$C_2(x=0) = 1$$

at $x = 1$: $\quad \frac{dC_1}{dx} = 0$

at $x = 1$: $\quad \frac{dC_2}{dx} = 0$

Here, D is the diffusivity ($D_1 = D_2 = 1.5$) and k the rate constant ($k_1 = k_2 = 10$).

5.61 In a counter-current double pipe heat exchanger, a process stream flows through the tube (at temperature, T_t) and a heating medium through the shell side (at temperature, T_s). At steady state, the energy balance on the tube side and shell side yields:

$$\upsilon_t \frac{dT_t}{dx} = C_t(T_s - T_t)$$

$$-\upsilon_s \frac{dT_s}{dx} = C_s(T_t - T_s)$$

with

$$C_t = \frac{US}{\rho_t C_{pt} A_t}$$

$$C_s = \frac{US}{\rho_s C_{ps} A_s}$$

where,

υ = velocity

U = overall heat transfer coefficient

S = perimeter of inner pipe

A = cross-sectional area

ρ = density

C_p = heat capacity

Note that the subscript t is used for tube and s for shell.

Boundary conditions:

at $x = 0$: $\quad T_t(0) = 40\,°C$, $T_s(0) = 80\,°C$

at $x = l$: $\quad T_t(l) = 120\,°C$, $T_s(l) = 150\,°C$

Given parameter values:

l = length of the heat exchanger = 10 m

$C_t = 0.2\ \text{min}^{-1}$

$C_s = 0.5\ \text{min}^{-1}$

$\upsilon_t = 2.5$ m/min

$\upsilon_s = 5$ m/min

Solve the coupled ODEs for $M = 25$.

5.62 Axial conduction and diffusion in a tubular reactor yield:

$$\frac{1}{2}\frac{d^2C}{dx^2} - \frac{dC}{dx} - kC\exp\left[\eta\left(1-\frac{1}{T}\right)\right] = 0$$

$$\frac{1}{2}\frac{d^2T}{dx^2} - \frac{dT}{dx} - \beta\, kC\exp\left[\eta\left(1-\frac{1}{T}\right)\right] = 0$$

over $0 \le x \le 1$ subject to boundary conditions:

at $x = 0$: $\quad \frac{1}{2}\frac{dC}{dx} = C - 1 \quad$ and $\quad \frac{1}{2}\frac{dT}{dx} = T - 1$

at $x = 1$: $\quad \frac{dC}{dx} = \frac{dT}{dx} = 0$

Here, k (= 2), η (= 1) and β (= 1) are known. Solve this coupled boundary value problem using the finite difference method.

5.63 The material and energy balance equations for the porous catalyst particles include (McGinnis, 1965):

$$\frac{d^2\bar{C}}{d\bar{x}^2} + \frac{2}{\bar{x}}\frac{d\bar{C}}{d\bar{x}} = \phi^2\bar{C}\exp\left[\gamma\left(1-\frac{1}{\bar{T}}\right)\right]$$

$$\frac{d^2\bar{T}}{d\bar{x}^2} + \frac{2}{\bar{x}}\frac{d\bar{T}}{d\bar{x}} = -\beta\phi^2\bar{C}\exp\left[\gamma\left(1-\frac{1}{\bar{T}}\right)\right]$$

subject to:

at $\bar{x} = 0$: $\quad \frac{d\bar{C}}{d\bar{x}} = \frac{d\bar{T}}{d\bar{x}} = 0$

at $\bar{x} = 1$: $\quad \bar{C} = \bar{T} = 1$

Here,

$\bar{C}$ = dimensionless concentration

$\bar{T}$ = dimensionless temperature

$\bar{x}$ = dimensionless radial coordinate (spherical geometry)

ϕ = Thiele modulus

γ = Arrhenius number

β = Prater number

Solve this BVP system for $\gamma = 25$, $\beta = 0.2$ and $\phi = 1$ using the:

i) Finite difference method

ii) Shooting method

Hint: The given material and energy balance equations can be combined into:

$$\frac{d^2\bar{C}}{d\bar{x}^2}+\frac{2}{\bar{x}}\frac{d\bar{C}}{d\bar{x}}=\phi^2\bar{C}\exp\left[\frac{\gamma\beta(1-\bar{C})}{1+\beta(1-\bar{C})}\right]$$

subject to:

at $\bar{x}=0$: $$\frac{d\bar{C}}{d\bar{x}}=0$$

at $\bar{x}=1$: $$\bar{C}=1$$

5.64 Consider a reaction–diffusion system (tubular reactor), in which the following reactions take place:

$$A+B \xrightarrow{k_1} C$$

$$A+C \xrightarrow{k_2} D$$

Based on Equation (A5.1), the mole balance for the four species yields:

Species A

$$D\frac{d^2C_A}{dx^2}-\upsilon_x\frac{dC_A}{dx}-k_1C_AC_B-k_2C_AC_C=0$$

Species B

$$D\frac{d^2C_B}{dx^2}-\upsilon_x\frac{dC_B}{dx}-k_1C_AC_B=0$$

Species C

$$D\frac{d^2C_C}{dx^2}-\upsilon_x\frac{dC_C}{dx}+k_1C_AC_B-k_2C_AC_C=0$$

Species D

$$D\frac{d^2C_D}{dx^2}-\upsilon_x\frac{dC_D}{dx}+k_2C_AC_C=0$$

For this system (over the domain $0\le x\le 5$), the boundary conditions include:

$$\text{at } x=0\text{ m:}\quad\begin{cases}D\dfrac{dC_A}{dx}=\upsilon_x(C_A-C_{Ai})\\ D\dfrac{dC_B}{dx}=\upsilon_x(C_B-C_{Bi})\\ D\dfrac{dC_C}{dx}=\upsilon_x(C_C-C_{Ci})\\ D\dfrac{dC_D}{dx}=\upsilon_x(C_D-C_{Di})\end{cases}$$

$$\text{at } x = 5\text{ m:}\quad \begin{cases} \dfrac{dC_A}{dx} = 0 \\ \dfrac{dC_B}{dx} = 0 \\ \dfrac{dC_C}{dx} = 0 \\ \dfrac{dC_D}{dx} = 0 \end{cases}$$

Given:

rate constants, $k_1 = 1.2\text{ m}^3/\text{mol.sec}$

$k_2 = 1\text{ m}^3/\text{mol.sec}$

diffusivity, $D = 1.2 \times 10^{-4}\ \text{m}^2/\text{sec}$

axial velocity, $\upsilon_x = 1$ m/sec

inlet concentrations, $C_{Ai} = C_{Bi} = 1\text{ mol/m}^3$

$C_{Ci} = C_{Di} = 0\text{ mol/m}^3$

$M = 50$

Solve the problem and produce the concentration profile (C versus x).

References

Chapra, S. C. (2012). *Applied Numerical Methods with MATLAB for Engineers and Scientists*, 3rd ed., New Delhi: Tata McGraw-Hill.

Jain, M. K., Iyengar, S. R. K., and Jain, R. K. (1993). *Numerical Methods for Scientific and Engineering Computation*, 3rd ed., New Delhi: New Age International.

Kharab, A. and Guenther, R. B. (2019). *An Introduction to Numerical Methods: A MATLAB Approach*, 4th ed., New York: CRC Press.

McGinnis, P. H. (1965). *Chem. Eng. Prog. Symp. Ser.*, No. 55, **61**, 1.

Salariya, A. K. (1980). Numerical solution of a differential equation in fluid mechanics. *Comput. Methods Appl. Mech. Eng.*, **21**, 211.

Schlichting, H. (1987). *Boundary Layer Theory*, 7th ed., New York: McGraw Hill.

6

PARTIAL DIFFERENTIAL EQUATION

"Science is the differential calculus of the mind. Art the integral calculus; they may be beautiful when apart, but are greatest only when combined." – Ronald Ross

Key Learning Objectives

- Learning some basics of partial differential equation (PDE)
- Knowing several finite difference methods for PDE
- Solving parabolic, hyperbolic and elliptic PDEs
- Developing solution methodology with the use of global index
- Numerically solving various linear and non-linear process examples

6.1 INTRODUCTION

There are many systems which are typically modeled with more than one independent variable [e.g., space (x) and time (t)] to characterize their states. These models thus contain partial derivatives and are constituted with *partial differential equations* (PDEs).[1] Indeed, the PDEs are used to describe a wide variety of phenomena, such as heat, diffusion, fluid dynamics, gravitation, quantum mechanics, sound, elasticity, among others.

Let us consider a general form of a linear PDE, which has two independent variables, namely x and y, as:

$$A\frac{\partial^2 u}{\partial x^2}+B\frac{\partial^2 u}{\partial x\partial y}+C\frac{\partial^2 u}{\partial y^2}+D\frac{\partial u}{\partial x}+E\frac{\partial u}{\partial y}+Fu=G \tag{6.1}$$

1 Formally, a PDE is defined as a differential equation having partial derivatives of one or more dependent variables with respect to more than one independent variable.

Here, u is the dependent variable. Obviously, this is a second-order[2] PDE.

Rewriting,

$$A\,u_{xx}+B\,u_{xy}+C\,u_{yy}+D\,u_x+E\,u_y+Fu=G \tag{6.2}$$

Note that here, A, B, C, D, E, F and G are all functions of x and y, not of u or its derivatives.

6.1.1 Basic Differences between ODE and PDE

A couple of basic differences between the ODE and PDE are highlighted in Table 6.1.

Table 6.1 Differences between ODE and PDE

ODE	PDE
1. An ODE is based on a single independent variable.	1. More than one independent variable.
2. Solving ODE requires either initial condition (for IVP) or boundary condition (for BVP).	2. Solving PDE requires either boundary condition[3] or both initial and boundary conditions.[4]
3. Numerical solution is sought at the time steps (for IVP) or grid points (for BVP).	3. Numerical solution is sought at the grid points.
4. Numerical solutions, if plotted, typically represent a curve.	4. For a PDE having two independent variables (and one dependent variable), the numerical solutions represent a surface (e.g., see Figure 6.5).

Note that in this chapter, we will confine ourselves mostly to two-dimensional (2D) differential systems, which have two independent variables (i.e., either time and space or two spatial dimensions). However, we will also have a few cases with three independent variables (see, for example, Section 6.10), namely time and two spatial dimensions.

6.1.2 Initial and Boundary Conditions

The partial differential equation systems often include derivative terms on spatial variables. In many cases, time derivative term is also additionally contained. For example, along with a second-order derivative with respect to space variable, the PDE may contain a first-order time derivative as:

$$\frac{\partial u}{\partial t}=\alpha\frac{\partial^2 u}{\partial x^2}$$

2 Recall that the order of the highest derivative present in a differential equation is called the *order* of that equation. Based on this definition, Equation (6.1) is a second-order PDE.

3 In this chapter, we will solve Laplace and Poisson equations having two spatial dimensions and they involve only boundary condition (no initial condition).

4 For solving a class of PDE systems, along with boundary condition (i.e., Dirichlet, Neumann or Robin type), one may need initial condition as well. Thus, these types of systems lead to form *initial boundary value problem* (IBVP).

For such a case, along with two boundary conditions (BCs), an initial condition (IC) needs to be prescribed. It thus becomes clear that we specify one IC for a first-order time derivative contained in a PDE, two ICs for a second-order time derivative, and so on. For a PDE having two spatial second-order derivatives [e.g., Equation (6.69)], two BCs should be provided for each spatial variable.

6.1.3 Classifying Partial Differential Equations

Like ODEs, the PDEs can also be classified as:

- Linear PDE versus non-linear PDE
- Homogeneous PDE versus non-homogeneous PDE

Note that the PDEs are also categorized based on their order.

Linear PDE versus non-linear PDE

Linear PDE

The definition of linear ODE given in Chapter 4 (Subsection 4.2.1) is also extendable to linear PDE. Accordingly, in a linear PDE, all of its coefficients are functions of the independent variables or constants. For example,

$$\frac{\partial u}{\partial x}+x^{3}\frac{\partial u}{\partial y}=\sin x$$

is linear because both E and G (i.e., $E=x^3$ and $G=\sin x$) are functions of independent variable, x only, and the rest one is a constant (i.e., $D = 1$).

Let us consider Equation (6.1) as our second example. As stated, all $A,\ B,\ C,\ D,\ E,\ F$ and G are functions of x and y, thus Equation (6.1) is also a linear PDE.

Non-linear PDE

Making a little modification in the above equation leads to form a non-linear PDE:

$$\frac{\partial u}{\partial x}+u\frac{\partial u}{\partial y}=\sin x$$

This is non-linear because the coefficient E depends on the dependent variable u (i.e., $E = u$).

Another example of the non-linear PDE includes:

$$\frac{\partial u}{\partial x}+\left(\frac{\partial u}{\partial y}\right)^{2}=G$$

This is non-linear because the power of u_y is 2.

The Navier–Stokes equation is a popular example of the non-linear PDE that typically describes the motion of fluids.

Homogeneous PDE versus non-homogeneous PDE

Similar to ODE, a PDE is said to be *homogeneous* if every term contains the dependent variable or its partial derivative. If it is violated, then the PDE is *non-homogeneous*.

Accordingly, Equation (6.1) with $G = 0$ leads to a homogeneous PDE as:

$$A\frac{\partial^2 u}{\partial x^2}+B\frac{\partial^2 u}{\partial x\partial y}+C\frac{\partial^2 u}{\partial y^2}+D\frac{\partial u}{\partial x}+E\frac{\partial u}{\partial y}+Fu=0$$

and that equation with $G \neq 0$ leads to a non-homogeneous PDE.

6.1.4 Classifying Second-Order PDEs

We are interested for a class of PDEs given in Equation (6.1) because they are the most common in practice. This second-order linear PDE:

$$A\frac{\partial^2 u}{\partial x^2}+B\frac{\partial^2 u}{\partial x\partial y}+C\frac{\partial^2 u}{\partial y^2}+D\frac{\partial u}{\partial x}+E\frac{\partial u}{\partial y}+Fu=G \tag{6.1}$$

can be classified as:

i) Parabolic: $B^2-4AC=0$
ii) Hyperbolic: $B^2-4AC>0$
iii) Elliptic: $B^2-4AC<0$

Here, the term, B^2-4AC is called the *discriminant* of the solution, sign of which dictates the behavior of the solution of Equation (6.1).

Example

A few popular examples include:

- Parabolic PDE

$$\frac{\partial u}{\partial t}=\alpha\frac{\partial^2 u}{\partial x^2} \qquad \alpha>0$$

This is known as the *heat* or *diffusion* equation. It is parabolic since $B^2-4AC=0$ ($A=-\alpha$ and $B=C=0$).

- Hyperbolic PDE

$$\frac{\partial^2 u}{\partial t^2}=a^2\frac{\partial^2 u}{\partial x^2}$$

This is known as the *wave* equation. It is hyperbolic since $B^2-4AC>0$ ($A=-a^2$, $B=0$ and $C=1$).

- Elliptic PDE

$$\frac{\partial^2 u}{\partial x^2}+\frac{\partial^2 u}{\partial y^2}=0$$

This is known as the *Laplace* equation. It is elliptic since $B^2 - 4AC < 0$ ($A = 1$, $B = 0$ and $C = 1$).

A similar classification is also there for quasi-linear PDEs (Hildebrand, 1962).

6.2 Numerical Methods

To solve the systems of PDEs, there are three most widely used numerical methods:

- Finite difference method (FDM)
- Finite volume method (FVM)
- Finite element method (FEM)

In this chapter, we will be limited to the finite difference based methods. They mainly include:

- Leap-Frog scheme
- Method of lines
- Lax–Friedrichs scheme
- Dufort–Frankel method
- Crank–Nicholson method
- Alternating direction implicit (ADI) scheme

These apart, a few other combinations of any two of the forward difference (FD), backward difference (BD) and central difference (CD) are to be studied under explicit and implicit schemes.

6.3 Finite Difference Approximation: A Review

The central idea of the finite difference based techniques is to approximate the differential terms by their finite difference equivalents. By this way, one can reduce a differential equation to a difference equation (discretized form). Let us review a few of these approximations.

First-Order Approximations
Central difference

$$\left(\frac{\partial u}{\partial x}\right)_m = \frac{u_{m+1} - u_{m-1}}{2k} + o\left(k^2\right) \tag{6.3}$$

$$\left(\frac{\partial^2 u}{\partial x^2}\right)_m = \frac{u_{m+1} - 2u_m + u_{m-1}}{k^2} + o\left(k^2\right) \tag{6.4}$$

Forward difference

$$\left(\frac{\partial u}{\partial x}\right)_m = \frac{u_{m+1} - u_m}{k} + o(k) \tag{6.5}$$

$$\left(\frac{\partial^2 u}{\partial x^2}\right)_m = \frac{u_{m+2} - 2u_{m+1} + u_m}{k^2} + o(k) \tag{6.6}$$

Backward difference

$$\left(\frac{\partial u}{\partial x}\right)_m = \frac{u_m - u_{m-1}}{k} + o(k) \tag{6.7}$$

$$\left(\frac{\partial^2 u}{\partial x^2}\right)_m = \frac{u_m - 2u_{m-1} + u_{m-2}}{k^2} + o(k) \tag{6.8}$$

Second-Order Approximations
Central difference

$$\left(\frac{\partial u}{\partial x}\right)_m = \frac{-u_{m+2} + 8u_{m+1} - 8u_{m-1} + u_{m-2}}{12k} + o\left(k^4\right) \tag{6.9}$$

$$\left(\frac{\partial^2 u}{\partial x^2}\right)_m = \frac{-u_{m+2} + 16u_{m+1} - 30u_m + 16u_{m-1} - u_{m-2}}{12k^2} + o\left(k^4\right) \tag{6.10}$$

Solution Methodology

The solution methodology for a system of partial differential equation is framed with the following steps in general by using the FDM:

Step 1: Divide the solution region into a grid of nodes.

Step 2: Approximate the differential terms of the given PDE by their finite difference equivalents. This results in relating the dependent variable at a point in the solution region with its values at the neighboring points.

Step 3: Solve the resulting discretized equations with the use of specified boundary (and initial) conditions.

Let us apply this solution methodology to a simple PDE system before going into details of the FDM methods and their applications to a couple of industrially relevant processes for final solutions.

Example 6.1 Advection equation (one-way wave equation): hyperbolic PDE[5]

Let us consider the simple advection equation,

$$u_t + au_x = 0 \tag{6.11}$$

That is,

$$\underbrace{\frac{\partial u}{\partial t}}_{\text{accumulation}} + a\underbrace{\frac{\partial u}{\partial x}}_{\text{convection/advection}} = 0 \tag{6.12}$$

for $0 < t < \infty$ (i.e., $t \in (0, \infty)$) and $x \in (0, 1)$, and here, a is a positive constant.[6] This system equation involves two independent variables, namely time (t) and space (x). For them,

$$t_n = t_0 + n\,h \qquad \text{where, } h = \Delta t \text{ and } n = 0,\ 1,\ 2, \ldots\ldots,\ N$$

$$x_m = x_0 + m\,k \qquad \text{where, } k = \Delta x \text{ and } m = 0,\ 1,\ 2, \ldots\ldots,\ M$$

This is illustrated in Figure 6.1. Here, h and k denote the *time* and *spatial increment*, respectively. Obviously, $(x_m,\ t_n)$ is the coordinate of a typical *grid point* or *node*.[7]

Now, we would like to know the initial and boundary conditions for this linear hyperbolic PDE.

Boundary condition (BC):[8]

$$u(0,\ t) = g(t) \qquad \text{for } x = 0 \text{ and } 0 \le t < \infty \text{ (i.e., } t \in [0,\ \infty))$$

Initial condition (IC):

$$u(x,\ 0) = \bar{g}(x) \qquad \text{for } t = 0 \text{ and } 0 \le x \le 1 \text{ (i.e., } x \in [0,\ 1])$$

---

5 Question is how this is a hyperbolic PDE!
Taking time derivative in Equation (6.12),

$$\frac{\partial^2 u}{\partial t^2} + a\frac{\partial^2 u}{\partial t \partial x} = 0$$

Rearranging:

$$\frac{\partial^2 u}{\partial t^2} + a\frac{\partial}{\partial x}\left(\frac{\partial u}{\partial t}\right) = 0$$

Substituting $\frac{\partial u}{\partial t} = -a\frac{\partial u}{\partial x}$, we get:

$$\frac{\partial^2 u}{\partial t^2} - a^2\frac{\partial^2 u}{\partial x^2} = 0$$

As shown before, it is hyperbolic since $B^2 - 4AC > 0$ ($A = -a^2$, $B = 0$ and $C = 1$).

6 Equation (6.12) can be thought of as being representative of a wave moving with a constant velocity of a.

7 Also called as the *mesh point* or *nodal point*.

8 Note that if $a < 0$, the BC is to be specified in terms of $u(1, t)$ instead of $u(0, t)$ [e.g., $u(1, t) = g'(t)$].

At this point it should be noted that for this wave equation, only the left (or inflow) boundary at $x = 0$ exists; the flow is from left to right, and the outflow boundary at $x = 1$ is left open.

Now, let us apply the finite difference approximations discussed before to solve this partial differential equation.

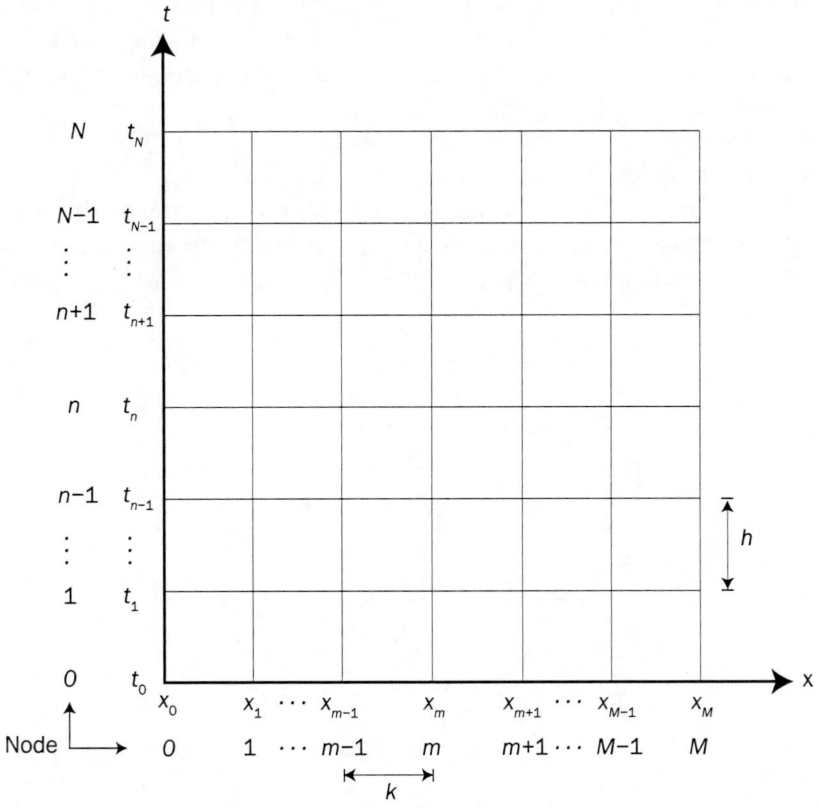

FIGURE 6.1 A grid consisting of total M by N rectangles or meshes with sides $\Delta x = k$ and $\Delta t = h$

Forward in time forward in space (FTFS)

Using Equation (6.5) for forward difference in time and forward difference in space,

$$\frac{\partial u}{\partial t}=u_t(x_m, t_n)=\frac{u(x_m, t_{n+1})-u(x_m, t_n)}{h}=\frac{u_{m,\,n+1}-u_{m,\,n}}{h}$$

Similarly,

$$\frac{\partial u}{\partial x}=u_x(x_m, t_n)=\frac{u(x_{m+1}, t_n)-u(x_m, t_n)}{k}=\frac{u_{m+1,\,n}-u_{m,\,n}}{k}$$

Plugging in Equation (6.12),

$$\frac{u_{m,\,n+1}-u_{m,\,n}}{h}=-a\frac{u_{m+1,\,n}-u_{m,\,n}}{k}+o(k,\ h) \tag{6.13}$$

Rearranging,

$$u_{m,\,n+1} = \left(1 + a\frac{h}{k}\right) u_{m,\,n} - a\frac{h}{k} u_{m+1,\,n} \tag{6.14}$$

for $m = 1,\ 2, \ldots\ldots, M-1$. This is the final discretized form (explicit) of advection equation obtained by the application of FTFS.[9] This algebraic form is used to find time row $n + 1$ across the grid, with that, approximation in row n is known. For this, start from the bottommost time row ($n = 0$, where $t = t_0 = 0$) that corresponds to the specified initial condition [i.e., $u(x,\ 0) = \bar{g}(x)$].

Computational Molecule

In Figure 6.2, Equation (6.14) is illustrated in a graphical way through the *computational molecule.* This graphical depiction is also called as a *stencil.* Here, the 'square' is sometimes used to represent a node where u is presumed known and a 'circle' where u is unknown.

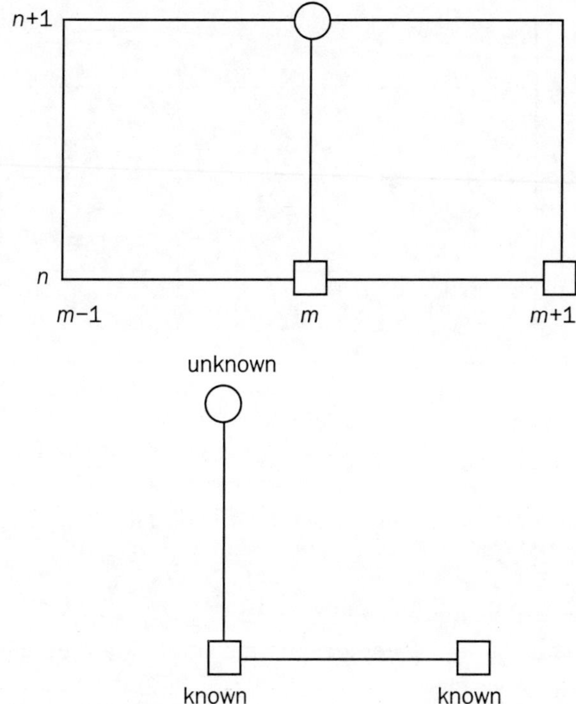

Figure 6.2 Computational molecule for the advection example (FTFS)

Remarks

1. As shown in Equation (6.13), the explicit FTFS scheme for the advection equation is first-order accurate in both space and time.
2. This is one-step finite difference scheme in that finding u at n+1th time step requires knowing u only at nth time step (as shown in Equation (6.14) and Figure 6.2).

9 $a\frac{h}{k}$ is often called the *Courant number.*

Forward in time backward in space (FTBS)

Using forward difference [Equation (6.5)] in time and backward difference [Equation (6.7)] in space, we have, respectively:

$$\frac{\partial u}{\partial t} = \frac{u_{m,\, n+1} - u_{m,\, n}}{h}$$
$$\frac{\partial u}{\partial x} = \frac{u_{m,\, n} - u_{m-1,\, n}}{k}$$

Substituting in Equation (6.12),

$$\frac{u_{m,\, n+1} - u_{m,\, n}}{h} = -a\frac{u_{m,\, n} - u_{m-1,\, n}}{k} + o(k,\ h) \tag{6.15}$$

Like FTFS, this is first-order accurate in both space and time. Rearranging,

$$u_{m,\, n+1} = \left(1 - a\frac{h}{k}\right)u_{m,\, n} + a\frac{h}{k}u_{m-1,\, n} \tag{6.16}$$

This is the final discretized form (explicit) of advection equation obtained by the application of FTBS. The three-point stencil is illustrated in Figure 6.3 through the computational molecule.

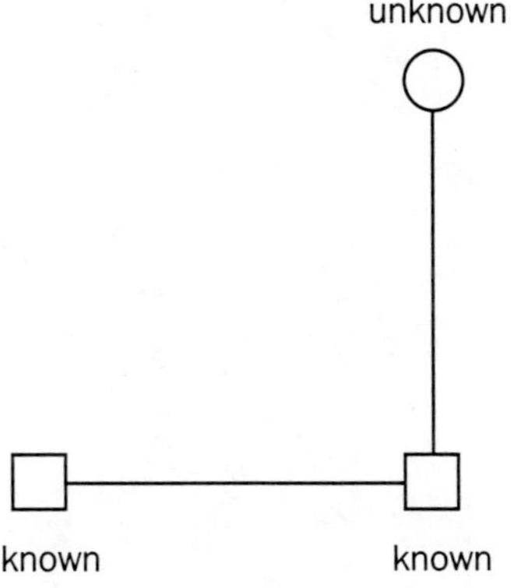

FIGURE 6.3 Computational molecule for the advection example (FTBS)

Remark

Both the remarks made a little before for the advection problem with FTFS are valid for that with this FTBS scheme.

Central difference in time central difference in space (CTCS)

Using Equation (6.3) for central difference in time as well as in space, one obtains:

$$\frac{\partial u}{\partial t} = \frac{u_{m,\, n+1} - u_{m,\, n-1}}{2h}$$
$$\frac{\partial u}{\partial x} = \frac{u_{m+1,\, n} - u_{m-1,\, n}}{2k}$$

Equation (6.12) yields,

$$\frac{u_{m,\,n+1}-u_{m,\,n-1}}{2h}=-\,a\frac{u_{m+1,\,n}-u_{m-1,\,n}}{2k}+o(k^2,\ h^2) \tag{6.17}$$

Rearranging,

$$u_{m,\,n+1}=u_{m,\,n-1}-a\frac{h}{k}(u_{m+1,\,n}-u_{m-1,\,n}) \tag{6.18}$$

This is the final form (explicit) obtained by the application of CTCS. It is illustrated in Figure 6.4 through the computational molecule.

Remarks

1. This CTCS scheme for the advection equation is second-order accurate in both space and time.
2. This is also called the *Leap-Frog method*,[10] which is basically a two-step finite difference scheme in that it uses the results from two previous time steps (nth and $n-1$th steps) in calculating u for the subsequent step ($n+1$th step). This is quite obvious in Figure 6.4 as well.

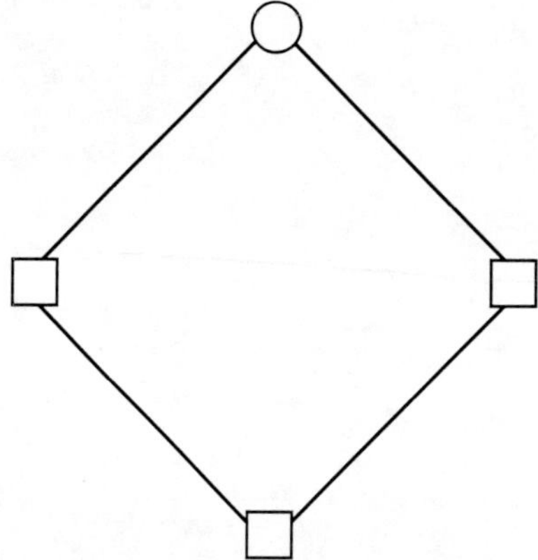

FIGURE 6.4 Computational molecule for the advection example (CTCS)

Apart from FTFS, FTBS, and CTCS, one may use other combinations, including FTCS, BTCS and so on, without maintaining the order of accuracy in a consistent manner. Further remember, all such combinations are not useful, mainly because of their stability concern.

10 It is obvious from Equation (6.18) that the value at time $(n + 1)h$ is obtained by adding a term to the value at $(n - 1)h$. Because of this *leap* over the time nh, it is called the Leap-frog or Leapfrog method.

6.4 Method of Lines (MoL)

This numerical method is employed first to convert the PDE to a set of ODEs using the finite difference approximation. Then those transformed ODEs are solved by employing an ODE solver. With this, we develop the computational steps below.

Computational Steps

Step 1: Discretize the spatial domain in $M + 1$ nodes, $x_0, x_1, \ldots, x_M$.

Step 2: Use the finite difference approximation for the spatial derivative in the PDE.

Step 3: Integrate the resulting set of ODEs with the use of specified boundary and initial conditions.

Figure 6.5 demonstrates the method of lines for a typical initial boundary value problem (IBVP), for which, u is the dependent variable, and x and t are the two independent variables. As shown, the time trajectories are produced along lines starting from the different spatial nodes and in this sense, this scheme is called the *method of lines.*

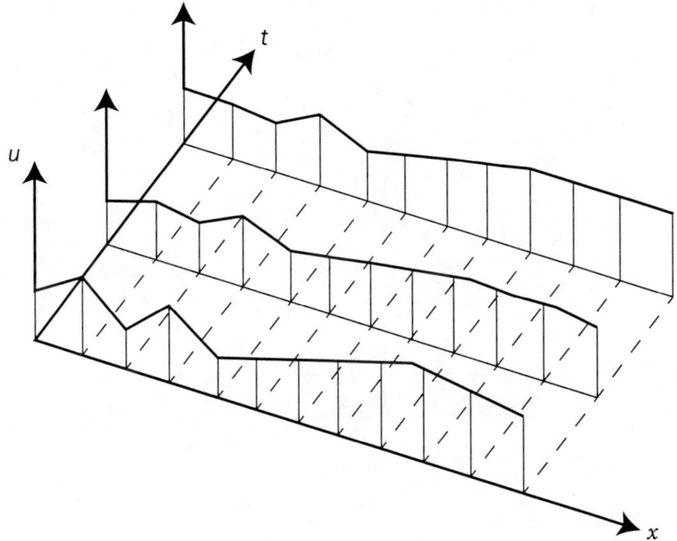

Figure 6.5 Illustrating the method of lines for a typical IBVP system

A Sample Application

To understand this finite difference scheme, let us consider the same advection equation used in Example 6.1:

$$\frac{\partial u}{\partial t} + a\,\frac{\partial u}{\partial x} = 0 \tag{6.12}$$

With the central difference approximation [Equation (6.3)] of the spatial derivative (i.e., $\frac{\partial u}{\partial x}$),

$$\frac{du_m}{dt}+a\ \frac{u_{m+1}-u_{m-1}}{2k}=0 \tag{6.19}$$

This is obviously representing a set of ODE-IVP equations, in which, $m=1,\ 2,.....,\ M-1$ (i.e., interior nodes). Now, one can use an ODE solver discussed in Chapter 4 (e.g., Euler or Runge–Kutta method).

In the following, this method is further illustrated by a complete process example.

Case Study 6.1 Method of lines: Parabolic PDE

Unsteady state heat conduction in a slab: One dimension

Let us extend the steady-state one-dimensional heat conduction problem discussed in Chapter 5 (see Case Study 5.4) to its unsteady state form. For this, we again consider an elemental volume in a slab shown in Figure 6.6.

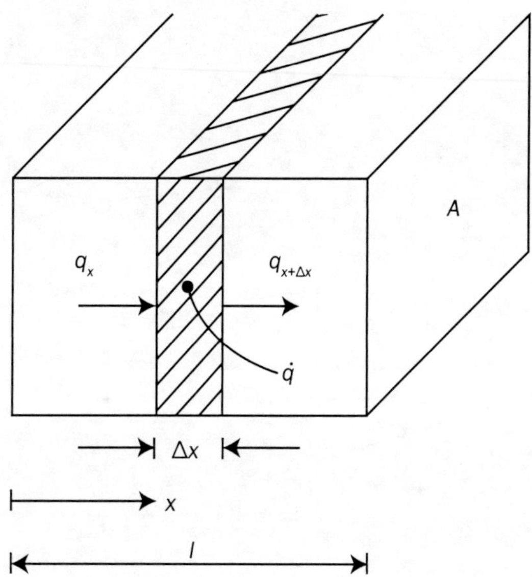

FIGURE 6.6 An elemental volume for one-dimensional heat conduction problem

Here,

A = heat transfer area

C_p = specific heat capacity

K = thermal conductivity

l = slab length

q = heat transfer rate

$\dot{q}$ = rate of heat generation per unit volume

T = temperature

t = time

ρ = density

$\alpha = \dfrac{K}{\rho\, C_p}$ = thermal diffusivity

Deriving model

We make energy balance based on Equation (A4.3), which is rewritten as:

$$\begin{pmatrix}\text{Rate of heat}\\ \text{accumulation}\end{pmatrix} = \begin{pmatrix}\text{Rate of heat}\\ \text{input}\end{pmatrix} - \begin{pmatrix}\text{Rate of heat}\\ \text{output}\end{pmatrix} + \begin{pmatrix}\text{Rate of heat}\\ \text{generation}\end{pmatrix} \tag{6.20a}$$

This means,

$$\begin{pmatrix}\text{Amount of heat accumulated}\\ \text{during time period } \Delta t\end{pmatrix} = \begin{pmatrix}\text{Amount of heat}\\ \text{in during } \Delta t\end{pmatrix} - \begin{pmatrix}\text{Amount of heat}\\ \text{out during } \Delta t\end{pmatrix} + \begin{pmatrix}\text{Amount of heat}\\ \text{generated during } \Delta t\end{pmatrix}$$

With this, we have:

$$A\Delta x \rho\, C_p [T_{t+\Delta t} - T_t] = q_x \Delta t - q_{x+\Delta x} \Delta t + \dot{q} A \Delta x \Delta t$$

Using the Fourier law of heat conduction:

$$A\rho\, C_p \Delta x [T_{t+\Delta t} - T_t] = -KA\left(\frac{\partial T}{\partial x}\right)_x \Delta t + KA\left(\frac{\partial T}{\partial x}\right)_{x+\Delta x} \Delta t + \dot{q} A \Delta x \Delta t$$

Dividing both sides by $A\Delta x\ \Delta t$, and taking limit $\Delta x \to 0$ and $\Delta t \to 0$,

$$\rho C_p \frac{\partial T}{\partial t} = \frac{\partial}{\partial x}\left[K \frac{\partial T}{\partial x}\right] + \dot{q} \tag{6.20b}$$

Assuming constant thermal conductivity,

$$\frac{\partial T}{\partial t} = \alpha \frac{\partial^2 T}{\partial x^2} + \frac{\dot{q}}{\rho C_p} \tag{6.20c}$$

This gives,

$$\frac{1}{\alpha}\frac{\partial T}{\partial t} = \frac{\partial^2 T}{\partial x^2} + \frac{\dot{q}}{K} \tag{6.20d}$$

Note that this equation can also be derived from Equation (A5.3).

Further, neglecting heat generation (i.e., $\dot{q} = 0$),

$$\frac{\partial T}{\partial t} = \alpha \frac{\partial^2 T}{\partial x^2} \tag{6.20e}$$

This is the unsteady state heat conduction model, which is obviously a parabolic PDE.

Problem statement

Let us solve the following parabolic PDE:

$$\frac{\partial T}{\partial t} = \alpha \frac{\partial^2 T}{\partial x^2} \tag{6.20e}$$

subject to the Dirichlet boundary conditions (BCs):

$$\begin{cases} T(0,\ t) = 0 \\ T(1,\ t) = 1 \end{cases} \tag{6.21}$$

and initial condition (IC):

$$T(x,\ 0) = 2 \qquad x \in (0,\ 1) \tag{6.22}$$

Here, two independent variables are involved, namely time (t) and space (x), and it is obviously an IBVP system.

Solution

Let us first discretize the spatial domain into $M + 1$ equidistant nodes (see Figure 6.7).

FIGURE 6.7 Discretization of spatial domain into $M + 1$ equidistant nodes

Using the central difference spatial approximation [Equation (6.4)],

$$\frac{dT_m}{dt} = \alpha \left[\frac{T_{m+1}(t) - 2T_m(t) + T_{m-1}(t)}{k^2} \right] \tag{6.23}$$

Now, we need to determine the time-dependent temperature at each node m [i.e., $T_m(t)$]. Since $T_0(t)$ and $T_M(t)$ are known (given BCs) as:

$$\begin{cases} T(0,\ t) = 0 = T_0(t) \\ T(1,\ t) = 1 = T_M(t) \end{cases}$$

so, the rest temperatures, which include $T_1(t)$, $T_2(t)$, ..., $T_{M-1}(t)$, need to be determined. For this, we use Equation (6.23) as follows:

- $m = 1$

$$\frac{dT_1}{dt} = \frac{\alpha}{k^2} \left[T_2(t) - 2T_1(t) + T_0(t) \right]$$

- $m = 2$

$$\frac{dT_2}{dt} = \frac{\alpha}{k^2} \left[T_3(t) - 2T_2(t) + T_1(t) \right] \tag{6.24}$$

$\vdots$

- $m = M-1$

$$\frac{dT_{M-1}}{dt} = \frac{\alpha}{k^2}\left[T_M(t) - 2T_{M-1}(t) + T_{M-2}(t)\right]$$

In compact form,

$$\frac{d\bar{T}}{dt} = F(\bar{T}), \qquad \bar{T}(t=0) = \bar{T}(0) \tag{6.25}$$

where,

$$\bar{T} = \begin{bmatrix} T_1 \\ T_2 \\ . \\ . \\ T_{M-1} \end{bmatrix}, \qquad F(\bar{T}) = \frac{\alpha}{k^2}\begin{bmatrix} T_2 - 2T_1 + T_0 \\ T_3 - 2T_2 + T_1 \\ . \\ . \\ T_M - 2\,T_{M-1} + T_{M-2} \end{bmatrix} \quad \text{and} \qquad \bar{T}(0) = \begin{bmatrix} 2 \\ 2 \\ . \\ . \\ 2 \end{bmatrix}$$

Obviously, Equation (6.25) represents a set of coupled ODE-IVP equations (total $M-1$ equations with $M-1$ unknowns).

Let us solve Equation (6.25) using the explicit Euler method (an ODE solver). For this, we select $M = 4$ and thus, $k = 0.25$, and thermal diffusivity $\alpha = 0.00138$. With this, we have from Equation (6.24),

$$\frac{dT_1}{dt} = \frac{\alpha}{k^2}\left[T_2 - 2T_1 + T_0\right] \qquad \text{since } T_0(t) = 0$$

$$= 0.02208(T_2 - 2T_1)$$

$$\frac{dT_2}{dt} = 0.02208\left[T_3 - 2T_2 + T_1\right]$$

$$\frac{dT_3}{dt} = \frac{\alpha}{k^2}\left[T_4 - 2T_3 + T_2\right] \qquad \text{since } T_4(t) = 1$$

$$= 0.02208\left[1 - 2T_3 + T_2\right]$$

Given initial condition:

$$T_1(0) = T_2(0) = T_3(0) = 2$$

For the explicit Euler, let us adopt $h = 0.025$ and the numerical solutions of T (at $t = 25$) are obtained for this IBVP as:

$T_0(25) = 0.0$

$T_1(25) = 1.2927$

$T_2(25) = 1.7632$

$T_3(25) = 1.6270$

$T_4(25) = 1.0$

These solutions are further plotted in Figure 6.8 to visualize the slab temperature at a glance along its length at time $t = 25$.

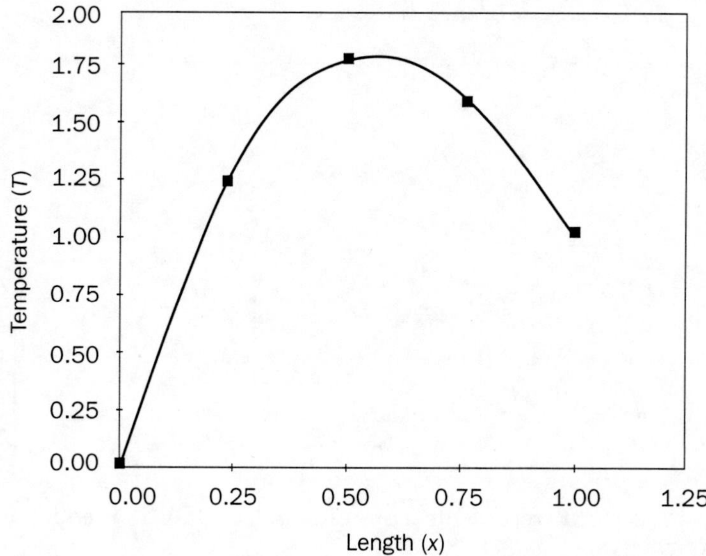

Figure 6.8 Illustrating numerical solutions of the unsteady state heat conduction problem

Remarks

1. Notice that method of lines employed here in solving the unsteady state heat conduction problem first uses the central difference in space to convert the PDE to a coupled ODE-IVP system and then Euler (forward difference) method to solve those ODE-IVP equations.
2. Applying the explicit FTCS method, one can get the following form for the parabolic PDE [Equation (6.20e)] as:

$$T_{m,\,n+1} = r\,T_{m-1,\,n} + (1-2r)T_{m,\,n} + rT_{m+1,\,n}$$

where, $r = \dfrac{\alpha h}{k^2}$. This is sometimes called the *Bender–Schmidt* (explicit) method.

Numerical Stability Analysis: Unsteady state heat conduction in a slab (revisited)

We would like to analyze the numerical stability of this IBVP system solved above by the method of lines that uses centered differencing in space followed by forward Euler for coupled IVP equations. As we know, whenever we employ the explicit Euler scheme, there is a probability of numerical instability problem. For this, care must be exercised in selecting the time step to guarantee that the numerical solution will be bounded (provided that the actual (analytical) solution in time is bounded). Accordingly, we fix our goal to find the analytical solution along with a set of conditions for numerical stability of the heat conduction problem when solved with explicit Euler. Here, we will closely follow the analysis made in Chapter 4 (see Section 4.13).

Equation (6.25) can be presented in the following form:

$$\frac{d\bar{T}}{dt}=\frac{\alpha}{k^2}\begin{bmatrix} 0 & 0 & 0 & 0 & \cdots\cdots\cdots & 0 & 0 & 0 & 0 \\ 1 & -2 & 1 & 0 & \cdots\cdots\cdots & 0 & 0 & 0 & 0 \\ 0 & 1 & -2 & 1 & \cdots\cdots\cdots & 0 & 0 & 0 & 0 \\ \cdot & \cdot & \cdot & \cdot & \cdot & \cdot & \cdot & \cdot & \cdot \\ \cdot & \cdot & \cdot & \cdot & \cdot & \cdot & \cdot & \cdot & \cdot \\ \cdot & \cdot & \cdot & \cdot & \cdot & \cdot & \cdot & \cdot & \cdot \\ 0 & 0 & 0 & 0 & \cdots\cdots\cdots & 0 & 1 & -2 & 1 \\ 0 & 0 & 0 & 0 & \cdots\cdots\cdots & 0 & 0 & 0 & 0 \end{bmatrix}\begin{bmatrix} T_0(t) \\ T_1(t) \\ T_2(t) \\ \cdot \\ \cdot \\ T_{M-2}(t) \\ T_{M-1}(t) \\ T_M(t) \end{bmatrix} \tag{6.26}$$

Note that here T_0 and T_M (both are known) are additionally included in vector $\bar{T}$.

We can represent this linear system in compact form:

$$\frac{d\bar{T}}{dt}=\frac{\alpha}{k^2}A\bar{T} \tag{6.27}$$

for which, A matrix is clearly defined in Equation (6.26). It can further be transformed into [see Equation (4.71)],

$$\frac{d\hat{T}}{dt}=\frac{\alpha}{k^2}\Lambda\hat{T}$$

Here,

$$\Lambda=\begin{bmatrix} \lambda_0 & 0 & \cdots\cdots & 0 \\ 0 & \lambda_1 & \cdots\cdots & 0 \\ \cdot & \cdot & \cdot & \cdot \\ \cdot & \cdot & \cdot & \cdot \\ 0 & 0 & \cdots\cdots & \lambda_M \end{bmatrix}$$

in which, $\lambda_0, \lambda_1, \ldots, \lambda_M$ are the eigenvalues of the A matrix.[11]

For any node m,

$$\frac{d\hat{T}_m}{dt}=\frac{\alpha\lambda_m}{k^2}\hat{T}_m \tag{6.28}$$

where, λ_m is the eigenvalue corresponding to node m. Recall that the analytical solution of Equation (6.27) is to be bounded (stable) if $\text{Re}(\lambda_m)\leq 0$.

11 One can substitute for T_0 and T_M to avoid null rows in matrix A [see Equation (6.26)] that leads to $M-2$ eigenvalues.

Along with eigenvalues having non-positive real parts, for numerical stability of the explicit Euler, the time step size h should satisfy:

$$h \leq \frac{2k^2}{\alpha\,|\lambda_{max}|} \tag{6.29}$$

This yields,

$$\frac{h}{k^2} \leq \frac{2}{\alpha\,|\lambda_{max}|} \tag{6.30}$$

Remarks

1. Similar stability conditions are there for other explicit methods. For example, for the RK4,

$$\frac{h}{k^2} \leq \frac{2.785}{\alpha\,|\lambda_{max}|}$$

2. It becomes obvious that the MoL used for the parabolic PDE [i.e., Equation (6.20e)] is conditionally stable.

Case Study 6.2 Method of lines: Parabolic PDE

Non-linear reaction–diffusion system

The *reaction–diffusion* problem is governed by the following mass balance equation for one dimensional flow (i.e., only x-direction is considered):

$$\underbrace{\frac{\partial C}{\partial t}}_{\text{accumulation}} = \underbrace{D\frac{\partial^2 C}{\partial x^2}}_{\text{diffusion/conduction}} - \underbrace{\frac{\partial (\upsilon_x C)}{\partial x}}_{\text{convection}} + \underbrace{R(C, x)}_{\text{reaction}}$$

For deriving this equation, one can revisit Appendix A5.2.

Neglecting the convective term,

$$\frac{\partial C}{\partial t} = D\frac{\partial^2 C}{\partial x^2} + R(C, x) \tag{6.31}$$

This equation represents a pure diffusion system if the reaction term gets vanished (i.e., $R(C, x) = 0$),

$$\frac{\partial C}{\partial t} = D\frac{\partial^2 C}{\partial x^2} \tag{6.32}$$

This is the form of *Fick's second law*.

This apart, there are a few more special cases based on the reaction term R as highlighted below.

Special cases

- When $R = C(1-C)$, Equation (6.31) yields the *Fisher equation* that is typically used to describe the growth of biological population
- When $R = C(1-C^2)$, Equation (6.31) leads to the *Newell–Whitehead–Segel* equation used to describe the Rayleigh–Benard convection
- When $R = C(1-C)(C-\alpha)$, Equation (6.31) becomes the *Zeldovich equation* that arises in combustion theory with $0<\alpha<1$

Problem statement

Let us adopt the following non-linear reaction–diffusion system [based on Equation (6.31)]:

$$\frac{\partial C}{\partial t} = D\frac{\partial^2 C}{\partial x^2} + C(1-C)(C-\alpha) \tag{6.33}$$

with

Boundary conditions:

$$C(0,\ t) = 1 \qquad 0 \le t \le 10 \tag{6.34a}$$

$$C(10,\ t) = 0 \tag{6.34b}$$

Initial condition:

$$C(x,\ 0) = \begin{cases} 1 & \text{when } x < 5 \\ 0 & \text{when } x \ge 5 \end{cases} \qquad 0 \le x \le 10 \tag{6.35}$$

Given parameters:

$D = 1$

$\alpha = 0.5$

$k = 1$ (i.e., $M = 10$)

$h = 0.005$

Use the method of lines (MoL) and obtain the numerical solutions. Illustrate these results in graphical form [C versus x at different times (i.e., t = 2, 5, 8 and 10)].

Solution

Using the central difference spatial approximation (with $D = 1$),

$$\frac{dC_m}{dt} = \frac{C_{m+1}(t) - 2C_m(t) + C_{m-1}(t)}{k^2} + C_m(t)[1 - C_m(t)][C_m(t) - \alpha] \tag{6.36}$$

Rearranging (with $k = 1$),

$$\frac{dC_m}{dt} = C_{m+1} - 2C_m + C_{m-1} - [\alpha\, C_m - C_m^2(1+\alpha) + C_m^3] \tag{6.37}$$

where, $m = 1, 2, \ldots\ldots 9$.

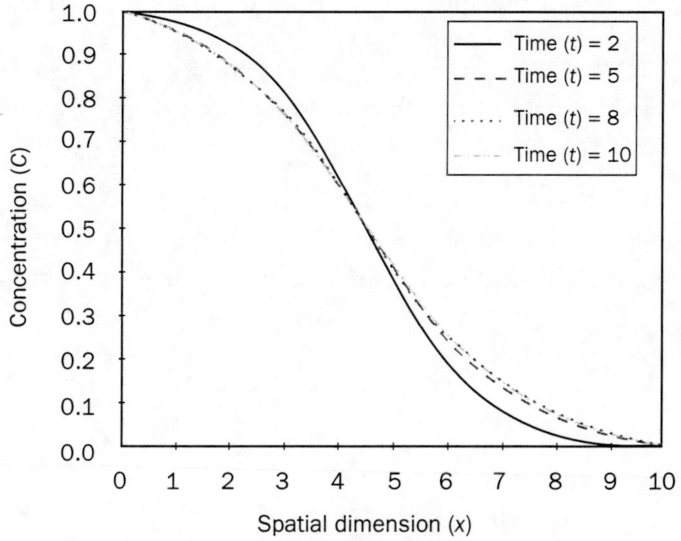

FIGURE 6.9 Concentration profile of the example reaction–diffusion problem

We solve this set of ODEs using the forward Euler through computer programming. Produced numerical solutions are provided in Table 6.2 and presented in Figure 6.9 as concentration (C) versus spatial dimension (x) at four different times (i.e., $t = 2, 5, 8$ and 10).

TABLE 6.2 Numerical solutions of the example reaction–diffusion system

x	C (t = 2)	C (t = 5)	C (t = 8)	C (t = 10)
0	1	1	1	1
1	0.97798	0.95338	0.94882	0.94838
2	0.93043	0.88334	0.87536	0.87477
3	0.82268	0.76797	0.76015	0.76017
4	0.62517	0.59873	0.59698	0.59852
5	0.37486	0.40339	0.41096	0.41418
6	0.17744	0.23511	0.2483	0.25212
7	0.07005	0.12215	0.13505	0.13834
8	0.02392	0.0574	0.06666	0.06888
9	0.0068	0.02214	0.02678	0.02786
10	0	0	0	0

6.5 Lax–Friedrichs Method

To discuss the Lax–Friedrichs approach,[12] we consider the one-way wave equation:

$$\frac{\partial u}{\partial t}+a\frac{\partial u}{\partial x}=0 \tag{6.12}$$

Now, let us apply forward in time and central in space (FTCS),

$$\frac{u_{m,\,n+1}-u_{m,\,n}}{h}=-a\frac{u_{m+1,\,n}-u_{m-1,\,n}}{2k}+o(k^2,h) \tag{6.38}$$

The Lax–Friedrichs scheme takes a spatial average of the neighbors of $u_{m,n}$ (i.e., $u_{m+1,n}$ and $u_{m-1,n}$) in finding the time derivative at the corresponding point as:

$$u_{m,n}=\frac{1}{2}(u_{m+1,n}+u_{m-1,n})+o(k^2)$$

It can easily be derived from the Taylor series.

Substituting in Equation (6.38),

$$\frac{u_{m,n+1}-\frac{1}{2}(u_{m+1,n}+u_{m-1,n})}{h}=-a\frac{u_{m+1,n}-u_{m-1,n}}{2k}$$

Rearranging,

$$u_{m,n+1}=\frac{1}{2}\left(1-a\frac{h}{k}\right)u_{m+1,n}+\frac{1}{2}\left(1+a\frac{h}{k}\right)u_{m-1,n} \tag{6.39}$$

This is the final form of the Lax–Friedrichs scheme (explicit) for the example advection system. It is illustrated in Figure 6.10 through the computational molecule.

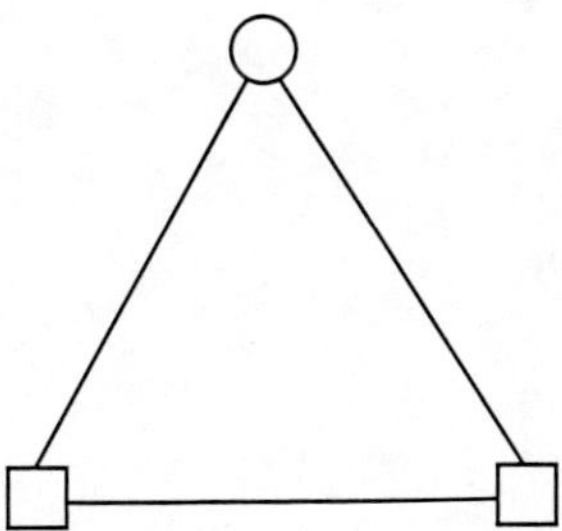

Figure 6.10 Concentration molecule for the Lax–Friedrichs approach applied to the advection example

12 Named after Hungarian-born American mathematician Péter Dávid Lax (born: May 1, 1926) and a noted German American mathematician Kurt O. Friedrichs (1901–1982).

Remarks

1. Unlike the Leap-Frog method [Equation (6.18)] that requires information at the two previous time steps (i.e., at nth and $(n-1)$th step), the Lax–Friedrichs method requires information only at the last time step (i.e., at nth step). Thus, it is a one-step finite difference scheme.
2. The FTCS scheme is unconditionally unstable for hyperbolic PDEs. The Lax–Friedrichs method makes the FTCS scheme stable if and only if the following condition is satisfied:[13]

$$\left|a\frac{h}{k}\right| \leq 1$$

6.6 Dufort–Frankel Method

To formulate the Dufort–Frankel scheme, let us adopt the following parabolic PDE:

$$\frac{\partial u}{\partial t} = \alpha \frac{\partial^2 u}{\partial x^2} \tag{6.40}$$

Notice that it is the one-dimensional unsteady state heat conduction Equation (6.20e) (T is only replaced here by u).

Applying the CTCS method, we get:

$$\frac{u_{m,\,n+1} - u_{m,\,n-1}}{2h} = \alpha \left[\frac{u_{m+1,\,n} - 2u_{m,\,n} + u_{m-1,\,n}}{k^2}\right] + o(k^2, h^2)$$

Rearranging,

$$u_{m,\,n+1} = u_{m,\,n-1} + 2r\left[u_{m+1,\,n} - 2u_{m,\,n} + u_{m-1,\,n}\right]$$

where, $r = \frac{\alpha h}{k^2}$. Introducing,

$$u_{m,n} = \frac{1}{2}(u_{m,\,n+1} + u_{m,\,n-1})$$

we get:

$$u_{m,\,n+1} = u_{m,\,n-1} + 2r\left[u_{m+1,\,n} - (u_{m,\,n+1} + u_{m,\,n-1}) + u_{m-1,\,n}\right]$$

It gives:

$$u_{m,\,n+1} = \frac{1-2r}{1+2r} u_{m,\,n-1} + \frac{2r}{1+2r}(u_{m+1,\,n} + u_{m-1,\,n}) \tag{6.41}$$

---

13 Finding this stability condition is quite straightforward by using the so called *von Neumann stability analysis*; but it is beyond the scope of this book.

This is the final discretized form of the *Dufort–Frankel method* for the one-dimensional unsteady state heat conduction Equation (6.40). It is illustrated in Figure 6.11 through the computational molecule.

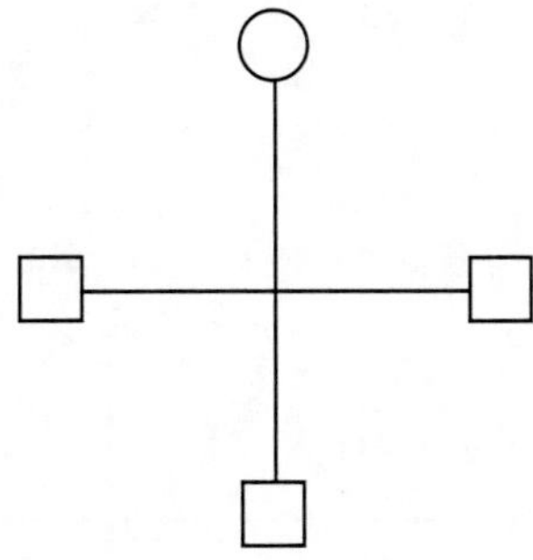

FIGURE 6.11 Computational molecule for the Dufort–Frankel method applied to the unsteady state heat conduction problem

Remark
Although this Dufort–Frankel approach is an explicit method, it is unconditionally stable.

6.7 Crank–Nicholson Method

The Crank–Nicholson (CN) method[14] is an improved finite difference technique, which works by performing the numerical approximations at the point $\left(x_m, t_{n+\frac{1}{2}}\right)$ that typically lies between the rows in the grid. This is illustrated in Figure 6.12. To understand this, let us continue with the advection problem (Example 6.1) considered before with the following form:

$$\frac{\partial u}{\partial t}+a\frac{\partial u}{\partial x}=0 \tag{6.12}$$

The C–N method evaluates the derivatives at point A′ (Figure 6.12). Using the central difference in time with step size $\frac{h}{2}$, we have:

14 Named after the British mathematical physicist John Crank (1916–2006) and British mathematician Phyllis Nicholson (1917–1968).

$$\left(\frac{\partial u}{\partial t}\right)_{(x_m,\, t_{n+1/2})} = \frac{u_{m,n+1} - u_{m,n}}{h} + o(h^2)$$

Let us write it in a little simplified form:

$$\left(\frac{\partial u}{\partial t}\right)_{m,\, n+\frac{1}{2}} = \frac{u_{m,n+1} - u_{m,n}}{h} + o(h^2)$$

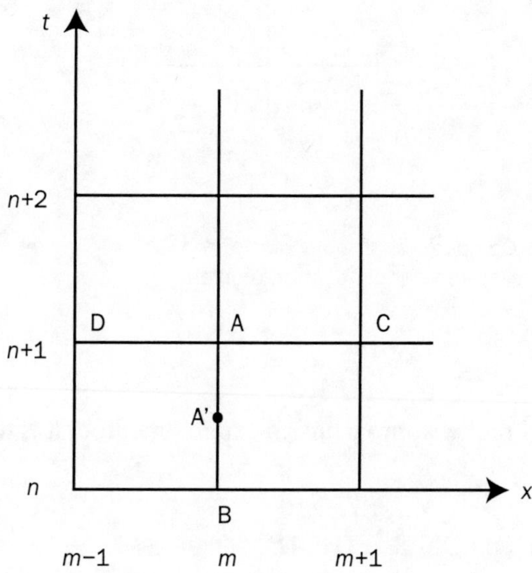

Figure 6.12 Discretization made for Crank–Nicholson

Again, this C–N method approximates $u_x\left(m, n+\frac{1}{2}\right)$ by averaging $u_x(m, n+1)$ and $u_x(m, n)$ with the use of central difference as:

$$\left(\frac{\partial u}{\partial x}\right)_{m,\, n+\frac{1}{2}} = \frac{1}{2}\left[\frac{u_{m+1,\, n+1} - u_{m-1,\, n+1}}{2k} + \frac{u_{m+1,\, n} - u_{m-1,\, n}}{2k}\right] + o(k^2)$$

Inserting into Equation (6.12),

$$\frac{u_{m,\, n+1} - u_{m,\, n}}{h} = -\frac{a}{2}\left[\frac{u_{m+1,\, n+1} - u_{m-1,\, n+1}}{2k} + \frac{u_{m+1,\, n} - u_{m-1,\, n}}{2k}\right] + o(k^2,\, h^2)$$

It gives:

$$-\frac{ah}{4k}\, u_{m-1,n+1} + u_{m,n+1} + \frac{ah}{4k} u_{m+1,n+1} = \frac{ah}{4k}\, u_{m-1,n} + u_{m,n} - \frac{ah}{4k} u_{m+1,n}$$

This is the final form of the C–N method and it is illustrated in Figure 6.13 through the computational molecule.

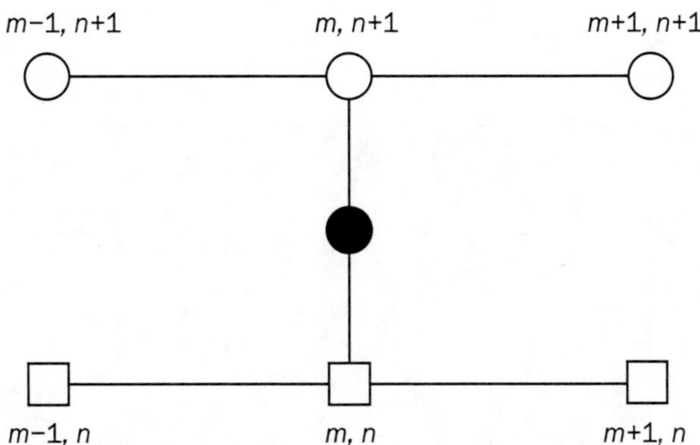

FIGURE 6.13 Computational molecule for the C–N method applied to the advection example

Remarks

1. As shown above, the C–N method is second-order accurate in both space and time.
2. It is also evident that this is an implicit technique and thus, unconditionally stable.

In the next phase, the C–N method is illustrated by the heat conduction problem.

Case Study 6.3 Crank–Nicholson (C–N) method

Unsteady state heat conduction in a slab: One dimension (revisited)

Let us apply the Crank–Nicholson method on the one-dimensional unsteady state heat conduction problem governed by:

$$\frac{\partial u}{\partial t}=\alpha\frac{\partial^2 u}{\partial x^2} \tag{6.40}$$

subject to the following boundary conditions (BCs):

$$\begin{cases} u(0,\ t)=0 \\ u(1,\ t)=1 \end{cases} \qquad 0\le t\le 1 \tag{6.42a}$$

and initial condition (IC):

$$u(x,\ 0)=2 \qquad x\in(0,\ 1) \tag{6.42b}$$

Solution

According to the C–N method, Equation (6.40) is rewritten as:

$$\left(\frac{\partial u}{\partial t}\right)_{m,\, n+\frac{1}{2}}=\alpha\left(\frac{\partial^2 u}{\partial x^2}\right)_{m,\, n+\frac{1}{2}}$$

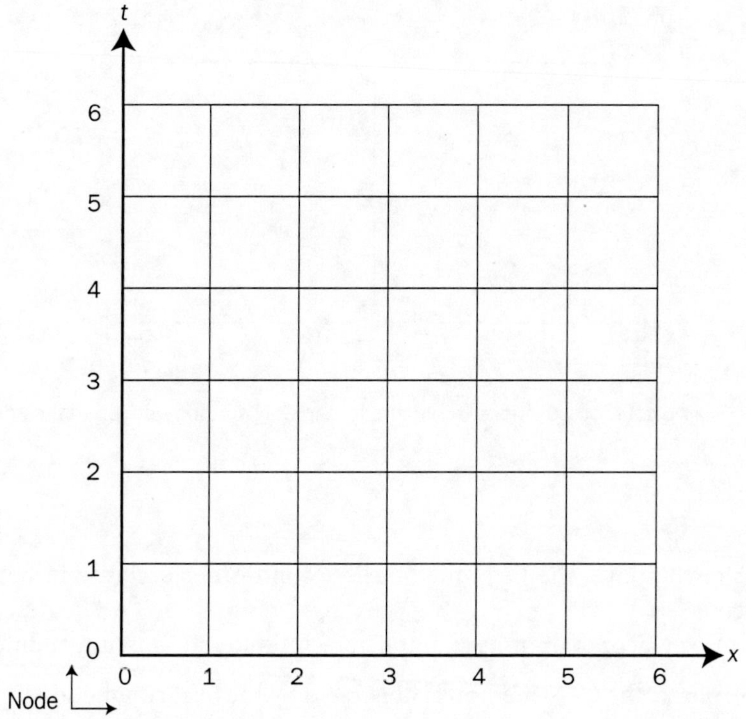

FIGURE 6.14 Discretization made for unsteady state heat conduction problem

It gives:

$$\frac{u_{m,\,n+1}-u_{m,\,n}}{h}=\frac{\alpha}{2}\left[\frac{u_{m+1,\,n+1}-2u_{m,\,n+1}+u_{m-1,\,n+1}}{k^2}+\frac{u_{m+1,\,n}-2u_{m,\,n}+u_{m-1,\,n}}{k^2}\right]+o(k^2,\ h^2) \tag{6.43}$$

or,

$$u_{m,\,n+1}-u_{m,\,n}=\frac{1}{2}\frac{\alpha\,h}{k^2}\left[u_{m+1,\,n+1}-2u_{m,\,n+1}+u_{m-1,\,n+1}+u_{m+1,\,n}-2u_{m,\,n}+u_{m-1,\,n}\right]$$

It gives:

$$2u_{m,\,n+1}-2u_{m,\,n}=r\left[u_{m+1,\,n+1}-2u_{m,\,n+1}+u_{m-1,\,n+1}+u_{m+1,\,n}-2u_{m,\,n}+u_{m-1,\,n}\right]$$

where, $r=\frac{\alpha\,h}{k^2}$. Rearranging,

$$r\,u_{m+1,n+1}-(2+2r)\,u_{m,n+1}+r\,u_{m-1,\,n+1}=-r\,u_{m+1,\,n}+(2r-2)u_{m,n}-r\,u_{m-1,n} \tag{6.44}$$

↑ Point C ↑ Point A ↑ Point D

for $m=1, 2,\ldots, M-1$. Obviously, point D corresponds to the left boundary (see Figure 6.12), indicating $u_{m-1,n+1}$ is known. Further, all terms on the right hand side of Equation (6.44) are also known (specified IC). So, Equation (6.44) involves two unknowns (at points A and C or equivalently, $u_{m,n+1}$ and $u_{m+1,n+1}$) at future time step. Thus, the Crank–Nicholson is an implicit method and it is unconditionally stable.

To solve this problem, let us discretize the spatial and time domain as shown in Figure 6.14. The grid consists of 6 by 6 rectangles with sides h and k. Here h and k are considered equal, and thus there are total 36 squares.

With this, let us note down the following known terms:

Known: $u_{1,0}, u_{2,0}, u_{3,0}, u_{4,0}, u_{5,0}$ IC

$u_{0,0}, u_{0,1}, u_{0,2}, u_{0,3}, u_{0,4}, u_{0,5}, u_{0,6}$ .... Left boundary (BC)

$u_{6,0}, u_{6,1}, u_{6,2}, u_{6,3}, u_{6,4}, u_{6,5}, u_{6,6}$ .... Right boundary (BC)

At this point, we should note that for the corner node (e.g., left bottom and right bottom corner), sometimes there are two known values; one based on IC and other one on BC. In such a case,[15] one would take the arithmetic average of those two values for a corner node and it is elaborated later.

- For $m = 1$: Equation (6.44) yields,

$$r u_{2,n+1} - (2+2r)u_{1,n+1} + r u_{0,n+1} = -r u_{2,n} + (2r-2)u_{1,n} - r u_{0,n}$$

Obviously, for $n = 0$, all right hand terms are known. Additionally, $u_{0,n+1}$ belongs to the left boundary (thus, known). Hence, there are total two unknowns present in the left hand side of the above equation. Rearranging,

$$-(2+2r)u_{1,n+1} + r u_{2,n+1} = b_1$$

or,

$$a_{11}\, u_{1,n+1} + a_{12}\, u_{2,n+1} = b_1 \tag{6.45}$$

where, $a_{11} = -(2+2r)$ and $a_{12} = r$.

- For $m = 2$

$$r\, u_{3,n+1} - (2+2r)\, u_{2,n+1} + r u_{1,n+1} = -r\, u_{3,n} + (2r-2)u_{2,n} - r u_{1,n}$$

which yields,

$$a_{21}\, u_{1,n+1} + a_{22}\, u_{2,n+1} + a_{23} u_{3,n+1} = b_2 \tag{6.46}$$

where, $a_{21} = a_{23} = r$ and $a_{22} = -(2+2r)$.

- Likewise, for $m = 5$ (not shown for $m = 3$ and 4)

$$r u_{6,n+1} - (2+2r)u_{5,n+1} + r u_{4,n+1} = -r u_{6,n} + (2r-2)u_{5,n} - r u_{4,n}$$

15 For example, the IC of this unsteady state heat conduction problem is given for $x \in [0, 1]$, instead of $x \in (0, 1)$.

Rearranging,

$$a_{54}\, u_{4,\,n+1} + a_{55}\, u_{5,\,n+1} = b_5 \tag{6.47}$$

where, $a_{54} = r$, $a_{55} = -(2+2r)$ and $u_{6,n+1}$ belongs to the right boundary (thus, known).

There is one equation for each discrete spatial point [for example, Equation (6.45) for $m = 1$]. Thus, we have total 5 equations (each for $m = 1,\ 2, \ldots,\ 5$). Combining them,

$$\begin{bmatrix} a_{11} & a_{12} & 0 & 0 & 0 \\ a_{21} & a_{22} & a_{23} & 0 & 0 \\ 0 & a_{32} & a_{33} & a_{34} & 0 \\ 0 & 0 & a_{43} & a_{44} & a_{45} \\ 0 & 0 & 0 & a_{54} & a_{55} \end{bmatrix} \begin{bmatrix} u_{1,n+1} \\ u_{2,n+1} \\ u_{3,n+1} \\ u_{4,n+1} \\ u_{5,n+1} \end{bmatrix} = \begin{bmatrix} b_1 \\ b_2 \\ b_3 \\ b_4 \\ b_5 \end{bmatrix} \tag{6.48}$$

for $n = 0, 1, \ldots, 5$. If the time domain is specified as $t \geq 0$ (instead of $0 \leq t \leq 1$), one can continue to solve Equation (6.48) till any large value of n (instead of 5) as required.

Note that here, $a_{ii} = -(2+2r)$ and other elements are equal to either 0 or r. Obviously, Equation (6.48) forms a tridiagonal linear system:

$$AU = B$$

Notice that here B is different at different time steps. Now to find temperature ($= u$), one can solve it by either direct means or by iteration. Choose a suitable method detailed in Chapter 2 and the numerical solution of this problem is left as an exercise.

Remarks

1. Here we obtained a set of linear algebraic equations (i.e., $AU = B$) since we started with a linear PDE. If we start with a non-linear PDE, it will lead to non-linear algebraic equations (e.g., Case Study 6.2), which can be solved by using a suitable iterative method discussed in Chapter 3 (e.g., the Newton–Raphson method).
2. It should further be noted that since the C–N method is unconditionally stable, one can use any step size, h. But recall that the large size affects the accuracy.

6.8 General Implicit Scheme: Extension of C–N Method

Let us continue with the 1D heat conduction equation (parabolic):

$$\frac{\partial u}{\partial t} = \alpha \frac{\partial^2 u}{\partial x^2} \tag{6.40}$$

Notice that the C–N method is applied in Equation (6.43) giving equal weightage (i.e., $\frac{1}{2} \equiv 50\%$) to both n and $n + 1$ levels. Now we would like to formulate a general implicit scheme by extending the concept of C–N method for the parabolic Equation (6.40). For this, a different weight is introduced to different levels (say, θ to level n and $(1 - \theta)$ to level $n + 1$). Accordingly, Equation (6.40) yields:

$$\frac{u_{m,\,n+1}-u_{m,\,n}}{h}=\alpha\left[\theta\frac{u_{m+1,\,n}-2u_{m,\,n}+u_{m-1,\,n}}{k^2}+(1-\theta)\frac{u_{m+1,\,n+1}-2u_{m,\,n+1}+u_{m-1,\,n+1}}{k^2}\right]+o(k^2,\,h^2)$$

Denoting $r=\frac{\alpha h}{k^2}$, we get:

$$r(1-\theta)\,u_{m+1,\,n+1}-[1+2r(1-\theta)]\,u_{m,\,n+1}+r(1-\theta)\,u_{m-1,\,n+1}=-r\theta\,u_{m+1,\,n}-(1-2r\theta)\,u_{m,n}-r\theta u_{m-1,\,n} \quad (6.49)$$

This represents the *general implicit scheme.*

Remarks

1. As shown above, this scheme is second-order accurate in both space and time.
2. This implicit method leads to the C–N method [Equation (6.44)] at $\theta=\frac{1}{2}$.

So far we have discussed the initial boundary value problems (IBVPs), having space and time as two independent variables. Now, let us turn our attention to solving the PDE that has two spatial dimensions as independent variables.

6.9 Finite Differencing of Elliptic PDE

Let us consider a general form of a typical elliptic PDE:

$$\frac{\partial^2 u}{\partial x^2}+\frac{\partial^2 u}{\partial y^2}=H(x,\,y) \quad (6.50)$$

This is popularly known as the *Poisson equation.*[16] Here, H represents the source term.

For this, each edge is typically specified in Figure 6.15 with:

$u(x,\,0)=u_1$	Bottom boundary	(6.51a)
$u(0,\,y)=u_2$	Left boundary	(6.51b)
$u(x,\,1)=u_3$	Top boundary	(6.51c)
$u(1,\,y)=u_4$	Right boundary	(6.51d)

16 Named after the French mathematician and physicist Siméon Denis Poisson (1781–1840).

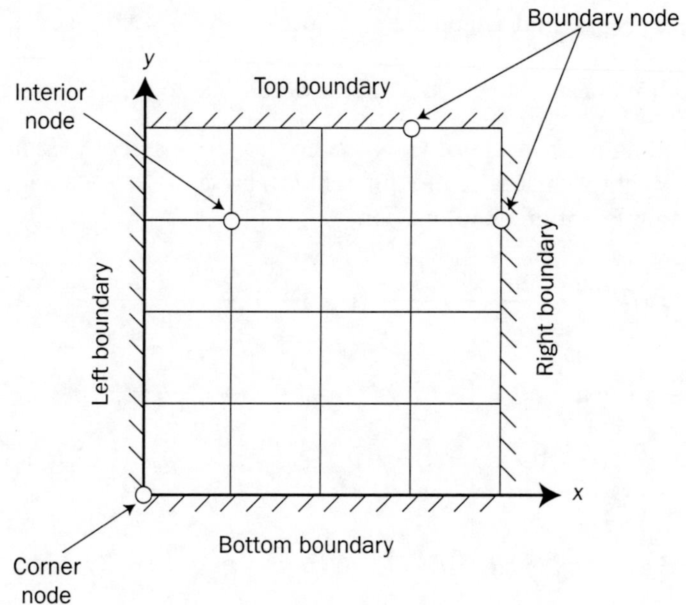

FIGURE 6.15 Showing four boundaries and different nodes

Notice that all these are Dirichlet boundary conditions.

With this, let us develop a solution methodology for this elliptic problem.

Developing a four-steps approach

Step 1: Discretize the two domains as:

$$k_1 = \Delta x = x_{i+1} - x_i \qquad (x\text{-axis}) \tag{6.52a}$$

$$k_2 = \Delta y = y_{j+1} - y_j \qquad (y\text{-axis}) \tag{6.52b}$$

Step 2: Discretize the PDE using a finite difference scheme.

Step 3: Translate the boundary conditions into the form of algebraic equations.

Step 4: Numerically solve the resulting system of algebraic equations to find the unknowns.

Notice that this methodology is quite similar to that followed for the one-dimensional BVP in Chapter 5.

Applying the four-steps approach

Let us follow these four steps to solve the representative elliptic PDE [Equation (6.50)] subject to the BCs given in Equations (6.51).

Step 1: Here we discretize both the x- and y-axes into the same number of points (i.e., $M + 1$ points) with $\Delta x = \Delta y = k$. This is illustrated shortly in a square domain in Figure 6.16. Notice that there are total $(M + 1)^2$ grid points, at which, we need to determine the respective u values. It reveals that there are total $(M + 1)^2$ unknowns.

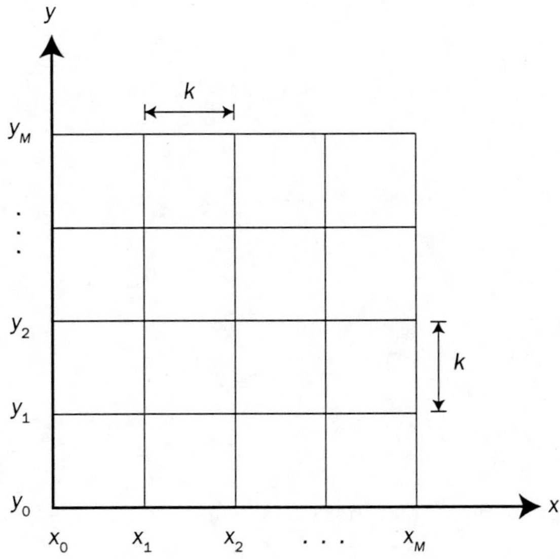

FIGURE 6.16 Uniform discretization[17] of x- and y-axes

Step 2: Discretizing Equation (6.50) by the central finite difference scheme:

$$\frac{u_{i+1,j}-2u_{i,j}+u_{i-1,j}}{k_1^2}+\frac{u_{i,j+1}-2u_{i,j}+u_{i,j-1}}{k_2^2}=H_{i,j}$$

Considering same mesh size in x- and y-axes (i.e., $k_1=k_2=k$):

$$u_{i,j}=\frac{1}{4}\left[u_{i+1,j}+u_{i-1,j}+u_{i,j+1}+u_{i,j-1}-k^2H_{i,j}\right] \tag{6.53}$$

When the source term vanishes (i.e., $H = 0$),

$$u_{i,j}=\frac{1}{4}\left[u_{i+1,j}+u_{i-1,j}+u_{i,j+1}+u_{i,j-1}\right] \tag{6.54}$$

This is the representation of *Laplace equation.*[18] It is also called as the *standard 5-points formula*. This equation is valid for $1 \le i, j \le M-1$, indicating that there are total $(M-1)^2$ such equations. This approach [Equation (6.54)] is illustrated through the computational molecule in Figure 6.17.

17 Non-uniform discretization is quite useful to take care of the accuracy level when sharp gradients exist in the function.

18 Named after the French scholar Pierre-Simon marquis de Laplace (1749–1827). It is a special case of the more general Poisson Equation (6.50) with $H = 0$.

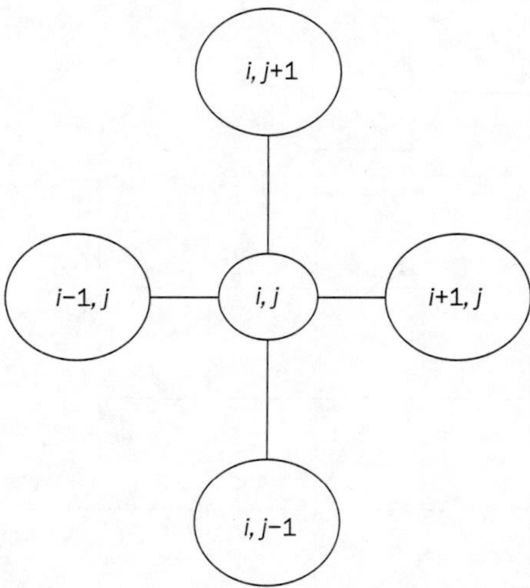

FIGURE 6.17 Computational molecule for standard 5-points formula

It is obvious from Equation (6.54) and Figure 6.17 that the value of u for each node is the average of them at the four surrounding nodes.

Step 3: Translate the boundary conditions into the form of algebraic equations. Note that the given Dirichlet BCs [i.e., Equations (6.51a–d)] always have algebraic forms and let them simply write here in the following way:[19]

$$u_{i,0} = u_1 \qquad 1 \le i \le M-1 \qquad \text{Bottom boundary} \tag{6.55a}$$

$$u_{0,j} = u_2 \qquad 1 \le j \le M-1 \qquad \text{Left boundary} \tag{6.55b}$$

$$u_{i,M} = u_3 \qquad 1 \le i \le M-1 \qquad \text{Top boundary} \tag{6.55c}$$

$$u_{M,j} = u_4 \qquad 1 \le j \le M-1 \qquad \text{Right boundary} \tag{6.55d}$$

There are obviously total $4(M-1)$ equations. These equations correspond to the edges, except the corner nodes.

For a corner node, one can have two values based on the two adjacent edges having different boundary conditions. By convention, one can take the arithmetic average of those two values. Accordingly, for the four corner nodes, we have:

$$u_{0,0} = \frac{u_1 + u_2}{2} \tag{6.56a}$$

$$u_{0,M} = \frac{u_2 + u_3}{2} \tag{6.56b}$$

19 For the Neumann or Robin BC, one needs to translate them into algebraic forms through finite difference approximations.

$$u_{M,0} = \frac{u_1 + u_4}{2} \tag{6.56c}$$

$$u_{M,M} = \frac{u_3 + u_4}{2} \tag{6.56d}$$

Step 4: At this moment, we have total $(M+1)^2$ equations, which include:

i) $(M-1)^2$ equations formed in Step 2 [i.e., all correspond to $(M-1)^2$ interior nodes]
ii) $4M$ equations formed in Step 3 [$4(M-1)$ boundary nodes + 4 corner nodes]

These equations now need to be solved to find the same number of unknowns [i.e., $(M+1)^2$]. Note that since the boundary nodes (including, corner nodes) are directly obtained from the specified Dirichlet BCs, we need to actually solve $(M-1)^2$ equations developed in Step 2 for finding unknown u's for all $(M-1)^2$ interior nodes.

Remarks

1. Obviously, finite differencing of elliptic PDE leads to a system of large number of algebraic equations. Their solution is truly a major issue. Anyway, these algebraic equations can be solved using the linear method [e.g., the Jacobi and Gauss–Seidel (detailed in Chapter 2)] or non-linear root-finding method [e.g., the Newton–Raphson and Secant (detailed in Chapter 3)], as suitable.
2. Similar to the standard 5-points formula, there is an alternative formula for the Laplace equation that can be obtained by considering four diagonal points for finite differences as:

$$u_{i,j} = \frac{1}{4}\left[u_{i+1,j+1} + u_{i-1,j-1} + u_{i-1,j+1} + u_{i+1,j-1}\right]$$

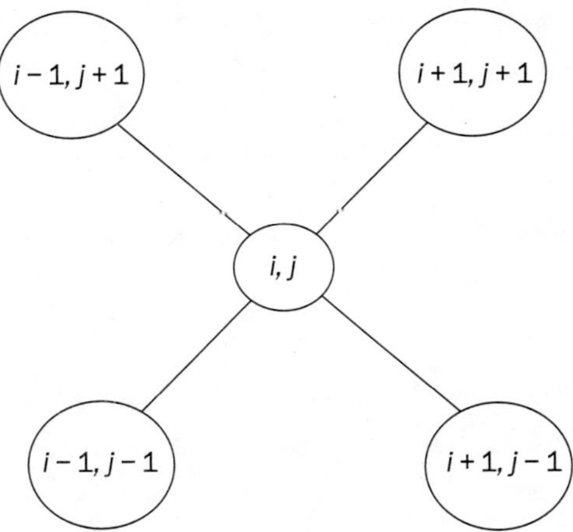

FIGURE 6.18 Computational molecule for diagonal 5-points formula

It is often called the *diagonal 5-points formula*, according to which, the value of u at any point is the average of its values at the 4 neighboring diagonal nodes. It is shortly illustrated in Figure 6.18.

It should be noted that the standard 5-points formula is more accurate over the diagonal 5-points formula. This is because the neighboring four points used in the former formula are generally located at less distance compared to the neighboring four diagonal points required in the later scheme. However, this diagonal formula is sometimes used:

- when the standard formula is not applicable to the grid;
- for the ease of computation; and
- for approximating the starting values at the grid points required for an iterative method to solve the resulting set of algebraic equations.

Example 6.2 Laplace equation: elliptic PDE

Solve the Laplace equation represented by:

$$\frac{\partial^2 u}{\partial x^2}+\frac{\partial^2 u}{\partial y^2}=0 \tag{6.57}$$

over the domain $0 \le x,\ y \le 1$ subject to:

$$\begin{aligned} &u(x,\ 0)=0\\ &u(0,\ y)=0\\ &u(x,\ 1)=9(1-x)\\ &u(1,\ y)=0 \end{aligned} \tag{6.58}$$

Consider square domain with $M = 3$ and thus, $\Delta x = \Delta y = k = \frac{1}{3}$.

Solution

Figure 6.19 shows the finite difference grid with 4-columns wide by 4-rows high. Values of all boundary nodes are given and they are placed in that figure. Except the top left corner node $\left(=\frac{1}{2}(9+0)=4.5\right)$, all other three corner nodes have the same value of 0. With this, we need to find the values of u at 4 interior nodes (unknown) as marked in Figure 6.19.

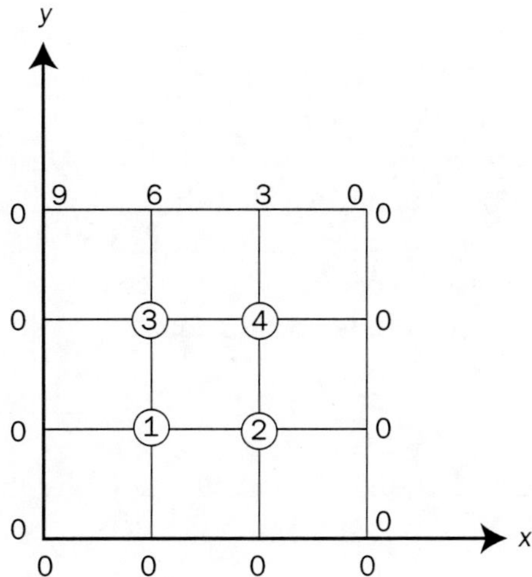

FIGURE 6.19 Finite difference grid

Equation (6.57) is discretized before and obtained in Equation (6.54) as:

$$u_{i,j} = \frac{1}{4}\left[u_{i+1,j} + u_{i-1,j} + u_{i,j+1} + u_{i,j-1}\right] \tag{6.54}$$

Representing node (i, j) as (follow Figure 6.19):

Point 1: $u_1 = \frac{1}{4}[u_2 + 0 + u_3 + 0]$

Point 2: $u_2 = \frac{1}{4}[0 + u_1 + u_4 + 0]$

Point 3: $u_3 = \frac{1}{4}[u_4 + 0 + 6 + u_1]$

Point 4: $u_4 = \frac{1}{4}[0 + u_3 + 3 + u_2]$

Rearranging them for the four points, one obtains, respectively:

$$4u_1 - u_2 - u_3 = 0 \tag{6.59a}$$

$$u_1 - 4u_2 + u_4 = 0 \tag{6.59b}$$

$$-u_1 + 4u_3 - u_4 = 6 \tag{6.59c}$$

$$-u_2 - u_3 + 4u_4 = 3 \tag{6.59d}$$

Solving this linear set of Equations (6.59),

$u_1 = 0.625$

$u_2 = 0.5$

$u_3 = 2.0$

$u_4 = 1.375$

Remarks

1. There are total $(M-1)^2$ (= 4) equations that are solved for the same number of unknowns, namely u_1, u_2, u_3 and u_4.
2. For solving the Laplace Equation (6.57) with the Dirichlet boundary conditions [Equation set (6.58)], it is noticed that there is no use of corner nodes (which are approximated by making average of the two neighboring boundary nodes). It reveals that there is no effect of these four corner nodes on the solution of remaining variables. But remember, this is not the case when we have the Neumann or Robin boundary conditions.

A short note

With this background study, we are now in a position to solve the wide varieties of elliptic PDEs, including:

$\nabla^2 u = 0$	Laplace equation
$\nabla^2 u = H(x, y)$	Poisson equation
$\nabla^2 u + u\,f(x, y) = H(x, y)$	Helmholtz equation

where,

$$\nabla^2 u = u_{xx} + u_{yy}$$

In the similar fashion, we can also solve the Laplace or Poisson equation in 3-dimensions; but it is avoided here due to cumbersome computation.

6.10 Finite Differencing of Elliptic PDE: Concept of Global Index

The concept of global variable along with global index we have studied before in solving the coupled ODE-BVPs in Chapter 5 (see Case Study 5.6). There we have used a common index (called the global index), *m* for the two dependent variables after clubbing them into a global variable. Here, we will use the same index for the elliptic PDE problems having two independent variables and one dependent variable.

Let us start with the following equation:

$$\frac{\partial^2 u}{\partial x^2} + \frac{\partial^2 u}{\partial y^2} = H(x, y) \tag{6.50}$$

Boundary conditions:

$u(x, 0) = u'_1$	Bottom boundary
$u(0, y) = u'_2$	Left boundary
$u(x, 1) = u'_3$	Top boundary
$u(1, y) = u'_4$	Right boundary

For solving this PDE system, a four-steps approach developed before (Section 6.9) is to be followed. To use the concept of global index in that approach, let us reformulate Steps 2 and 3, keeping Steps 1 and 4 unaltered. For completeness, however, we have presented the all four-steps below in order.

Step 1: As stated before, discretize the two domains as:

$$k_1 = \Delta x = x_{i+1} - x_i \qquad (x\text{-axis}) \tag{6.52a}$$

$$k_2 = \Delta y = y_{j+1} - y_j \qquad (y\text{-axis}) \tag{6.52b}$$

Step 2: This step is concerned with the discretization of PDE. Instead of using two local indices, namely i and j for x- and y-axes, respectively, we would like to adopt a common index, m that is called the *global index*.

For this, let us produce Figure 6.20 with $M = 3$ in both x- and y-axes. Note that the numbers written within the circular molecules belong to the global index, m. Accordingly, one can write:

$$m = i + j(M+1) \tag{6.60}$$

where, $m = 0, 1, 2, \ldots. (M+1)^2 - 1$.

For example, Equation (6.60) yields:

If $i = j = 0$ $\qquad m = 0$

$i = 1, j = 0$ $\qquad m = 1$

$i = 2, j = 3$ $\qquad m = 3M + 5 = 14$ (since $M = 3$)

$i = M, j = M$ $\qquad m = M^2 + 2M = (M+1)^2 - 1 = 15$

One can also use Figure 6.20 to verify these values of global index calculated above against various i-j pairs.

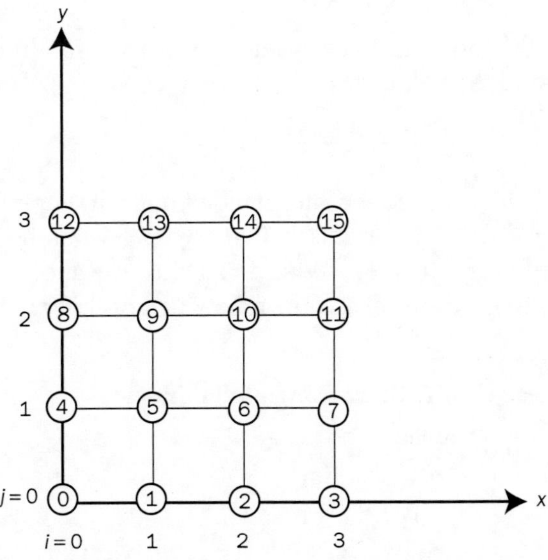

FIGURE 6.20 Converting local indices (i, j) to a global index (m)

Let us now discretize Equation (6.50) with the use of global index, m, considering $H = 0$,

$$\frac{u_{m+1} - 2u_m + u_{m-1}}{k^2} + \frac{u_{m+M+1} - 2u_m + u_{m-(M+1)}}{k^2} = 0 \quad (6.61)$$

Step 3: Write equations for the given boundary conditions and then use the global index with those equations.

Bottom boundary: $u_m = u'_1$ $\quad m = i$ with $i = 0, 1, \ldots., M$ (6.62a)

[put $j = 0$ in Equation (6.60) and vary $i = 0, 1, \ldots., M$]

According to Equation (6.62a), the bottom boundary includes nodes 0, 1, 2 and 3, and the same is also depicted in Figure 6.20.

Left boundary: $u_m = u'_2$ $\quad m = j(M+1)$ with $j = 0, 1, \ldots., M$ (6.62b)

[put $i = 0$ in Equation (6.60) and vary $j = 0, 1, \ldots., M$]

It becomes obvious that the left boundary includes nodes 0, 4, 8 and 12 as shown in Figure 6.20 as well.

Top boundary: $u_m = u'_3$ $\quad m = i + M(M+1)$ with $i = 0, 1, \ldots., M$ (6.62c)

[put $j = M$ in Equation (6.60) and vary $i = 0, 1, \ldots., M$]

It leads to nodes 12, 13, 14 and 15.

Right boundary: $u_m = u'_4$ $\quad m = M + j(M+1)$ with $j = 0, 1, \ldots., M$ (6.62d)

[put $i = M$ in Equation (6.60) and vary $j = 0, 1, \ldots., M$]

With this, we get nodes 3, 7, 11 and 15 located on the right boundary.

Then find the four corner nodes (which are approximated by making average of the two adjacent boundary nodes).

Step 4: As stated, in this final step, one needs to solve the resulting system of algebraic equations to find the unknowns.

Remark

For solving an elliptic PDE with the use of global index, one can simply substitute the specified Dirichlet boundary conditions in Step 2 itself. In that case, the number of available equations and unknowns get reduced to $(M-1)^2$. With doing this, the above solution methodology looks simpler and it is followed in solving the following non-linear PDE system.

Example 6.3 Non-linear elliptic PDE: using global index

Solve the following non-linear PDE:

$$\frac{\partial^2 u}{\partial x^2} + \frac{\partial^2 u}{\partial y^2} - \alpha u^2 = 0 \qquad 0 \le x,\ y \le 1 \quad (6.63)$$

All boundaries are fixed at the value of u equals 1, which reveals:

$$\begin{aligned} &u(x,\ 0)=1 && \text{Bottom boundary} \\ &u(0,\ y)=1 && \text{Left boundary} \\ &u(x,\ 1)=1 && \text{Top boundary} \\ &u(1,\ y)=1 && \text{Right boundary} \end{aligned} \tag{6.64}$$

Consider a square domain with $M = 4$ (i.e., $\Delta x = \Delta y = 0.25$) and $\alpha = 1$.

Solution

Let us follow the four-steps approach described above in solving this elliptic PDE with the use of global index.

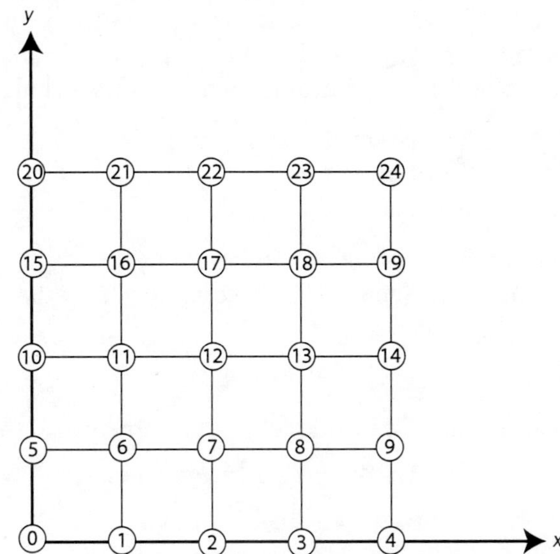

FIGURE 6.21 Discretization of x and y domains with global index

Step 1: We discretize both the x and y spatial domains into same number of points (i.e., $M + 1 = 5$ points) with $\Delta x = \Delta y = k = 0.25$. With this, we have a complete picture shown in Figure 6.21.

Step 2: Discretizing Equation (6.63),

$$\frac{u_{i+1,j}-2u_{i,j}+u_{i-1,j}}{k^2}+\frac{u_{i,j+1}-2u_{i,j}+u_{i,j-1}}{k^2}-\alpha\, u_{i,j}^2=0 \tag{6.65}$$

In terms of global index, m

$$\frac{u_{m+1}-2u_m+u_{m-1}}{k^2}+\frac{u_{m+M+1}-2u_m+u_{m-(M+1)}}{k^2}-\alpha\, u_m^2=0$$

Rearranging,

$$u_{m+1} - 4u_m + u_{m-1} + u_{m+M+1} + u_{m-(M+1)} - \alpha k^2 u_m^2 = 0 \quad (6.66)$$

This equation is to be used for all interior nodes (total 9) and thus, there are total 9 $[=(M-1)^2]$ equations. This is also obvious from Figure 6.21.

Step 3: For all of the boundary nodes, including corner nodes:

$$u_m - 1 = 0 \qquad m \in \text{boundary}$$

With this, there are total 16 (=4*M*) equations.

Step 4: At this moment, we have total 25 equations, which need to be solved to find the same number of unknown variables.

However, there is a scope of making the resulting set of algebraic equations smaller, which, in turn, is easier to solve. For this, as stated before, one can exclude the equations related to all known boundary nodes (total 16). Consequently, the number of unknowns gets reduced to nine, all of which are specific to the non-linear Equation (6.66). Those nine unknown variables include (see Figure 6.21):

u_6, u_7, u_8	Second row (from bottom)
u_{11}, u_{12}, u_{13}	Third row
u_{16}, u_{17}, u_{18}	Fourth row

Using N–R method to solve the non-linear Equation (6.66)

For the Newton–Raphson (N–R) method, we need to construct the Jacobian matrix. For this, Equation (6.66) is represented as:

$$u_{m-1} - 4u_m - \alpha k^2 u_m^2 + u_{m+1} + u_{m-M-1} + u_{m+M+1} = g_m \quad (6.67)$$

Now let us use this equation for the nine interior nodes, namely 6, 7, 8, 11, 12, 13, 16, 17, 18.

- $m = 6$ $\qquad u_5 - 4u_6 - \alpha k^2 u_6^2 + u_7 + u_1 + u_{11} = g_6$

Using boundary conditions $\qquad 1 - 4u_6 - \alpha k^2 u_6^2 + u_7 + 1 + u_{11} = g_6$

Rearranging $\qquad -4u_6 - \alpha k^2 u_6^2 + u_7 + u_{11} + 2 = g_6$

Differentiating $\qquad \dfrac{\partial g_6}{\partial u_6} = -4 - 2\alpha k^2 u_6 = g_{6,6}$

$$\frac{\partial g_6}{\partial u_7} = 1 = g_{6,7} \qquad \frac{\partial g_6}{\partial u_{11}} = 1 = g_{6,11}$$

- $m = 7$ $\qquad u_6 - 4u_7 - \alpha k^2 u_7^2 + u_8 + u_2 + u_{12} = g_7$

Using boundary condition $\qquad u_6 - 4u_7 - \alpha k^2 u_7^2 + u_8 + u_{12} + 1 = g_7$

Differentiating $\qquad \dfrac{\partial g_7}{\partial u_6} = 1 = g_{7,6} \qquad \dfrac{\partial g_7}{\partial u_7} = -4 - 2\alpha k^2 u_7 = g_{7,7}$

$$\frac{\partial g_7}{\partial u_8} = 1 = g_{7,8} \qquad \frac{\partial g_7}{\partial u_{12}} = 1 = g_{7,12}$$

- $m = 8$ $\quad u_7 - 4u_8 - \alpha k^2 u_8^2 + u_9 + u_3 + u_{13} = g_8$

Using boundary conditions $\quad u_7 - 4u_8 - \alpha k^2 u_8^2 + u_{13} + 2 = g_8$

Differentiating $\quad \dfrac{\partial g_8}{\partial u_7} = 1 = g_{8,7} \qquad \dfrac{\partial g_8}{\partial u_8} = -4 - 2\alpha k^2 u_8 = g_{8,8}$

$$\frac{\partial g_8}{\partial u_{13}} = 1 = g_{8,13}$$

- $m = 11$ $\quad u_{10} - 4u_{11} - \alpha k^2 u_{11}^2 + u_{12} + u_6 + u_{16} = g_{11}$

Using boundary condition $\quad u_6 - 4u_{11} - \alpha k^2 u_{11}^2 + u_{12} + u_{16} + 1 = g_{11}$

Differentiating $\quad \dfrac{\partial g_{11}}{\partial u_6} = 1 = g_{11,6} \qquad \dfrac{\partial g_{11}}{\partial u_{11}} = -4 - 2\alpha k^2 u_{11} = g_{11,11}$

$$\frac{\partial g_{11}}{\partial u_{12}} = 1 = g_{11,12} \qquad \frac{\partial g_{11}}{\partial u_{16}} = 1 = g_{11,16}$$

- $m = 12$ $\quad u_{11} - 4u_{12} - \alpha k^2 u_{12}^2 + u_{13} + u_7 + u_{17} = g_{12}$

Differentiating $\quad \dfrac{\partial g_{12}}{\partial u_7} = 1 = g_{12,7} \qquad \dfrac{\partial g_{12}}{\partial u_{11}} = 1 = g_{12,11}$

$$\frac{\partial g_{12}}{\partial u_{12}} = -4 - 2\alpha k^2 u_{12} = g_{12,12}$$

$$\frac{\partial g_{12}}{\partial u_{13}} = 1 = g_{12,13} \qquad \frac{\partial g_{12}}{\partial u_{17}} = 1 = g_{12,17}$$

- $m = 13$ $\quad u_{12} - 4u_{13} - \alpha k^2 u_{13}^2 + u_{14} + u_8 + u_{18} = g_{13}$

Using boundary condition $\quad u_8 + u_{12} - 4u_{13} - \alpha k^2 u_{13}^2 + u_{18} + 1 = g_{13}$

Differentiating $\quad \dfrac{\partial g_{13}}{\partial u_8} = 1 = g_{13,8} \qquad \dfrac{\partial g_{13}}{\partial u_{12}} = 1 = g_{13,12}$

$$\frac{\partial g_{13}}{\partial u_{13}} = -4 - 2\alpha k^2 u_{13} = g_{13,13}$$

$$\frac{\partial g_{13}}{\partial u_{18}} = 1 = g_{13,18}$$

- $m = 16$ $\quad u_{15} - 4u_{16} - \alpha k^2 u_{16}^2 + u_{17} + u_{11} + u_{21} = g_{16}$

Using boundary conditions $\quad u_{11} - 4u_{16} - \alpha k^2 u_{16}^2 + u_{17} + 2 = g_{16}$

Differentiating $\frac{\partial g_{16}}{\partial u_{11}}=1=g_{16,\,11}$ $\frac{\partial g_{16}}{\partial u_{16}}=-4-2\alpha\, k^2 u_{16}=g_{16,\,16}$

$\frac{\partial g_{16}}{\partial u_{17}}=1=g_{16,\,17}$

- $m = 17$ $u_{16}-4u_{17}-\alpha\, k^2 u_{17}^2+u_{18}+u_{12}+u_{22}=g_{17}$

Using boundary condition $u_{12}+u_{16}-4u_{17}-\alpha\, k^2 u_{17}^2+u_{18}+1=g_{17}$

Differentiating $\frac{\partial g_{17}}{\partial u_{12}}=1=g_{17,\,12}$ $\frac{\partial g_{17}}{\partial u_{16}}=1=g_{17,\,16}$

$\frac{\partial g_{17}}{\partial u_{17}}=-4-2\alpha\, k^2 u_{17}=g_{17,\,17}$

$\frac{\partial g_{17}}{\partial u_{18}}=1=g_{17,\,18}$

- $m = 18$ $u_{17}-4u_{18}-\alpha\, k^2 u_{18}^2+u_{19}+u_{13}+u_{23}=g_{18}$

Using boundary conditions $u_{13}+u_{17}-4u_{18}-\alpha\, k^2 u_{18}^2+2=g_{18}$

Differentiating $\frac{\partial g_{18}}{\partial u_{13}}=1=g_{18,\,13}$ $\frac{\partial g_{18}}{\partial u_{17}}=1=g_{18,\,17}$

$\frac{\partial g_{18}}{\partial u_{18}}=-4-2\alpha\, k^2 u_{18}=g_{18,\,18}$

Now one can write with the application of the N–R method,

$$U^{n+1}=U^{n}-\left(J^{n}\right)^{-1}G^{n} \tag{6.68}$$

where,

$$U=[u_6 \quad u_7 \quad u_8 \quad u_{11} \quad u_{12} \quad u_{13} \quad u_{16} \quad u_{17} \quad u_{18}]^{T}$$

$$G=\left[g_6 \quad g_7 \quad g_8 \quad g_{11} \quad g_{12} \quad g_{13} \quad g_{16} \quad g_{17} \quad g_{18}\right]^{T}$$

$$J=\begin{bmatrix} g_{6,6} & g_{6,7} & g_{6,8} & g_{6,11} & g_{6,12} & g_{6,13} & g_{6,16} & g_{6,17} & g_{6,18} \\ g_{7,6} & g_{7,7} & g_{7,8} & g_{7,11} & g_{7,12} & g_{7,13} & g_{7,16} & g_{7,17} & g_{7,18} \\ g_{8,6} & g_{8,7} & g_{8,8} & g_{8,11} & g_{8,12} & g_{8,13} & g_{8,16} & g_{8,17} & g_{8,18} \\ g_{11,6} & g_{11,7} & g_{11,8} & g_{11,11} & g_{11,12} & g_{11,13} & g_{11,16} & g_{11,17} & g_{11,18} \\ g_{12,6} & g_{12,7} & g_{12,8} & g_{12,11} & g_{12,12} & g_{12,13} & g_{12,16} & g_{12,17} & g_{12,18} \\ g_{13,6} & g_{13,7} & g_{13,8} & g_{13,11} & g_{13,12} & g_{13,13} & g_{13,16} & g_{13,17} & g_{13,18} \\ g_{16,6} & g_{16,7} & g_{16,8} & g_{16,11} & g_{16,12} & g_{16,13} & g_{16,16} & g_{16,17} & g_{16,18} \\ g_{17,6} & g_{17,7} & g_{17,8} & g_{17,11} & g_{17,12} & g_{17,13} & g_{17,16} & g_{17,17} & g_{17,18} \\ g_{18,6} & g_{18,7} & g_{18,8} & g_{18,11} & g_{18,12} & g_{18,13} & g_{18,16} & g_{18,17} & g_{18,18} \end{bmatrix}_{9\times 9}$$

This gives,

$$J = \begin{bmatrix} g_{6,6} & 1 & 0 & 1 & 0 & 0 & 0 & 0 & 0 \\ 1 & g_{7,7} & 1 & 0 & 1 & 0 & 0 & 0 & 0 \\ 0 & 1 & g_{8,8} & 0 & 0 & 1 & 0 & 0 & 0 \\ 1 & 0 & 0 & g_{11,11} & 1 & 0 & 1 & 0 & 0 \\ 0 & 1 & 0 & 1 & g_{12,12} & 1 & 0 & 1 & 0 \\ 0 & 0 & 1 & 0 & 1 & g_{13,13} & 0 & 0 & 1 \\ 0 & 0 & 0 & 1 & 0 & 0 & g_{16,16} & 1 & 0 \\ 0 & 0 & 0 & 0 & 1 & 0 & 1 & g_{17,17} & 1 \\ 0 & 0 & 0 & 0 & 0 & 1 & 0 & 1 & g_{18,18} \end{bmatrix}_{9\times 9}$$

In the above Jacobian matrix, there are five types of non-zero entries as:

$j_{m,\, m-(M+1)} = 1$ for $m = 11, 12, 13, 16, 17, 18$

$j_{m,\, m-1} = 1$ for $m = 7, 8, 12, 13, 17, 18$

$j_{m,\, m} = -4 - 2\alpha\, k^2 u_m$ for all interior nodes (i.e., $m = 6, 7, 8, 11, 12, 13, 16, 17, 18$)

$j_{m,\, m+1} = 1$ for $m = 6, 7, 11, 12, 16, 17$

$j_{m,\, m+(M+1)} = 1$ for $m = 6, 7, 8, 11, 12, 13$

in which, $j_{m,\, m} = g_{m,\, m}$. Recall that here $M = 4$.

Using Equation (6.68), we get the unknowns as (with a tolerance of 10^{-5}):

$u_6 = 0.960819$ $u_7 = 0.950487$ $u_8 = 0.960819$

$u_{11} = 0.950487$ $u_{12} = 0.936776$ $u_{13} = 0.950487$

$u_{16} = 0.960819$ $u_{17} = 0.950487$ $u_{18} = 0.960819$

6.11 Alternating Direction Implicit (ADI) Method

A Little about ADI

When we solve the elliptic PDE equations using an implicit finite difference method (e.g., the Crank–Nicholson), the number of algebraic equations to be solved is quite large. It is typically seen that although the coefficient matrix has many zeros, that is not a banded system. To deal with such problems, Peaceman and Rachford (1955) proposed the alternating direction implicit (ADI) method. It is briefly highlighted with the following points:

- The ADI method is an iterative scheme used in numerical linear algebra to solve large matrix equations.
- It involves two-stage iteration in the solution methodology for alternately updating the column and row spaces.

- This ADI method is an implicit scheme and thus, unconditionally stable.
- It is usually employed for numerical solution of parabolic and elliptic PDEs.

Let us understand the motivation behind this ADI method followed by its formulation and solution mechanism through an example system.

Case Study 6.4 ADI method: Elliptic PDE

Unsteady state heat conduction: Two dimension

Previously, we have developed the unsteady state heat conduction model for 1D case (Case Study 6.1). Extending to 2D case, one can have:

$$\frac{\partial T}{\partial t} = \alpha \left(\frac{\partial^2 T}{\partial x^2} + \frac{\partial^2 T}{\partial y^2} \right) \tag{6.69}$$

with specified initial and boundary conditions. Notice that this is again an IBVP system, which includes total three independent variables (i.e., time and two spatial dimensions).

Motivation

Let us first identify the difficulty associated in solving Equation (6.69) by the Crank–Nicholson method. Here we define:

$$T_{i,j}^{n} \equiv T(x_i,\ y_j,\ t_n)$$

According to the C–N method, Equation (6.69) yields:

$$\left(\frac{\partial T}{\partial t} \right)_{i,\,j,\,n+\frac{1}{2}} = \alpha \left(\frac{\partial^2 T}{\partial x^2} + \frac{\partial^2 T}{\partial y^2} \right)_{i,\,j,\,n+\frac{1}{2}}$$

It gives:

$$\frac{T_{i,j}^{n+1} - T_{i,j}^{n}}{\Delta t} = \frac{\alpha}{2} \left[\frac{T_{i+1,j}^{n+1} - 2T_{i,j}^{n+1} + T_{i-1,j}^{n+1}}{(\Delta x)^2} + \frac{T_{i+1,j}^{n} - 2T_{i,j}^{n} + T_{i-1,j}^{n}}{(\Delta x)^2} + \frac{T_{i,j+1}^{n+1} - 2T_{i,j}^{n+1} + T_{i,j-1}^{n+1}}{(\Delta y)^2} + \frac{T_{i,j+1}^{n} - 2T_{i,j}^{n} + T_{i,j-1}^{n}}{(\Delta y)^2} \right]$$

With $\Delta x = \Delta y = k$ and $\beta = \dfrac{\alpha \Delta t}{k^2}$, we obtain:

$$\left[-\frac{\beta}{2} T_{i-1,j} - \frac{\beta}{2} T_{i,j-1} + (1+2\beta) T_{i,j} - \frac{\beta}{2} T_{i,j+1} - \frac{\beta}{2} T_{i+1,j} \right]^{n+1}$$
$$= \left[\frac{\beta}{2} T_{i-1,j} + \frac{\beta}{2} T_{i,j-1} + (1-2\beta) T_{i,j} + \frac{\beta}{2} T_{i,j+1} + \frac{\beta}{2} T_{i+1,j} \right]^{n}$$

This discretized form obtained by the implicit C–N method involves a large and sparse *pentadiagonal* matrix (Finlayson, 1980). For example, if we have 10 interior nodes (excluding known boundary nodes) in both x- and y-directions separately, then total 100 simultaneous equations are obtained in 100 variables. As a result, we will have a 100 × 100 matrix associated with 100 variables or equations. It requires a huge computational effort and round-off error becomes very large. Thus, we are motivated to use a special approach, namely the ADI method.

Instead of solving 100 simultaneous equations as in C–N method, we will see in the next phase of our study that the ADI scheme involves the solution of 10 sets of 10 simultaneous equations in each iteration stage. By this way, the ADI scheme motivates us to solve the concerned PDE systems with reducing the computational effort at large extent and thus, the round-off error.

Formulation and Solution Mechanism

Our next assignment is to develop the formulation and solution strategy of the ADI method. For this, let us proceed through formulating the computational steps that would be followed to solve the example PDE problem.

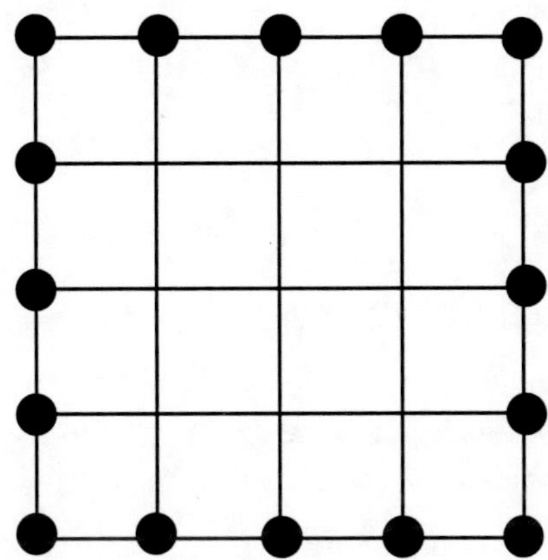

FIGURE 6.22 Showing boundary nodes (marked)

Computational steps

Step 1: The boundary nodes are shown in Figure 6.22. For those nodes, all temperatures are directly given (Dirichlet BC) or obtainable through the given boundary condition (e.g., Neumann BC). As far as temperatures in interior nodes are concerned, one can assume them all at the beginning (i.e., at $t = 0$).

Step 2: This step involves the first iteration stage of the ADI method as briefed below. Before performing this, let us divide the time step, Δt into two half-steps.

Iteration stage 1

In the first half-time step, an implicit scheme (backward Euler in time) is employed for the x-terms and the y-terms are treated explicitly. Accordingly, Equation (6.69) yields the following expression in the first half-step in time:

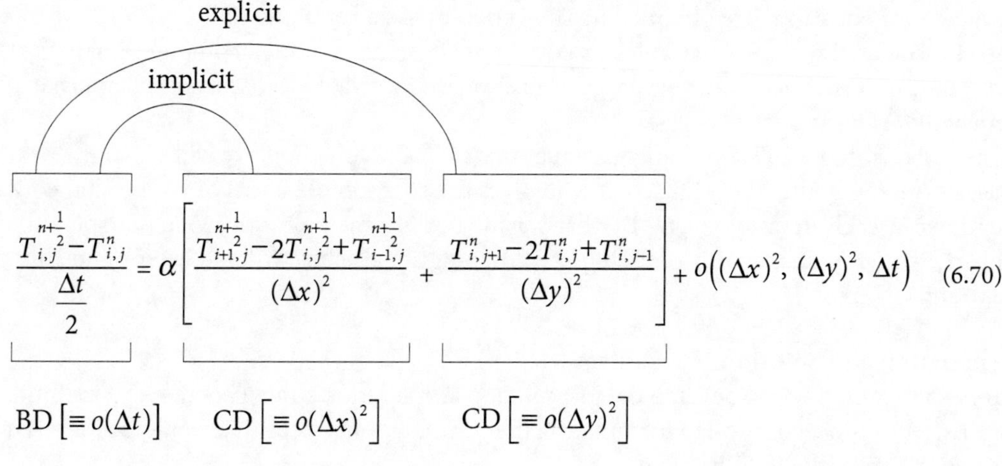

$$\frac{T_{i,j}^{n+\frac{1}{2}}-T_{i,j}^{n}}{\frac{\Delta t}{2}} = \alpha\left[\frac{T_{i+1,j}^{n+\frac{1}{2}}-2T_{i,j}^{n+\frac{1}{2}}+T_{i-1,j}^{n+\frac{1}{2}}}{(\Delta x)^2} + \frac{T_{i,j+1}^{n}-2T_{i,j}^{n}+T_{i,j-1}^{n}}{(\Delta y)^2}\right] + o\left((\Delta x)^2, (\Delta y)^2, \Delta t\right) \quad (6.70)$$

BD $[\equiv o(\Delta t)]$ CD $\left[\equiv o(\Delta x)^2\right]$ CD $\left[\equiv o(\Delta y)^2\right]$

With $\Delta x = \Delta y = k$ and $\beta = \frac{\alpha\,\Delta t}{k^2}$,

$$\frac{2}{\beta}\left(T_{i,j}^{n+\frac{1}{2}}-T_{i,j}^{n}\right)=T_{i+1,j}^{n+\frac{1}{2}}-2T_{i,j}^{n+\frac{1}{2}}+T_{i-1,j}^{n+\frac{1}{2}}+T_{i,j+1}^{n}-2T_{i,j}^{n}+T_{i,j-1}^{n}$$

Rearranging,

$$-T_{i-1,j}^{n+\frac{1}{2}}+\left(2+\frac{2}{\beta}\right)T_{i,j}^{n+\frac{1}{2}}-T_{i+1,j}^{n+\frac{1}{2}}=T_{i,j+1}^{n}+\left(\frac{2}{\beta}-2\right)T_{i,j}^{n}+T_{i,j-1}^{n} \quad (6.71)$$

Note that here T^n is the temperature at time, t and $T^{n+\frac{1}{2}}$ the updated temperature obtained in the first half-time step of iteration. Actually, using Equation (6.71), one can compute $T^{n+\frac{1}{2}}$ based on the known T^n (given or obtainable for boundary nodes and assumed for interior nodes at the beginning).

This Equation (6.71) is to be solved row-wise from bottom to top. It means, we need to first consider:

$j = 0$ and $i = 0, 1, 2, \ldots, M$

This may be a specified bottom boundary and if so, there is no calculation involved. Additionally, if the left and right boundaries are specified, then proceed for:

$j = 1$ and $i = 1, 2, \ldots, M-1$,

$j = 2$ and $i = 1, 2, \ldots, M-1$

and so on. It is illustrated in Figure 6.23 for $M = 4$.

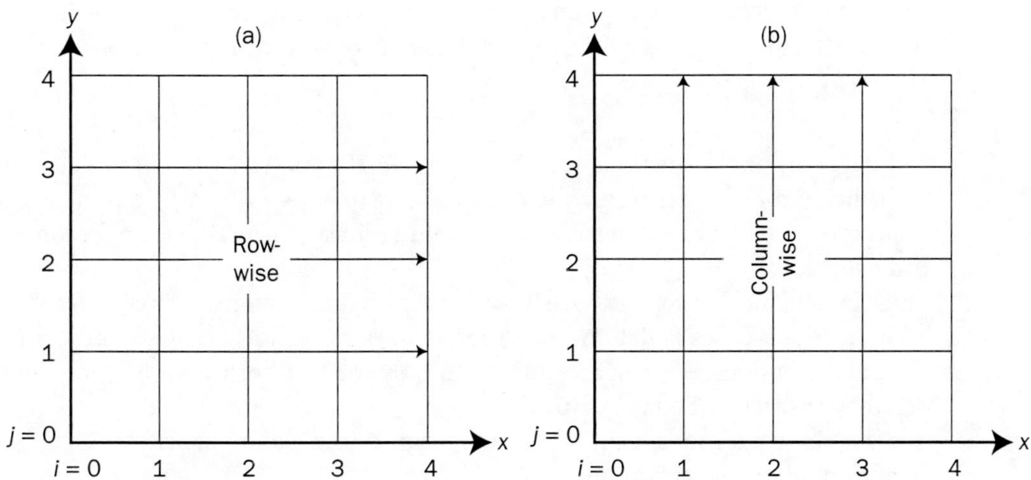

Figure 6.23 Two-stage iteration in ADI method: (a) stage 1, and (b) stage 2

Step 3: This step involves the second iteration stage of the ADI method.

Iteration stage 2

In the second half-step in time, the y-terms are treated implicitly and the x-terms explicitly. Accordingly, Equation (6.69) yields:

$$\frac{T_{i,j}^{n+1}-T_{i,j}^{n+\frac{1}{2}}}{\frac{\Delta t}{2}}=\alpha\left[\frac{T_{i+1,j}^{n+\frac{1}{2}}-2T_{i,j}^{n+\frac{1}{2}}+T_{i-1,j}^{n+\frac{1}{2}}}{(\Delta x)^2}+\frac{T_{i,j+1}^{n+1}-2T_{i,j}^{n+1}+T_{i,j-1}^{n+1}}{(\Delta y)^2}\right]+o\left((\Delta x)^2,\ (\Delta y)^2,\ \Delta t\right) \tag{6.72}$$

Rearranging,

$$-T_{i,j-1}^{n+1}+\left(2+\frac{2}{\beta}\right)T_{i,j}^{n+1}-T_{i,j+1}^{n+1}=T_{i+1,j}^{n+\frac{1}{2}}+\left(\frac{2}{\beta}-2\right)T_{i,j}^{n+\frac{1}{2}}+T_{i-1,j}^{n+\frac{1}{2}} \tag{6.73}$$

Here, $T^{n+\frac{1}{2}}$ is known [obtained in first iteration stage (Step 2)] and T^{n+1} is to be computed. As shown in Figure 6.23 with $M = 4$, this Equation (6.73) needs to be solved column-wise from left to right. It means, one should start with:

$i = 0$ and $j = 0, 1, 2, \ldots, M$ (this may be a specified left boundary)

Next,

$i = 1$ and $j = 1, 2, \ldots, M-1$,

$i = 2$ and $j = 1, 2, \ldots, M-1$

and so on.

Step 4: Check convergence by computing the absolute error between the temperatures obtained at current and last time steps. If it is not converged, go back to Step 2.

Remarks

1. Although the above relations [i.e., Equations (6.70) and (6.72)] are implicit but with their proper ordering [made in Equations (6.71) and (6.73), respectively], the respective coefficient matrix of the resulting linear algebraic system becomes tridiagonal.
2. The ADI scheme developed above is first-order accurate in time and second-order accurate in space [see Equations (6.70) and (6.72)]. Note that to have second-order accurate in both time and space, one can simply use CD (instead of BD) for time derivative as well in Equations (6.70) and (6.72).
3. There is some additional gain we have in terms of computational effort since the coefficient matrices remain unchanged.

Problem statement

Solve the following PDE:

$$\frac{\partial T}{\partial t}=\alpha\left(\frac{\partial^2 T}{\partial x^2}+\frac{\partial^2 T}{\partial y^2}\right) \tag{6.69}$$

with the specified conditions shown in Figure 6.24. Considering $\Delta x = \Delta y = 0.5$ m, $\alpha = 1.0$ m^2/sec and $\Delta t = 0.5$ sec, it is asked to find the final temperature at all interior nodes with a desired tolerance of 10^{-6}.

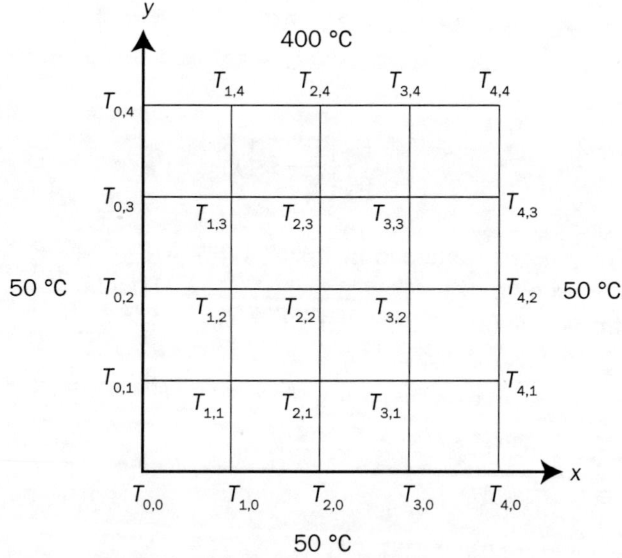

FIGURE 6.24 Specifying conditions for the slab

Solution

To solve this problem, we would like to follow the computational steps presented above.

Step 1: Temperatures at all boundary nodes are specified and fixed (see Figure 6.24). For the rest nodes (i.e., all interior nodes), we assume the initial temperatures as:

$$T_{1,1} = T_{2,1} = T_{3,1} = 60\ °C$$

$$T_{1,2} = T_{2,2} = T_{3,2} = 80\ °C$$

$$T_{1,3} = T_{2,3} = T_{3,3} = 100\ °C$$

Note that for the corner nodes, take the mean temperature values.[20]

Step 2: In this step, the row-wise operation is to be performed based on Equation (6.71).

- For various i nodes at $j = 1$
- Fix $j = 1$ and $i = 1$ [i.e., node (1, 1)]

$$-T_{0,1}^{n+\frac{1}{2}} + \left(2+\frac{2}{\beta}\right)T_{1,1}^{n+\frac{1}{2}} - T_{2,1}^{n+\frac{1}{2}} = T_{1,2}^{n} + \left(\frac{2}{\beta}-2\right)T_{1,1}^{n} + T_{1,0}^{n}$$

Here,

$T_{0,1}^{n+\frac{1}{2}} = 50$ (left boundary node)

$T_{1,0}^{n} = 50$ (bottom boundary node)

$T_{1,1}^{n} = 60$ (assumed at $t = 0$)

$T_{1,2}^{n} = 80$ (assumed at $t = 0$)

Substituting,

$$\left(2+\frac{2}{\beta}\right)T_{1,1}^{n+\frac{1}{2}} - T_{2,1}^{n+\frac{1}{2}} = \left(\frac{2}{\beta}-2\right)T_{1,1}^{n} + T_{1,2}^{n} + 100 \qquad (6.74a)$$

- Fix $j = 1$ and $i = 2$ [i.e., node (2, 1)]

$$-T_{1,1}^{n+\frac{1}{2}} + \left(2+\frac{2}{\beta}\right)T_{2,1}^{n+\frac{1}{2}} - T_{3,1}^{n+\frac{1}{2}} = \left(\frac{2}{\beta}-2\right)T_{2,1}^{n} + T_{2,2}^{n} + 50 \qquad (6.74b)$$

- Fix $j = 1$ and $i = 3$ [i.e., node (3, 1)]

$$-T_{2,1}^{n+\frac{1}{2}} + \left(2+\frac{2}{\beta}\right)T_{3,1}^{n+\frac{1}{2}} = \left(\frac{2}{\beta}-2\right)T_{3,1}^{n} + T_{3,2}^{n} + 100 \qquad (6.74c)$$

So for various $(i, 1)$ nodes, we have:

20 In this problem, the temperatures at the corner nodes are not required.

$$\begin{bmatrix} a & -1 & 0 \\ -1 & a & -1 \\ 0 & -1 & a \end{bmatrix} \begin{bmatrix} T_{1,1}^{n+\frac{1}{2}} \\ T_{2,1}^{n+\frac{1}{2}} \\ T_{3,1}^{n+\frac{1}{2}} \end{bmatrix} = \begin{bmatrix} b\,T_{1,1}^{n} + T_{1,2}^{n} + 100 \\ b\,T_{2,1}^{n} + T_{2,2}^{n} + 50 \\ b\,T_{3,1}^{n} + T_{3,2}^{n} + 100 \end{bmatrix} \tag{6.75}$$

where,

$$\beta = \frac{\alpha\,\Delta t}{k^2} = \frac{1 \times 0.5}{(0.5)^2} = 2$$

and thus,

$$a = \left(2 + \frac{2}{\beta}\right) = 3$$

$$b = \left(\frac{2}{\beta} - 2\right) = -1$$

Assumed values include:

$$T_{1,1} = T_{2,1} = T_{3,1} = 60$$

$$T_{1,2} = T_{2,2} = T_{3,2} = 80$$

Notice that Equation (6.75) represents a system with a tridiagonal set of linear algebraic equations. Solving it,

$$T_{1,1}^{n+\frac{1}{2}} = \frac{430}{7} = 61.4286$$

$$T_{2,1}^{n+\frac{1}{2}} = \frac{450}{7} = 64.2857$$

$$T_{3,1}^{n+\frac{1}{2}} = \frac{430}{7} = 61.4286$$

- For various i nodes at $j = 2$

Similarly, varying i at a fixed $j = 2$, we get:

$$\begin{bmatrix} a & -1 & 0 \\ -1 & a & -1 \\ 0 & -1 & a \end{bmatrix} \begin{bmatrix} T_{1,2}^{n+\frac{1}{2}} \\ T_{2,2}^{n+\frac{1}{2}} \\ T_{3,2}^{n+\frac{1}{2}} \end{bmatrix} = \begin{bmatrix} T_{1,1}^{n} + bT_{1,2}^{n} + T_{1,3}^{n} + 50 \\ T_{2,1}^{n} + bT_{2,2}^{n} + T_{2,3}^{n} \\ T_{3,1}^{n} + bT_{3,2}^{n} + T_{3,3}^{n} + 50 \end{bmatrix} \tag{6.76}$$

Use the temperature values assumed in Step 1. Solving Equation (6.76) with $a = 3$ and $b = -1$, we get:

$$T_{1,2}^{n+\frac{1}{2}} = \frac{470}{7} = 67.1428$$

$$T_{2,2}^{n+\frac{1}{2}} = \frac{500}{7} = 71.4286$$

$$T_{3,2}^{n+\frac{1}{2}} = \frac{470}{7} = 67.1428$$

- For various i nodes at $j = 3$

Again, varying i at a fixed $j = 3$,

$$\begin{bmatrix} a & -1 & 0 \\ -1 & a & -1 \\ 0 & -1 & a \end{bmatrix} \begin{bmatrix} T_{1,3}^{n+\frac{1}{2}} \\ T_{2,3}^{n+\frac{1}{2}} \\ T_{3,3}^{n+\frac{1}{2}} \end{bmatrix} = \begin{bmatrix} T_{1,2}^{n} + bT_{1,3}^{n} + 450 \\ T_{2,2}^{n} + bT_{2,3}^{n} + 400 \\ T_{3,2}^{n} + bT_{3,3}^{n} + 450 \end{bmatrix} \tag{6.77}$$

Solving this set of equations with $a = 3$ and $b = -1$, we get:

$$T_{1,3}^{n+\frac{1}{2}} = \frac{1670}{7} = 238.5714$$

$$T_{2,3}^{n+\frac{1}{2}} = \frac{2000}{7} = 285.7143$$

$$T_{3,3}^{n+\frac{1}{2}} = \frac{1670}{7} = 238.5714$$

Step 3: In this step, we need to perform the column-wise operation with the use of Equation (6.73).

- For various j nodes at $i = 1$
- Fix $i = 1$ and $j = 1$ [i.e., node (1, 1)]

$$-T_{1,0}^{n+1} + aT_{1,1}^{n+1} - T_{1,2}^{n+1} = T_{2,1}^{n+\frac{1}{2}} + bT_{1,1}^{n+\frac{1}{2}} + T_{0,1}^{n+\frac{1}{2}}$$

Here,

$T_{1,0}^{n+1} = 50$ (bottom boundary node)

$T_{0,1}^{n+\frac{1}{2}} = 50$ (left boundary node)

Substituting,

$$aT_{1,1}^{n+1} - T_{1,2}^{n+1} = bT_{1,1}^{n+\frac{1}{2}} + T_{2,1}^{n+\frac{1}{2}} + 100 \tag{6.78a}$$

- Fix $i = 1$ and $j = 2$ [i.e., node (1, 2)]

$$-T_{1,1}^{n+1}+aT_{1,2}^{n+1}-T_{1,3}^{n+1}=T_{2,2}^{n+\frac{1}{2}}+bT_{1,2}^{n+\frac{1}{2}}+T_{0,2}^{n+\frac{1}{2}}$$

This gives,

$$-T_{1,1}^{n+1}+aT_{1,2}^{n+1}-T_{1,3}^{n+1}=bT_{1,2}^{n+\frac{1}{2}}+T_{2,2}^{n+\frac{1}{2}}+50 \tag{6.78b}$$

- Fix $i = 1$ and $j = 3$ [i.e., node (1, 3)]

$$-T_{1,2}^{n+1}+aT_{1,3}^{n+1}-T_{1,4}^{n+1}=T_{2,3}^{n+\frac{1}{2}}+bT_{1,3}^{n+\frac{1}{2}}+T_{0,3}^{n+\frac{1}{2}}$$

This gives,

$$-T_{1,2}^{n+1}+aT_{1,3}^{n+1}=bT_{1,3}^{n+\frac{1}{2}}+T_{2,3}^{n+\frac{1}{2}}+450 \tag{6.78c}$$

So for various j nodes at $i = 1$,

$$\begin{bmatrix} a & -1 & 0 \\ -1 & a & -1 \\ 0 & -1 & a \end{bmatrix}\begin{bmatrix} T_{1,1}^{n+1} \\ T_{1,2}^{n+1} \\ T_{1,3}^{n+1} \end{bmatrix}=\begin{bmatrix} bT_{1,1}^{n+\frac{1}{2}}+T_{2,1}^{n+\frac{1}{2}}+100 \\ bT_{1,2}^{n+\frac{1}{2}}+T_{2,2}^{n+\frac{1}{2}}+50 \\ bT_{1,3}^{n+\frac{1}{2}}+T_{2,3}^{n+\frac{1}{2}}+450 \end{bmatrix} \tag{6.79}$$

Along with $a = 3$ and $b = -1$, use the values of the following temperatures obtained in Step 2:

$T_{1,1}^{n+\frac{1}{2}}=61.4286$ $\quad T_{2,1}^{n+\frac{1}{2}}=64.2857$ $\quad T_{1,2}^{n+\frac{1}{2}}=67.1428$

$T_{2,2}^{n+\frac{1}{2}}=71.4286$ $\quad T_{1,3}^{n+\frac{1}{2}}=238.5714$ $\quad T_{2,3}^{n+\frac{1}{2}}=285.7143$

Solving Equation (6.79), one obtains:

$$T_{1,1}^{n+1}=\frac{3460}{49}=70.6122$$

$$T_{1,2}^{n+1}=\frac{5340}{49}=108.9796$$

$$T_{1,3}^{n+1}=\frac{9900}{49}=202.0408$$

- For various j nodes at fixed $i = 2$

For this, we have:

$$\begin{bmatrix} a & -1 & 0 \\ -1 & a & -1 \\ 0 & -1 & a \end{bmatrix} \begin{bmatrix} T_{2,1}^{n+1} \\ T_{2,2}^{n+1} \\ T_{2,3}^{n+1} \end{bmatrix} = \begin{bmatrix} T_{1,1}^{n+\frac{1}{2}} + bT_{2,1}^{n+\frac{1}{2}} + T_{3,1}^{n+\frac{1}{2}} + 50 \\ T_{1,2}^{n+\frac{1}{2}} + bT_{2,2}^{n+\frac{1}{2}} + T_{3,2}^{n+\frac{1}{2}} \\ T_{1,3}^{n+\frac{1}{2}} + bT_{2,3}^{n+\frac{1}{2}} + T_{3,3}^{n+\frac{1}{2}} + 400 \end{bmatrix} \tag{6.80}$$

Solving this,

$$T_{2,1}^{n+1} = \frac{11540}{147} = 78.5034$$

$$T_{2,2}^{n+1} = \frac{6220}{49} = 126.9388$$

$$T_{2,3}^{n+1} = \frac{35200}{147} = 239.4558$$

- For various j nodes at fixed $i = 3$

Similarly, for this we have:

$$\begin{bmatrix} a & -1 & 0 \\ -1 & a & -1 \\ 0 & -1 & a \end{bmatrix} \begin{bmatrix} T_{3,1}^{n+1} \\ T_{3,2}^{n+1} \\ T_{3,3}^{n+1} \end{bmatrix} = \begin{bmatrix} T_{2,1}^{n+\frac{1}{2}} + bT_{3,1}^{n+\frac{1}{2}} + 100 \\ T_{2,2}^{n+\frac{1}{2}} + bT_{3,2}^{n+\frac{1}{2}} + 50 \\ T_{2,3}^{n+\frac{1}{2}} + bT_{3,3}^{n+\frac{1}{2}} + 450 \end{bmatrix} \tag{6.81}$$

Solving this, we obtain:

$$T_{3,1}^{n+1} = \frac{3460}{49} = 70.6122$$

$$T_{3,2}^{n+1} = \frac{5340}{49} = 108.9796$$

$$T_{3,3}^{n+1} = \frac{9900}{49} = 202.0408$$

Step 4: This step involves the convergence check. In the last two iteration stages, we got the following values (Table 6.3) for 9 interior nodes.

Table 6.3 Computed temperatures in two iterative stages

Iteration stage 1	*Iteration stage 2*
$T_{1,1}^{n+\frac{1}{2}} = 61.4286$	$T_{1,1}^{n+1} = 70.6122$
$T_{2,1}^{n+\frac{1}{2}} = 64.2857$	$T_{2,1}^{n+1} = 78.5034$
$T_{3,1}^{n+\frac{1}{2}} = 61.4286$	$T_{3,1}^{n+1} = 70.6122$
$T_{1,2}^{n+\frac{1}{2}} = 67.1428$	$T_{1,2}^{n+1} = 108.9796$
$T_{2,2}^{n+\frac{1}{2}} = 71.4286$	$T_{2,2}^{n+1} = 126.9388$
$T_{3,2}^{n+\frac{1}{2}} = 67.1428$	$T_{3,2}^{n+1} = 108.9796$
$T_{1,3}^{n+\frac{1}{2}} = 238.5714$	$T_{1,3}^{n+1} = 202.0408$
$T_{2,3}^{n+\frac{1}{2}} = 285.7143$	$T_{2,3}^{n+1} = 239.4558$
$T_{3,3}^{n+\frac{1}{2}} = 238.5714$	$T_{3,3}^{n+1} = 202.0408$

Obviously,

$$\left| \left(T_{i,j}^{n+1}\right)_{t+\Delta t} - \left(T_{i,j}^{n+1}\right)_{t} \right| > \text{tol } (=10^{-6})$$

and thus, repeat the computation with going back to Step 2. Now use the temperature profile at $t+\Delta t$ to obtain that at $t+2\Delta t$.

A sample calculation is shown here. Following this, you perform the rest of the computations as your homework and report the final temperature for all 9 interior nodes with satisfying the given tolerance.

6.12 Coupled PDEs

So far we discussed the numerical solution of single PDE systems. Now, we would like to extend the application of finite difference method to the system of coupled PDEs. To elaborate this, let us take an example of a non-isothermal plug flow reactor (PFR). Here, we will use the method of lines to solve the PFR. The details of MoL, including its computational steps, are presented before with an application.

Case Study 6.5 Coupled PDEs

Non-isothermal plug flow reactor (PFR)

Here, we will study how to develop a plug flow reactor (also called the *continuous tubular reactor*) model and to perform its numerical simulation. For this, let us consider a non-isothermal liquid phase PFR depicted in Figure 6.25. A fluid stream traveling in the axial direction in a PFR may be modeled as a series of infinite thin coherent *plugs*. Although the composition in each plug is assumed uniform, but it varies from plug-to-plug. Each plug of differential volume (shown with length Δz in Figure 6.25) is deemed as an infinitesimally small continuous stirred tank reactor (CSTR), limiting to zero volume.

With this, the following assumptions are adopted here to develop the PFR model.

Assumptions

1. Perfect mixing in the radial direction, not in the axial direction (i.e., no variation of component composition and temperature in radial direction of the reactor) takes place.
2. There is no heat loss to the surroundings (i.e., perfectly insulated).
3. Only reaction, no diffusion (mass and thermal diffusion), is involved.
4. All physical properties (density, ρ and heat capacity, C_p) remain unchanged.
5. Convective velocity is assumed constant.
6. One-dimensional flow (i.e., in z direction) is considered in the PFR.
7. Second-order endothermic irreversible reaction takes place.

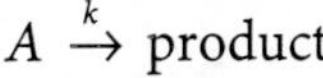

$$A \xrightarrow{k} \text{product}$$

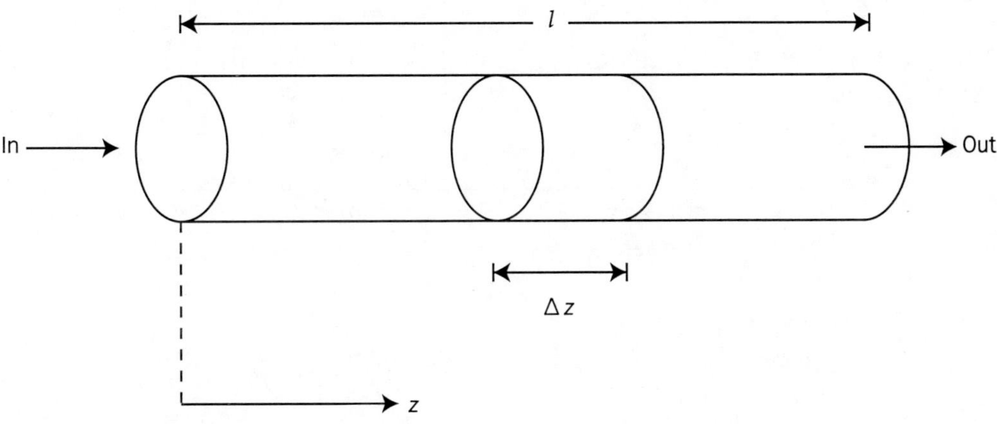

FIGURE 6.25 Schematic of a PFR

Deriving model

With these assumptions, Equations (A5.7) and (A5.8) reduce, respectively, to:

$$\frac{\partial C_A}{\partial t} + \upsilon \frac{\partial C_A}{\partial z} + (-r_A) = 0 \tag{6.82}$$

$$\frac{\partial T}{\partial t} + \upsilon \frac{\partial T}{\partial z} + \frac{(\Delta H_r)(-r_A)}{\rho C_p} = 0 \tag{6.83}$$

with reaction rate,

$$-r_A = kC_A^2 \tag{6.84}$$

where,

$$k = k_0 \exp\left[-\frac{\Delta E}{R}\left(\frac{1}{T}-\frac{1}{T_r}\right)\right] \tag{6.85}$$

All unknown notations are defined in Table 6.4.

Problem statement

The transient PFR model developed above consists of two hyperbolic PDEs. Let us proceed to solve these coupled equations with the given conditions and parameter values by using the MoL scheme.

Initial conditions:

$C_A(z,\ 0) = 1\ \text{mol/m}^3 \qquad 0 \le z \le l$

$T(z,\ 0) = 500\ \text{K}$

where, the reactor length, l is equal to 2 m.

Boundary conditions:

$C_A(0, t) = 1\ \text{mol/m}^3 \qquad t \ge 0$

$T(0, t) = 500\ \text{K}$

Consider $M = 25$ (i.e., $\Delta z = 0.08$ m) and $\Delta t = 0.001$ min. Given parameter values are reported in Table 6.4.

Table 6.4 Process parameters

Pre-exponential factor	$k_0 = 5\ \text{m}^3/\text{mol.min}$
Reference temperature	$T_r = 500$ K
Activation energy	$\Delta E = 50 \times 10^3$ J/mol
Endothermic heat of reaction	$\Delta H_r = 10 \times 10^3$ J/mol
Convective velocity	$\upsilon = 0.5$ m/min
Reactor length	$l = 2$ m
	$\rho C_p = 1000\ \text{J/m}^3.\text{K}$
	$R = 8.314$ J/mol.K

Solution

Let us employ the MoL with following its computational steps detailed before.

Step 1: Discretizing the spatial domain (= 2 m), we have total 26 nodes.

Step 2: Converting the PDEs into a set of ODEs. For this, we apply the backward difference approximation for the spatial derivative. Accordingly, Equations (6.82) and (6.83) yield, respectively:

$$\frac{dC_{Am}}{dt} + \upsilon\frac{C_{Am} - C_{A\ m-1}}{\Delta z} = -(-r_{Am}) \tag{6.86}$$

$$\frac{dT_m}{dt}+\upsilon\frac{T_m-T_{m-1}}{\Delta z}=-\frac{(\Delta H_r)(-r_{Am})}{\rho C_p} \tag{6.87}$$

Here, m = 1, 2, 3, ….., 25. Thus, there are total 50 (= $2M$) equations and the same number of unknown variables, namely C_m and T_m.

Step 3: In the next phase, we integrate the set of Equations (6.86) and (6.87) by using the forward Euler formula.

With this, Equations (6.86) and (6.87) are integrated with the use of given conditions and parameter values. The resulting concentration and temperature profiles along the reactor length are depicted in Figure 6.26. The non-isothermal PFR takes about six minutes or so to reach the steady state.

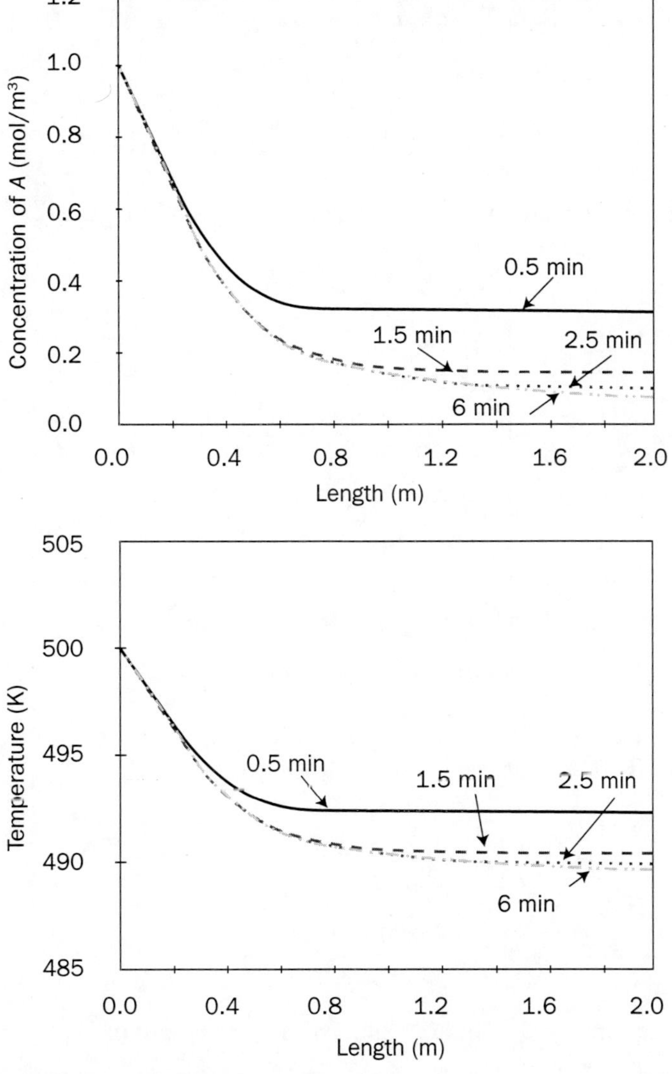

Figure 6.26 Concentration and temperature profile of the non-isothermal PFR along the reactor length

Remark

The PFR model is converted to an ODE-IVP system that consists of Equations (6.86) and (6.87). The given initial conditions at all spatial nodes (i.e., boundary and interior nodes) are enough to solve it. But sometimes the system boundary is maintained at a certain condition (i.e., BC). Like in this PFR problem, the left boundary is maintained with $C = 1$ mol/m^3 and $T = 500$ K for all times, keeping the right boundary open. It is also reflected in Figure 6.26 that C and T at $z = 0$ are fixed, but they are different with time at $z = \ell = 2$ m. Finally, we reach the steady state at all nodes, and no further change of any C and T is noticed.

6.13 Solved Example and Case Studies

Example 6.4 Elliptic PDE on Triangular Domain

Let us solve the Laplace equation:

$$\frac{\partial^2 u}{\partial x^2}+\frac{\partial^2 u}{\partial y^2}=0$$

that is defined over the right angle triangle region as shown in Figure 6.27.

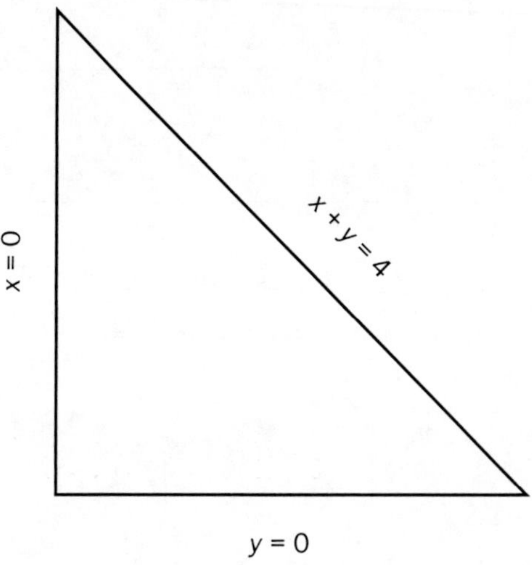

Figure 6.27 Defining right angle triangular domain

Boundary conditions:[21]

$u = 0$	over the edge $x = 0$
$u = 0$	over the edge $y = 0$
$u=16-x^2-y^2$	at the edge $x+y=4$

Find u at the grid points of the square region with mesh length of 1.

21 For the triangular boundary, there are 3 BCs (instead of 4 in rectangular domain).

Solution

Based on the given information, Figure 6.28 is produced that shows all nodes, including the known boundary nodes as:

$$u_{0,0} = u_{1,0} = u_{2,0} = u_{3,0} = u_{4,0} = 0$$

$$u_{0,1} = u_{0,2} = u_{0,3} = u_{0,4} = 0$$

$$u_{3,1} = 16 - x^2 - y^2 = 16 - 3^2 - 1^2 = 6 = u_{1,3}$$

$$u_{2,2} = 16 - 2^2 - 2^2 = 8$$

For the three interior nodes, standard 5-points formula [i.e., Equation (6.54)] yields:

- Node (1,1) $\quad u_{2,1} + u_{1,2} - 4u_{1,1} = 0$
- Node (2,1) $\quad u_{1,1} + 6 + 8 - 4u_{2,1} = 0$
- Node (1,2) $\quad 8 + 6 + u_{1,1} - 4u_{1,2} = 0$

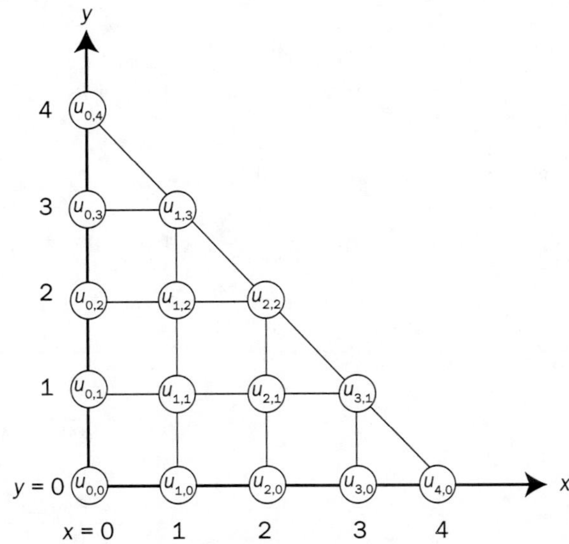

Figure 6.28 Finite difference grid for Laplace equation over triangular region

Solving these three linear algebraic equations, we get the u at interior nodes as:

$$u_{1,1} = 2$$
$$u_{1,2} = u_{2,1} = 4$$

Case Study 6.6 Porous media (non-linear model)

The IBVP system is described by the following non-linear model equation:

$$\frac{\partial u}{\partial t} = \left(\frac{\partial u}{\partial x}\right)^2 + u\frac{\partial^2 u}{\partial x^2} \tag{6.88}$$

Initial condition:

$$u(x,\ 0) = x \qquad 0 \le x \le 1$$

Boundary conditions:

$$u(0,\ t) = t \qquad t \ge 0$$

$$u(1,\ t) = 1 + t$$

Solve this problem using the FTCS method with $M = 10$ and $h = \Delta t = 0.0001$ to find the u trajectories at time, $t = 0$, 5 and 10.

Solution

Applying the FTCS scheme,

$$\frac{u_{m,\,n+1} - u_{m,\,n}}{h} = \left(\frac{u_{m+1,\,n} - u_{m-1,\,n}}{2k}\right)^2 + u_{m,\,n}\frac{u_{m+1,\,n} - 2u_{m,\,n} + u_{m-1,\,n}}{k^2}$$

Rearranging,

$$u_{m,\,n+1} = \frac{h}{4k^2}(u_{m+1,n} - u_{m-1,n})^2 + u_{m,\,n} + \frac{h}{k^2}u_{m,\,n}(u_{m+1,\,n} - 2u_{m,\,n} + u_{m-1,\,n}) \qquad (6.89)$$

for $m = 1, 2, \ldots, 9$, and $n = 1, 2, \ldots\ldots$ so on. Note that here, $k = \frac{1}{M} = 0.1$ and $h = 0.0001$ (given).

With this, we solve Equation (6.89) and produce Figure 6.29 that depicts the profile of u against x at three different times (i.e., $t = 0$, 5 and 10).

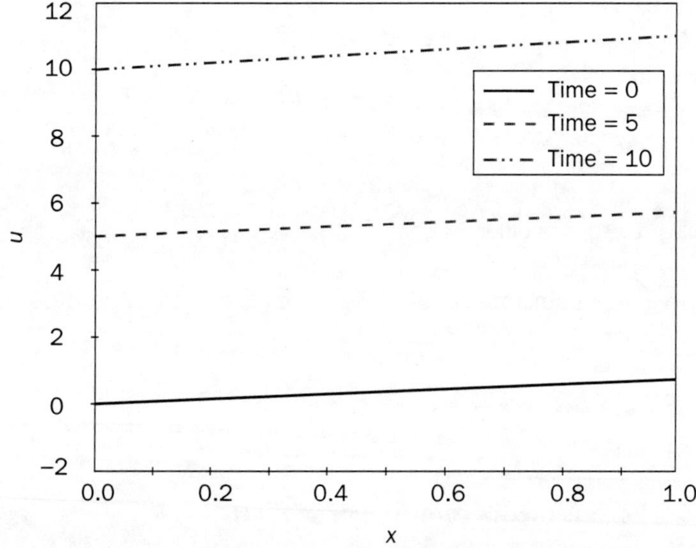

FIGURE 6.29 Predicting u against x for the non-linear porous media problem

Case Study 6.7 Unsteady state double pipe heat exchanger

A double pipe counter-current heat exchanger is schematically shown in Figure 6.30. For numerical simulation and analysis of this unsteady state process, first we will develop its mathematical model by means of the energy balance equation. For this, let us adopt the following assumptions.

Assumptions

1. There is no heat loss to the surroundings and the saturated steam flowed counter-currently through the shell side remains at a constant temperature, T_s (i.e., only phase change occurs).
2. Process stream (liquid) flows through the tube side with a constant velocity of υ.
3. Temperature of process stream (T) changes only in axial direction, not radially.
4. Physical properties (i.e., ρ and C_p) remain unchanged.
5. There is no heat accumulation in the tube and shell walls.

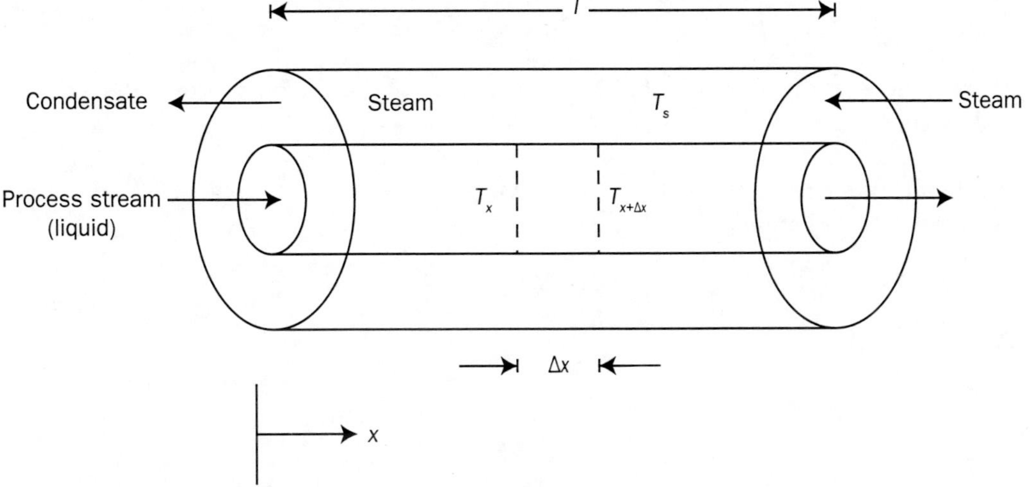

FIGURE 6.30 A double pipe heat exchanger

Deriving model

Applying the principle of energy conservation,

Heat accumulated = Heat input − Heat output + Heat generation

with zero heat generation, one obtains:

$$A\Delta x\ \rho\ C_p[T_{t+\Delta t}-T_t]=\upsilon\ A\rho\ C_p\Delta t T_x-\upsilon\ A\rho\ C_p\Delta t T_{x+\Delta x}+Q\Delta t(\pi\ d\ \Delta x) \tag{6.90}$$

Note that there are two heat input terms, which are combined as:

Heat input at x through flowing liquid during Δt + Heat input from shell side steam during Δt

$$=\upsilon\ A\rho\ C_p\Delta t T_x+Q\Delta t(\pi\ d\ \Delta x)$$

Here,

A = flow area for tube side liquid
Q = heat flux
d = tube diameter

Dividing both sides by $A\rho\ C_p \Delta x\ \Delta t$, and taking limits $\Delta x \to 0$ and $\Delta t \to 0$,

$$\frac{\partial T}{\partial t}+\upsilon\frac{\partial T}{\partial x}=\frac{Q(\pi\ d)}{A\rho\ C_p} \tag{6.91}$$

Considering,

$$A=\frac{\pi}{4}d^2$$
$$Q=U(T_s-T)$$

we have,

$$\frac{\partial T}{\partial t}+\upsilon\frac{\partial T}{\partial x}=\frac{4U}{\rho C_p d}(T_s-T) \tag{6.92}$$

where, U denotes the overall heat transfer coefficient. This is the final form of the dynamic heat balance equation.

Steady state model

At steady state, Equation (6.92) yields:

$$\upsilon\frac{dT}{dx}=\frac{4U}{\rho C_p d}(T_s-T) \tag{6.93}$$

This gives,

$$\frac{dT}{dx}=\phi(T_s-T)$$

where,

$$\phi=\frac{4U}{\upsilon\rho C_p d}$$

Integrating,

$$\int_{T(0,0)}^{T(x,0)}\frac{dT}{T-T_s}=-\phi\int_0^x dx$$

we get,

$$T(x,\ 0)=T_s+(T(0,\ 0)-T_s)\exp(-\phi\ x) \tag{6.94}$$

Problem statement

The unsteady state heat exchanger is modeled above by:

$$\frac{\partial T}{\partial t}+v\frac{\partial T}{\partial x}=\frac{4U}{\rho C_p d}(T_s-T) \tag{6.92}$$

Let us consider that this heat exchanger operates initially (i.e., at t = 0) at steady state. Accordingly, we fix the following conditions:

Initial condition (i.e., steady state model):

$$T(x,\ 0)=T_s+(T(0,\ 0)-T_s)\exp(-\phi\ x) \tag{6.94}$$

at t = 0 over the domain $0 \le x \le l$, where $T(0,\ 0)=75$ °F.

Boundary condition:

$T(0, t)$ = 60 °F $\qquad\qquad t > 0$

The above set of conditions indicate that the heat exchanger is forced to deviate from its steady state by introducing a step decrease in inlet temperature of process liquid (that is indeed a disturbance variable) from 75 to 60 °F. This is shortly demonstrated in Figure 6.31.

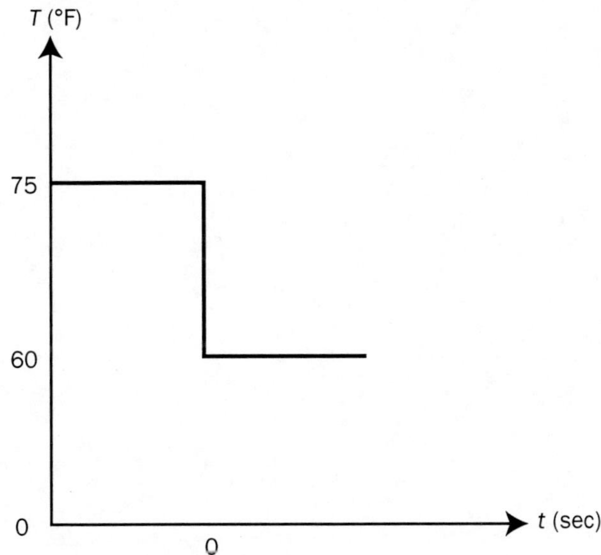

FIGURE **6.31** Step decrease of T at x = 0 and t > 0

Relevant process information is given in Table 6.5 (Ramirez, 1997). Adopting M = 40, we have:

$$k=\frac{l}{M}=0.3 \text{ ft}$$

$$h=\frac{k}{v}=0.1 \text{ sec}$$

Table 6.5 Heat exchanger information

Process stream	water
Steam temperature	T_s = 250 °F
Specific heat capacity of water	C_p = 1 Btu/lb$_m$.°F
Tube diameter	d = 0.5 ft
Density of water	ρ = 62.4 lb/ft^3
Overall heat transfer coefficient	U = 0.2 Btu/sec.ft^2.°F
Water velocity	υ = 3 ft/sec
Tube length	l = 12 ft

Solve this IBVP system using the FTBS method and produce the temperature profile at time, t = 0, 1 and 5 sec.

Solution

From Equation (6.92),

$$\frac{1}{\upsilon}\frac{\partial T}{\partial t}=-\frac{\partial T}{\partial x}+\phi(T_s-T)$$

Applying FTBS scheme,

$$\frac{1}{\upsilon}\frac{T_{m,\,n+1}-T_{m,\,n}}{h}=-\frac{T_{m,\,n}-T_{m-1,\,n}}{k}+\phi(T_s-T_{m,\,n})$$

or,

$$T_{m,\,n+1}=T_{m,\,n}-\frac{h\upsilon}{k}(T_{m,\,n}-T_{m-1,\,n})+\phi\, h\upsilon(T_s-T_{m,\,n})$$

$$=\phi\, h\upsilon T_s+\left(1-\frac{h\upsilon}{k}-\phi\, h\upsilon\right)T_{m,\,n}+\frac{h\upsilon}{k}T_{m-1,\,n}$$

This gives:

$$T_{m,\,n+1}=aT_s+bT_{m,\,n}+cT_{m-1,\,n} \tag{6.95}$$

where,

$$a=\phi\, h\upsilon=0.0025641$$

$$b=\left(1-\frac{h\upsilon}{k}-\phi\, h\upsilon\right)=-0.0025641$$

$$c=\frac{h\upsilon}{k}=1$$

Having these information, we solve Equation (6.95) and produce the numerical solution for temperature at different grid points. Plotting these simulation data, the temperature profile is obtained in Figure 6.32 at time, $t = 0$, 1 and 5 sec.

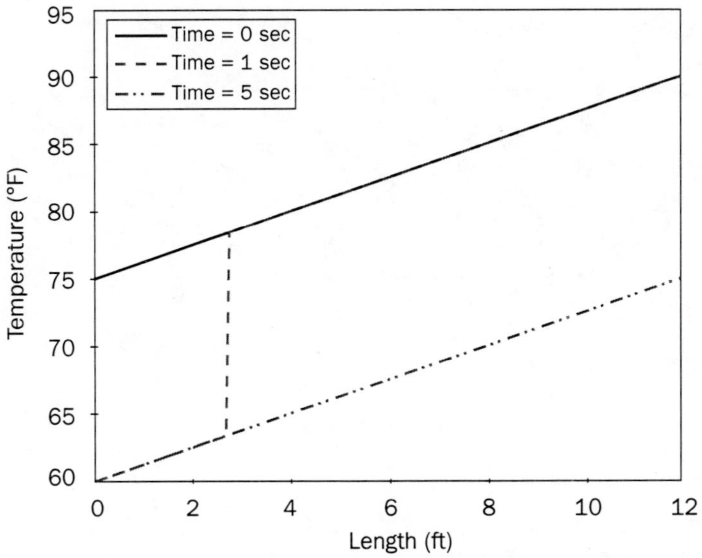

Figure 6.32 Temperature profile of the heat exchanger

Case Study 6.8 An implicit FTCS scheme

Unsteady state heat conduction in a cylinder

Here our goal is to analyze the unsteady state heat conduction in a cylindrical shell by using an implicit finite difference scheme. For this, let us first develop the heat transfer model. Considering an elemental volume shown in Figure 6.33, we adopt the following assumptions:

Assumptions

1. There is no heat loss.
2. Heat conduction occurs in radial direction only. This is quite reasonable when the length, l is much larger than the diameter, d ($=2R$).
3. Volumetric rate of heat generation, $\dot{q}$ remains uniform throughout the medium.
4. Physical properties [i.e., density (ρ), specific heat capacity (C_p) and thermal conductivity (K)] remain constant.

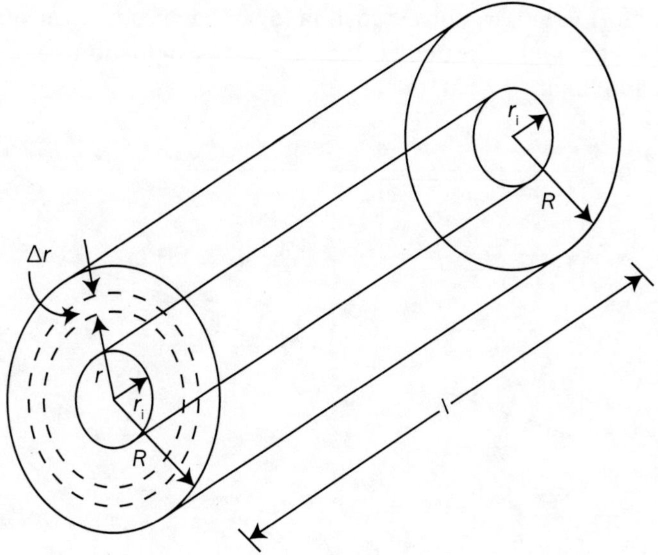

FIGURE 6.33 An elemental volume in a cylindrical shell

Deriving model

Making energy balance,

$$\text{Heat accumulated} = \text{Heat input} - \text{Heat output} + \text{Heat generation}$$

we get,

$$2\pi\ rl\Delta r\rho C_p[T_{t+\Delta t} - T_t] = \Delta t\ q_r - \Delta t\ q_{r+\Delta r} + \dot{q}\ (2\pi\ rl\Delta r\Delta t)$$

Here, r denotes the radius and q the rate of heat transfer. Dividing both sides by $\Delta t\ \Delta r$, and taking $\Delta t \to 0$ and $\Delta r \to 0$:

$$2\pi\ rl\rho C_p\frac{\partial T}{\partial t} = -\frac{\partial}{\partial r}\left(-K2\pi\ rl\frac{\partial T}{\partial r}\right) + \dot{q}\ (2\pi\ rl)$$

Dividing further by $2\pi\ lK$,

$$\frac{r}{\alpha}\frac{\partial T}{\partial t} = \frac{\partial}{\partial r}\left(r\frac{\partial T}{\partial r}\right) + \frac{r}{K}\dot{q}$$

This gives,

$$\frac{1}{\alpha}\frac{\partial T}{\partial t} = \frac{\partial^2 T}{\partial r^2} + \frac{1}{r}\frac{\partial T}{\partial r} + \frac{\dot{q}}{K} \tag{6.96}$$

This is the final model equation and it can also be derived from Equation (A5.8).

Problem statement

Neglecting heat generation (i.e., $\dot{q} = 0$), the unsteady state heat conduction model for a cylindrical shell [i.e., Equation (6.96)] gets the following form:

$$\frac{\partial T}{\partial t} = \alpha \left(\frac{\partial^2 T}{\partial r^2} + \frac{1}{r} \frac{\partial T}{\partial r} \right) \tag{6.97}$$

Boundary conditions (BCs):

at $r = 0$: $\quad \frac{\partial T}{\partial r}(0, t) = 0 \quad t \geq 0$

at $r = R$: $\quad T\ (R, t) = T_s = 150\ °C$

Initial condition (IC):

at $t = 0$: $\quad T(r, 0) = 50\ °C \quad 0 \leq r < R$

Here, T_s denotes the surface temperature. Given information include:

$R = 0.4$ cm

$\Delta r = k = 0.1$ cm

$\Delta t = h = 0.1$ sec

$\alpha = 2.3 \times 10^{-5}\ m^2/sec$

Solve this PDE system by using the forward in time and central in space (FTCS) implicit difference scheme. Report the temperature at different nodes against time up to 2 sec.

Solution

For the given system, we make discretization and the resulting grid points in the radius of the cylindrical shell are shown in Figure 6.34.

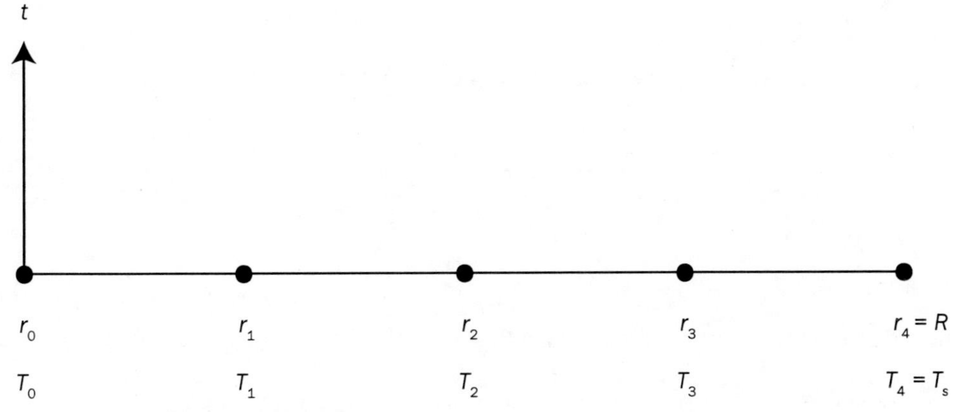

FIGURE 6.34 Finite difference grid points

Applying forward in time and central in space implicit scheme,

$$\frac{T_{m,\,n+1}-T_{m,\,n}}{\Delta t}=\alpha\left[\frac{T_{m+1,\,n+1}-2T_{m,\,n+1}+T_{m-1,\,n+1}}{(\Delta r)^2}+\frac{1}{r_m}\frac{T_{m+1,\,n+1}-T_{m-1,\,n+1}}{2\Delta r}\right]$$

where,

$$r_m=r_0+m\,\Delta r \qquad r_0=0 \text{ and } m=0,\ 1,\ 2,\ \ldots.$$

and

$$\beta=\frac{\alpha\,\Delta t}{(\Delta r)^2}=\frac{(2.3\times10^{-5})\times0.1}{(0.1\times10^{-2})^2}=2.3$$

Rearranging,

$$\left(\beta-\frac{\beta}{2m}\right)T_{m-1,\,n+1}-(2\beta+1)T_{m,\,n+1}+\left(\beta+\frac{\beta}{2m}\right)T_{m+1,\,n+1}=-T_{m,n} \tag{6.98}$$

where, $m=1,\ 2,\ \ldots.,\ M-1$ (here, $M=\frac{R}{\Delta r}=4$).

Now the question is how to formulate the discretized form of Equation (6.97) at $m = 0$! It is shown below.

- $m = 0$ (i.e., $r = r_0 = 0$)

At $m = 0$, the term $\frac{1}{r}\frac{\partial T}{\partial r}$ in Equation (6.97) gives $\frac{0}{0}$. Thus, we apply the L'Hospital's rule as:

$$\lim_{r\to0}\frac{1}{r}\frac{\partial T}{\partial r}=\lim_{r\to0}\frac{\partial^2 T}{\partial r^2}$$

Accordingly, Equation (6.97) yields:

$$\frac{\partial T}{\partial t}=2\alpha\frac{\partial^2 T}{\partial r^2}$$

Discretizing by the use of the same implicit FTCS method,

$$\frac{T_{m,\,n+1}-T_{m,\,n}}{\Delta t}=2\alpha\left[\frac{T_{m+1,\,n+1}-2T_{m,\,n+1}+T_{m-1,\,n+1}}{(\Delta r)^2}\right] \tag{6.99}$$

Fixing $m = 0$,

$$T_{0,\,n+1}-T_{0,\,n}=2\beta\,(T_{1,\,n+1}-2T_{0,\,n+1}+T_{-1,\,n+1})$$

From the given boundary condition at $r = 0$,

$$\frac{T_{m+1,\,n+1}-T_{m-1,\,n+1}}{2\Delta r}=0$$

This gives,

$$T_{-1,\,n+1} = T_{1,\,n+1}$$

Substituting this, we get the final discretized form at $m = 0$ as:

$$-(4\beta+1)T_{0,\,n+1} + 4\beta T_{1,\,n+1} = -T_{0,\,n} \tag{6.100a}$$

For $m = 1, 2, 3$, let us use Equation (6.98) derived before as follows.

- $m = 1$

$$\left(\beta - \frac{\beta}{2}\right)T_{0,\,n+1} - (2\beta+1)T_{1,\,n+1} + \left(\beta + \frac{\beta}{2}\right)T_{2,\,n+1} = -T_{1,\,n}$$

Simplifying,

$$\frac{\beta}{2}T_{0,\,n+1} - (2\beta+1)T_{1,\,n+1} + \frac{3\beta}{2}T_{2,\,n+1} = -T_{1,\,n} \tag{6.100b}$$

- $m = 2$

$$\frac{3\beta}{4}T_{1,\,n+1} - (2\beta+1)T_{2,\,n+1} + \frac{5\beta}{4}T_{3,\,n+1} = -T_{2,\,n} \tag{6.100c}$$

- $m = 3$

$$\frac{5\beta}{6}T_{2,\,n+1} - (2\beta+1)T_{3,\,n+1} + \frac{7\beta}{6}T_{4,\,n+1} = -T_{3,\,n}$$

This gives,

$$\frac{5\beta}{6}T_{2,\,n+1} - (2\beta+1)T_{3,\,n+1} = -T_{3,\,n} - \frac{7\beta}{6}\times 150 \tag{6.100d}$$

In matrix form, we obtain the following tridiagonal set of linear algebraic equations based on Equation (6.100),

$$\begin{bmatrix} -(4\beta+1) & 4\beta & 0 & 0 \\ \frac{\beta}{2} & -(2\beta+1) & \frac{3\beta}{2} & 0 \\ 0 & \frac{3\beta}{4} & -(2\beta+1) & \frac{5\beta}{4} \\ 0 & 0 & \frac{5\beta}{6} & -(2\beta+1) \end{bmatrix} \begin{bmatrix} T_{0,\,n+1} \\ T_{1,\,n+1} \\ T_{2,\,n+1} \\ T_{3,\,n+1} \end{bmatrix} = \begin{bmatrix} -T_{0,\,n} \\ -T_{1,\,n} \\ -T_{2,\,n} \\ -T_{3,\,n} - 175\beta \end{bmatrix} \tag{6.101}$$

Solving Equation (6.101), we get the results and they are reported for the first 2 sec in Table 6.6. Figure 6.35 is further produced by using the same simulation data. Recall that the surface temperature, T_4 is fixed at 150 °C. It is obvious that the temperatures at all four nodes reach the steady state (almost) within 2 sec time with attaining 150 °C.

Table 6.6 Temperature (°C) variation with time (sec) at different nodes of the cylindrical shell

Time (sec)	T_0 (°C)	T_1 (°C)	T_2 (°C)	T_3 (°C)
0.0	50.00	50.00	50.00	50.00
0.1	78.36979	81.45346	91.5983	112.15418
0.2	105.8369	108.82245	117.75069	132.2041
0.3	124.42967	126.45062	132.2339	140.7415
0.4	135.57251	136.78369	140.18253	144.98656
0.5	141.94395	142.63651	144.56383	147.24416
0.6	145.52014	145.90886	146.98695	148.47664
0.7	147.51282	147.7294	148.32932	149.15616
0.8	148.61998	148.74033	149.07347	149.5322
0.9	149.23448	149.30127	149.48615	149.7406
1.0	149.57539	149.61244	149.715	149.85614
1.1	149.76448	149.78503	149.84193	149.92021
1.2	149.86937	149.88077	149.91232	149.95575
1.3	149.92754	149.93385	149.95137	149.97545
1.4	149.95981	149.9633	149.97302	149.98639
1.5	149.97769	149.97964	149.98505	149.99245
1.6	149.98763	149.98869	149.9917	149.9958
1.7	149.99312	149.99373	149.99539	149.99767
1.8	149.99619	149.99652	149.99744	149.99872
1.9	149.99788	149.99806	149.99858	149.99928
2.0	149.99883	149.99892	149.99921	149.9996

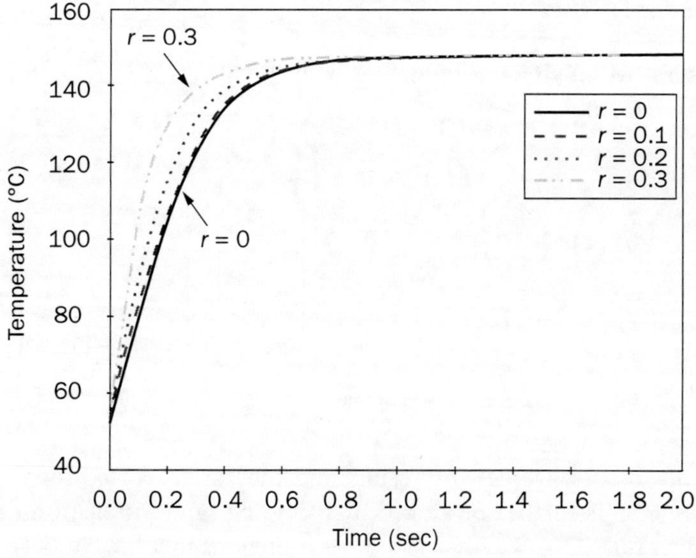

Figure 6.35 Comparative temperature profile at the four different nodes of the cylindrical shell

6.14 Summary and Concluding Remarks

In this chapter, we learn several elegant and advanced numerical techniques, and their applications in solving the systems of partial differential equations, including various industrially relevant processes. Several combinations of the finite difference approximations are made for the time and spatial derivatives under explicit and implicit schemes. Further, we have employed the implicit schemes (e.g., the Crank–Nicholson and alternating direction implicit (ADI) method) keeping the stability issue in mind. Solving the standard parabolic, hyperbolic and elliptic PDEs, we have moved to model the various processes having single PDE and coupled PDEs for their numerical simulation. In most of the cases, the plots (along with tabulated results) are produced for physical understanding of the system behavior.

General Recommendations

Method of lines is the most widely used technique for numerical solution of the PDE systems. In this scheme, the partial differential equation is first converted to a set of ODE-IVP equations. Then those coupled equations are solved by using an IVP solver, for which, the numerical stability is briefly discussed for Case Study 6.1 and general recommendations are made in more details in Chapter 4 (Section 4.16). There is a certain constraint imposed on time step, h; it must be followed when one is going to employ an explicit method (e.g., the forward Euler or Runge–Kutta) to solve the transformed set of IVP equations. However, using an implicit IVP solver (e.g., the backward Euler) can overcome this problem.

In this direction, we often prefer the implicit Crank–Nicholson (C–N) method. This is, in fact, a very popular technique for solving the PDE systems without having any concern on stability (unconditionally stable). Although there is no effect of step size (h) on numerical stability, still we need to be cautious about the use of a reasonably large step size because of its well-known effect on the accuracy of solution.

However, this C–N method sometimes results in a large system of linear algebraic equations for parabolic and elliptic PDEs. In such a case, it is recommended to prefer the alternating direction implicit (ADI) method that offers much lower computational work (thus, the round-off error) than the C–N scheme. Further, the ADI is an implicit scheme and thus, unconditionally stable.

Exercise

Review Questions

6.1 List the basic differences between:
- algebraic and differential equation.
- ODE-IVP, ODE-BVP and PDE.

6.2 Is the short note given under 'initial and boundary conditions' (see Subsection 6.1.2) applicable to ODE-IVP and ODE-BVP systems?

6.3 How many boundary and initial conditions need to be ideally specified for a PDE that contains a first-order time derivative as well as a first-order spatial derivative?

6.4 What is initial BVP? Give examples.

6.5 Give examples of the physical systems that have three independent variables.

6.6 What is the basic difference between advection and convection?

6.7 Why are the following methods called so?
– method of lines
– Leap-Frog

6.8 Is it right to say that the FTCS, BTCS and CTCS are the various forms of the MoL scheme?

6.9 Develop the computational molecule for the implicit FTCS method (discussed in Case Study 6.8 for a cylindrical shell).

6.10 Determine the type of the following PDEs (parabolic/hyperbolic/elliptic):

i) $$\frac{\partial T}{\partial t} = \alpha\left(\frac{\partial^2 T}{\partial x^2} + \frac{\partial^2 T}{\partial y^2}\right)$$

ii) $$\frac{\partial^2 u}{\partial x^2} + y\frac{\partial^2 u}{\partial y^2} + x\,u = 0$$

iii) $$\frac{\partial^2 u}{\partial x^2} - 2\frac{\partial^2 u}{\partial x \partial y} + 3\frac{\partial^2 u}{\partial y^2} = 0$$

iv) $$(1+x)\frac{\partial^2 T}{\partial x^2} - 2(2+x)\frac{\partial^2 T}{\partial x \partial y} + (3+x)\frac{\partial^2 T}{\partial y^2} = 0$$

6.11 In the Leap-Frog scheme [Equation (6.18)], why and how $a\frac{h}{k}$ is found to be of critical importance?

6.12 Show that the Lax–Friedrichs method [Equation (6.39)] is stable if and only if:

$$\left|a\frac{h}{k}\right| \le 1$$

6.13 Does Equation (6.49) remain implicit if $\theta = 1$?

6.14 i) What are the relative merit and demerit of ADI and C–N method?

ii) Develop the discretized form of Equation (6.69) using the CD method in time as well as in space in the ADI scheme.

Practice Problems

6.15 Consider the following parabolic PDE:

$$\frac{\partial u}{\partial t} = D\frac{\partial^2 u}{\partial x^2}$$

which describes the diffusion of a chemical species in a porous media with respect to time and space.

BCs: $u(0,\ t) = 0 \qquad t \ge 0$

$u(1,\ t) = 0$

IC: $u(x,\ 0) = \sin \pi x \qquad 0 \le x \le 1$

i) Solve this problem using the FTCS with $D = 1.0$ (choose h and k).

ii) Discuss its stability with respect to the value of $\beta\left(= \frac{hD}{k^2} = \frac{\Delta t D}{(\Delta x)^2}\right)$.

6.16 Repeat the above problem and solve it using the Crank–Nicholson method with $\Delta x = 0.2$, $\Delta t = 0.05$ and $D = 1$.

6.17 Apply the Dufort–Frankel method on the following parabolic equation:

$$\frac{\partial u}{\partial t} = \frac{1}{\mu}\frac{\partial^2 u}{\partial x^2} \qquad \mu > 0$$

Develop the solution strategy.

6.18 The following parabolic PDE is used to describe the heat transport by conduction:

$$\frac{\partial T}{\partial t} = D\frac{\partial^2 T}{\partial x^2}$$

subject to:

BCs: $T(0,\ t) = 0 \qquad 0 \le t \le 1$

$T(1,\ t) = 0$

IC: $T(x,\ 0) = \sin^2 4\pi x \qquad 0 \le x \le 1$

Solve this problem using the Crank–Nicholson method with $D = 1$ (choose h and k).

6.19 Solve Equation (6.40) with the given boundary [Equation (6.42a)] and initial condition [Equation (6.42b)] by using the Crank–Nicholson method [complete solution methodology is developed in Case Study 6.3, see Equation (6.48)] with $r = 1$.

6.20 Consider the following parabolic PDE:

$$\frac{\partial u}{\partial t} = \alpha\frac{\partial^2 u}{\partial x^2}$$

subject to the boundary conditions:

$$\left\{\begin{array}{ll} \dfrac{\partial u}{\partial x} = u & \text{at } x = 0 \\ \dfrac{\partial u}{\partial x} = -u & \text{at } x = 1 \end{array}\right. \qquad \text{for } t > 0$$

and initial conditions:

$$u(x,\ 0) = \begin{cases} 2x & 0 \le x \le \dfrac{1}{2} \\ 2(1-x) & \dfrac{1}{2} \le x \le 1 \end{cases}$$

Formulate the solution methodology and solve it.

6.21 Solve the following PDE:

$$\frac{\partial u}{\partial t} = 8\frac{\partial^2 u}{\partial x^2}$$

subject to:

BCs: $u(0,\ t) = 0 \qquad t \ge 0$

$u(80,\ t) = 0$

IC:
$$u(x,\ 0) = \begin{cases} 0 & 0 \le x < 20 \\ 40 & 20 \le x \le 60 \\ 0 & 60 < x \le 80 \end{cases}$$

Use the CTCS method with $\Delta x = 8$ and $\Delta t = 0.008$.

6.22 Repeat the above problem and solve it by using the Dufort–Frankel scheme.

6.23 Solve the following one-dimensional heat conduction equation:

$$\frac{\partial u}{\partial t} = 5\frac{\partial^2 u}{\partial x^2}$$

subject to:

$$BCs: \quad u(0,\ t) = 0 \qquad t \geq 0$$
$$u(1,\ t) = 4$$

$$IC: \quad u(x,\ 0) = x^2(5 - x) \qquad 0 \leq x \leq 1$$

Use an explicit finite difference scheme of your choice with $k = 0.2$ and $h = 0.005$ to find $u(x,\ 0.02)$.

6.24 The following PDE is used to describe the desorption of a gas from a liquid stream in a wetted-wall column (Davis, 1984):

$$(1 - \eta^2)\frac{\partial f}{\partial \xi} = \frac{\partial^2 f}{\partial \eta^2}$$

subject to:

$$\text{at } \eta = 0: \qquad f = 0$$
$$\eta = 1: \qquad \frac{\partial f}{\partial \eta} = 0$$
$$\xi = 0: \qquad f = 1$$

Solve it using a suitable numerical method.

6.25 Consider the following non-linear hyperbolic PDE (simplified form of Burger equation[22]):

$$\frac{\partial u}{\partial t} + \frac{\partial F(u)}{\partial x} = 0$$

where, $F(u) = \dfrac{u^2}{2}$.

$$BCs: \quad u(0,\ t) = 0 \qquad 0 \leq t \leq 1$$
$$u(1,\ t) = 0$$

$$IC: \quad u(x,\ 0) = \sin \pi x \qquad 0 \leq x \leq 1$$

Solve this problem using a suitable FDM.

6.26 Consider the following wave equation:

$$\frac{\partial^2 u}{\partial t^2} = a^2\frac{\partial^2 u}{\partial x^2}$$

which describes the *transverse vibration* of the string.

$$ICs: \quad u(x,\ 0) = f(x) \qquad 0 \leq x \leq 1$$

$$\frac{\partial u}{\partial t}(x,\ 0) = g(x)$$

---

22 Detailed form is given in Problem 6.35 [Equation (6.102)].

BCs: $u(0, t) = 0 \qquad t \geq 0$

$u(1, t) = 0$

Formulate the solution methodology using the CTCS method.

6.27 Repeat the above problem with:

$$a^2 = 1$$

ICs: $u(x, 0) = x(x-2)^2 \qquad 0 \leq x \leq 2$

$\frac{\partial u}{\partial t}(x, 0) = 0$

BCs: $u(0, t) = 0 \qquad t \geq 0$

$u(2, t) = 0$

Solve this problem using the FTCS method with $M = 10$ and $\Delta t = 0.1$.

6.28 Consider the following hyperbolic PDE (one-dimensional wave equation):

$$\frac{\partial^2 T}{\partial t^2} = \frac{\partial^2 T}{\partial x^2}$$

subject to:

BCs: $T(0, t) = 0 \qquad t \geq 0$

$T(1, t) = 0$

ICs: $T(x, 0) = \sin \pi x \qquad 0 \leq x \leq 1$

$\frac{\partial T}{\partial t}(x, 0) = 0$

i) Use the FTFS scheme, and produce numerical results with adopting $h = 0.2$ and $k = 0.25$

ii) Compare numerical solution with the analytical one given as:

$$T(x, t) = \sin \pi x \cos \pi t$$

6.29 Consider the following IBVP system:

$$\frac{\partial^2 u}{\partial t^2} = \frac{\partial^2 u}{\partial x^2}$$

subject to:

BCs: $u(0, t) = 0 \qquad t \geq 0$

$u(1, t) = 0$

ICs: $u(x, 0) = 1 - \exp\left[-20(x-1)^2\right] \qquad 0 \leq x \leq 1$

$\frac{\partial u}{\partial t}(x, 0) = 0$

Solve this problem using the FTCS with $h \leq k$.

6.30 Consider the following hyperbolic PDE:

$$\frac{\partial^2 u}{\partial t^2} = 25 \frac{\partial^2 u}{\partial x^2}$$

subject to:

BCs: $u(0,\ t) = 0 \qquad t \geq 0$

$u(1,\ t) = 0$

ICs: $u(x,\ 0) = \sin \pi x \qquad 0 \leq x \leq 1$

$\dfrac{\partial u}{\partial t}(x,\ 0) = 0$

Solve this problem with $\Delta x = 0.2$ and $\Delta t = 0.05$ using an explicit FDM.

6.31 Solve the following IBVP system:

$$\frac{\partial^2 u}{\partial t^2} = \frac{\partial^2 u}{\partial x^2}$$

subject to:

ICs: $u(x,\ 0) = \begin{cases} 25x^2(1-x)^2 & 0 \leq x \leq 1 \\ 0 & 1 < x \leq 5 \end{cases}$

$\dfrac{\partial u}{\partial t}(x,\ 0) = 0 \qquad 0 \leq x \leq 5$

BCs: $u(0,\ t) = 0 \qquad 0 \leq t \leq 1$

$u(5,\ t) = 0$

Use $\Delta x = 0.5$ and $\Delta t = 0.2$.

6.32 Repeat the above problem with the following conditions:

BCs: $u(0,\ t) = 0 \qquad t \geq 0$

$u(50,\ t) = 0$

ICs: $u(x,\ 0) = \begin{cases} 0.1x & 0 \leq x \leq 15 \\ 0.25 - \dfrac{x-15}{140} & 15 < x \leq 50 \end{cases}$

$\dfrac{\partial u}{\partial t}(x,\ 0) = 0 \qquad 0 \leq x \leq 50$

Solve the problem using the implicit FTCS method with $M = 50$ and $h = 1$.

6.33 The transient diffusion–reaction problem is modeled as:

$$\frac{\partial C}{\partial t} = D\frac{\partial^2 C}{\partial x^2} - kC^2$$

where, C denotes the concentration, D the mass diffusivity and k the reaction rate constant. This model is subject to:

BCs: $C(0,\ t) = 1$

$C(1,\ t) = 1$

IC: $C(x,\ 0) = 1$

Solve this problem using the method of lines with $D = k = 1$.

6.34 Consider a PDE system:

$$\frac{\partial u}{\partial t}=\frac{\partial^2 u}{\partial x^2}+x\frac{\partial u}{\partial x}$$

subject to:

IC: $u(x,\ 0)=2x-x^3 \qquad 0\le x\le 1$

BCs: $u(0,\ t)=0 \qquad t\ge 0$

$u(1,\ t)=1$

Use FTCS to find $u(x,\ 0.05)$ when $k = 0.25$ and $h = 0.005$.

6.35 One-dimensional gas dynamics is modeled by the following quasi-linear parabolic PDE:[23]

$$\frac{\partial u}{\partial t}+u\frac{\partial u}{\partial x}=\frac{1}{\text{Re}}\frac{\partial^2 u}{\partial x^2} \tag{6.102}$$

subject to:

BCs: $u(0,\ t)=0 \qquad 0\le t\le 1$

$u(1,\ t)=0$

IC: $u(x,\ 0)=\sin\pi x \qquad 0\le x\le 1$

Solve this problem using the FTCS with Reynolds number (Re) = 20, $\Delta x=0.1$ and $\Delta t=0.001$.

6.36 Solve the following non-linear PDE:

$$\frac{\partial C}{\partial t}=D\frac{\partial^2 C}{\partial x^2}+C(1-C^2)$$

subject to:

BCs: $C(0,\ t)=1 \qquad t\ge 0$

$C(10,\ t)=0$

IC: $C(x,\ 0)=\begin{cases}1 & x<5\\ 0 & x\ge 5\end{cases} \qquad 0\le x\le 10$

Use MoL with $k = 1$, $h = 0.005$ and $D = 1$.

6.37 Repeat the above problem, keeping everything same except $k = 2.5$ and $h = 0.1$. Solve it and analyze the numerical stability.

6.38 Consider the Laplace equation:

$$\frac{\partial^2 u}{\partial x^2}+\frac{\partial^2 u}{\partial y^2}=0$$

subject to:

$u(x,\ 0)=0$

$u(0,\ y)=0$

$u(x,\ 1)=200x$

$u(1,\ y)=200y$

Adopting $M = 5$ with $\Delta x=\Delta y$, solve the problem for all the internal nodes.

---

23 This is called Burger equation used in describing many systems (e.g., gas dynamics, fluid mechanics and traffic flow).

6.39 Find a solution of the Laplace equation as:

$$u = A\sinh(x)\sin(y) + B\sin(x)\sinh(y)$$

Compare the analytical solutions with the numerical results obtained in the last problem.

6.40 Solve the Laplace equation subject to the following conditions:

$$u(x,\ 0) = 0 \qquad 0 \le x,\ y \le \pi$$
$$u(0,\ y) = 0$$
$$u(x,\ \pi) = \sin x$$
$$u(\pi,\ y) = 0$$

Use the concept of global index detailed in Section 6.10.

6.41 The Laplace equation is defined over the right angle triangle region shown in Figure 6.36.

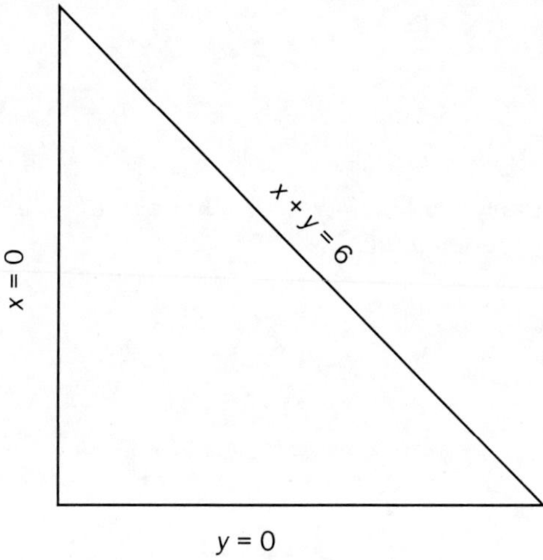

FIGURE 6.36 Defining right angle triangle region

Boundary conditions:

$u = 0$ over the edge $x = 0$

$u = 0$ over the edge $y = 0$

$u = 36 - x^2 - y^2$ at the edge $x + y = 6$

Compute u at all interior nodes of the square region with mesh length equals 1.

6.42 Consider the following Poisson equation:

$$\frac{\partial^2 u}{\partial x^2} + \frac{\partial^2 u}{\partial y^2} = -4 \qquad 0 \le x \le 9$$
$$0 \le y \le 12$$

In all four boundaries, $u = 0$. Choosing $\Delta x = \Delta y = 3$, solve the problem for the 6 interior nodes shown in Figure 6.37.

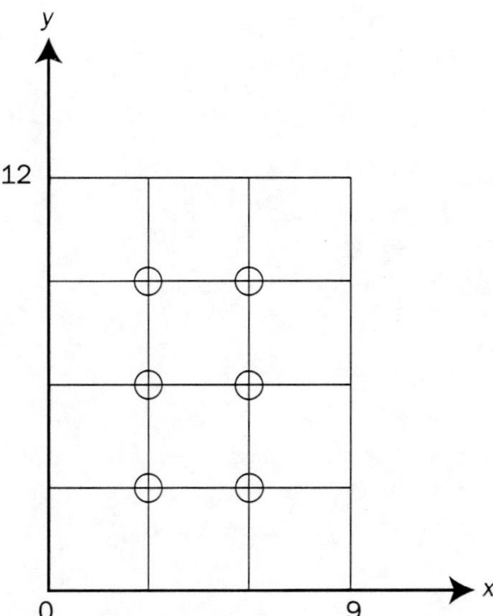

FIGURE 6.37 Illustrating 6 interior nodes

6.43 Consider the following form of Poisson equation:

$$\frac{\partial^2 u}{\partial x^2}+\frac{\partial^2 u}{\partial y^2}=x+y$$

over the domain $0 \le x,\ y \le 1$ subject to:

$u(x, 0) = 50$	Bottom boundary
$u(0, y) = 200$	Left boundary
$u(x, 1) = 50$	Top boundary
$u(1, y) = 200$	Right boundary

Solve this PDE system for $M = 10$ with $\Delta x = \Delta y$.

6.44 Consider the following form of Poisson equation:

$$\frac{\partial^2 u}{\partial x^2}+\frac{\partial^2 u}{\partial y^2}+xy-x=0$$

over the domain $0 \le x,\ y \le 1$ with boundary conditions:

$u(x,\ 0)-0$	Bottom boundary
$u(0,\ y)=-20$	Left boundary
$u(x,\ 1)=40$	Top boundary
$u(1,\ y)=20$	Right boundary

Solve this PDE system for $M = 4$ with $\Delta x = \Delta y$. Use the successive over-relaxation method to solve the resulting set of algebraic equations.

6.45 Consider the following form of Poisson equation:

$$\frac{\partial^2 u}{\partial x^2}+\frac{\partial^2 u}{\partial y^2}=\sin(x+y)$$

over the domain $0 \le x,\ y \le 1$ with boundary conditions:

$u(x, 0) = 0$ Bottom boundary

$u(0, y) = 200$ Left boundary

$u(x, 1) = 0$ Top boundary

$u(1, y) = 200$ Right boundary

Solve this PDE system for $M = 5$ with $\Delta x = \Delta y$.

6.46 Consider the following form of Poisson equation:

$$\frac{\partial^2 u}{\partial x^2} + \frac{\partial^2 u}{\partial y^2} = \exp(xy)$$

over the domain $0 \le x,\ y \le 1$ with boundary conditions:

$u = x + y$ for all four boundaries

Solve this PDE system for $M = 4$ with $\Delta x = \Delta y$.

6.47 Consider the following form of Poisson equation:

$$\frac{\partial^2 u}{\partial x^2} + \frac{\partial^2 u}{\partial y^2} = \sin x - \cos y$$

over the domain $0 \le x,\ y \le 1$ with boundary conditions:

$u(x, 0) = 0.5$ Bottom boundary

$u(0, y) = 0.2$ Left boundary

$u(x, 1) = 0.7$ Top boundary

$u(1, y) = 0.8$ Right boundary

Solve this PDE system for $M = 5$ with $\Delta x = \Delta y$.

6.48 Consider a system having the following PDE:

$$\frac{\partial^2 T}{\partial x^2} + \frac{\partial^2 T}{\partial y^2} = 0$$

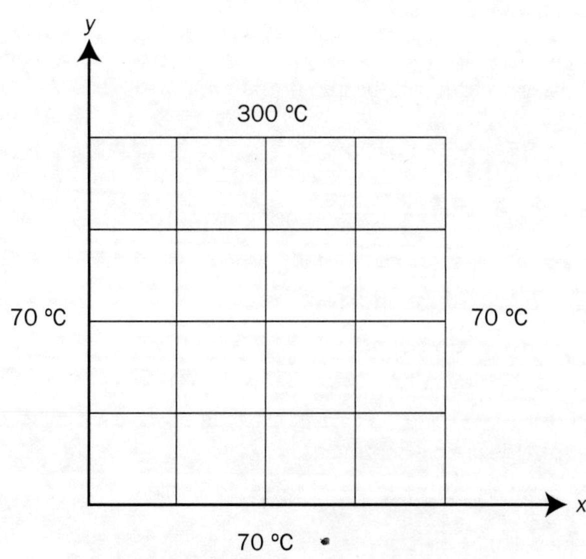

FIGURE **6.38** Showing a square slab with temperatures in all four boundaries

Find the temperature at all interior nodes shown in Figure 6.38 with $\Delta x = \Delta y = 0.5$ m and $x = y = 2$ m by the use of the ADI method.

6.49 i) Develop the ADI scheme for the 2D heat conduction problem [Equation (6.69)] by employing the central difference approximation for both the time and space.

ii) Using this ADI scheme, solve Equation (6.69) with the given BCs in Figure 6.24. Other relevant information is given as: $\Delta x = \Delta y = \Delta t = 0.25$ and $\alpha = 1$.

6.50 i) For the following hyperbolic PDE (wave equation):

$$\frac{\partial^2 u}{\partial t^2} = a^2 \frac{\partial^2 u}{\partial x^2}$$

formulate an implicit scheme by using the CD for the time derivative and the average of central differences at $n + 1$ and $n - 1$ levels for space derivative.

ii) Note that in the implicit scheme asked to develop above in Part (i), equal weightage (i.e., 0.5) is given to $n + 1$ and $n - 1$ levels, and zero weightage to n level. Now you are further asked to develop a general implicit scheme by giving θ_1 weightage to $n + 1$ level, θ_2 to $n - 1$ level and $(1 - \theta_1 - \theta_2)$ to n level.

6.51 Consider the following PDE:

$$\frac{\partial^2 u}{\partial x^2} + \frac{\partial^2 u}{\partial y^2} + 5u = 0 \qquad 0 \le x,\ y \le 1$$

with the boundary condition:

$$u(x,\ y) = \cos 2.5x + \sin 2.5y$$

Solve the problem for $M = 5$ with $\Delta x = \Delta y$.

6.52 Consider the following PDE:

$$\frac{\partial^2 T}{\partial t^2} = \frac{\partial^2 T}{\partial x^2} - T$$

subject to:

BCs: $T(0,\ t) = 0 \qquad t \ge 0$

$T(\pi,\ t) = 0$

ICs: $$T(x,\ 0) = \begin{cases} x & 0 \le x < \dfrac{\pi}{2} \\ \pi - x & \dfrac{\pi}{2} \le x \le \pi \end{cases}$$

$$\left(\frac{\partial T}{\partial t}\right)_{t=0} = 0 \qquad 0 \le x \le \pi$$

Solve the problem using a suitable FDM.

6.53 Unsteady state one-dimensional diffusion in a sphere is modeled as:

$$\frac{\partial C}{\partial t} = D\frac{\partial^2 C}{\partial r^2} + \frac{2D}{r}\frac{\partial C}{\partial r}$$

subject to:

IC: at $t = 0 \qquad C = C_0 = 0.07\ \text{gm/cm}^3 \qquad 0 \le r \le R$

BCs: at $r = 0 \qquad \dfrac{\partial C}{\partial r} = 0 \qquad t \ge 0$

at $r = R \qquad C = 0 \qquad t > 0$

Given information:

radius, $R = 0.35$ cm

diffusivity, $D = 2.5 \times 10^{-7}$ cm²/sec

Solve the IBVP system with $\Delta r = 0.35$ mm and $\Delta t = 1$ sec.

6.54 In a cylindrical coordinate system, we have:

$$\frac{\partial T}{\partial t} = \sin r \frac{\partial^2 T}{\partial r^2} + r^2 \frac{\partial T}{\partial r}$$

subject to:

$$IC: \quad T(r, 0) = r(1-r) \qquad 0 \le r \le 1$$

$$BCs: \quad T(0, t) = 0 \qquad t \ge 0$$

$$T(1, t) = 0$$

Solve this problem using the FTCS with $\Delta t = 0.01$ and $\Delta r = 0.1$. Show the results up to $t = 0.1$.

6.55 An unsteady state plug flow reactor (PFR) is modeled with considering a first-order irreversible reaction and constant volume as:

$$\frac{\partial C_A}{\partial t} + \upsilon \frac{\partial C_A}{\partial x} + k\, C_A = 0 \qquad 0 \text{ m} \le x \le 0.5 \text{ m}$$

The initial condition can be obtained from the following steady state model:

$$C_A(x, 0) = C_A(0, 0) \exp\left(-\frac{k}{\upsilon} x\right)$$

with $C_A(0, 0) = 1$ mol/L.

The boundary condition includes:

$$C_A(0, t) = 1.5 \text{ mol/L} \qquad t > 0$$

(i.e., a step increase in C_A is introduced at $x = 0$ and $t > 0$)

Given information:

$\upsilon = 0.5$ m/min

$k = 0.3$ min^{-1}

$M = 10$

Solve this IBVP system using the FTBS method and produce the dynamic C_A profile at time, $t = 0$, 2 and 5 min. You may revisit the solution methodology discussed in Case Study 6.7.

6.56 Nutrient transport occurs from the surrounding soil mass (assumed as an annular cylinder of large outer diameter) to a plant root (assumed cylindrical). The root typically absorbs the nutrient at the surface (radius = a), which involves a first-order chemical reaction.

With this, the nutrient transport is modeled by:

$$\frac{\partial C}{\partial t} = \frac{D}{r} \frac{\partial}{\partial r}\left(r \frac{\partial C}{\partial r}\right)$$

IC: $C(r, 0) = C_0$

BCs: $D\frac{\partial C}{\partial r}(a, t) = kC$

$C(\infty, t) = C_0$

Develop the solution methodology for this PDE system.

6.57 A fluid is flowing through a tube, in which, the fluid gets progressively heated because of the hot boundary. Heat transfer occurs in the axial direction by convection and in radial direction by diffusion. This is modeled at steady state by the following parabolic PDE:

$$\upsilon\frac{\partial T}{\partial x} = \frac{\alpha}{r}\frac{\partial}{\partial r}\left(r\frac{\partial T}{\partial r}\right)$$

Here, the thermal diffusivity, $\alpha = 1.5 \times 10^{-4}$ m^2/sec. The parabolic velocity profile is given as:

$$\upsilon(r) = \upsilon_{max}\left(1 - \left(\frac{r}{R}\right)^2\right)$$

in which, $\upsilon_{max} = 0.4$ m/sec and radius, $R = 5$ cm. For this 2 m long tube, the following conditions are valid:

Inlet condition:

$$T(x = 0, r) = 350 \text{ K}$$

Boundary conditions:

$$T(x, R) = 450 \text{ K}$$

$$\frac{\partial T}{\partial r}(x, 0) = 0$$

Using the MoL, solve the PDE with $\Delta r = 0.05$ cm and $\Delta x = 2$ cm.

6.58 Steady state two-dimensional atmospheric dispersion involves convective dispersion of a species in the axial direction and eddy diffusion in the transverse direction. There is a uniform area (source density = ρ kg/m^2.min) where the species originates. In the axial direction (x-direction), the wind flows and the dispersion is assumed negligible over convection. However, in the vertical direction (z-direction), the dispersion is quite significant.

With this, the steady state dispersion model of the species is developed (Lebedeff and Hameed, 1975) as:

$$\upsilon(z)\frac{\partial C}{\partial x} = \frac{\partial}{\partial z}\left(D_z\frac{\partial C}{\partial z}\right)$$

with

$$\upsilon(z) = \upsilon_0 z^\beta$$
$$D_z = D_0 z^\gamma$$

where,

υ = axial velocity
D_z = transverse eddy diffusivity
β, γ = constants

BCs: at the ground level: $D_z \frac{\partial C}{\partial z}(x, z=0) = -\rho$

at a large height: $C(x, z \to \infty) = 0$

at $x = 0$: $C(0, z) = 0$

(since dispersion in axial direction is assumed negligible)
Formulate a suitable solution methodology for this PDE system.

6.59 For the following 2D unsteady state heat conduction equation:

$$\frac{\partial T}{\partial t} = \alpha\left(\frac{\partial^2 T}{\partial x^2} + \frac{\partial^2 T}{\partial y^2}\right)$$

develop the solution methodology using the C–N method.

6.60 Solve the following PDE with the use of ADI scheme:

$$\frac{\partial T}{\partial t} = \alpha\left(\frac{\partial^2 T}{\partial x^2} + \frac{\partial^2 T}{\partial y^2}\right)$$

with the specified conditions shown in Figure 6.39. Considering $\Delta x = \Delta y = 0.25$ m, $\alpha = 1.2$ m²/sec and $\Delta t = 0.1$ sec, find the final temperature at all interior nodes with a desired tolerance of 10^{-5}.

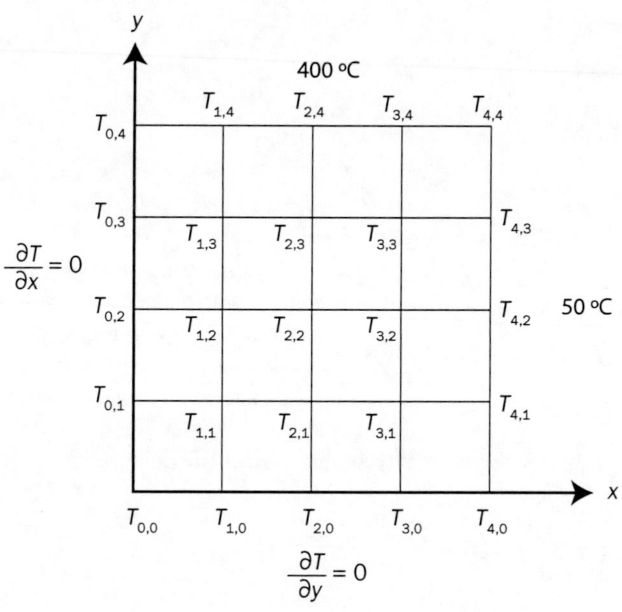

FIGURE 6.39 Specifying boundary conditions

6.61 Repeat the above problem with considering central difference formula, instead of backward Euler used for $\frac{\partial T}{\partial t}$ with step size $\frac{\Delta t}{2}$. Keeping all other given information same, find the temperature at all interior nodes.

6.62 Solve the following PDE with the use of ADI scheme:

$$\frac{\partial T}{\partial t} = \alpha\left(\frac{\partial^2 T}{\partial x^2} + \frac{\partial^2 T}{\partial y^2}\right)$$

with the specified conditions shown in Figure 6.40. Considering $\Delta x = \Delta y = 0.5$ m, $\alpha = 1.2$ m²/sec and $\Delta t = 0.1$ sec, find the final temperature at all interior nodes with a desired tolerance of 10^{-5}. Use the following initial condition for the interior nodes:

IC (at $t = 0$): $$T(x,\ y) = \cos\left(\frac{\pi}{3}x\right)\cos\left(\frac{\pi}{3}y\right) \qquad -1 \le x,\ y \le 1$$

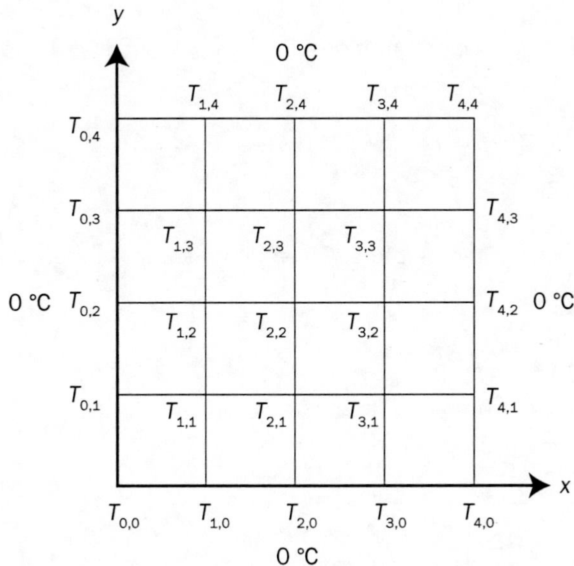

FIGURE **6.40** Specifying boundary conditions for the slab

6.63 Consider an unsteady state isothermal tubular reactor with axial mixing, in which, a first-order irreversible reaction takes place. It is modeled in dimensionless form as:

$$\frac{\partial c}{\partial \tau} = \frac{1}{\text{Pe}}\frac{\partial^2 c}{\partial z^2} - \frac{\partial c}{\partial z} - \text{Da}\ c$$

subject to:

IC: $c(z,\ 0) = 0 \qquad 0 \le z \le 1$

BCs: $\frac{\partial c}{\partial z}(0,\ \tau) = \text{Pe}(c-1) \qquad 0 \le \tau \le 1$

$$\frac{\partial c}{\partial z}(1,\ \tau) = 0$$

Given information include:

Peclet number, Pe = 5

Damköhler number, Da = 2

Solve this problem using the MoL.

6.64 Revisit Case Study 6.6 [porous media (non-linear model)]. Find the analytical solution of Equation (6.88) and compare with the numerical results.

6.65 Revisit Case Study 6.7 (unsteady state double pipe heat exchanger). Considering a step increase in inlet temperature of process liquid from 75 to 100 °F, produce the temperature profile (similar to Figure 6.32) at time, t = 0, 2, 5, 10 sec. For this, keep all other information given there unchanged.

6.66 Consider the following PDE:

$$\frac{\partial T}{\partial t} = \frac{\partial^2 T}{\partial r^2} + \frac{5}{r}\frac{\partial T}{\partial r}$$

subject to the following boundary conditions (BCs):

at $r = 0$: $T(0,\ t) = 100$ °C $t \geq 0$

at $r = R$: $T(R,\ t) = T_s = 250$ °C

and initial condition (IC):

at $t = 0$: $T(r,\ 0) = 100$ °C $0 \leq r < R$

Here, T_s denotes the surface temperature. Given information include:

$R = 2$ cm

$\Delta r = k = 0.25$ cm

$\Delta t = h = 0.1$ sec

Solve this PDE system by using the FTBS scheme.

6.67 Consider the following PDE:

$$\frac{\partial T}{\partial t} = \alpha\left(\frac{\partial^2 T}{\partial r^2} + \frac{3}{r}\frac{\partial T}{\partial r}\right)$$

subject to the following boundary conditions (BCs):

at $r = 0$: $\frac{\partial T}{\partial r} = 0$ $t \geq 0$

at $r = R$: $T(R,\ t) = 400$ °C

and initial condition (IC):

at $t = 0$: $T(r,\ 0) = 100$ °C $0 \leq r < R$

Given information include:

$R = 1$ cm

$\Delta r = k = 0.1$ cm

$\Delta t = h = 0.1$ sec

$\alpha = 1.2 \times 10^{-5}$ m²/sec

Solve this PDE system by using the forward in time and central in space (FTCS) implicit difference scheme.

6.68 Solve the above problem by developing and then using an implicit BTCS scheme.

6.69 Diffusion of CO_2 in a leaf for photosynthesis is modeled by (Gross, 1981):

$$\frac{\partial C}{\partial t} = D_e\frac{\partial^2 C}{\partial x^2} - k\ C + \mu$$

Notice that the rate of photosynthesis is described by a first-order reaction term ($\equiv k\ C$). Here, μ is a zero-order sink term and D_e the effective diffusivity.

IC: Find $C(x, 0)$ at steady state from the given PDE and use it as an IC [$C(0, 0) = 0.005$ mol/m³]

BCs: $C(0, t) = 0$

$$\frac{\partial C}{\partial x}(l, t) = 0$$

Formulate the solution methodology and solve it for:

$D_e = 1.67 \times 10^{-7}$ m²/sec

$k = 0.04$ sec^{-1}

$\mu = 30$ μmol/m³.sec

$l = 5$ cm

6.70 Solve the following system of PDE:

$$\frac{\partial^2 u}{\partial t^2} = \frac{1}{r}\frac{\partial u}{\partial r} + \frac{\partial^2 u}{\partial r^2}$$

subject to:

BCs: $u(0, t) = 0.8$ $\quad t \geq 0$

$u(a, t) = 0$

ICs: $u(r, 0) = 1 - \dfrac{r^2}{a^2}$ $\quad 0 \leq r \leq a$

$$\frac{\partial u}{\partial t}(r, 0) = 0$$

Solve this problem using the Lax–Friedrichs method for $a = 1$.

6.71 In an isothermal laminar flow reactor, a first-order chemical reaction takes place. The convective diffusion model for this reactor has the following form:

$$a(1-r^2)\frac{\partial C}{\partial x} = \frac{1}{r}\frac{\partial C}{\partial r} + \frac{\partial^2 C}{\partial r^2} - b\,C$$

subject to:

BCs: $C(r, 0) = 0.8$ $\quad 0 \leq r \leq 1$

$\dfrac{\partial C}{\partial r}(0, x) = 0$ $\quad 0 \leq x \leq 1$

$$\frac{\partial C}{\partial r}(1, x) = 0$$

Given information include:

$a = 20$

$b = 10$

Solve this problem using a suitable finite difference method with selecting Δx and Δr.

6.72 Steady state reaction–diffusion in a cylindrical catalyst pellet is described by a dimensionless PDE as:

$$\frac{1}{\bar{r}}\frac{\partial}{\partial \bar{r}}\left(\bar{r}\frac{\partial \bar{c}}{\partial \bar{r}}\right) + \frac{1}{\gamma^2}\frac{\partial^2 \bar{c}}{\partial \bar{z}^2} - \phi^2\,\bar{c} = 0$$

subject to:

BCs: $\bar{c}(\bar{z},\ \bar{r}=1)=1$

$$\frac{\partial \bar{c}}{\partial \bar{r}}(\bar{z},\ \bar{r}=0)=0$$

$$\bar{c}(\bar{z}=1,\ \bar{r})=1$$

$$\frac{\partial \bar{c}}{\partial \bar{z}}(\bar{z}=0,\ \bar{r})=0$$

Solve this problem using the Crank–Nicholson method with adopting: $\gamma=\phi=1$.

6.73 Considering thermal conductivity as a linear function of temperature, we developed the model equation for steady state one-dimensional heat conduction in a slab (see Case Study 5.4). Extending that model to the unsteady state case, we have:

$$\frac{\partial \theta}{\partial t}=(1+a\theta)\frac{\partial^2 \theta}{\partial \zeta^2}+a\left(\frac{\partial \theta}{\partial \zeta}\right)^2 \qquad 0\le\zeta\le 1$$

subject to:

IC: $\theta(\zeta,\ 0)=0$

BCs: $\theta(0,\ t)=1$

$$\theta(1,\ t)=0$$

Note that here the non-dimensionalization is slightly different than that in Case Study 5.4. Formulate the solution methodology using the Lax–Friedrichs and MoL methods for $a = 1$.

6.74 The steady state temperature distribution in a rectangular block having sides *a*, *b* and *c* (Figure 6.41) is described by:

$$\frac{\partial^2 T}{\partial x^2}+\frac{\partial^2 T}{\partial y^2}+\frac{\partial^2 T}{\partial z^2}=0$$

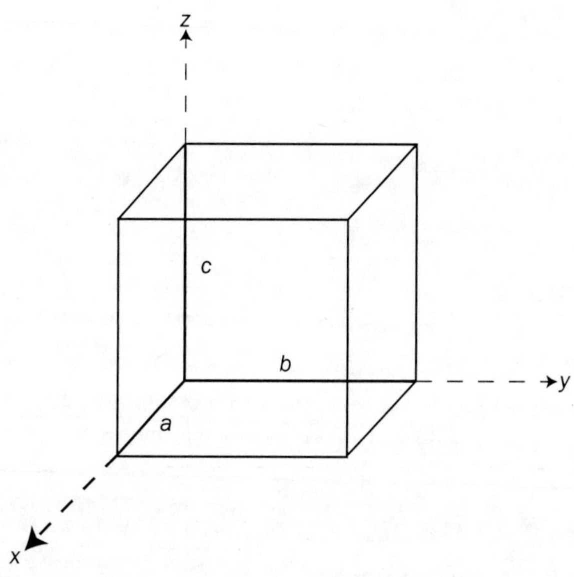

FIGURE **6.41** A rectangular block having sides *a*, *b* and *c*

All the five surfaces are maintained at 25 °C, and the face ($z = c$) is at 200 °C. Accordingly, the boundary conditions are written as:

$$T(0, y, z) = 25\,°C$$
$$T(a, y, z) = 25\,°C$$
$$T(x, 0, z) = 25\,°C$$
$$T(x, b, z) = 25\,°C$$
$$T(x, y, 0) = 25\,°C$$
$$T(x, y, c) = 200\,°C$$

Solve this PDE system and produce the temperature profile for $a = b = c = 1$ m.

6.75 Consider a system of coupled PDEs:

$$\frac{\partial C_A}{\partial t} = D_A \frac{\partial^2 C_A}{\partial x^2} \qquad 0 \le x \le 1$$

$$\frac{\partial C_B}{\partial t} = D_B \frac{\partial^2 C_B}{\partial x^2} \qquad x \ge 1$$

subject to:

ICs: $C_A(x, 0) = C_B(x, 0) = 0.2$

BCs: $C_A(0, t) = 1$

$C_A(1, t) = 0$

$C_B(1, t) = 0$

$C_B(\infty, t) = 0.2$

Solve this problem for $D_A = D_B = 10^{-6}$.

6.76 A counter-current double pipe heat exchanger is modeled before (see Case Study 6.7) with a single PDE, assuming constant shell side temperature. Now, we would like to extend that model to consider the dynamics of both tube side (T_t) and shell side (T_s) temperature. The energy balance on the tube side and shell side yields, respectively:

$$\frac{\partial T_t}{\partial t} + \upsilon_t \frac{\partial T_t}{\partial x} = C_t(T_s - T_t)$$

$$\frac{\partial T_s}{\partial t} - \upsilon_s \frac{\partial T_s}{\partial x} = C_s(T_t - T_s)$$

with

$$C_t = \frac{US}{\rho_t C_{pt} A_t}$$

$$C_s = \frac{US}{\rho_s C_{ps} A_s}$$

where,

υ = velocity

U = overall heat transfer coefficient

S = perimeter of inner pipe

A = cross sectional area

ρ = density

C_p = specific heat capacity

ICs: known steady state profile[24] (i.e., at $t = 0$) of $T_t(x)$ and $T_s(x)$, for which, you adopt:

$$T_t(0,\ 0) = 50\ °C$$
$$T_s(l,\ 0) = 80\ °C$$

Note that the shell side stream enters at $x = l$, while tube side stream at $x = 0$.

BCs: $T_t(0,\ t) = 40\ °C \qquad t > 0$

$T_s(l,\ t) = 80\ °C$

Given parameter values:

l = length of the heat exchanger = 10 m

$C_t = 0.2\ \text{min}^{-1}$

$C_s = 0.5\ \text{min}^{-1}$

$\upsilon_t = 2.5$ m/min

$\upsilon_s = 5$ m/min

i) Find steady state profile (i.e., at $t = 0$) of $T_t(x)$ and $T_s(x)$.

ii) Solve the coupled PDEs by using the FTBS method with $M = 100$.

6.77 A non-isothermal liquid phase plug flow reactor (PFR) with a constant inlet velocity (υ) and complex reaction kinetics is modeled as:

Mass balance

$$\frac{\partial C_A}{\partial t} + \upsilon\frac{\partial C_A}{\partial z} + (-r_A) = 0$$

Energy balance

$$\frac{\partial T}{\partial t} + \upsilon\frac{\partial T}{\partial z} - \frac{(\Delta H_r)(-r_A)}{\rho C_p} = 0$$

Obviously, the PFR model comprises two coupled hyperbolic PDEs. In this reactor, an exothermic irreversible chemical reaction takes place. The Langmuir–Hinshelwood reaction kinetics is given as:

$$-r_A = \frac{k\ C_A}{k_R + C_A}$$

24 You need to develop them based on the given energy balance equations.

where,

$$k = k_0 \exp\left[-\frac{\Delta E}{R}\left(\frac{1}{T}-\frac{1}{T_r}\right)\right]$$

$$k_R = k_{R0} \exp\left[-\frac{\Delta E_R}{R}\left(\frac{1}{T}-\frac{1}{T_r}\right)\right]$$

ICs: $C_A(z,\ 0) = 1$ mol/m³ $\quad 0 \le z \le l$

$T(z,\ 0) = 320$ K

BCs: $C_A(0,\ t) = 1$ mol/m³ $\quad t \ge 0$

$T(0,\ t) = 320$ K

Given parameter values:

Kinetic factor,	$k_0 = 0.15$ mol/m³.min
	$k_{R0} = 1$ mol/m³
Reference temperature,	$T_r = 320$ K
Density × specific heat capacity,	$\rho C_p = 500$ J/m³.K
Activation energy,	$\Delta E = 40 \times 10^3$ J/mol
	$\Delta E_R = -9 \times 10^3$ J/mol
Heat of reaction,	$\Delta H_r = 80 \times 10^3$ J/mol
Velocity,	$\upsilon = 0.3$ m/min
Reactor length,	$l = 2$ m
Universal gas constant,	$R = 8.314$ J/mol.K

Solve this problem using the FTBS method for $M = 100$ and $h = 0.0005$ min.

6.78 Consider the following biological reactions:

$$A + B_1 \xrightarrow{k_1} B_2 + P$$

$$P + B_3 \xrightarrow{k_2} A + B_4$$

The concerned reactor is modeled as:

$$\frac{\partial A}{\partial t} = D_A \frac{\partial^2 A}{\partial x^2} - k_1 A + k_2 P$$

$$\frac{\partial P}{\partial t} = D_P \frac{\partial^2 P}{\partial x^2} + k_1 A - k_2 P$$

$$\frac{\partial B_1}{\partial t} = D_{B1} \frac{\partial^2 B_1}{\partial x^2} - k_1 A$$

Here, A, P and B_1 are the local concentrations of the corresponding species. The diffusivities, $D_A = D_P = D_{B1} = D$. The following conditions are specified:

ICs: $A(x,\ 0) = 0 \qquad 0 \le x \le l$

$P(x,\ 0) = 0$

$B_1(x,\ 0) = B_{10}\left(1 - \frac{x}{l}\right)$

BCs: $A(0,\ t) = A_0 \qquad t \ge 0$

$P(0,\ t) = 0$

$B_1(0,\ t) = B_{10}$

$\frac{\partial A}{\partial x}(l,\ t) = 0$

$\frac{\partial P}{\partial x}(l,\ t) = 0$

$\frac{\partial B_1}{\partial x}(l,\ t) = 0$

i) Formulate the solution methodology with the MoL method.

ii) Produce results for $B_{10} = 1$, $A_0 = 2$, $l = 1$, $D = 1$ and $k_1 = k_2 = 0.2$.

References

Davis, M. E. (1984). *Numerical Methods and Modeling for Chemical Engineers*, 2nd ed., New York: John Wiley & Sons.

Finlayson, B. A. (1980). *Non-linear Analysis in Chemical Engineering*, 1st ed., New York: McGraw Hill.

Gross, L. J. (1981). On the dynamics of internal leaf carbon dioxide intake. *J. Math. Biol.*, 11, 181–191.

Hildebrand, F. B. (1962). *Advanced Calculus for Applications*, 1st ed., Englewood Cliffs, New Jersey: Prentice Hall.

Lebedeff, S. A. and Hameed, S. (1975). Steady state solution of the semi-empirical diffusion equation for area sources. *J. Appl. Meteor.*, 14, 546–549.

Ramirez, W. F. (1997). *Computational Methods for Process Simulation*, 2nd ed., Oxford: Reed Educational and Professional Publishing Ltd.

Peaceman, D. W. and Rachford Jr., H. H. (1955). The numerical solution of parabolic and elliptic differential equations. *Journal of the Society for Industrial and Applied Mathematics*, 3 (1), 28–41.

Part IV

Function Approximations

Part IV of this book consists of:

- Data fitting
- Numerical integration

Data fitting

Functions are used to model many real world problems, for which, we have discrete data points that are either directly measured or derived quantities. To understand the behavior of these functions, they need to be continuous or differentiable. Constructing such a continuous function based on the discrete data is called *data fitting*. Here, we will cover two approaches for data fitting:

- Least-squares data fitting (Chapter 7)
- Polynomial interpolation (Chapter 8)

In the subsequent Chapter 9, we will discuss numerical integration approaches using the functions fitted in Chapter 8.

7

Method of Least-Squares

"Thus, in a sense, mathematics has been most advanced by those who distinguished themselves by intuition rather than by rigorous proofs." – Felix Klein

Key Learning Objectives

- Knowing the basics of data fitting
- Learning various error criteria
- Understanding the fundamental of method of least-squares
- Applying least-squares on linear and non-linear examples

7.1 Introduction

The availability of data (in terms of dependent variable, y) at several discrete locations (values of the independent variable, x) is quite common. Often, we call it as *input-output data set* in different fields of our study. It may be laboratory-scale, pilot-scale, industrial-scale, or field-scale data. To deal with different scientific and engineering problems at various occasions, one may need to:

- Interpolate[1] or extrapolate[2] these data.
- Find slope (i.e., $\frac{dy}{dx}$) of the function f described by that given data set.
- Find integral (i.e., $\int y \, dx$) of that function.

For these mathematical operations, one needs to first construct an analytical function, f based on the available y versus x data. Thus, prior to performing the interpolation, differentiation or any kind of integration, one has to know an appropriate technique to

1 Means estimate intermediate values of y between known discrete data points.

2 Means estimate values of y outside the range of known data.

fit the data sets by function *f*. Accordingly, here we fix our goal to determine a formula, $y = f(x)$ that best fits the available experimental data set. To do so, the most common technique used is *method of least-squares* or *least-squares curve fitting*. Actually, there is a class of formulas (depending on the process characteristics) and studying them will be the subject matter of this chapter.

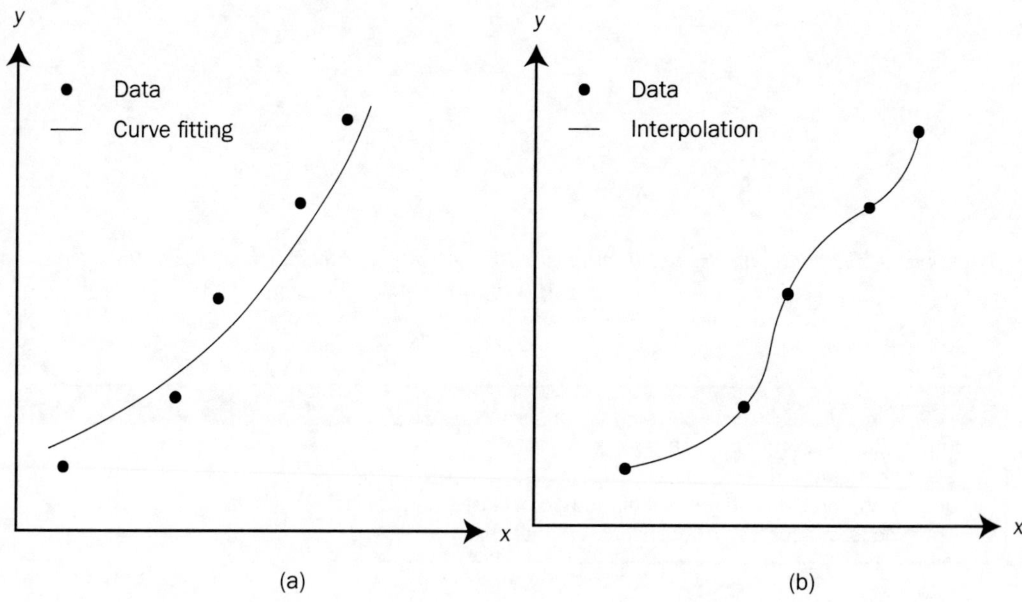

FIGURE 7.1 Difference of (a) curve fitting, and (b) interpolation

At this point of time, let us know shortly the distinction between the interpolation and curve fitting. Any individual data point may be incorrect and thus, without intersecting every data point, we can derive a curve to follow the pattern (or general trend) of the points as a group. This is often called *curve fitting* (or sometimes regression[3]). Accordingly, Figure 7.1(a) is drawn to show that the function determined through curve fitting does not necessarily pass through any data point. On the other hand, in case of *interpolation* [Figure 7.1(b)], the derived function typically passes through every data point (to be covered in Chapter 8) with the presumption that the data are known to be very precise.[4]

7.2 Least-Squares Data Fitting: Some Basics

Let us go through some basic concepts involved in the method of least-squares used for data fitting. There are a few specialized equipment that generate experimental data points with

3 It is employed to obtain an approximate function and handle the uncertainty existed in the data set (unobserved random error).

4 For example, heat capacity of water is precisely available against temperature.

about five digits of accuracy or so.[5] However, a number of equipment is reliable to three or even fewer digits of accuracy. It is also fact that there is some experimental error involved in measurement for the real-world problems. With this, one can write, in general:

$$f(x_k) = y_k + e_k \tag{7.1}$$

where for kth data point,

$f(x_k)$ = true value of y

e_k = error (also called *residual* or *deviation*)

Notice that the residual, e_k is used to quantify the discrepancy between the true value of y and its approximate value at x_k.

An example

Let us fit a straight line:

$$y = f(x) = a_0 + a_1 x \tag{7.2}$$

through y versus x data. Here, a_0 and a_1 are the parameters representing the intercept and the slope, respectively. With this, we have:

$$e_k = a_0 + a_1 x_k - y_k$$

This e_k is geometrically interpreted in Figure 7.2 and it is nothing but the vertical distance between the data point (x_k, y_k) and the least-squares line $y = a_0 + a_1 x$. Now we target to find a_0 and a_1 such that this distance (i.e., discrepancy) becomes minimal. This minimization is typically performed based on some error criteria, which are discussed below.

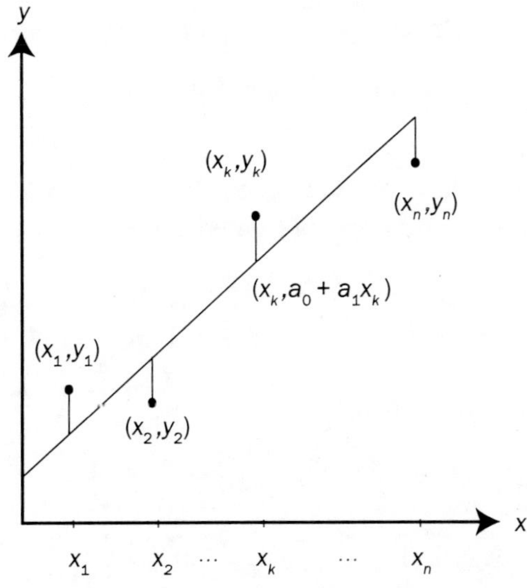

FIGURE 7.2 Vertical distances between the points (x_k, y_k) and least-squares line $y = a_0 + a_1 x$

5 As indicated, when the data sets are available to several significant digits of accuracy, polynomial interpolation is preferably used (discussed in Chapter 8).

Error Criteria

For a general system, Equation (7.1) yields:

$$e_k = f(x_k) - y_k \tag{7.3}$$

that is true for $0 \leq k \leq n$ (i.e., there are total $n+1$ data points). As indicated, this residual is typically used to quantify how far the curve $y = f(x)$ lies from the experimental data. To minimize the total distance, there are various norms employed.

Maximum error: $$E(f) = \max_{0 \leq k \leq n} [|f(x_k) - y_k|] \tag{7.4}$$

Average error: $$E(f) = \frac{1}{n+1} \sum_{k=0}^{n} |f(x_k) - y_k| \tag{7.5}$$

Sum of square errors: $$E(f) = \sum_{k=0}^{n} [f(x_k) - y_k]^2 \tag{7.6}$$

Root mean square (RMS) error: $$E(f) = \left[\frac{1}{n+1} \sum_{k=0}^{n} [f(x_k) - y_k]^2 \right]^{1/2} \tag{7.7}$$

Remark

These various forms of error are applicable to both linear and non-linear least-squares. The criterion shown in Equation (7.6) is called *least-squares*.

Let us apply these error criteria to a simple example.

Example 7.1 Applying error criteria

A straight line,

$$y = f(x) = 5 + 2.6x$$

is constructed through the following data points: (−2, 0), (−1, 2.5), (0, 5), (1, 7.5), (2, 10). Find *E(f)* applying the four error criteria presented above.

Solution

To find *E(f)*, let us first calculate $f(x_k)$, $|e_k|$ and e_k^2 in Table 7.1. With this, we get:

Maximum error = max [0.2, 0.1, 0.0, 0.1, 0.2] = 0.2

$$\textit{Average error} = \frac{1}{5}[0.2 + 0.1 + 0.0 + 0.1 + 0.2] = 0.12$$

Sum of square errors = [0.04 + 0.01 + 0.0 + 0.01 + 0.04] = 0.1

$$\textit{RMS error} = \left[\frac{1}{5}(0.1)\right]^{1/2} = 0.1414$$

TABLE 7.1 Calculated values of $f(x_k)$, $|e_k|$ and e_k^2

k	x_k	y_k	$f(x_k) = 5 + 2.6x_k$	$\|e_k\|$	e_k^2
0	−2	0	−0.2	0.2	0.04
1	−1	2.5	2.4	0.1	0.01
2	0	5	5	0.0	0.0
3	1	7.5	7.6	0.1	0.01
4	2	10	10.2	0.2	0.04

7.3 METHOD OF LEAST-SQUARES

Here, we would like to cover the following methods:

- Linear least-squares
- Least-squares for non-linear functions
- Least-squares for polynomial functions
- Non-linear least-squares

Let us discuss them one-by-one with their relevant applications to various types of systems.

7.4 LINEAR LEAST-SQUARES

The simplest technique is to fit a straight line:

$$y = f(x) = a_0 + a_1 x \tag{7.2}$$

through y versus x data set. Recall that a_0 and a_1 are the constant parameters. To find the best fitting line, one needs to minimize one of the E terms defined before in Equations (7.4)–(7.7). For this, we adopt the sum of the squares of the residuals [Equation (7.6)] as our first choice:[6]

$$E = \sum_{k=0}^{n} [f(x_k) - y_k]^2 \tag{7.6}$$

As stated, this is also called *least-squares criterion*.

Inserting Equation (7.2),

$$E = \sum_{k=0}^{n} [a_0 + a_1 x_k - y_k]^2 \tag{7.8}$$

Notice that E is a function of two parameters, a_0 and a_1. Now we would like to determine the values of a_0 and a_1, which would lead to minimize this total error. Accordingly, we find:

6 To avoid cancellation of positive and negative values of residuals and easy to deal with computationally.

$$\frac{\partial E}{\partial a_0} = 2\sum_{k=0}^{n} [a_0 + a_1 x_k - y_k] = 0 \tag{7.9a}$$

$$\frac{\partial E}{\partial a_1} = 2\sum_{k=0}^{n} \{x_k[a_0 + a_1 x_k - y_k]\} = 0 \tag{7.9b}$$

Equation (7.9) yields:

$$a_0(n+1) + a_1 \sum_{k=0}^{n} x_k = \sum_{k=0}^{n} y_k \tag{7.10a}$$

$$a_0 \sum_{k=0}^{n} x_k + a_1 \sum_{k=0}^{n} x_k^2 = \sum_{k=0}^{n} y_k x_k \tag{7.10b}$$

Solving these two simultaneous equations:

$$a_1 = \frac{(n+1)\sum_{k=0}^{n} y_k x_k - \sum_{k=0}^{n} x_k \sum_{k=0}^{n} y_k}{(n+1)\sum_{k=0}^{n} x_k^2 - \left(\sum_{k=0}^{n} x_k\right)^2} \tag{7.11}$$

$$a_0 = \frac{1}{n+1}\left[\sum_{k=0}^{n} y_k - a_1 \sum_{k=0}^{n} x_k\right] \tag{7.12}$$

which basically constitute Equation (7.2) [i.e., $y = a_0 + a_1 x$].

Alternatively, one can find the unknown parameters, a_0 and a_1, by first presenting Equation (7.10) in matrix form as:

$$Aa = B \tag{7.13}$$

where,

$$A = X^T X$$

$$B = X^T Y$$

with

$$X = \begin{bmatrix} 1 & x_0 \\ 1 & x_1 \\ . & . \\ . & . \\ 1 & x_n \end{bmatrix}$$

$$a = \begin{bmatrix} a_0 \\ a_1 \end{bmatrix}$$

$$Y = \begin{bmatrix} y_0 \\ y_1 \\ . \\ . \\ y_n \end{bmatrix}$$

Then from Equation (7.13), one gets the parameters:

$$a = A^{-1}B \tag{7.14}$$

Remarks

1. Notice that Equation (7.13) has the same form with $AX = B$ [i.e., Equation (2.2) in Chapter 2]. In order to solve this system of linear equations, one can use a suitable linear technique described in Chapter 2.
2. It is perhaps easier to obtain the solution analytically by using Equations (7.11) and (7.12).
3. This linear technique can easily be extended to *multiple linear least-squares* approach. To understand this, let us consider a two-dimensional case, in which, y is a linear function of x_1 and x_2 as:

   $$y = f(x) = a_0 + a_1 x_1 + a_2 x_2$$

   We all know that the *line* (formed in one-dimensional case) becomes a *plane* in two-dimensional case. Now, one can similarly use Equation (7.6) and get the three unknown parameters, a_0, a_1 and a_2, with consideration of:

   $$\frac{\partial E}{\partial a_0} = \frac{\partial E}{\partial a_1} = \frac{\partial E}{\partial a_2} = 0$$

This linear least-squares method is illustrated in an example below.

Example 7.2 Linear least-squares

Find the linear function that best fits the data set given in Table 7.2 and evaluate the performance of the regression function.

TABLE 7.2 y versus x data points

x_k	y_k
1.0	6.1
1.5	7.5
2.0	8.7
2.5	9.9
3.0	11.3

Solution

The given y versus x data are to be used to fit the following linear equation:

$$y_k = a_0 + a_1 x_k \tag{7.15}$$

For this,

$$n+1=5$$

$$\sum x_k = 1+1.5+2+2.5+3=10$$

$$\sum x_k^2 = 1^2+1.5^2+2^2+2.5^2+3^2=22.5$$

$$\sum y_k = 6.1+7.5+8.7+9.9+11.3=43.5$$

$$\sum y_k x_k = (6.1\times 1)+(7.5\times 1.5)+(8.7\times 2)+(9.9\times 2.5)+(11.3\times 3)=93.4$$

From Equations (7.11) and (7.12), respectively, we get:

$$a_1 = \frac{5\times 93.4 - 10\times 43.5}{5\times 22.5 - 10^2} = 2.56$$

$$a_0 = \frac{1}{5}\left[43.5 - 2.56\times 10\right] = 3.58$$

The linear Equation (7.15) finally gets the following form:

$$y = 3.58 + 2.56x$$

Table 7.3 and Figure 7.3 compare the given data with the best least-squares line, $y = 3.58 + 2.56x$. Note that here the same data set is used for both parameter identification and testing, and it is followed in a few more examples later.

Table 7.3 Comparing the given and predicted data

x_k	y_k	f_k
1.0	6.1	6.14
1.5	7.5	7.42
2.0	8.7	8.7
2.5	9.9	9.98
3.0	11.3	11.26

It is evident from Figure 7.3 that the developed model predicts the data pretty well. Similar kind of qualitative evaluation can also be performed by making a *parity plot* between the given y (abscissa) and predicted y (ordinate). We obtain a good fit when the data lies along the 45° line.

Let us now quantify the performance of the regression model below.

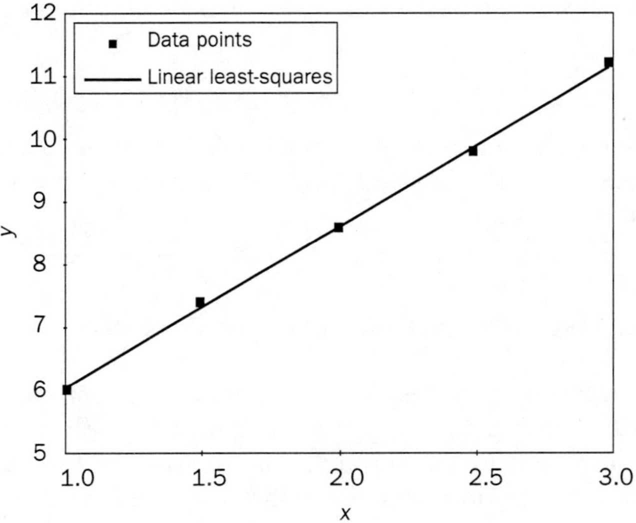

FIGURE 7.3 Linear least-squares fit

Evaluating performance of regression function: quantification of error

To evaluate the performance of the regression model (i.e., $y = 3.58 + 2.56x$), one needs to do the following analysis with producing a set of data as shown in Table 7.4.

Using Equation (A7.2), we get the standard deviation as:

$$S_d = \sqrt{\frac{16.4}{4}} = 2.025$$

The standard error of the estimate we get from Equation (A7.4) as:

$$S_e = \sqrt{\frac{0.016}{5-2}} = 0.073$$

Since S_e is reasonably small,[7] the data points are spread around the regression line, indicating that we have a good fit. In a little different way, we can say that the regression model has merit since we got $S_d > S_e$.

TABLE 7.4 Data to compute the goodness-of-fit statistics (details in Appendix 7A)

	x_k	y_k	f_k	$(y_k-\bar{y})^2$	$(f_k-y_k)^2$
	1.0	6.1	6.14	6.76	0.0016
	1.5	7.5	7.42	1.44	0.0064
	2.0	8.7	8.7	0.0	0.0
	2.5	9.9	9.98	1.44	0.0064
	3.0	11.3	11.26	6.76	0.0016
Σ	10	43.5		16.4	0.016

7 Around 1.4% of the range of y_k (i.e., the maximum value of $|y_k - y_l|$, where $k \neq l$ and $k, l = 0, 1, 2, 3, 4$).

Further, to quantify the goodness-of-fit, we compute r^2-value from Equation (A7.5):

$$r^2 = \frac{16.4 - 0.016}{16.4} = 0.999$$

It reveals that the regression model (i.e., $y = 3.58 + 2.56x$) is able to explain 99.9% of the original uncertainty (i.e., variability of the data). This indicates very close to *perfect fitting*, which is also evident in Figure 7.3.

7.5 Least-Squares for Non-linear Functions

Linear least-squares method is not restricted to linear functions only. Here, we will consider a few non-linear functions as well for curve fitting by data linearization. In the solution methodology, at first the non-linear function that we wish to fit through the given data points needs to be converted to a linear form; then use the linear least-squares technique discussed before for finding the concerned parameters.

With this mechanism, let us go into details of the least-squares for two non-linear functions that have wide applications.

7.5.1 Power Law Model

Consider the following non-linear function:

$$y = ax^b \tag{7.16}$$

where, a and b are constant parameters. This is the form of *power law*. This model has wide applicability in fitting the experimental data when the underlying model is not known. It includes the examples of planetary motion and pressure–volume relation of a gas (i.e., $PV^\gamma =$ constant).

Taking natural logarithm of both sides of Equation (7.16):

$$\ln y = \ln a + b \ln x \tag{7.17}$$

Considering,

$$\ln y = \bar{y} \qquad \ln x = \bar{x}$$
$$\ln a = \bar{a}_0 \qquad b = \bar{a}_1$$

we get,

$$\bar{y} = \bar{a}_0 + \bar{a}_1 \bar{x} \tag{7.18}$$

This is the converted linear form that is similar to the straight line [Equation (7.2)]. This process is often called *data linearization* and is depicted in Figure 7.4. Note that to linearize this power model, one can use any base logarithm. Further, here the y versus x data set may be obtained through direct measurement but $\bar{y}$ versus $\bar{x}$ data points are derived quantities.

In the next phase, we need to use the linear data fitting technique discussed before to solve the problem. Accordingly, from Equations (7.11) and (7.12), we have:

$$\bar{a}_1 = \frac{(n+1)\sum_{k=0}^{n} \bar{y}_k \bar{x}_k - \sum_{k=0}^{n} \bar{x}_k \sum_{k=0}^{n} \bar{y}_k}{(n+1)\sum_{k=0}^{n} \bar{x}_k^2 - \left(\sum_{k=0}^{n} \bar{x}_k\right)^2} \tag{7.19}$$

$$\bar{a}_0 = \frac{1}{n+1}\left[\sum_{k=0}^{n} \bar{y}_k - \bar{a}_1 \sum_{k=0}^{n} \bar{x}_k\right] \tag{7.20}$$

Having obtained $\bar{a}_0$ and $\bar{a}_1$, we can get:

$$a = \exp(\bar{a}_0)$$

$$b = \bar{a}_1$$

for equation $y = ax^b$ and our work gets done.

Remark

To fit this non-linear power equation to the experimental data directly, one can use the non-linear least-squares method. But we will see later that the non-linear scheme is not so simple compared to the approach discussed above through data linearization.

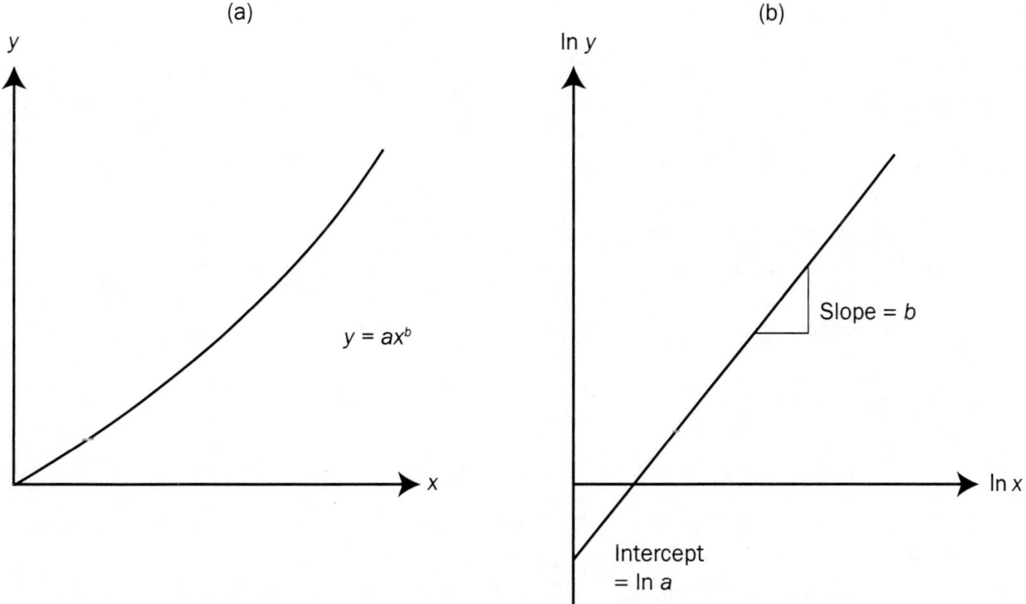

Figure 7.4 Illustrating the (a) power equation, and (b) its linearized version

Example 7.3 Power fit $y = ax^b$

Find the best fit values of a and b for Equation (7.16) based on the data set given in Table 7.5.

Table 7.5 Given data set for $y = ax^b$

x_k	y_k
1	2
2	11.3137
3	31.1769
4	64
5	111.8034

Solution

In this problem,

$$n+1=5$$

$$\Sigma\bar{x}_k = \ln(1) + \ln(2) + \ln(3) + \ln(4) + \ln(5)$$

$$= 0 + 0.69315 + 1.09861 + 1.38629 + 1.60944 = 4.78749$$

$$\Sigma\bar{x}_k^2 = 6.19950$$

$$\Sigma\bar{y}_k = 15.43446$$

$$\Sigma\bar{y}_k\bar{x}_k = 18.81719$$

From Equations (7.19) and (7.20), we get, respectively:

$$\bar{a}_1 = \frac{5\times18.81719-4.78749\times15.43446}{5\times6.19950-(4.78749)^2} = 2.5$$

$$\bar{a}_0 = \frac{1}{5}\left[15.43446-2.5\times4.78749\right] = 0.693147$$

So we get Equation (7.18) as:

$$\bar{y} = 0.693147+2.5\,\bar{x}$$

which gives:

$$y = 2x^{2.5}$$

This is the final form obtained by least-squares fitting.

Example 7.4 Power fit $y = ax^b$

Solve another problem to find the best fit values of a and b for Equation (7.16) based on the data set given in Table 7.6.

TABLE 7.6 Given data set for $y = ax^b$

x_k	y_k
50	28
450	30
780	32
1200	36
4400	51
4800	58
5300	69

Solution

Let us solve this problem in a slightly different way with taking common logarithm (i.e., base-10 logarithm), instead of natural logarithm, of $y = ax^b$. Accordingly,

$$\log_{10}^{y} = \log_{10}^{a} + b\log_{10}^{x} \tag{7.21}$$

Considering,

$$\log_{10}^{y} = \bar{y} \qquad \log_{10}^{x} = \bar{x}$$

$$\log_{10}^{a} = \bar{a}_0 \qquad b = \bar{a}_1$$

we get,

$$\bar{y} = \bar{a}_0 + \bar{a}_1\bar{x}$$

Then we can find the parameters $\bar{a}_0$ and $\bar{a}_1$ by using Equations (7.20) and (7.19), respectively, based on the data given in Table 7.6.

In this problem,

$$n+1=7$$

$$\sum \bar{x}_k = \log_{10}^{50} + \log_{10}^{450} + \log_{10}^{780} + \log_{10}^{1200} + \log_{10}^{4400} + \log_{10}^{4800} + \log_{10}^{5300}$$

$$= 1.699 + 2.653 + 2.892 + 3.079 + 3.643 + 3.681 + 3.724$$

$$= 21.371$$

$$\sum \bar{x}_k^2 = 68.458$$

$$\sum \bar{y}_k = \log_{10}^{28} + \log_{10}^{30} + \log_{10}^{32} + \log_{10}^{36} + \log_{10}^{51} + \log_{10}^{58} + \log_{10}^{69}$$

$$= 1.447 + 1.477 + 1.505 + 1.556 + 1.707 + 1.763 + 1.839$$

$$= 11.294$$

$$\sum \bar{y}_k \bar{x}_k = 35.077$$

From Equations (7.19) and (7.20), we get, respectively:

$$\bar{a}_1 = \frac{7\times 35.077 - 21.371\times 11.294}{7\times 68.458 - (21.371)^2} = 0.185665$$

$$\bar{a}_0 = \frac{1}{7}\left[11.294 - 0.185665\times 21.371\right] = 1.0466$$

So we get Equation (7.18) as:

$$\bar{y} = 1.0466 + 0.185665\,\bar{x}$$

which gives:

$$y = 11.1327x^{0.185665}$$

To qualitatively verify the analytical function constructed above, Figure 7.5 is produced. It is evident that the function predicts the data set pretty well, except the last point that corresponds to $x_k = 5300$.

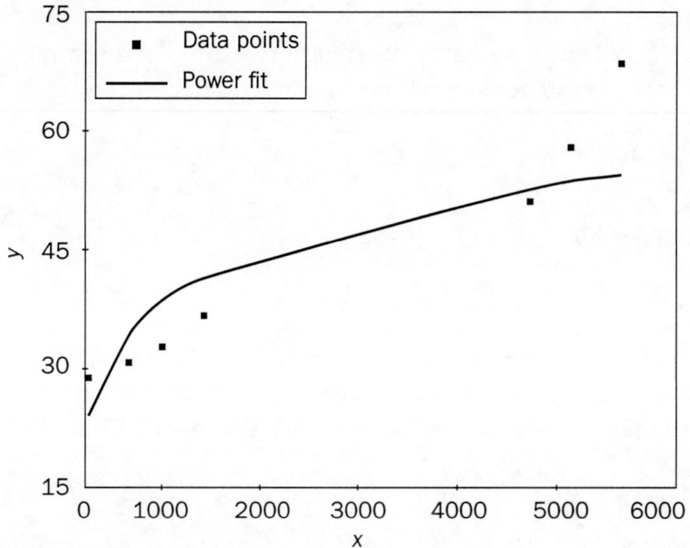

FIGURE 7.5 Least-squares power fitting

7.5.2 Exponential Model

Let us consider the following non-linear exponential function in the form of:

$$y = ae^{bx} \tag{7.22}$$

where, a and b are constants. This form of model is widely used in various areas of science and engineering, including in chemical and biological reaction fields.

Taking natural logarithm of both sides:

$$\ln y = \ln a + bx \tag{7.23}$$

Considering,

$$\ln y = \bar{y} \qquad x = \bar{x}$$

$$\ln a = \bar{a}_0 \qquad b = \bar{a}_1$$

we get,

$$\bar{y} = \bar{a}_0 + \bar{a}_1 \bar{x} \tag{7.18}$$

For this straight line, one can compute $\bar{a}_1$ and $\bar{a}_0$ from Equations (7.19) and (7.20), respectively. Having obtained $\bar{a}_0$ and $\bar{a}_1$, we can get:

$$a = \exp(\bar{a}_0)$$

$$b = \bar{a}_1$$

for $y = ae^{bx}$.

In the following examples, this procedure is illustrated.

Example 7.5 Least-squares fit $y = ae^{bx}$

Find the exponential curve of the form $y = ae^{bx}$ that best fits the data set given in Table 7.7.

Table 7.7 Data to fit $y = ae^{bx}$

x_k	y_k
1	14.6190
1.5	51.0253
2	178.0958
2.5	621.6154

Solution

Based on the given data in Table 7.7,

$$n + 1 = 4$$

$$\sum \bar{x}_k = 1 + 1.5 + 2 + 2.5 = 7$$

$$\sum \bar{x}_k^2 = 1^2 + 1.5^2 + 2^2 + 2.5^2 = 13.5$$

$$\sum \bar{y}_k = \ln(14.619) + \ln(51.0253) + \ln(178.0958) + \ln(621.6154)$$

$$= 2.6823 + 3.9323 + 5.1823 + 6.4323$$

$$= 18.2292$$

$$\sum \bar{y}_k \bar{x}_k = 35.0261$$

From Equations (7.19) and (7.20), we get respectively:

$$\bar{a}_1 = \frac{4\times 35.0261 - 7\times 18.2292}{4\times 13.5 - 49} = 2.5$$

$$\bar{a}_0 = \frac{1}{4}\left[18.2292 - 2.5\times 7\right] = 0.1823$$

It gives,

$$\bar{y} = 0.1823 + 2.5\,\bar{x}$$

In exponential form,

$$y = 1.2\, e^{2.5x}$$

This is the final form obtained by least-squares fitting.

Example 7.6 Least-squares fit $y = ae^{bx}$

Repeat Example 7.3 to find the least-squares fit $y = ae^{bx}$ based on a set of paired observations given in Table 7.5.

Solution

In this problem,

$$n + 1 = 5$$

$$\sum \bar{x}_k = \sum x_k = 1 + 2 + 3 + 4 + 5 = 15$$

$$\sum \bar{x}_k^2 = \sum x_k^2 = 1^2 + 2^2 + 3^2 + 4^2 + 5^2 = 55$$

$$\sum \bar{y}_k = \ln(2) + \ln(11.3137) + \ln(31.1769) + \ln(64) + \ln(111.8034)$$

$$= 0.69315 + 2.42601 + 3.43968 + 4.15888 + 4.71674$$

$$= 15.43446$$

$$\sum \bar{y}_k \bar{x}_k = 56.08343$$

So,

$$\bar{a}_1 = \frac{5\times 56.08343 - 15\times 15.43446}{5\times 55 - (15)^2} = 0.978$$

$$\bar{a}_0 = \frac{1}{5}\left[15.43446 - 0.978\times 15\right] = 0.15289$$

It gives the following linearized form:

$$\bar{y} = 0.15289 + 0.978\,\bar{x}$$

In exponential form:

$$y = 1.1652\, e^{0.978\,x}$$

Remark

Here, we got the following exponential fit:

$$y = 1.1652\, e^{0.978\, x}$$

For the same data set (Table 7.5), we got previously (Example 7.3) the power fit:

$$y = 2x^{2.5}$$

Now we compare these two functions in Table 7.8 with reference to the given data set. Additionally, Figure 7.6 is produced to visualize the difference at a glance. It is obvious that for the sample data set, the least-squares power fitting leads to provide a far better prediction than the exponential fitting. As your homework, it is asked to confirm this outperformance of power fitting through statistical analysis.

Table 7.8 Comparing two non-linear functions constructed through least-squares data fitting

x_k	y_k	$y_k\,[=2x_k^{2.5}]$	$y_k\,[=1.1652\,\exp(0.978x_k)]$
1	2	2	3.0984
2	11.3137	11.3137	8.2391
3	31.1769	31.1769	21.9089
4	64	64	58.2587
5	111.8034	111.8034	154.9175

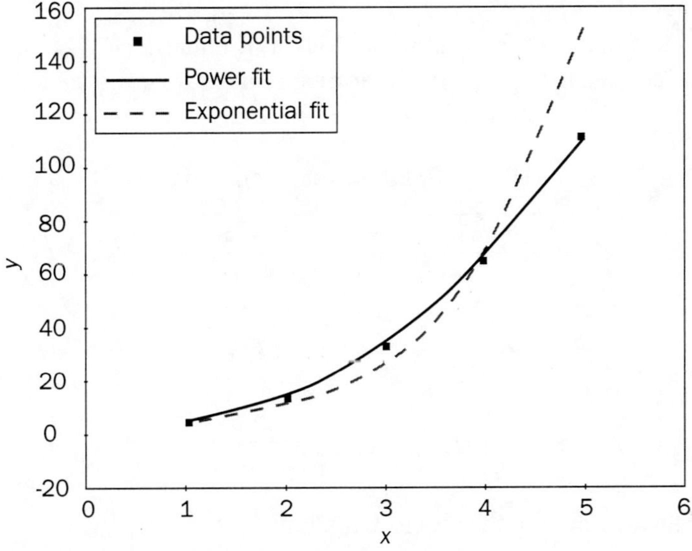

Figure 7.6 Comparing two non-linear functions

Case Study 7.1 A batch reactor

A first-order reaction,

$$A \rightarrow P$$

takes place in a batch reactor. Here, A is the reactant and P the product. The reactor model has the following form:

$$\frac{dC_A}{dt} = -k\,C_A$$

where,

C_A = concentration of reactant A, kmol/m^3
k = reaction rate constant, min^{-1}

Integrating,

$$\int_{C_{A0}}^{C_A} \frac{dC_A}{C_A} = -k\int_0^t dt$$

we get,

$$\ln\frac{C_A}{C_{A0}} = -k\,t$$

It gives:

$$C_A = C_{A0}e^{-k\,t} \tag{7.24}$$

This is the derived function, which is similar to that in Equation (7.22).

For the transient batch reactor, the experimental data are given in Table 7.9 (Bequette, 1998). Find C_{A0} and k for Equation (7.24).

Table 7.9 Given experimental data

t (min)	C_A (kmol/m^3)
1	5.0
2	2.95
3	1.82
4	1.05
5	0.71

Solution

Taking natural logarithm of both sides in Equation (7.24),

$$\ln C_A = \ln C_{A0} - k\,t$$

We can represent it as:

$$\bar{y} = \bar{a}_0 + \bar{a}_1\bar{x} \tag{7.18}$$

where,

$$\bar{y} = \ln C_A \qquad \bar{a}_0 = \ln C_{A0}$$
$$\bar{a}_1 = -k \qquad \bar{x} = t$$

For finding $\bar{a}_0$ and $\bar{a}_1$, let us do the following calculations:

$n+1=5$

$\sum \bar{x}_k = 1 + 2 + 3 + 4 + 5 = 15$

$\sum \bar{x}_k^2 = 1^2 + 2^2 + 3^2 + 4^2 + 5^2 = 55$

$\sum \bar{y}_k = \ln(5) + \ln(2.95) + \ln(1.82) + \ln(1.05) + \ln(0.71)$

$= 1.6094 + 1.0818 + 0.5988 + 0.04879 - 0.3425$

$= 2.99629$

$\sum \bar{y}_k \bar{x}_k = 4.05206$

From Equations (7.19) and (7.20),

$$\bar{a}_1 = \frac{5 \times 4.05206 - 15 \times 2.99629}{5 \times 55 - (15)^2} = -0.49368$$

$$\bar{a}_0 = \frac{1}{5}\left[2.99629 + 0.49368 \times 15\right] = 2.0803$$

From these parameter values, we get:

$C_{A0} = e^{\bar{a}_0} = 8.0069$ kmol/m^3

$k = -\bar{a}_1 = 0.49368$ min^{-1}

Accordingly, Equation (7.24) gets the following final form:

$$C_A = 8.0069\, e^{-0.49368\, t}$$

A Short Note

So far we have discussed least-squares principle for linear and non-linear functions with different forms. There are still many more non-linear functions, which are not truly covered here. However, one can deal them in the similar way as we have treated the power and exponential functions through data linearization. For completeness, many of them are listed shortly with their linearization methodology in Table A7.1 (see Appendix A7.2). As shown, after linearizing to the following form:

$$\bar{y} = \bar{a}_0 + \bar{a}_1 \bar{x}$$

one can employ Equations (7.19) and (7.20) to find $\bar{a}_1$ and $\bar{a}_0$, respectively. Then the regression model, $y = f(x)$ is finally formed through proper conversion of $\bar{a}_0$ and $\bar{a}_1$ to the actual function parameters.

7.6 Least-Squares for Polynomial Functions

Let us first know some basics of the polynomial functions. A polynomial[8] of degree m (non-negative integer) is commonly represented by:

$$y = f(x) = a_0 + a_1 x + a_2 x^2 + \dots\dots + a_{m-1} x^{m-1} + a_m x^m \quad (7.25)$$

where, a is the coefficient. It can be rewritten as:

$$f(x) = \sum_{i=0}^{m} a_i x^i \quad (7.26)$$

When the degree,

$m = 0$, it is called *constant* (non-zero) polynomial
$m = 1$, it is called *linear* polynomial
$m = 2$, it is called *quadratic* polynomial
$m = 3$, it is called *cubic* polynomial
$m = 4$, it is called *quartic* polynomial

and so on. Note that when $m \geq 2$, it is basically *non-linear* in nature.

Polynomials have various applications, including in curve fitting, interpolation, and in characterizing the dynamic systems (e.g., reactors, electrical circuits and mechanical devices).

Deriving least-squares formula for polynomials

The polynomial model has the following form:

$$f(x_k) = a_0 + a_1 x_k + a_2 x_k^2 + \dots\dots + a_{m-1} x_k^{m-1} + a_m x_k^m \quad (7.27)$$

where, $k = 0, 1, 2, \dots, n$. Note that here, $n \geq m$ since for finding $(m + 1)$ number of coefficients involved in the above equation (i.e., a_0, a_1, ..., a_m), we need at least $(m + 1)$ or more data points.

Now we need to find those coefficients, a_0, a_1, ..., a_m, which minimize:

$$E = \sum_{k=0}^{n} \left[\sum_{i=0}^{m} a_i x_k^i - y_k \right]^2 \quad (7.28)$$

It yields:

$$\frac{\partial E}{\partial a_0} = \sum_{k=0}^{n} 2 \left[\sum_{i=0}^{m} a_i x_k^i - y_k \right] = 0$$

$$\frac{\partial E}{\partial a_1} = \sum_{k=0}^{n} 2 \left[\sum_{i=0}^{m} a_i x_k^i - y_k \right] x_k = 0$$

$$\vdots$$

8 It is an expression that typically consists of coefficients along with variables with non-negative integer exponents. The highest power of the variable in a polynomial is called the *degree* of that polynomial expression.

$$\frac{\partial E}{\partial a_m} = \sum_{k=0}^{n} 2\left[\sum_{i=0}^{m} a_i x_k^i - y_k\right] x_k^m = 0$$

Rearranging these $(m + 1)$ equations for the $(m + 1)$ unknown coefficients,

$$\begin{aligned}
&a_0(n+1) + a_1 \sum_{k=0}^{n} x_k + \dots\dots\dots + a_m \sum_{k=0}^{n} x_k^m = \sum_{k=0}^{n} y_k \\
&a_0 \sum_{k=0}^{n} x_k + a_1 \sum_{k=0}^{n} x_k^2 + \dots\dots\dots + a_m \sum_{k=0}^{n} x_k^{m+1} = \sum_{k=0}^{n} y_k x_k \\
&\vdots \\
&a_0 \sum_{k=0}^{n} x_k^m + a_1 \sum_{k=0}^{n} x_k^{m+1} + \dots\dots\dots + a_m \sum_{k=0}^{n} x_k^{2m} = \sum_{k=0}^{n} y_k x_k^m
\end{aligned} \tag{7.29}$$

In matrix form,

$$\begin{bmatrix} (n+1) & \sum x_k & \dots & \sum x_k^m \\ \sum x_k & \sum x_k^2 & \dots & \sum x_k^{m+1} \\ \cdot & \cdot & & \cdot \\ \cdot & \cdot & & \cdot \\ \sum x_k^m & \sum x_k^{m+1} & \dots & \sum x_k^{2m} \end{bmatrix} \begin{bmatrix} a_0 \\ a_1 \\ \cdot \\ \cdot \\ a_m \end{bmatrix} = \begin{bmatrix} \sum y_k \\ \sum y_k x_k \\ \cdot \\ \cdot \\ \sum y_k x_k^m \end{bmatrix} \tag{7.30}$$

This is obviously a linear form in the unknown coefficients that are to be estimated; here $\sum x_k, \sum x_k^2, \dots, \sum y_k x_k^m$ are all known numerical values. The set of Equations (7.29) is often called *normal equations* for linear least-squares. To solve it, one can use a suitable method detailed in Chapter 2.[9]

This least-squares scheme is illustrated below through an example.

Example 7.7 Least-squares polynomial

Find the least-squares polynomial of degree 2,

$$f(x) = a_0 + a_1 x + a_2 x^2$$

based on the data set given in Table 7.10. Evaluate the performance of the regression polynomial.

---

9 When the number of coefficients increases, a larger amount of data points is required that leads to make the concerned matrix in Equation (7.30) ill-conditioned. Inverting this matrix is not recommended since this operation is sensitive to round-off errors. In such a situation, we need to use a computer with a reasonably high precision. This apart, there is an advanced method available, namely *singular value decomposition* (SVD). Interested readers may consult any advanced numerical book for details.

Table 7.10 Given data for least-squares polynomial

x	y
1	6.1
2	17.1
3	34.2
4	57.2
5	86.1

Solution

In this problem,

$n+1=5$ $\qquad \sum x_k = 15$ $\qquad \sum x_k^2 = 55$

$\sum x_k^3 = 225$ $\qquad \sum x_k^4 = 979$ $\qquad \sum y_k = 200.7$

$\sum y_k x_k = 802.2$ $\qquad \sum y_k x_k^2 = 3450$

From Equation (7.30),

$$\begin{bmatrix} 5 & 15 & 55 \\ 15 & 55 & 225 \\ 55 & 225 & 979 \end{bmatrix} \begin{bmatrix} a_0 \\ a_1 \\ a_2 \end{bmatrix} = \begin{bmatrix} 200.7 \\ 802.2 \\ 3450 \end{bmatrix}$$

Solving (Gauss elimination with backward sweep),

$a_0 = 0.956\ (\approx 1)$

$a_1 = 2.142\ (\approx 2)$

$a_2 = 2.978\ (\approx 3)$

So the polynomial function gets the following form:

$$f(x) = 1 + 2x + 3x^2$$

Performance evaluation of regression model

In order to quantify the performance of the regression polynomial that is constructed above to predict the data set, let us do the following statistical analysis with producing the data in Table 7.11.

Table 7.11 Data for statistical analysis

	x_k	y_k	f_k	$(y_k - \bar{y})^2$	$(f_k - y_k)^2$
	1	6.1	6	1158.72	0.01
	2	17.1	17	530.84	0.01
	3	34.2	34	35.28	0.04
	4	57.2	57	291.04	0.04
	5	86.1	86	2112.32	0.01
Σ	15	200.7		4128.20	0.11

Using Equation (A7.2), we get the standard deviation as:

$$S_d = \sqrt{\frac{4128.20}{4}} = 32.12$$

The standard error of the estimate is obtained from Equation (A7.4) as:

$$S_e = \sqrt{\frac{0.11}{5-3}} = 0.2345$$

Notice that here,
the total number of data points, $n+1=5$
the total number of unknown parameters, $m+1=3$

The coefficient of determination is obtained from Equation (A7.5) as:

$$r^2 = \frac{4128.20-0.11}{4128.20} = 0.9999$$

These results reveal that the quadratic equation (i.e., $f(x)=1+2x+3x^2$) represents an excellent fit with explaining 99.99% of the uncertainty in the original data.

7.7 Non-linear Least-Squares

Let us consider the following non-linear exponential curve:

$$y = f(x) = ae^{bx} \tag{7.22}$$

for which, the data points are given as: (x_0, y_0), (x_1, y_1), …, (x_n, y_n).

For the non-linear least-squares method, similarly we require to minimize the E:

$$E = \sum_{k=0}^{n}\left[ae^{bx_k} - y_k\right]^2 \tag{7.31}$$

Notice that here we have two unknown parameters, a and b, with respect to which, the partial derivatives of E yield:

$$\frac{\partial E}{\partial a} = 2\sum_{k=0}^{n}[ae^{bx_k} - y_k]e^{bx_k} \tag{7.32a}$$

$$\frac{\partial E}{\partial b} = 2\sum_{k=0}^{n}[ae^{bx_k} - y_k]ax_k e^{bx_k} \tag{7.32b}$$

Setting,

$$\frac{\partial E}{\partial a} = \frac{\partial E}{\partial b} = 0$$

we get the normal equations as:

$$a\sum_{k=0}^{n} e^{2bx_k} - \sum_{k=0}^{n} y_k e^{bx_k} = 0 \tag{7.33a}$$

$$a\sum_{k=0}^{n} x_k e^{2bx_k} - \sum_{k=0}^{n} x_k y_k e^{bx_k} = 0 \tag{7.33b}$$

It is obvious that the set of Equations (7.33) is non-linear in the unknowns a and b. To solve these coupled equations, one can use an iterative convergence method (e.g., Newton–Raphson approach).

This method is illustrated in the following example.

Example 7.8 Non-linear least-squares

Use the non-linear least-squares method to find the exponential fit $y = ae^{bx}$ for the data set given in Table 7.12.

Table 7.12 Data for $y = ae^{bx}$

x_k	y_k
0	1.0
1	2.0
2	3.5
3	5.0
4	7.5

Solution

For this problem, Equation (7.31) yields:

$$E = (a-1)^2 + (ae^b-2)^2 + (ae^{2b}-3.5)^2 + (ae^{3b}-5)^2 + (ae^{4b}-7.5)^2 \tag{7.34}$$

Differentiating it, we get based on Equations (7.32a) and (7.32b), respectively, as:

$$\begin{aligned}\frac{\partial E}{\partial a} &= 2[(a-1)+(ae^b-2)e^b+(ae^{2b}-3.5)e^{2b}+(ae^{3b}-5)e^{3b}+(ae^{4b}-7.5)e^{4b}] \\ &= 0 = f_1 \text{ (say)}\end{aligned} \tag{7.35a}$$

$$\begin{aligned}\frac{\partial E}{\partial b} &= 2[(ae^b-2)ae^b+(ae^{2b}-3.5)2ae^{2b}+(ae^{3b}-5)3ae^{3b}+(ae^{4b}-7.5)4ae^{4b}] \\ &= 0 = f_2 \text{ (say)}\end{aligned} \tag{7.35b}$$

Notice that both the above Equations (7.35) are non-linear. Applying the Newton–Raphson (N–R) method [Equation (3.18)],

$$\begin{bmatrix} a_{j+1} \\ b_{j+1} \end{bmatrix} = \begin{bmatrix} a_j \\ b_j \end{bmatrix} - \begin{bmatrix} \dfrac{\partial f_{1j}}{\partial a} & \dfrac{\partial f_{1j}}{\partial b} \\ \dfrac{\partial f_{2j}}{\partial a} & \dfrac{\partial f_{2j}}{\partial b} \end{bmatrix}^{-1} \begin{bmatrix} f_{1j} \\ f_{2j} \end{bmatrix} \tag{7.36}$$

where,

$$\frac{\partial f_1}{\partial a} = \frac{\partial^2 E}{\partial a^2} \qquad \frac{\partial f_1}{\partial b} = \frac{\partial^2 E}{\partial b \partial a}$$

$$\frac{\partial f_2}{\partial a} = \frac{\partial^2 E}{\partial a \partial b} \qquad \frac{\partial f_2}{\partial b} = \frac{\partial^2 E}{\partial b^2}$$

With adopting the initial guess for a and b as 1.6 and 0.38, respectively, we get the solution from Equation (7.36) as:

$a = 1.3446397639$

$b = 0.4328346202$

This gives the final form:

$$y = 1.3446397639\, e^{0.4328346202x}$$

You can find the exponential fit $y = ae^{bx}$ through data linearization for the same problem and compare with the form obtained above by using the non-linear least-squares.

Remark

It is true that the N–R method requires a good starting values for a and b, and it involves time consuming computation. Because of this, one may prefer to use a software package that has in-built minimization algorithm.

7.8 Summary and Concluding Remarks

Here we cover the fundamentals of the method of least-squares for data fitting. Based on the given data points, we learned how to fit least-squares line. For non-linear functions, both the linear and non-linear least-squares methods are used subsequently. Several relevant examples are used to illustrate these methods. This chapter is devoted not only to find the function f (with its coefficients/parameters) but also to quantify the goodness-of-fit.

Appendix 7A

A7.1 Quantification of Error and Goodness-of-Fit

To gain insight into the characteristics of the given data set, we need to do some statistical analysis. In fact, we are also curious to know how good our model is in predicting the experimental data points. But it is fairly true that figuring out the goodness of the regression model is not a trivial task. Here we will discuss these issues with knowing the statistics at the minimum level.

Arithmetic mean ($\bar{y}$)

We all know that it is the sum of the individual data points (i.e., y_k, where $k = 0, 1, \ldots, n$) divided by the total number of those points (i.e., $n+1$) as:

$$\bar{y} = \frac{1}{n+1} \sum_{k=0}^{n} y_k \tag{A7.1}$$

Standard deviation (S_d)

It is used as a *measure of spread* for a sample about the mean as:

$$S_d = \sqrt{\frac{\bar{E}}{n}} \tag{A7.2}$$

where, $\bar{E}$ is the sum of the squares of the residuals between the data points and their mean as:

$$\bar{E} = \sum_{k=0}^{n} [y_k - \bar{y}]^2 \tag{A7.3}$$

If the data set is too scattered around the mean, then the S_d will be large. In contrast, if the data points are tightly grouped, S_d will be small. One can also measure the spread by S_d^2, which is called the *variance*.

Standard error of the estimate (S_e)

It is defined by:

$$S_e = \sqrt{\frac{E(f)}{(n+1)-(m+1)}} \tag{A7.4}$$

where,

E = sum of square errors [Equation (7.6)]

$n + 1$ = total number of data points

$m + 1$ = total number of unknown parameters

Note that the denominator in Equation (A7.4) [i.e., $(n + 1) - (m + 1)$] is referred to as the *degrees of freedom*. To understand it better, let us consider a case, in which, a linear model (i.e., $f(x) = a_0 + a_1 x$) is used to fit two experimental data points. In such a case, the degree of freedom is zero since $n+1 = m+1 = 2$.

Further, let us know how $m + 1$ varies in Equation (A7.4) in different models.

Example 1: For linear least-squares regression with:

$$f(x) = a_0 + a_1 x$$

we have $m + 1 = 2$ (i.e., a_0 and a_1).

Example 2: For multiple linear regression with:

$$f(x) = a_0 + a_1 x_1 + a_2 x_2 + ... + a_m x_m$$

we have total $m + 1$ unknown parameters (i.e., a_0, a_1, …, a_m).

Example 3: For a polynomial of degree m [Equation (7.25)]:

$$f(x) = a_0 + a_1 x + a_2 x^2 + ... + a_m x^m$$

we have again $m + 1$ data-derived parameters.

The fitting is said to be *good* if S_e is about 1% of the range of y_k (i.e., the maximum value of $|y_k - y_l|$, where $k \neq l$ and $k, l = 0,\ 1,\ ....,\ n$).

Notice that the standard deviation measures the spread around the mean, whereas the standard error quantifies the spread around the regression curve.

Coefficient of determination (r^2)

It is the difference, $\bar{E} - E$, which measures the performance improvement or error reduction secured through the prediction by the regression model rather than the data described by their mean value. Now point is that the magnitude of $\bar{E} - E$ is scale-dependent and thus, it is defined with normalization as:

$$r^2 = \frac{\bar{E} - E}{\bar{E}}$$

$$= \frac{\sum_{k=0}^{n} [y_k - \bar{y}]^2 - \sum_{k=0}^{n} [f(x_k) - y_k]^2}{\sum_{k=0}^{n} [y_k - \bar{y}]^2} \quad \text{(A7.5)}$$

This coefficient of determination is popularly called the r^2-value.

From this, one can also calculate r, which is called the *correlation coefficient*. Finding r^2, one can quantify how well the model fits the data set. Like, when the regression model provides *perfect* fitting (i.e., $E = 0$), we get $r^2 = 1$. Similarly, when $r^2 = 0$ (i.e., $\bar{E} = E$), there is no error reduction achieved at all through the regression model.

A7.2 Least-Squares for Various Non-linear Functions

Table A7.1 Data linearization, and resulting variable and parameter

Model $y=f(x)$	Linearized model $\bar{y}=\bar{a}_0+\bar{a}_1\bar{x}$	Changed variable and parameter
$y=a+\frac{b}{x}$	$y=a+b\frac{1}{x}$	$\bar{y}=y,\ \bar{x}=\frac{1}{x}$ $\bar{a}_0=a,\ \bar{a}_1=b$
$y=\frac{a}{b+x}$	$y=\frac{a}{b}-\frac{1}{b}xy$	$\bar{y}=y,\ \bar{x}=xy$ $\bar{a}_0=\frac{a}{b},\ \bar{a}_1=-\frac{1}{b}$
$y=\frac{1}{a+bx}$	$\frac{1}{y}=a+bx$	$\bar{y}=\frac{1}{y},\ \bar{x}=x$ $\bar{a}_0=a,\ \bar{a}_1=b$
$y=(a+bx)^{-n}$	$y^{-\frac{1}{n}}=a+bx$	$\bar{y}=y^{-\frac{1}{n}},\ \bar{x}=x$ $\bar{a}_0=a,\ \bar{a}_1=b$
$y=\frac{x}{a+bx}$	$\frac{1}{y}=b+a\frac{1}{x}$	$\bar{y}=\frac{1}{y},\ \bar{x}=\frac{1}{x}$ $\bar{a}_0=b,\ \bar{a}_1=a$
$y=a+b\ln x$	$y=a+b\ln x$	$\bar{y}=y,\ \bar{x}=\ln x$ $\bar{a}_0=a,\ \bar{a}_1=b$
$y=ax^b$	$\ln y=\ln a+b\ln x$	$\bar{y}=\ln y,\ \bar{x}=\ln x$ $\bar{a}_0=\ln a,\ \bar{a}_1=b$
$y=a\exp(bx)$	$\ln y=\ln a+bx$	$\bar{y}=\ln y,\ \bar{x}=x$ $\bar{a}_0=\ln a,\ \bar{a}_1=b$
$y=ax\exp(-bx)$	$\ln\frac{y}{x}=\ln a-bx$	$\bar{y}=\ln\frac{y}{x},\ \bar{x}=x$ $\bar{a}_0=\ln a,\ \bar{a}_1=-b$
$y=a\exp\left(-\frac{b}{x}\right)$	$\ln y=\ln a-b\frac{1}{x}$	$\bar{y}=\ln y,\ \bar{x}=\frac{1}{x}$ $\bar{a}_0=\ln a,\ \bar{a}_1=-b$

Contd

Contd

Model $y = f(x)$	Linearized model $\bar{y} = \bar{a}_0 + \bar{a}_1\bar{x}$	Changed variable and parameter
$y = \dfrac{c}{d + a\exp(bx)}$	$\ln\left(\dfrac{c}{y} - d\right) = \ln a + bx$	$\bar{y} = \ln\left(\dfrac{c}{y} - d\right)$, $\bar{x} = x$ $\bar{a}_0 = \ln a$, $\bar{a}_1 = b$
$\ln y = a - \dfrac{b}{c + x}$ (Antoine equation)	$\ln y = a - b\dfrac{1}{c + x}$	$\bar{y} = \ln y$, $\bar{x} = \dfrac{1}{c + x}$ $\bar{a}_0 = a$, $\bar{a}_1 = -b$
$-r_A = -\dfrac{dC_A}{dt} = kC_A^{\alpha}C_B^{\beta}$	$\ln\left(-\dfrac{dC_A}{dt}\right) = \ln k + \alpha\ln C_A + \beta\ln C_B$ i.e., $\bar{y} = \bar{a}_0 + \bar{a}_1\bar{x}_1 + \bar{a}_2\bar{x}_2$	$\bar{y} = \ln\left(-\dfrac{dC_A}{dt}\right)$, $\bar{x}_1 = \ln C_A$, $\bar{x}_2 = \ln C_B$, $\bar{a}_0 = \ln k$, $\bar{a}_1 = \alpha$, $\bar{a}_2 = \beta$

EXERCISE

Review Questions

7.1 List the basic differences between:
- data regression and interpolation
- linear and non-linear least-squares

7.2 Develop the solution methodology for least-squares line by minimizing the
- average error
- root mean square (RMS) error

7.3 What are the gains of minimizing the sum of square errors over the other error criteria in least-squares method?

7.4 Draw the graphs for the following functions:

i) $y = a + \dfrac{b}{x}$ with $a = 1$ and $b = 2$

ii) $y = \dfrac{x}{a + bx}$ with $a = 2$ and $b = 3$

iii) $y = a\exp\left(-\dfrac{b}{x}\right)$ with $a = 1$ and $b = 3$

iv) $\ln y = a - \dfrac{b}{c + x}$ with $a = 1$, $b = 3$, $c = 5$

7.5 What are the merits and demerits of increasing the degree of a polynomial in data fitting?

7.6 Do you think that the non-linear least-squares leads to provide better prediction than the linear least-squares for a given data set? Justify your answer.

7.7 How to select an analytical function for least-squares based on the given data set?

Practice Problems

7.8 Develop the linear least-squares method for a parabola:

$$f(x) = a_0 + a_1x + a_2x^2$$

with minimizing the RMS error.

7.9 Find the least-squares line [Equation (7.2)] for the data points given in Table 7.1 (y vs. x).

7.10 i) Find the linear function that best fits the data set given in Table 7.13. Neglect the first data point since it seems erroneous.

Table 7.13 y versus x data points

x_k	y_k
1.0	6.0
1.5	70.5
2.0	80.7
2.5	90.9
3.0	110.3
3.5	120.2

ii) Develop the parity plot and comment on the performance of your developed function.

7.11 i) Find the least-squares line based on the data points given in Table 7.14.

Table 7.14 y versus x data points

x_k	y_k
−3	−17
−2	−14.5
−1	−8.8
0	−3
1	2.5
2	8.5
3	14

ii) Quantify the performance of the regression model through statistical analysis.

7.12 Find a_0 and a_1 for Equation (7.2) based on the data set given in Table 7.15.

Table 7.15 y versus x data points

x_k	y_k
1.1	9.4
1.5	12.5
1.9	15.2
2.3	18.0
2.8	21.8
3.2	24.3

7.13 Repeat Example 7.3 and solve it by using Equation (7.21) based on the data given in Table 7.5.

7.14 Find the best fit values of a_0 and a_1 for Equation (7.2) based on the data set given in Table 7.16.

Table 7.16 *y* versus *x* data points

x_k	y_k
5	−4
10	−11.752
15	−19.325
20	−27.120
25	−34.685
30	−42.35

7.15 Find the best fit values of *a* and *b* for Equation (7.16) based on the data set given in Table 7.17.

Table 7.17 *y* versus *x* data points

x_k	y_k
−2	−640
−1	−5
1	5
2	640
3	10935
4	81920

7.16 Find the best power fit through the data given in Table 7.18.

Table 7.18 Homework versus grades data

Homework	Grades
1	60
2	70.5
3	75.7
4	70.9
5	80.3
6	52.2
7	96

7.17 i) Find the power fit $y = ax^b$ for the data set given in Table 7.19.

Table 7.19 *y* versus *x* data points

x_k	y_k
0.1	500
0.3	55.56
0.8	7.813
1.2	3.47
1.5	2.22
2.0	1.22

ii) Evaluate the goodness of the regression model by calculating r^2-value.

7.18 Fit the exponential curve $y = ae^{bx}$ using the data given in Table 7.20.

Table 7.20 Day time versus temperature data

Day time	Temperature (°C)
7	15
9	16
11	18
13	20
15	19
17	17
19	15

7.19 The following form of equation relates the pressure (*P*) and volume (*V*) of a gas:

$$P\,V^{\gamma} = C$$

in which, γ and *C* are constants. Find the values of these two constants based on the data given in Table 7.21.

Table 7.21 Given data for $PV^{\gamma} = C$

P (kg/cm²)	V (L)
0.5	1.5
1.0	1.0
1.5	0.7
2.0	0.6
2.5	0.5

7.20 Consider the following form of equation (parabola):

$$y = a_0 + a_1 x + a_2 x^2$$

Find the values of a_0, a_1 and a_2 based on the data given in Table 7.22. How good is this model?

Table 7.22 Given data for $y = a_0 + a_1x + a_2x^2$

x_k	y_k
1	8.8
2	24.1
3	47
4	78.4
5	117

7.21 Use the data given in Table 7.22 to find the parameters of the following polynomial:

$$y = a_0 + a_1x + a_2x^2 + a_3x^3$$

7.22 Arrhenius law is used to determine the reaction rate constant, k that varies with temperature, T as:

$$k = k_0 \exp\left(-\frac{E}{RT}\right)$$

where, the universal gas constant $R = 8.314$ J/gmol.K. Find the pre-exponential factor, k_0 and activation energy, E based on the data given in Table 7.23.

Table 7.23 Rate constant versus temperature data

T (K)	k (min^{-1})
312	0.025
320	0.060
325	0.10
330	0.215
333	0.40

7.23 Number of iterations (n) in the Bisection root finding method varies with the desired tolerance limit (tol) as given in Table 7.24. Find a suitable least-squares approximation for this data set.

Table 7.24 Number of iterations versus tol in Bisection method

tol	n
10^{-2}	7
10^{-3}	10
10^{-4}	14
10^{-5}	17
10^{-6}	20
10^{-7}	24
10^{-8}	27

7.24 For a biochemical reaction, the Monod model is used to estimate the specific growth rate (μ):

$$\mu = \frac{\mu_{max} s}{k_s + s}$$

in which,

s = substrate concentration

μ_{max}, k_s = constant parameters

The above equation can be written as:

$$\frac{1}{\mu} = \frac{1}{\mu_{max}} + \frac{k_s}{\mu_{max}} \frac{1}{s}$$

i) Using the linear least-squares method, find k_s and μ_{max} based on the data given in Table 7.25.

ii) Evaluate the performance of the regression model by calculating r^2-value.

Table 7.25 μ versus s data

s (gm/L)	μ (hr^{-1})
0.1	0.25
0.15	0.31
0.25	0.36
0.5	0.43
0.75	0.45
1.0	0.47
1.5	0.50
3.0	0.52

7.25 Find the least-squares polynomial of degree 4 in time, t:

$$y = a_0 + a_1 t + a_2 t^2 + a_3 t^3 + a_4 t^4$$

that best fits the data given in Table 7.26.

Table 7.26 Given data for a polynomial of degree 4

t	y
0.0	1.31
2.5	1.0
5.0	0.78
10.0	0.51
15.0	0.37
20.0	0.21
25.0	0.17

7.26 The gas phase decomposition of di-tert-butyl peroxide is carried out in an isothermal, constant volume batch reactor. It is modeled as:[10]

$$\frac{dP}{dt} = k(3P_0 - P)^n$$

The pressure is recorded at different times during the reaction as in Table 7.27. Find the order of the reaction, n.

TABLE 7.27 Given data for the batch reactor

t (min)	P (mm Hg)
0.0	7.5 (= P_0)
2.5	10.5
5.0	12.5
10.0	15.8
15.0	17.9
20.0	19.4

Hint: Integrate the model equation with limits $P = P_0$ when $t = t_0$. With doing this, for example, if a plot of $\ln\frac{2P_0}{3P_0 - P}$ versus t becomes linear, then the reaction is first-order.

7.27 Fit the following form:

$$\ln P^s = A - \frac{B}{T + C}$$

for the data given in Table 7.28 for acetone. Here,

P^s = vapor pressure

T = temperature

A, B, C = constants

and C is given as −35.93.

TABLE 7.28 Vapor pressure data for acetone

T (K)	P^s (mm Hg)
250	18.4548
260	34.0676
280	99.8522
300	248.6748
320	544.6367

10 Trotman–Dickenson, A. F. (1969). *J. Chem. Edu.*, **46**, 396.

7.28 i) Find the power fit $T = ax^b$ for the data given in Table 7.29.

TABLE 7.29 Given data

x	T
58	88
108	225
150	365
228	687

ii) Find the exponential fit $T = ae^{bx}$ and compare the results with those of power fit performing statistical analysis.

7.29 The heat capacity (C_p) of chloroform is estimated from:

$$C_p = a_0 + a_1T + a_2T^2 + a_3T^3$$

Find the coefficients of this polynomial using the data given in Table 7.30.

TABLE 7.30 Heat capacity data

T (K)	C_p (J/mol.K)
300	66.01839
310	66.9742
330	68.8128
350	70.5569
380	73.0028

7.30 Determine the curve fit:

$$y = ax\exp(-bx)$$

for the data given in Table 7.31.

TABLE 7.31 Given data points

x	y
1	0.170268
1.5	6.2981e-2
2.5	6.3832e-3
4.0	1.53151e-4
5.0	1.16414e-5
6.0	8.49497e-7

7.31 Determine the curve fit:

$$y = (a+bx)^{-1.3}$$

for the data given in Table 7.32 through data linearization.

Table 7.32 Given data points

x	y
–1	1.0
1.2	0.111657
2.3	7.16048e-2
3.4	5.14525e-2
4.5	3.95425e-2
5.6	3.177058e-2

7.32 Find the least-squares fit:

$$y = \frac{2}{1.5 + ae^{bx}}$$

for the data given in Table 7.33 through data linearization.

Table 7.33 Given data points

x	y
0	0.8
1	0.63518
2	0.4741267
3	0.334354
4	0.224996

7.33 Determine the least-squares fit:

$$y = a + b\ln x$$

for the data given in Table 7.34 through data linearization.

Table 7.34 Given data points

x	y
1	2.3
5	4.67869e-2
10	–0.923619
20	–1.894025
50	–3.176832

7.34 The following glucose levels are recorded for a non-diabetic patient over 50 minutes in Table 7.35.

Table 7.35 Glucose level versus time

Time	Glucose level (fasting)
0	100
10	103
20	101
30	90
40	85
50	88

Find a suitable function with its parameters.

7.35 Life expectancy in India from 1881 to 1950 is recorded in Table 7.36.

Table 7.36 Year-wise data on life expectancy in India

Year	Life expectancy
1881	25.4
1891	24.3
1901	23.5
1905	24
1911	23.2
1915	24
1921	24.9
1925	27.6
1931	29.3
1935	31
1941	32.6
1950	35.4

Construct a suitable function that can predict the life expectancy data precisely.

7.36 Find the exponential fit:

$$y = a\exp(bx)$$

for the data given in Table 7.37 by the use of:

i) linear least-squares
ii) non-linear least-squares

Table 7.37 Given data points

x	y
0	1.5
1	2.5
2	3.5
3	5.1
4	7.2

7.37 Find the exponential fit:

$$y = a\exp(bx)$$

for the data given in Table 7.20 by the use of non-linear least-squares.

7.38 i) Find the exponential fit:

$$y = a\exp(bx)$$

for the experimental data given in Table 7.38 using the:

a) linear least-squares method
b) non-linear least-squares method

ii) Which method leads to provide better prediction?

TABLE 7.38 Given data points

x	y
–5	–2.28
–4	–2.13
6	2.37
7	3.45
8	4.74

7.39 Using the non-linear least-squares approach (with N–R method), find the exponential fit:

$$C = a\exp(bt)$$

for the experimental data given in Table 7.39. These data points are produced as concentration (C) versus time (t) for the photo-degradation of aqueous bromine (Chapra, 2012).

TABLE 7.39 Given data points

Time (min)	Concentration (ppm)
10	3.4
20	2.6
30	1.6
40	1.3
50	1.0
60	0.5

7.40 Find the power fit:

$$F = a\upsilon^b$$

for the data given in Table 7.40. These data points of force (F) versus velocity (υ) are specific to a wind tunnel (Chapra, 2012).

Table 7.40 Given data points

v (m/sec)	F (N)
10	25
20	70
30	380
40	550
50	610
60	1220
70	830
80	1450

7.41 There was a pandemic declared in 2020 by the World Health Organization (WHO) caused by the infectious Corona virus (COVID-19). The total cases from February 5 to March 10 in 2020 are reported in Table 7.41.

Table 7.41 Total worldwide Corona cases

Day	Corona cases
1	28266
5	40553
10	67100
15	75700
20	80087
25	86604
30	98425
35	118948

Using a suitable least-squares method, construct a function that can precisely predict the reported data.

7.42 Table 7.42 lists the data of the distance of nine planets from the sun versus their sidereal period (Mathews, 2000).

Table 7.42 Given data

Planet	Distance (km × 10^{-6})	Sidereal period (days)
Mercury	57.59	87.99
Venus	108.11	224.70
Earth	149.57	365.26
Mars	227.84	686.98
Jupiter	778.14	4332.4
Saturn	1427.00	10759
Uranus	2870.30	30684
Neptune	4499.90	60188
Pluto	5909.00	90710

Using a suitable least-squares method, fit a function that can precisely predict the reported data.

7.43 Using the non-linear least-squares approach (with the N–R method), find the exponential fit:

$$y = a\exp(bx)$$

for the population data of India given in Table 7.43.

Table 7.43 Population data of India

Year	Population (crores)
1961	43.92
1971	54.82
1981	68.33
1991	84.64
2001	102.87
2011	121.02

7.44 Table 7.44 reports the reaction rate ($-r_A$) versus concentration (C_A) data.

Table 7.44 $-r_A$ versus C_A data

C_A (mol/dm³)	$-r_A$ (mol/cm².sec)×10^7
0.1	0.36
0.5	0.74
1.0	1.25
2.0	1.32
4.0	2.07

Find the order of the reaction (n) using:

$$-r_A = k\, C_A^n$$

and the reaction rate constant, k.

7.45 Absorbance, A is related with wave number, m as:

$$A = a_0 + a_1 m + a_2 m^2 + a_3 m^3$$

Finding all coefficients based on the data given in Table 7.45, estimate the absorbance at $m = 1000$.

Table 7.45 Absorbance versus wave number data

Wave number	Absorbance
803.20	0.1581
828.33	0.0428
847.61	0.0048
871.75	0.0072
891.04	0.0437
893.90	0.0629
901.61	0.1201

Evaluate the goodness of the regression model by calculating r^2-value.

7.46 The reaction kinetics of polymer degradation is modeled as:

$$-\ln(1-\alpha) = k_0 \frac{T - T_0}{b} \exp\left(-\frac{E}{RT}\right)$$

where,

α = reaction conversion

k_0 = pre-exponential factor, min^{-1}

T = reaction temperature, K

T_0 = activation temperature = 340 K

b = heating rate = 10 K/min

E = activation energy, J/mol

R = universal gas constant = 8.314 J/mol.K

Using a suitable least-squares technique, find k_0 and E based on the data given in Table 7.46.

Table 7.46 Temperature (T) versus conversion (α) data

T (K)	α
360	0.1050
370	0.2010
380	0.3422
390	0.5145
400	0.6756
410	0.8025
420	0.8925
430	0.9550
440	0.9700

References

Bequette, B. W. (1998). *Process Dynamics, Modeling, Analysis, and Simulation*, 1st ed., Upper Saddle River, New Jersey: Prentice-Hall.

Chapra, S. C. (2012). *Applied Numerical Methods with MATLAB for Engineers and Scientists*, 3rd ed., New Delhi: Tata McGraw-Hill.

Mathews, J. H. (2000). *Numerical Methods for Mathematics, Science, and Engineering*, 2nd ed., New Delhi: Prentice-Hall.

8

Polynomial Interpolation

"The essence of mathematics is not to make simple things complicated, but to make complicated things simple." – S. Gudder

Key Learning Objectives

- Understanding the basics of interpolating polynomial
- Knowing a couple of commonly used polynomial interpolation approaches
- Learning how to estimate the errors (truncation errors)
- Knowing the piecewise polynomial interpolation

8.1 Introduction

In the last chapter, we have learned how to fit an analytical function, f based on the available y versus x data set. The *least-squares curve fitting* method was used there for this purpose to know the overall trend of our experimental observations. It reveals that the function is not necessarily going through any of the data points and it is expected so, since the data are not precise enough. Now here we would like to cover various techniques of *interpolation* to find the function that should pass through all of the data points with the presumption that the data are known to be very precise.[1]

For the given (tabulated) data points, (x_0, y_0), (x_1, y_1), ..., (x_m, y_m), the value of y at x is called:

- Interpolated value when $x_0 < x < x_m$
- Extrapolated value when either $x < x_0$ or $x > x_m$

Basically, we need to approximate y at non-tabulated abscissas over the interval $[x_0, x_m]$ for interpolation and outside that interval for extrapolation.

Let us consider a non-linear equation (linear in the parameters, a's) that we would like to use to predict our data points:

1 For example, heat capacity of methanol is precisely available against temperature.

$$y = f(x) = a_0 p_0(x) + a_1 p_1(x) + \ldots\ldots.. + a_m p_m(x) \qquad (8.1)$$

where, for example:

$p_i(x) = x^i$ for polynomial expansion

$p_i(x) = \sin(i\,x)$ or $\cos(i\,x)$ for Fourier expansion

$p_i(x) = \exp(\alpha_i\,x)$ for exponential function

Depending on the given data points, one can choose the polynomial, periodic (sine) or exponential function. However, here our interest is confined to polynomial function.

Now to make the above model Equation (8.1) ready for use, one needs to find the values of all coefficients/parameters, $a_0, a_1, \ldots.., a_m$. Accordingly, we fix our objective as the construction of the function $f(x)$ as a polynomial of minimum degree. It is fairly true that as m in the polynomial form of Equation (8.1) increases, the difficulty level in finding the associated parameters values gets intensified.

Before leaving the *introduction* section, let us briefly highlight the basic differences between the least-squares and polynomial interpolation in Table 8.1.

TABLE **8.1** Basic differences between least-squares and polynomial interpolation

Least-squares	**Polynomial interpolation**
1. We are free to adopt any form of analytical function (including polynomial) suitable for data fitting.	1. The interpolating polynomial is unique although it is obtained in various ways under different interpolation schemes.
2. It involves minimization of error for finding the coefficients/parameters.	2. No such minimization is involved here.
3. Number of data points is larger than the number of unknown parameters.	3. Number of data points = number of unknown parameters.
4. The fitted function does not necessarily go through any of the data points.	4. The fitted polynomial passes through all of the data points exactly.
5. This is used when the available data set is imprecise.	5. It is suitable when the available data set is precise enough.

8.2 Polynomial Interpolation and Various Approaches

The most common choice of interpolants is the polynomials because they have simple structures, and are easy to evaluate, differentiate and integrate. We are thus curious to learn the polynomial interpolation for various engineering and scientific problems.

Let us consider a polynomial of degree m, which has the following *power form* obtained from Equation (8.1):

$$y = f(x) = a_0 + a_1 x + a_2 x^2 + \ldots\ldots.. + a_m x^m \qquad (7.25)$$

To fit such a polynomial that has total $m+1$ unknown coefficients (i.e., a_0, a_1, ..., a_m), one needs a set of $m+1$ data points. With this, one can have a completely specified system [i.e., degrees of freedom = (number of data points – number of unknown parameters) = 0], thus guaranteeing a unique solution that fits the data exactly, rather than approximately. As stated, we would like to have this typical fitting through polynomial interpolation with that the available data points are known to a high degree of accuracy.

Let $y = f(x)$ be a real-valued function defined on some interval $[a,b]$ that contains $m+1$ distinct points, $x_0, x_1, ..., x_m$ (i.e., $a = x_1 < x_2 < .... < x_m = b$). If a polynomial of degree $\leq m$, namely $T(x)$ provides the same value with the function $f(x)$ at those $m+1$ points, then we can write $T(x_i) = f(x_i)$, $i = 0, 1,...., m$ (i.e., called the *unique polynomial*). It reveals that there exists a unique polynomial of degree $\leq m$, when it passes through $m+1$ data points [i.e., $(x_i, f(x_i))$, $i = 0, 1,...., m$]. Based on this definition, this unique polynomial is referred as the *interpolating polynomial.* Note that there are infinitely many polynomials of degree $> m$ that pass through those $m+1$ data points.

Methods of Polynomial Interpolation

There are a wide variety of forms available for representing an interpolating polynomial beyond the familiar format of Equation (7.25). With this, there are various schemes of polynomial interpolation developed. Here, a few important ones we would like to cover:

- Newton polynomial interpolation
  - Newton forward divided difference
  - Newton backward divided difference
- Polynomial interpolation with equally spaced data
  - Newton–Gregory forward difference interpolation
  - Newton–Gregory backward difference interpolation
- Lagrange polynomial interpolation
- Piecewise polynomial interpolation by spline functions
  - Linear spline
  - Quadratic spline
  - Cubic spline

Let us now pay our attention to learn these interpolation schemes one-by-one.

8.3 Newton Polynomial Interpolation

Let us start the Newton divided-difference polynomial interpolation (or shortly, Newton polynomial interpolation) with its first- and second-order versions[2] to understand the basic structure, and then we will extend it to its general form.

8.3.1 Linear Interpolation

Linear interpolation is the simplest type of interpolation scheme between two data points. In this, we would like to find a linear function (i.e., a straight line):

$$y = a_0 + a_1 x \tag{7.2}$$

2 Sometimes the *order* term is used to mean the *degree* of a polynomial.

For this first-order polynomial, two given data points, namely (x_0, y_0) and (x_1, y_1), are to be used to find the unknowns, a_0 and a_1. Accordingly, we develop the following matrix form based on Equation (7.2):

$$\begin{bmatrix} 1 & x_0 \\ 1 & x_1 \end{bmatrix} \begin{bmatrix} a_0 \\ a_1 \end{bmatrix} = \begin{bmatrix} y_0 \\ y_1 \end{bmatrix} \tag{8.2}$$

Finding the values of a_0 and a_1, one can use Equation (7.2) to estimate the intermediate values of y between the known discrete data points through interpolation. This linear technique is illustrated in Figure 8.1, showing how to find the dependent variable y against independent variable x (where $x_0 < x < x_1$).

Remarks

1. Computation of the coefficients, a_0 and a_1, requires solving the system of linear Equations (8.2). Similarly for Equation (7.25), we need to find all coefficients, $a_0, a_1, \ldots, a_m$, by solving a corresponding linear system represented by, say $Aa = Y$. This matrix A is called the *Vandermonde* matrix for the points $x_0, x_1, \ldots, x_m$.
2. Since $x_0, x_1, \ldots, x_m$ in A matrix are distinct, it is invertible (i.e., $|A| \neq 0$), which ensures that the linear system $Aa = Y$ has a unique solution.
3. Further note that when the points, $x_0, x_1, \ldots, x_m$, are close together, the Vandermonde matrix A becomes ill-conditioned. That is, the solution of $Aa = Y$ is very sensitive to round-off errors. We will illustrate this issue a little latter. Now, to overcome this shortcoming, alternative approaches, namely the Newton and Lagrange polynomial interpolation, are presented.

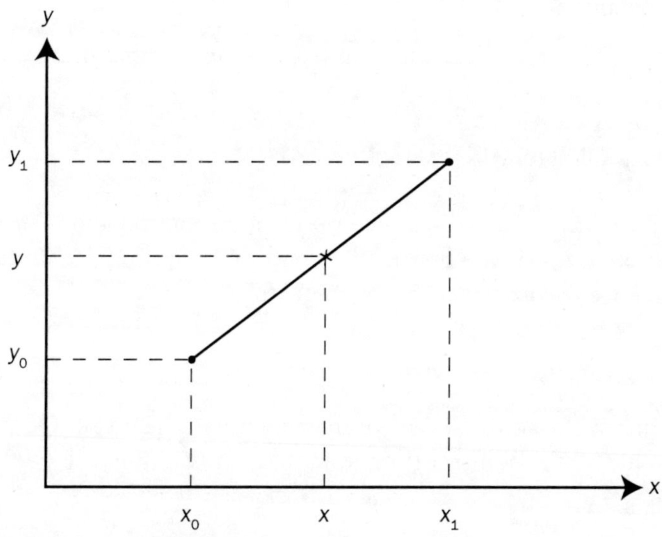

FIGURE 8.1 Linear interpolation

An alternative approach: Newton linear interpolation

Let us replace Equation (7.2) by the following interpolant:

$$y = a_0 + a_1(x - x_0) \tag{8.3}$$

This also involves two unknown coefficients, a_0 and a_1, which can be found out based on the given data points, (x_0, y_0) and (x_1, y_1). Let us do this job below.

Finding a_0

To find a_0, we substitute $x = x_0$ (i.e., $y = y_0$) in Equation (8.3) and get:

$$y_0 = a_0 \tag{8.4}$$

Accordingly, Equation (8.3) gets the following form:

$$y = y_0 + a_1(x - x_0) \tag{8.5}$$

Finding a_1

Similarly, we evaluate Equation (8.5) at $x = x_1$:

$$y_1 = y_0 + a_1(x_1 - x_0)$$

Rearranging,

$$a_1 = \frac{y_1 - y_0}{x_1 - x_0} \tag{8.6}$$

Plugging in Equation (8.5),

$$y = y_0 + \left[\frac{y_1 - y_0}{x_1 - x_0}\right](x - x_0) \tag{8.7}$$

This is the *Newton linear interpolation* formula in the interval (x_0, x_1). Outside this interval, this formula one can also use for linear extrapolation.

The term in the third bracket in Equation (8.7) is an approximation for the first derivative of the linear function as seen in Chapter 5 (boundary value problems). In the context of interpolation, it is referred as the *first divided difference* over the interval $x \in [x_0, x_1]$. On the other way, it is nothing but the slope between x_0 and x_1.

Geometric interpretation

In Figure 8.1, a straight line is drawn by connecting the two data points. Using this plot, one can then easily find y at some intermediate x. Theoretically, equating the two slopes (one from x_0 to x and other one from x_0 to x_1), we have:

$$\frac{y - y_0}{x - x_0} = \frac{y_1 - y_0}{x_1 - x_0} \tag{8.8}$$

Rearranging this equation for unknown y, one gets Equation (8.7) for the use at some intermediate x.

Remarks

1. We can have better approximation through linear interpolation if the interval between the data points is smaller. But the issue is that for the reported data points, we do not have any scope of changing the interval. Thus, in such a case, any improvement in approximation remains beyond our hand.
   Again the accuracy of approximation depends on how close we are [i.e., point (x, y)] to the given data points (x_0, y_0) and (x_1, y_1).
2. It is fairly true that the accuracy of linear interpolation depends on how straight/curved the function is originally. To improve the accuracy level, particularly for non-linear functions, the quadratic interpolation technique gets preference and it is discussed in the next section. But prior to that, we would like to know the answer of the following question:

Why Equation (8.3) [$y = a_0 + a_1(x - x_0)$] instead of (7.2) [$y = a_0 + a_1x$]?

Let us take a simple example with:

$$f(1110) = -\frac{6}{7} \text{ and } f(1111) = \frac{1}{7}$$

Using five significant digits floating point rounding arithmetic, we get from Equation (7.2):

$$a_0 + 1110a_1 = -0.85714$$

$$a_0 + 1111a_1 = 0.14286$$

Solving these equations, polynomial Equation (7.2) gets the following form:

$$y = -1110.8 + x$$

using which, we get:

$$f(1110) = -0.8 \qquad \left[\text{actually } -\frac{6}{7}\ (= -0.85714)\right]$$

$$f(1111) = 0.2 \qquad \left[\text{actually } \frac{1}{7}\ (= 0.14286)\right]$$

It reveals that the estimates are correct up to one significant digit only (i.e., loss of four significant digits).

To overcome this problem, let us now use Equation (8.3) as:

$$y = a_0 + a_1(x - 1110)$$

For this, we can easily find:

$$a_0 = -0.85714 \qquad \text{[from Equation (8.4)]}$$

$$a_1 = 1 \qquad \text{[from Equation (8.6)]}$$

Accordingly, we have:

$$y = -0.85714 + (x - 1110)$$

using which, we get:

$$f(1110) = -0.85714 \qquad \left[\text{actually } -\frac{6}{7} \ (= -0.85714)\right]$$

$$f(1111) = 0.14286 \qquad \left[\text{actually } \frac{1}{7} \ (= 0.14286)\right]$$

Obviously, the estimates are correct up to five significant digits (i.e., there is no loss of any significant digit). This is the advantage of using Equation (8.3) (correct up to five significant digits) over Equation (7.2) (correct up to one significant digit only). Notice that in Equation (7.2), the data set is analyzed from the origin at x = 0, whereas the origin is at x = 1110 in Equation (8.3); it leads to a significant improvement in approximation. Through this example, it becomes obvious that the solution of Equation (8.2), which is the matrix form of Equation (7.2), is very sensitive to round-off errors.

Another advantage of using Equation (8.3) is the ease of finding the coefficients, *a*'s, which is shown before for linear interpolation and will be shown later for quadratic interpolation.

Example 8.1 Linear interpolation

For compressed liquid water (pressure = 25 bar), data are given in Table 8.2.

TABLE 8.2 Given data

T (°C)	*u* (kJ/kg)
20	83.80
80	334.29

Find *T* at *u* = 167.30 kJ/kg through linear interpolation.

Solution

From Equation (8.7),

$$167.30 = 83.80 + \frac{334.29 - 83.80}{80 - 20}(T - 20)$$

Solving it, we get:

T = 40 °C

8.3.2 Quadratic Interpolation

Let us consider the following quadratic polynomial:

$$y = a_0 + a_1 x + a_2 x^2 \tag{8.9}$$

For this second-order polynomial, the three data points, namely (x_0, y_0), (x_1, y_1) and (x_2, y_2), are available and it is illustrated in Figure 8.2.

To find the unknowns, a_0, a_1 and a_2, accordingly we have:

$$\begin{bmatrix} 1 & x_0 & x_0^2 \\ 1 & x_1 & x_1^2 \\ 1 & x_2 & x_2^2 \end{bmatrix} \begin{bmatrix} a_0 \\ a_1 \\ a_2 \end{bmatrix} = \begin{bmatrix} y_0 \\ y_1 \\ y_2 \end{bmatrix} \tag{8.10}$$

Having the values of a_0, a_1 and a_2, one can use Equation (8.9) to estimate y at some intermediate x [i.e., (x, y)] through interpolation as shown in Figure 8.2.

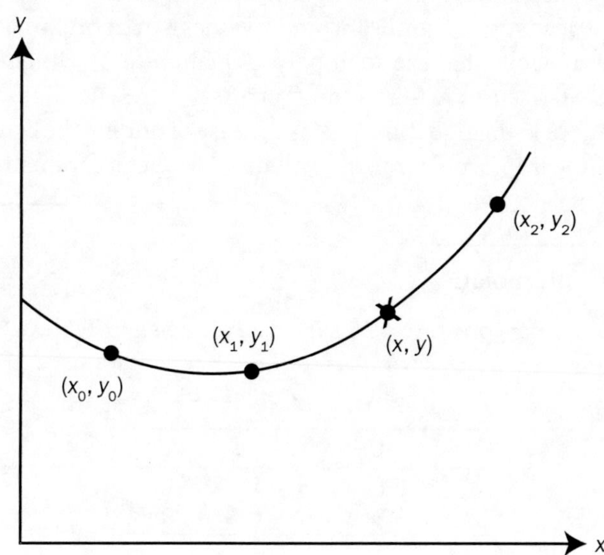

FIGURE 8.2 Quadratic interpolation

Remark

Here the known A matrix in Equation (8.10) is called the *Vandermonde matrix*, which is, as stated, very ill-conditioned. As shown in case of linear interpolation, similarly the solution of this Equation (8.10) in quadratic interpolation is also very sensitive to round-off errors. Thus, we move to the following alternative approach under the framework of Newton interpolation that does not manifest this shortcoming.

An alternative approach: Newton quadratic interpolation

Let us use the following interpolant:

$$y = a_0 + a_1(x - x_0) + a_2(x - x_0)(x - x_1) \tag{8.11}$$

For this, we need to determine the three unknown coefficients, a_0, a_1 and a_2, based on a set of three data points. The procedure is shown below.

Finding a_0

To find a_0, we substitute $x = x_0$ (i.e., $y = y_0$) in Equation (8.11) and get:

$$y_0 = a_0$$

Accordingly, Equation (8.11) yields:

$$y = y_0 + a_1(x - x_0) + a_2(x - x_0)(x - x_1) \tag{8.12}$$

Finding a_1

Similarly, evaluating Equation (8.12) at $x = x_1$:

$$y_1 = y_0 + a_1(x_1 - x_0)$$

Rearranging,

$$a_1 = \frac{y_1 - y_0}{x_1 - x_0}$$

Accordingly, Equation (8.12) gets the following form,

$$y = y_0 + \left[\frac{y_1 - y_0}{x_1 - x_0}\right](x - x_0) + a_2(x - x_0)(x - x_1) \tag{8.13}$$

Finding a_2

Evaluating Equation (8.13) at $x = x_2$:

$$y_2 = y_0 + \left[\frac{y_1 - y_0}{x_1 - x_0}\right](x_2 - x_0) + a_2(x_2 - x_0)(x_2 - x_1)$$

Rearranging,

$$a_2 = \frac{1}{x_2 - x_1}\left[\frac{y_2 - y_0}{x_2 - x_0} - \frac{y_1 - y_0}{x_1 - x_0}\right] \tag{8.14}$$

Alternatively,

$$a_2 = \frac{1}{x_2 - x_0}\left[\frac{y_2 - y_0}{x_2 - x_1} - \frac{(y_1 - y_0)(x_2 - x_0)}{(x_1 - x_0)(x_2 - x_1)}\right]$$

Considering,

$$x_2 - x_0 = (x_2 - x_1) + (x_1 - x_0)$$

we get,

$$a_2 = \frac{1}{x_2 - x_0}\left[\frac{y_2 - y_1}{x_2 - x_1} - \frac{y_1 - y_0}{x_1 - x_0}\right] \tag{8.15}$$

where,

$\dfrac{y_2 - y_1}{x_2 - x_1}$ = divided difference of the first derivative over the interval $x \in [x_1, x_2]$

$\dfrac{y_1 - y_0}{x_1 - x_0}$ = divided difference of the first derivative over the interval $x \in [x_0, x_1]$

Final expression

Plugging Equation (8.15) into Equation (8.13), we get:

$$y = y_0 + \left[\frac{y_1 - y_0}{x_1 - x_0}\right](x - x_0) + \frac{1}{x_2 - x_0}\left[\frac{y_2 - y_1}{x_2 - x_1} - \frac{y_1 - y_0}{x_1 - x_0}\right](x - x_0)(x - x_1) \tag{8.16}$$

This is the formula for the *Newton quadratic interpolation*. Notice that the first two right hand terms in Equation (8.16) are equivalent to linear interpolation [see Equation (8.7)] and the last term represents the second-order curvature in the formula.

Example 8.2 Quadratic interpolation

i) Construct the quadratic polynomial that passes through the unequally spaced data points (1, 1), (2, 1) and (4, 2).
ii) Using the fitted polynomial, find $y(3)$.

Solution

i) From Equation (8.16),

$$\begin{aligned} y &= 1 + 0 + \frac{1}{4-1}\left[\frac{2-1}{4-2} - 0\right](x-1)(x-2) \\ &= 1 + \frac{1}{6}(x^2 - 3x + 2) \\ &= \frac{1}{6}x^2 - \frac{1}{2}x + \frac{4}{3} \end{aligned}$$

This is the quadratic polynomial we obtained.

ii) Using this polynomial, we get:

$$\begin{aligned} y(3) &= \frac{1}{6}(3)^2 - \frac{1}{2}\times 3 + \frac{4}{3} \\ &= 1.333 \end{aligned}$$

Figure 8.3 depicts the point (3, 1.333), whose y-coordinate is approximated by the use of the fitted quadratic polynomial.

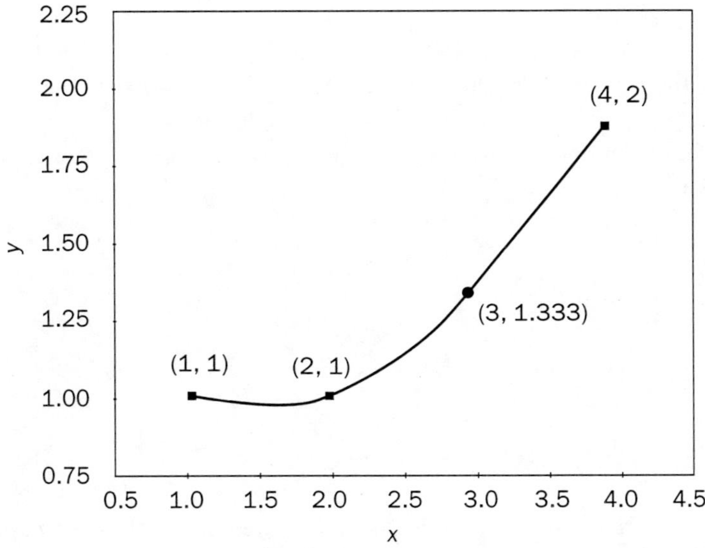

FIGURE 8.3 Fitted quadratic polynomial used for interpolation at the point (3, 1.333)

8.3.3 General Form of Newton Polynomial Interpolation

For higher-order interpolation, the formula derived in the preceding sections for the linear and quadratic interpolation can be extended. Following their pattern, one can have a general form of mth-order interpolant as:

$$y(x)=a_0+a_1(x-x_0)+a_2(x-x_0)(x-x_1)+.....+a_m(x-x_0)(x-x_1)...(x-x_{m-1})+R_m(\xi) \tag{8.17}$$

This is the general form of the *Newton*[3] *polynomial* that is constituted with the following *basis functions*[4] $\{N_j(x)\}_{j=0}^{m}$, where:

$$N_0(x)=1 \tag{8.18a}$$

$$N_j(x)=\prod_{i=0}^{j-1}(x-x_i) \tag{8.18b}$$

for $j=1, 2, ..., m$. In the Newton polynomial Equation (8.17), there are m *centers*,[5] namely $x_0, x_1, ..., x_{m-1}$. In this mth-order interpolant, $R_m(\xi)$ is called the *remainder* (i.e., truncation error), where, $x_0 \le \xi \le x_m$.

3 Named after the English mathematician, physicist, astronomer, theologian and author Sir Isaac Newton, PRS (1643–1727).

4 They are sometimes called as the *Newton basis polynomials* or the *Newton fundamental polynomials*.

5 The numbers, $x_0, x_1, \ldots, x_{m-1}$, which are subtracted from x in the linear factors that constitute the basis functions [Equation (8.18b)] are called the *centers* of the Newton polynomial.

Note that Equation (8.17) has total $(m+1)$ unknown coefficients (i.e., $a_0, a_1, \ldots, a_m$) and to estimate them, we need $(m+1)$ data points [i.e., $(x_0, y_0), (x_1, y_1), \ldots, (x_m, y_m)$] with assuming the remainder equals zero [i.e., $R_m(\xi)=0$]. We will further discuss this issue related to R_m later.

With the known data points, we use the following equations to estimate the coefficients of Equation (8.17):

$$\begin{aligned} a_0 &= y[x_0] \\ a_1 &= y[x_1, x_0] \\ a_2 &= y[x_2, x_1, x_0] \\ a_3 &= y[x_3, x_2, x_1, x_0] \\ &\vdots \\ a_m &= y[x_m, x_{m-1}, \ldots, x_1, x_0] \end{aligned} \tag{8.19}$$

Here the $y[]$ terms are often called *finite divided differences* (or simply, *divided differences*). Further we define them as:

$$\begin{aligned} y[x_0] &= y_0 \\ y[x_1, x_0] &= \frac{y_1 - y_0}{x_1 - x_0} \\ y[x_2, x_1, x_0] &= \frac{y[x_2, x_1] - y[x_1, x_0]}{x_2 - x_0} \\ &\vdots \\ y[x_m, x_{m-1}, \ldots, x_0] &= \frac{y[x_m, x_{m-1}, \ldots, x_1] - y[x_{m-1}, x_{m-2}, \ldots, x_0]}{x_m - x_0} \end{aligned} \tag{8.20}$$

These are called, respectively, 0th, 1st, 2nd, …, mth-order divided differences. Notice that the divided differences in Equations (8.20) are recursive in that the higher-order differences are determined by taking differences of lower-order differences.

It is worth emphasizing that

$$\begin{aligned} y[x_1, x_0] &= y[x_0, x_1] \\ y[x_2, x_1, x_0] &= y[x_0, x_1, x_2] \end{aligned}$$

and so on.

With this background study, we now discuss the Newton polynomial interpolation scheme with forward and backward divided difference.

Newton Forward Divided Difference

Inserting Equation (8.19) into Equation (8.17), we get:

$$y(x) = y[x_0] + y[x_1, x_0](x - x_0) + \ldots + y[x_m, x_{m-1}, \ldots, x_0](x - x_0)(x - x_1)\ldots(x - x_{m-1}) \tag{8.21}$$

This represents the *Newton divided difference* formula, more specifically, the *Newton forward divided difference formula*. Here, we consider the remainder,

$$R_m(\xi)=y\,[\xi,\ x_m,\ x_{m-1},\ \ldots,\ x_0](x-x_0)(x-x_1)\ldots(x-x_m) \qquad x_0\leq\xi\leq x_m \tag{8.22}$$

equal to 0.

Rewriting Equation (8.21) in compact form,

$$y(x)=\sum_{j=0}^{m}\left\{y\,[x_j,x_{j-1},\ldots,\ x_0]\prod_{i=0}^{j-1}(x-x_i)\right\} \tag{8.23}$$

Using the basis functions [Equation (8.18)], it gives:

$$y(x)=\sum_{j=0}^{m}a_jN_j(x)=\sum_{j=0}^{m}\{y\,[x_j,x_{j-1},\ldots,\ x_0]\ N_j(x)\}$$

In a sample case with $m=1$, this forward divided difference formula yields:

$$y(x)=y[x_0]+y[x_1,x_0](x-x_0) \tag{8.24}$$

Newton Backward Divided Difference

In the similar fashion, when the points are recorded in a reverse way as $x_m,\ x_{m-1},\ \ldots,\ x_0$, the polynomial is forced to pass through the following $(m+1)$ data points, $(x_m,\ y_m)$, $(x_{m-1},\ y_{m-1})$, …, $(x_0,\ y_0)$. Accordingly, Equation (8.21) gets an equivalent form as:

$$y(x)=y[x_m]+y[x_m,x_{m-1}](x-x_m)+\ldots\ldots+y[x_m,x_{m-1},\ldots,\ x_0](x-x_m)(x-x_{m-1})\ldots(x-x_1) \tag{8.25}$$

In more compact form,

$$y(x)=\sum_{j=0}^{m}\left\{y\,[x_m,x_{m-1},\ldots,\ x_{m-j}]\prod_{i=0}^{j-1}(x-x_{m-i})\right\} \tag{8.26}$$

This is often called the *Newton backward divided difference* formula. It has the following remainder term,

$$R_m(\xi)=y\,[\xi,\ x_m,\ x_{m-1},\ \ldots,\ x_0](x-x_m)(x-x_{m-1})\ldots(x-x_0) \qquad x_0\leq\xi\leq x_m$$

that is indeed Equation (8.22) obtained for the Newton forward divided difference.

For $m=1$, Equation (8.26) yields:

$$y(x)=y\,[x_1]+y\,[x_1,x_0](x-x_1) \tag{8.27}$$

Remarks

1. It is derived earlier (see Subsection 8.3.2) for *quadratic* interpolation that

$$a_0=y_0$$

$$a_1=\frac{y_1-y_0}{x_1-x_0}$$

$$a_2 = \frac{1}{x_2 - x_0}\left[\frac{y_2 - y_1}{x_2 - x_1} - \frac{y_1 - y_0}{x_1 - x_0}\right]$$

Now we would like to find these coefficients directly by using the definition of finite divided differences presented through Equations (8.19) and (8.20) for verification.

The equations for the first two coefficients, a_0 and a_1, are quite obvious from Equations (8.19) and (8.20). For the third coefficient a_2, let us show:

$$a_2 = y[x_2, x_1, x_0] = \frac{1}{x_2 - x_0}\left[\frac{y_2 - y_1}{x_2 - x_1} - \frac{y_1 - y_0}{x_1 - x_0}\right] \tag{8.28}$$

From Equation (8.20),

$$y[x_2, x_1, x_0] = \frac{y[x_2, x_1] - y[x_1, x_0]}{x_2 - x_0}$$

$$= \frac{\frac{y_2 - y_1}{x_2 - x_1} - \frac{y_1 - y_0}{x_1 - x_0}}{x_2 - x_0}$$

which is nothing but Equation (8.28).

2. In the similar fashion, we can directly employ the divided difference formulas to formulate the cubic interpolation, quartic interpolation, and so on.
3. We can show from Equation (8.21) that:

   At $x = x_0$

   $$y\,(x = x_0) = y_0$$

   At $x = x_1$

   $$y\,(x = x_1) = y_0 + y[x_1, x_0](x_1 - x_0) + 0$$

   $$= y_0 + \frac{y_1 - y_0}{x_1 - x_0}(x_1 - x_0)$$

   $$= y_1$$

   Similarly, $y\,(x = x_2) = y_2$ and $y\,(x = x_i) = y_i$, $i = 0, 1, 2, \ldots, m$. It means, the Newton polynomial passes through all $m+1$ data points, (x_0, y_0), (x_1, y_1), ..., (x_m, y_m), and thus, it is an interpolating polynomial determined uniquely.
4. The Newton polynomial interpolation (both forward and backward divided difference) is applicable for equally as well as unequally spaced data points.
5. The divided differences are calculated as shown in Table 8.3 as a sample case.

TABLE 8.3 Divided differences

x_i	$y_i = y[x_i]$	$y[x_{i+1}, x_i]$	$y[x_{i+2}, x_{i+1}, x_i]$	$y[x_{i+3}, x_{i+2}, x_{i+1}, x_i]$
x_0	$y[x_0]$			
		$y[x_1, x_0]$		
x_1	$y[x_1]$		$y[x_2, x_1, x_0]$	
		$y[x_2, x_1]$		$y[x_3, x_2, x_1, x_0]$
x_2	$y[x_2]$		$y[x_3, x_2, x_1]$	
		$y[x_3, x_2]$		
x_3	$y[x_3]$			

6. Interchanging the position of the points in the above divided difference table does not affect the results. For example, swapping x_0 and x_1 in $y[x_m, x_{m-1}, ..., x_2, x_1, x_0]$ of Equation (8.21), we have $y[x_m, x_{m-1}, ..., x_2, x_0, x_1]$. Since the polynomial is unique,
$$y[x_m, x_{m-1}, ..., x_2, x_1, x_0] = y[x_m, x_{m-1}, ..., x_2, x_0, x_1]$$
and so, the order of the arguments in $y[x_m, x_{m-1}, ..., x_0]$ is immaterial.
7. Based on the definition of divided differences and using interchangeability of arguments, Equation (8.22) gets the following form:
$$R_m(\xi) = y[\xi, x_m, x_{m-1}, ..., x_0](\xi - x_0)(\xi - x_1)...(\xi - x_m) \tag{8.29}$$
The derivation of this equation is shown in Appendix 8A.

Example 8.3 Forward divided difference: cubic interpolation

Using the Newton forward divided difference approach, construct the cubic polynomial that passes through the points given in Table 8.4.

TABLE 8.4 Data points

x	y
2	1
4	2
6	4
8	8

Solution

From Equation (8.21), we get the cubic polynomial as:
$$y(x) = y[x_0] + y[x_1, x_0](x - x_0) + y[x_2, x_1, x_0](x - x_0)(x - x_1) +$$
$$y[x_3, x_2, x_1, x_0](x - x_0)(x - x_1)(x - x_2)$$

For this, let us produce Table 8.5. The above equation yields:

Table 8.5 Divided differences

x_i	$y_i = y[x_i]$	$y[x_{i+1}, x_i]$	$y[x_{i+2}, x_{i+1}, x_i]$	$y[x_{i+3}, x_{i+2}, x_{i+1}, x_i]$
2	1			
		$\frac{2-1}{4-2} = \frac{1}{2}$		
4	2		$\frac{1-\frac{1}{2}}{6-2} = \frac{1}{8}$	
		$\frac{4-2}{6-4} = 1$		$\frac{\frac{1}{4}-\frac{1}{8}}{8-2} = \frac{1}{48}$
6	4		$\frac{2-1}{8-4} = \frac{1}{4}$	
		$\frac{8-4}{8-6} = 2$		
8	8			

$$y = 1 + \frac{1}{2}(x-2) + \frac{1}{8}(x-2)(x-4) + \frac{1}{48}(x-2)(x-4)(x-6)$$

$$= 1 + \frac{x}{2} - 1 + \frac{1}{8}(x^2 - 6x + 8) + \frac{1}{48}(x^3 - 12x^2 + 44x - 48)$$

Rearranging,

$$y = \frac{1}{48}x^3 - \frac{1}{8}x^2 + \frac{2}{3}x$$

Example 8.4 Backward divided difference: cubic interpolation (unequally spaced data)

Using the Newton backward divided difference method, construct the cubic polynomial that passes through the unequally spaced data points given in Table 8.6.

Table 8.6 Data points

x	y
2	1
4	2
6	4
10	8

Solution

From the Newton backward divided difference formula [i.e., Equation (8.25)], we get the following form of cubic polynomial:

$$\begin{aligned} y(x) &= y[x_3] + y[x_3, x_2](x - x_3) + y[x_3, x_2, x_1](x - x_3)(x - x_2) + \\ &\quad y[x_3, x_2, x_1, x_0](x - x_3)(x - x_2)(x - x_1) \end{aligned} \tag{8.30}$$

For this, let us produce Table 8.7.

TABLE 8.7 Divided differences

x_i	$y_i = y[x_i]$	$y[x_{i+1}, x_i]$	$y[x_{i+2}, x_{i+1}, x_i]$	$y[x_{i+3}, x_{i+2}, x_{i+1}, x_i]$
2	1			
		$\frac{2-1}{4-2} = \frac{1}{2}$		
4	2		$\frac{1-\frac{1}{2}}{6-2} = \frac{1}{8}$	
		$\frac{4-2}{6-4} = 1$		$\frac{0-\frac{1}{8}}{10-2} = -\frac{1}{64}$
6	4		$\frac{1-1}{10-4} = 0$	
		$\frac{8-4}{10-6} = 1$		
10	8			

Equation (8.30) yields:

$$y = 8 + 1(x-10) + 0(x-10)(x-6) - \frac{1}{64}(x-10)(x-6)(x-4)$$

$$= 8 + x - 10 - \frac{1}{64}(x^3 - 20x^2 + 124x - 240)$$

Rearranging,

$$y = -\frac{1}{64}x^3 + \frac{5}{16}x^2 - \frac{15}{16}x + \frac{7}{4}$$

Note that we obtain this same polynomial by applying the Newton forward divided difference formula.

8.4 Polynomial Interpolation with Equally Spaced Data: A Special Case

So far, we have discussed the interpolation formula with linear, quadratic and higher-order polynomials where the base points are equally as well as unequally spaced. Now we would like to explore the interpolation scheme to deal with a class of interpolation problems with only equally spaced data points. This scheme requires less number of arithmetic operations, thereby reduced round-off error. In this special case, the values, $y_0, y_1, \ldots, y_m$, are specified at equally spaced values, $x_0, x_1, \ldots, x_m$, respectively, of the base points with:

$$\Delta x = x_{i+1} - x_i \tag{8.31}$$

It gives:

$$\begin{aligned} x_1 &= x_0 + \Delta x \\ x_2 &= x_1 + \Delta x = x_0 + 2\Delta x \\ x_3 &= x_2 + \Delta x = x_0 + 3\Delta x \\ &\vdots \\ x_m &= x_{m-1} + \Delta x = x_0 + m\Delta x \end{aligned} \tag{8.32}$$

Note that here x_0 is the starting/reference point.

Remarks

1. For equally spaced data, the linear interpolation formula [Equation (8.7)] yields:

$$y = y_0 + \frac{y_1 - y_0}{\Delta x}(x - x_0) \tag{8.33}$$

where,

$$y[x_1, x_0] = \frac{y_1 - y_0}{x_1 - x_0} = \frac{y_1 - y_0}{\Delta x} \tag{8.34}$$

2. Similarly, for the quadratic interpolation formula [Equation (8.16)]:

$$y = y_0 + \frac{y_1 - y_0}{\Delta x}(x - x_0) + \frac{1}{2\Delta x}\left[\frac{y_2 - y_1}{\Delta x} - \frac{y_1 - y_0}{\Delta x}\right](x - x_0)(x - x_1) \tag{8.35}$$

This gives:

$$y = y_0 + \frac{y_1 - y_0}{\Delta x}(x - x_0) + \frac{y_2 - 2y_1 + y_0}{2(\Delta x)^2}(x - x_0)(x - x_1) \tag{8.36}$$

where,

$$y[x_2, x_1, x_0] = \frac{y_2 - 2y_1 + y_0}{2(\Delta x)^2} \tag{8.37}$$

In the following section, we will discuss the methods of polynomial interpolation with equally spaced data.

8.4.1 Newton–Gregory Forward Difference Interpolation

The *first-order* forward difference is defined as:

$$\Delta y_i = y_{i+1} - y_i \qquad i = 0, 1, 2, \ldots, m-1 \tag{8.38}$$

Here we use Δ to denote the first forward difference operator.

The *second-order* forward difference is:

$$\Delta^2 y_i = \Delta(\Delta y_i) = \Delta(y_{i+1} - y_i) = (y_{i+2} - y_{i+1}) - (y_{i+1} - y_i)$$

$$= y_{i+2} - 2y_{i+1} + y_i \tag{8.39}$$

Similarly, for *jth-order* forward difference:

$$\Delta^j y_i = \Delta^{j-1} y_{i+1} - \Delta^{j-1} y_i \tag{8.40}$$

- With the *first-order forward difference*, Equation (8.34) yields:

$$y[x_1, x_0] = \frac{y_1 - y_0}{x_1 - x_0} = \frac{\Delta y_0}{\Delta x}$$

Rearranging,

$$\Delta y_0 = \Delta x \; y[x_1, x_0] \tag{8.41}$$

Similarly,

$$\Delta y_1 = \Delta x \; y[x_2, x_1] \tag{8.42}$$

- For the *second-order forward difference*:

$$\Delta^2 y_0 = \Delta(\Delta y_0) = \Delta(y_1 - y_0)$$

$$= \Delta y_1 - \Delta y_0$$

$$= \Delta x \; y\,[x_2, x_1] - \Delta x \; y\,[x_1, x_0]$$

$$= \Delta x \; 2\Delta x \left[\frac{y\,[x_2, x_1]}{2\Delta x} - \frac{y\,[x_1, x_0]}{2\Delta x} \right]$$

Using Equation (8.20) for equally spaced data, we get:

$$\Delta^2 y_0 = 2(\Delta x)^2 y\,[x_2, x_1, x_0] \tag{8.43}$$

- For the *jth-order forward difference*, one can similarly show,

$$\Delta^j y_i = j!\,(\Delta x)^j y\,[x_{i+j}, x_{i+j-1}, \ldots\ldots, x_i] \tag{8.44}$$

Rearranging,

$$y\,[x_{i+j}, x_{i+j-1}, \ldots\ldots\ldots, x_i] = \frac{\Delta^j y_i}{j!\,(\Delta x)^j} \tag{8.45}$$

For $i = 0$,

$$y\,[x_j, x_{j-1}, \ldots\ldots\ldots, x_0] = \frac{\Delta^j y_0}{j!\,(\Delta x)^j} \tag{8.46}$$

Plugging this equation into Equation (8.23),

$$y(x) = \sum_{j=0}^{m} \left\{ \frac{\Delta^j y_0}{j!\,(\Delta x)^j} \prod_{i=0}^{j-1} (x - x_i) \right\} \tag{8.47}$$

This is the *Newton forward difference* formula for equally spaced data set.

Deriving Newton–Gregory forward difference

From Equation (8.32),

$$x_i = x_0 + i\Delta x \tag{8.48}$$

Setting,

$$x = x_0 + s\Delta x \tag{8.49}$$

we have,

$$x - x_i = (s - i)\Delta x \tag{8.50}$$

Notice that here s is introduced as a dimensionless value[6] of x.

Further setting,

$$y(x) = y(s) \tag{8.51}$$

we get from Equation (8.47):

$$y(s) = \sum_{j=0}^{m} \left\{ \frac{\Delta^j y_0}{j!\,(\Delta x)^j} \prod_{i=0}^{j-1} (s - i)\Delta x \right\}$$

$$= \sum_{j=0}^{m} \frac{\Delta^j y_0}{j!\,(\Delta x)^j} [s(s-1)(s-2)....(s-j+1)](\Delta x)^j \tag{8.52}$$

We use the binomial-coefficient notation:

$$\binom{s}{j} = \frac{s(s-1)......(s-j+1)}{j!} \tag{8.53}$$

Plugging in Equation (8.52),

$$y(s) = \sum_{j=0}^{m} \binom{s}{j} \Delta^j y_0 \tag{8.54}$$

This is the well-known *Newton–Gregory*[7] *forward difference* interpolation formula.

Remark

For a given data set (x_i, y_i), $i = 0, 1, 2, 3, 4$, the forward differences are produced in Table 8.8.

---

6 Since $s = \dfrac{x - x_0}{\Delta x}$.

7 Named after Sir Isaac Newton, PRS (1643–1727) and a Scottish mathematician and astronomer, James Gregory, FRS (1638–1675).

TABLE 8.8 Forward differences

i	x_i	y_i	Δy_i	$\Delta^2 y_i$	$\Delta^3 y_i$	$\Delta^4 y_i$
0	x_0	y_0				
			Δy_0			
				$\Delta^2 y_0$		
1	x_1	y_1				
					$\Delta^3 y_0$	
			Δy_1			
				$\Delta^2 y_1$		$\Delta^4 y_0$
2	x_2	y_2				
					$\Delta^3 y_1$	
			Δy_2			
				$\Delta^2 y_2$		
3	x_3	y_3				
			Δy_3			
4	x_4	y_4				

Example 8.5 Newton–Gregory forward difference

Using the Newton–Gregory forward difference interpolation scheme, find $y(2)$ based on the data points given in Table 8.9.

TABLE 8.9 Given data

i	x_i	y_i
0	1	0
1	3	1.1
2	5	1.6
3	7	1.9
4	9	2.1

Solution

Based on the given data, let us first estimate the forward differences in Table 8.10.

TABLE 8.10 Forward differences

i	x_i	y_i	Δy_i	$\Delta^2 y_i$	$\Delta^3 y_i$	$\Delta^4 y_i$
0	1	0				
			1.1			
				–0.6		
1	3	1.1				
					0.4	
			0.5			
				–0.2		–0.3
2	5	1.6				
					0.1	
			0.3			
				–0.1		
3	7	1.9				
			0.2			
4	9	2.1				

Here,

$$x_0 = 1,\ \Delta x = 2$$

and $x = 2$ at which, y is asked to determine. From Equation (8.49):

$$s = \frac{x - x_0}{\Delta x} = \frac{1}{2} = 0.5$$

From Equation (8.54):

$$y(s) = \sum_{j=0}^{4} \binom{s}{j} \Delta^j y_0$$

$$= \binom{0.5}{0} y_0 + \binom{0.5}{1} \Delta y_0 + \binom{0.5}{2} \Delta^2 y_0 + \binom{0.5}{3} \Delta^3 y_0 + \binom{0.5}{4} \Delta^4 y_0$$

$$= 0 + 0.5 \times 1.1 + \frac{0.5(0.5-1)}{2}(-0.6) + \frac{0.5(0.5-1)(0.5-2)}{6}(0.4)$$

$$+ \frac{0.5(0.5-1)(0.5-2)(0.5-3)}{24}(-0.3)$$

$$= 0 + 0.55 + 0.075 + 0.025 + 0.011719$$

$$= 0.661719$$

So, $y(2)$ is obtained as 0.661719.

8.4.2 Newton–Gregory Backward Difference Interpolation

The *first-order* backward difference is defined as:

$$\nabla y_i = y_i - y_{i-1} \qquad i = 1, 2, \ldots\ldots, m \tag{8.55}$$

Here we use ∇ to denote the first backward difference operator.

The *second-order* backward difference is:

$$\nabla^2 y_i = \nabla y_i - \nabla y_{i-1}$$

$$= y_i - 2y_{i-1} + y_{i-2} \tag{8.56}$$

Similarly, for the *jth-order* backward difference:

$$\nabla^j y_i = \nabla^{j-1} y_i - \nabla^{j-1} y_{i-1} \tag{8.57}$$

Deriving Newton–Gregory backward difference

For backward difference, we set the dimensionless value of x as:

$$s = \frac{x - x_m}{\Delta x} \tag{8.58}$$

in which, x_m is the reference point. It yields:

$$x = x_m + s\Delta x \tag{8.59}$$

Accordingly, we have,

$$s+1 = \frac{x - x_{m-1}}{\Delta x} \tag{8.60}$$

and so on. With this, Equation (8.25) gives:

$$y(s) = y_m + s\nabla y_m + \frac{s(s+1)}{2\,!}\nabla^2 y_m + \ldots\ldots + \frac{s(s+1)\ldots\ldots(s+m-1)}{m\,!}\nabla^m y_m \tag{8.61}$$

Or,

$$y(s) = \sum_{j=0}^{m} (-1)^j \binom{-s}{j} \nabla^j y_m \tag{8.62}$$

This is called the *Newton–Gregory backward difference* interpolation formula.

Remarks

1. Like the forward differences (Table 8.8), one can produce the backward differences in Table 8.11 for a given data set (x_i, y_i), $i = 0, 1, 2, 3, 4$.

Table 8.11 Backward differences

i	x_i	y_i	∇y_i	$\nabla^2 y_i$	$\nabla^3 y_i$	$\nabla^4 y_i$
0	x_0	y_0				
			∇y_1			
				$\nabla^2 y_2$		
1	x_1	y_1			$\nabla^3 y_3$	
			∇y_2			
				$\nabla^2 y_3$		$\nabla^4 y_4$
2	x_2	y_2			$\nabla^3 y_4$	
			∇y_3			
				$\nabla^2 y_4$		
3	x_3	y_3				
			∇y_4			
4	x_4	y_4				

2. For a set of evenly spaced data (x_0, y_0), (x_1, y_1), …., (x_m, y_m), one can fit an mth-degree polynomial by using either the forward difference [Equation (8.54)] or backward difference formula [Equation (8.62)]. Note that both the formulae will provide the same final results since a unique mth-degree polynomial passes through a set of $m+1$ data points.
3. The Newton–Gregory forward difference is preferred when the interpolation is sought near the starting of a table of values. When it is reverse (i.e., interpolation is sought near the end of a table), one can prefer the Newton–Gregory backward difference.

Example 8.6 Newton–Gregory backward difference

Repeat Example 8.5 with the given data in Table 8.9 and find $y(2)$ using the Newton–Gregory backward difference scheme.

Solution

The backward differences are estimated in Table 8.12.

TABLE 8.12 Backward differences

i	x_i	y_i	∇y_i	$\nabla^2 y_i$	$\nabla^3 y_i$	$\nabla^4 y_i$
0	1	0				
			1.1			
1	3	1.1		–0.6		
			0.5		0.4	
2	5	1.6		–0.2		–0.3
			0.3		0.1	
3	7	1.9		–0.1		
			0.2			
4	9	2.1				

Here,

$$x_m = 9, \Delta x = 2$$

and $x = 2$ at which, y is asked to determine. From Equation (8.58):

$$s = \frac{x - x_m}{\Delta x} = \frac{2-9}{2} = -3.5$$

From Equation (8.61):

$$\begin{aligned} y(s) &= y_4 + s\nabla y_4 + \frac{s(s+1)}{2!}\nabla^2 y_4 + \frac{s(s+1)(s+2)}{3!}\nabla^3 y_4 + \frac{s(s+1)(s+2)(s+3)}{4!}\nabla^4 y_4 \\ &= 2.1 + (-3.5)(0.2) + \frac{3.5\times 2.5}{2}(-0.1) - \frac{3.5\times 2.5\times 1.5}{6}(0.1) + \frac{3.5\times 2.5\times 1.5\times 0.5}{24}(-0.3) \\ &= 2.1 - 0.7 - 0.4375 - 0.21875 - 0.082031 \\ &= 0.661719 \end{aligned}$$

So, $y(2)$ is obtained as 0.661719.

Remark

It is obvious that we got the same value for $y(2)$ $(= 0.661719)$ from both the forward difference (Example 8.5) and backward difference (this Example 8.6) methods for the reason stated before.

8.4.3 Error in Polynomial Interpolation: Equally Spaced Data

The error (i.e., truncation error or remainder) involved in the Newton divided difference interpolation for unequally spaced data is defined before by Equation (8.29). Now we would like to discuss how to estimate the error for equally spaced data set.

Newton–Gregory forward difference

Earlier in Equation (8.52) we derived the Newton–Gregory forward difference formula for equispaced data points. Keeping all terms intact, one can rewrite it:

$$y(s)=\sum_{j=0}^{m}\frac{\Delta^j y_0}{j!}[s(s-1)(s-2)....(s-j+1)]+R_m(\xi) \tag{8.63}$$

In which,

$$R_m(\xi)=\frac{\Delta^{m+1}y(\xi)}{(m+1)!}s(s-1)......(s-m) \qquad x_0\leq\xi\leq x_m$$

It gives:

$$R_m(\xi)=\frac{y^{(m+1)}(\xi)}{(m+1)!}(\Delta x)^{m+1}s(s-1)......(s-m) \tag{8.64}$$

since the $(m + 1)$th derivative of y evaluated at $x=\xi$ is:

$$y^{(m+1)}(\xi)=\frac{\Delta^{m+1}y(\xi)}{(\Delta x)^{m+1}}$$

Equation (8.64) is the final form of remainder involved in the Newton–Gregory forward difference interpolation for evenly spaced data.

Remarks

1. We can easily compute the remainder using Equation (8.64) only when y is available in the functional form. The difficulty arises in computing the derivative of y when the tabulated data are only available, which is quite common in practice. In such a situation, we can estimate the remainder from Equation (8.29) at the point x:

   $$R_m(x)=y\,[x,\ x_m,\ x_{m-1},\ ....,\ x_0](x-x_0)(x-x_1)...(x-x_m) \tag{8.65}$$

   based on the formula of divided differences given in Equation (8.20).
2. Notice that Equation (8.65), which is an equivalent form of Equation (8.29), is valid for both equally and unequally spaced data points.

Now we would like to discuss an interesting issue concerning the similarity between the Newton–Gregory forward difference and Taylor series.

Similarity between Newton–Gregory forward difference and Taylor series

We derived the Newton forward difference formula in Equation (8.47) for equispaced points. Keeping all terms intact, one can rewrite it:

$$y = y_0 + \frac{\Delta y_0}{\Delta x}(x - x_0) + \frac{1}{2!}\frac{\Delta^2 y_0}{(\Delta x)^2}(x - x_0)(x - x_1) + \ldots + \frac{1}{m!}\frac{\Delta^m y_0}{(\Delta x)^m}(x - x_0)(x - x_1)\ldots(x - x_{m-1}) + R_m(\xi) \tag{8.66}$$

Obviously, the remainder is:

$$R_m(\xi) = \frac{1}{(m+1)!}\frac{\Delta^{m+1} y(\xi)}{(\Delta x)^{m+1}}(x - x_0)(x - x_1)\ldots(x - x_m) \qquad x_0 \le \xi \le x_m$$

It gives:

$$R_m(\xi) = \frac{y^{(m+1)}(\xi)}{(m+1)!}(x - x_0)(x - x_1)\ldots(x - x_m) \tag{8.67}$$

Notice that Equation (8.66) looks like the Taylor series at a point x_0:

$$y = y_0 + \left(\frac{dy}{dx}\right)_{x = x_0}(x - x_0) + \frac{1}{2!}\left(\frac{d^2 y}{dx^2}\right)_{x = x_0}(x - x_0)^2 + \ldots + \frac{1}{m!}\left(\frac{d^m y}{dx^m}\right)_{x = x_0}(x - x_0)^m + R_m(\xi)$$

Rewriting,

$$y = y_0 + y'(x_0)(x - x_0) + \frac{1}{2!}y''(x_0)(x - x_0)^2 + \ldots + \frac{1}{m!}y^{(m)}(x_0)(x - x_0)^m + R_m(\xi) \tag{8.68}$$

where, the remainder is:

$$R_m(\xi) = \frac{y^{(m+1)}(\xi)}{(m+1)!}(x - x_0)^{m+1} \qquad x_0 \le \xi \le x \tag{8.69}$$

Newton–Gregory backward difference

Based on Equation (8.61), we get:

$$R_m(\xi) = \frac{\nabla^{m+1} y(\xi)}{(m+1)\,!}s(s+1)\ldots\ldots(s+m)$$

It gives:

$$R_m(\xi) = \frac{s(s+1)\ldots\ldots(s+m)}{(m+1)\,!}(\Delta x)^{m+1} y^{(m+1)}(\xi) \qquad x_0 \le \xi \le x_m \tag{8.70}$$

This is the final form of the error involved in the Newton–Gregory backward difference interpolation for *equally spaced* data.

Remark

The first remark made above for Newton–Gregory forward difference is also valid for its backward difference counterpart. It reveals that we can only use Equation (8.70) to compute the remainder for the cases with equally spaced data when y is available in the functional form.

8.5 Lagrange Polynomial Interpolation

The Lagrange interpolation[8] formula is:

$$\begin{aligned} y(x) = {} & a_0(x-x_1)(x-x_2)\dots(x-x_m)+a_1(x-x_0)(x-x_2)\dots(x-x_m)+\dots\dots+ \\ & a_j(x-x_0)(x-x_1)\dots(x-x_{j-1})(x-x_{j+1})\dots(x-x_m)+\dots\dots+ \\ & a_m(x-x_0)(x-x_1)\dots(x-x_{m-1})+R_m \end{aligned} \tag{8.71}$$

where,

$$R_m(\xi)=\frac{y^{(m+1)}(\xi)}{(m+1)!}(x-x_0)(x-x_1)\dots(x-x_m) \qquad x_0\le\xi\le x_m \tag{8.67}$$

Notice that this remainder term is the same with that of Newton–Gregory forward difference interpolation. Now we need to determine the coefficients, a_0, a_1, …, a_m, by forcing the remainders at the $m + 1$ base points to be zero.

Finding $\boldsymbol{a_0}$

Evaluating Equation (8.71) at $x = x_0$, we get:

$$a_0=\frac{y_0}{(x_0-x_1)(x_0-x_2)\dots(x_0-x_m)} \tag{8.72}$$

Finding $\boldsymbol{a_j}$

Similarly at $x = x_j$, Equation (8.71) yields:

$$a_j=\frac{y_j}{(x_j-x_0)(x_j-x_1)\dots(x_j-x_{j-1})(x_j-x_{j+1})\dots(x_j-x_m)} \tag{8.73}$$

Plugging these equations in Equation (8.71),

$$\begin{aligned} y(x)= {} & \frac{(x-x_1)(x-x_2)\dots(x-x_m)}{(x_0-x_1)(x_0-x_2)\dots(x_0-x_m)}y_0+\frac{(x-x_0)(x-x_2)\dots(x-x_m)}{(x_1-x_0)(x_1-x_2)\dots(x_1-x_m)}y_1+\dots\dots+ \\ & \frac{(x-x_0)(x-x_1)\dots(x-x_{j-1})(x-x_{j+1})\dots(x-x_m)}{(x_j-x_0)(x_j-x_1)\dots(x_j-x_{j-1})(x_j-x_{j+1})\dots(x_j-x_m)}y_j+\dots\dots+\frac{(x-x_0)(x-x_1)\dots(x-x_{m-1})}{(x_m-x_0)(x_m-x_1)\dots(x_m-x_{m-1})}y_m \end{aligned} \tag{8.74}$$

8 Named after Italian mathematician and astronomer Joseph-Louis Lagrange (1736–1813).

It gives:

$$y(x)=\sum_{j=0}^{m}\left[\prod_{\substack{i=0\\i\neq j}}^{m}\frac{(x-x_i)}{(x_j-x_i)}\right]y_j \tag{8.75}$$

This is the final form of *Lagrange interpolating* polynomial.

One can represent Equation (8.74) as:

$$y(x)=L_0(x)y_0+L_1(x)y_1+.....+L_j(x)y_j+.....+L_m(x)y_m \tag{8.76}$$

In compact form,

$$y(x)=\sum_{j=0}^{m}L_j(x)y_j \tag{8.77}$$

where the Lagrange coefficient function[9] (L_j),

$$\begin{aligned}L_j(x)&=\frac{(x-x_0)(x-x_1)...(x-x_{j-1})(x-x_{j+1})...(x-x_m)}{(x_j-x_0)(x_j-x_1)...(x_j-x_{j-1})(x_j-x_{j+1})...(x_j-x_m)}\\&=\prod_{\substack{i=0\\i\neq j}}^{m}\frac{(x-x_i)}{(x_j-x_i)}\qquad j=0,1,2,....,m\end{aligned} \tag{8.78}$$

Notice that in the above equation, the product of the linear factors $(x-x_i)$ in the numerator and (x_j-x_i) in the denominator are to be produced, skipping the factor (x_j-x_j).

Obviously, the Lagrange coefficient function holds the following property:

$$L_j(x_i)=\begin{cases}1 & \text{if } i=j\\0 & \text{if } i\neq j\end{cases}$$

Remarks

1. From Equation (8.76), we have:
   At $x=x_0$

   $$\begin{aligned}y\ (x=x_0)&=L_0(x_0)y_0+L_1(x_0)y_1+.....+L_j(x_0)y_j+.....+L_m(x_0)y_m\\&=y_0 1+y_1 0+.....+y_j 0+.....+y_m 0=y_0\end{aligned}$$

   At $x=x_1$

   $$\begin{aligned}y\ (x=x_1)&=L_0(x_1)y_0+L_1(x_1)y_1+.....+L_j(x_1)y_j+.....+L_m(x_1)y_m\\&=y_0 0+y_1 1+.....+y_j 0+.....+y_m 0=y_1\end{aligned}$$

   Similarly $y\ (x=x_2)=y_2$ and so on. It becomes obvious that the Lagrange polynomial passes through all the data points, (x_0, y_0), (x_1, y_1), ..., (x_m, y_m), and thus it is an interpolating polynomial determined uniquely.

9 It is sometimes called the *Lagrange coefficient polynomial* or *Lagrange fundamental polynomial.*

2. Although the Newton and Lagrange interpolating polynomials, obtained in Equations (8.21) and (8.74), respectively, look different in structure but they are identical provided the $m + 1$ data points used for these two applications are the same. Moreover, both these approaches are applicable to equally as well as unequally spaced data points.
3. If we wish to fit a higher-order polynomial (say, $m + 1$th-order) with adding a new point to the data set, employing Newton formula [i.e., Equation (8.21)] is advantageous in that one can make use of the previous calculations (performed in finding mth-order polynomial). However, in case of Lagrange polynomial, one needs to start afresh and repeat all calculations.
4. The Lagrange interpolation scheme may involve large round-off error, which is generated due to the subtraction of numbers having similar magnitudes.

Example 8.7 Lagrange quadratic interpolation

Construct the quadratic polynomial that passes through the data points (1, 1), (2, 1) and (3, 2). Using the fitted polynomial, find $y(2.5)$.

Solution

From Equation (8.75) with $m = 2$,

$$y = \frac{(x - x_1)(x - x_2)}{(x_0 - x_1)(x_0 - x_2)} y_0 + \frac{(x - x_0)(x - x_2)}{(x_1 - x_0)(x_1 - x_2)} y_1 + \frac{(x - x_0)(x - x_1)}{(x_2 - x_0)(x_2 - x_1)} y_2$$

Using the given data points:

$$\begin{aligned} y &= \frac{(x-2)(x-3)}{(1-2)(1-3)} 1 + \frac{(x-1)(x-3)}{(2-1)(2-3)} 1 + \frac{(x-1)(x-2)}{(3-1)(3-2)} 2 \\ &= \frac{1}{2}(x-2)(x-3) - (x-1)(x-3) + (x-1)(x-2) \\ &= \frac{1}{2}x^2 - \frac{3}{2}x + 2 \end{aligned}$$

This is the fitted quadratic polynomial, using which, we get:

$$\begin{aligned} y(2.5) &= \frac{1}{2}(2.5)^2 - \frac{3}{2}2.5 + 2 \\ &= 1.375 \end{aligned}$$

Remark

Based on the same given data points, the Newton divided difference interpolation scheme [Equation (8.16)] yields:

$$y = 1 + 0 + \frac{1}{3-1}\left[\frac{2-1}{3-2} - 0\right](x-1)(x-2)$$

$$= \frac{1}{2}x^2 - \frac{3}{2}x + 2$$

Obviously, we got the same quadratic polynomial although the interpolation schemes (Lagrange and Newton) are different.

8.6 Piecewise Polynomial Interpolation by Spline Functions

For a large number of data points, the interpolating polynomial is of a reasonably high degree that may result in oscillatory characteristics. The higher the degree of this polynomial, the more oscillatory is its nature. This is somewhat pictorially illustrated in Figure 8.4.[10]

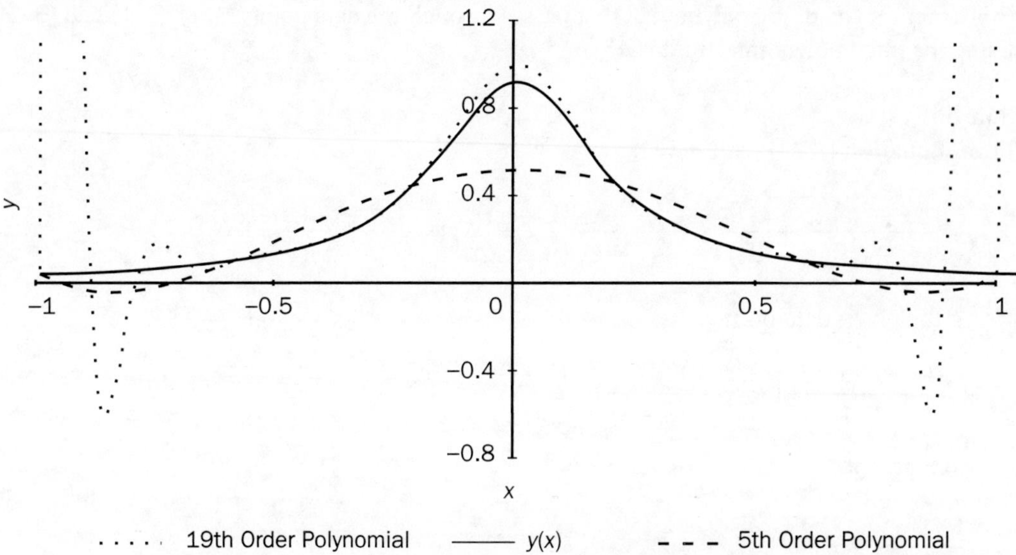

Figure 8.4 Illustrating oscillatory characteristics of interpolating polynomials

To circumvent this problem, one can do interpolation in a piecewise fashion by simply splitting the set of large data points into several small sets. Then represent them by a set of piecewise continuous curves rather than a single curve. By this way, one can divide the large interval into a set of subintervals and fit a lower degree approximating polynomial to each of these subintervals. This type of approximation is often referred to as the *piecewise polynomial interpolation*, which is demonstrated in Figure 8.5.

10 http://numericalmethods.eng.usf.edu.

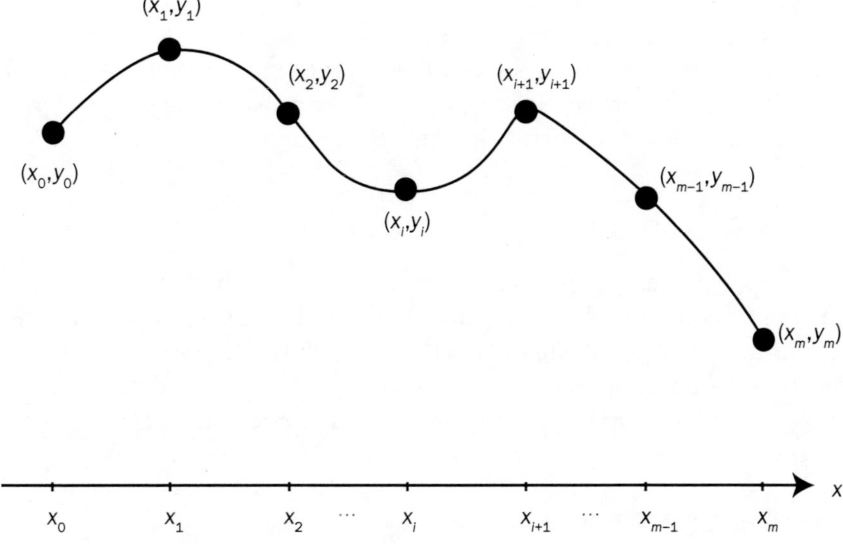

FIGURE 8.5 Piecewise polynomial interpolation

As shown, the curve between two successive nodes is referred to as a *spline* that is mathematically represented by a *spline function*. The common node between two adjacent splines is called a *knot or breakpoint.*[11] With this, for example, when the second-order curves used to connect each pair of data points, then those curves are referred to as the *quadratic splines*. One can construct the quadratic functions so that their interconnections look smooth. Connecting a sequence of such continuous curves forms a single continuous curve, which is often called the *spline curve*.

Here, we will use the notation $S_i(x)$ to denote the *spline function* (i.e., a piecewise polynomial function) between the successive nodes (x_i, y_i) and (x_{i+1}, y_{i+1}). As stated, a set of these functions, $S_i(x)$ leads to form a spline curve, which we are going to denote by $S(x)$.

General Definition of Spline Function

A function S is called a spline function of degree k if it holds the following properties:

(a) S is defined in the interval[12] $[a,b]$
(b) $S^{(r)}$ is continuous on $[a,b]$ for $0 \le r \le k-1$ (order of derivative, $r=0$ indicates that S is continuous)
(c) S is a polynomial of degree $\le k$ on each subinterval $[x_i, x_{i+1}]$, $i = 0, 1, 2, ..., m-1$

11 Thus, all interior nodes, $x_1, x_2, ..., x_{m-1}$, in the interval $[x_0, x_m]$ are considered knots or joints.

12 Recall that $a = x_1 < x_2 < ... < x_m = b$.

It is obvious that unlike conventional polynomial interpolation, the degree of the spline does not increase with the increase of data points. Basically here the degree remains fixed and one can rather use more splines. With this background study, let us focus on various types of spline functions for piecewise polynomial interpolation for a given large data set. For those functions, we will use here the *shifted power form.*[13]

8.6.1 Linear Spline

Piecewise interpolation using linear spline is the simplest method of piecewise polynomial approximation. This linear spline is nothing but a straight line that joins the two points. Figure 8.6 illustrates the connections of a set of data points by a series of such straight lines.

To construct the linear spline, let us consider a polynomial of degree 1 (= k) in each subinterval $[x_i, x_{i+1}]$. Accordingly, we use the following shifted power form:

$$S_i(x) = y(x) = a_{0i} + a_{1i}(x - x_i) \tag{8.79}$$

where a_0 and a_1 are constants to be determined.

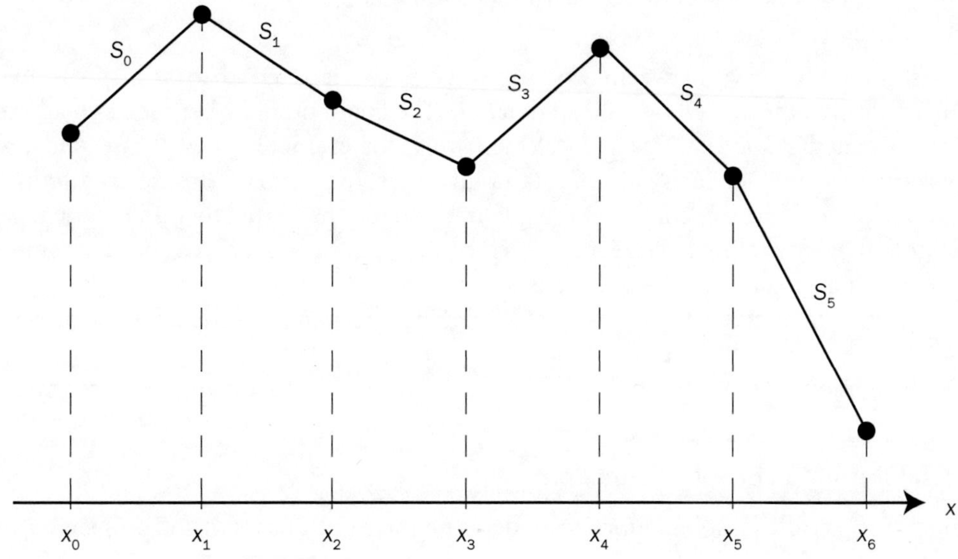

FIGURE 8.6 Piecewise linear interpolation

Finding a_{0i}

Put $x = x_i$ in Equation (8.79) and get:

$$a_{0i} = y(x_i) = y_i \tag{8.80}$$

13 It has the following general form: $y(x) = a_{0i} + a_{1i}(x - x_i) + a_{2i}(x - x_i)^2 + \ldots + a_{mi}(x - x_i)^m$.

Finding a_{1i}

Set $x = x_{i+1}$ in Equation (8.79) and get:

$$a_{1i} = \frac{y_{i+1} - y_i}{x_{i+1} - x_i} \tag{8.81}$$

Thus Equation (8.79) yields:

$$\begin{aligned} S_i(x) = y(x) &= y_i + \frac{y_{i+1} - y_i}{x_{i+1} - x_i}(x - x_i) \\ &= y(x_i) + y\,[x_{i+1}, x_i](x - x_i) \end{aligned} \tag{8.82}$$

Notice that this is the equivalent form of Equation (8.7).

With this, we get:

$$\begin{aligned} &S_0(x) = y_0 + \frac{y_1 - y_0}{x_1 - x_0}(x - x_0) && \text{for } i = 0 \\ &S_1(x) = y_1 + \frac{y_2 - y_1}{x_2 - x_1}(x - x_1) && \text{for } i = 1 \\ &\vdots \\ &S_{m-1}(x) = y_{m-1} + \frac{y_m - y_{m-1}}{x_m - x_{m-1}}(x - x_{m-1}) && \text{for } i = m-1 \end{aligned} \tag{8.83}$$

So the spline curve $S(x)$ is constructed by connecting all these individual linear spline functions as:

$$S(x) = \begin{cases} S_0(x) & \text{for } x \in [x_0, x_1] \\ S_1(x) & \text{for } x \in [x_1, x_2] \\ \cdot \\ \cdot \\ S_i(x) & \text{for } x \in [x_i, x_{i+1}] \\ \cdot \\ \cdot \\ S_{m-1}(x) & \text{for } x \in [x_{m-1}, x_m] \end{cases} \tag{8.84}$$

Remarks

1. Each spline goes through two consecutive data points, yielding two equations [based on Equation (8.79)] for two unknowns (i.e., a_0 and a_1). Computing the spline function for each subinterval and joining those splines together, one can construct an entire spline curve over the interval $[x_0, x_m]$.

2. Slope of the linear splines is discontinuous at the knots and thus, the resulting spline curve looks like a *broken line* (not smooth). Because of this reason the linear splines are seldom used.

To overcome this deficiency, we will formulate higher degree polynomial splines, which include quadratic splines (splines of degree 2) and cubic splines (splines of degree 3). Generalization to splines of any degree k is relatively straightforward and thus avoided here. Moreover, the higher degree splines (i.e., $k \geq 4$) are rarely used because of their increasing tendency of oscillation with k.

Example 8.8 Linear spline

Construct a first-degree spline $S(x)$ using the data given in Table 8.13.

TABLE 8.13 Data for linear spline

i	x_i	y_i
0	0.5	1
1	1.5	3
2	2.5	7
3	3.5	12
4	4.5	20

Find $y(3)$ by using the resulting spline and verify that all the spline functions are correctly derived.

Solution

From Equation (8.83),

$$S_0(x) = 1 + \frac{3-1}{1.5-0.5}(x-0.5) = 1 + 2x - 1 = 2x$$

$$S_1(x) = 3 + \frac{7-3}{2.5-1.5}(x-1.5) = 3 + 4x - 6 = 4x - 3$$

$$S_2(x) = 7 + \frac{12-7}{3.5-2.5}(x-2.5) = 7 + 5x - 12.5 = 5x - 5.5$$

$$S_3(x) = 12 + \frac{20-12}{4.5-3.5}(x-3.5) = 12 + 8x - 28 = 8x - 16$$

So we have:

$$S(x) = \begin{cases} 2x & \text{for } x \in [0.5,\ 1.5] \\ 4x-3 & \text{for } x \in [1.5,\ 2.5] \\ 5x-5.5 & \text{for } x \in [2.5,\ 3.5] \\ 8x-16 & \text{for } x \in [3.5,\ 4.5] \end{cases}$$

These linear spline functions are graphically illustrated in Figure 8.7.

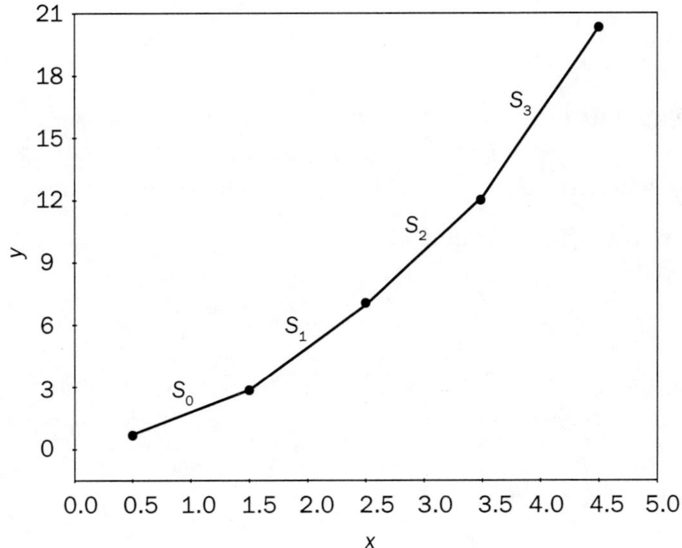

FIGURE 8.7 Piecewise linear interpolation

Finding y(3)
Notice that $x = 3$ lies in [2.5, 3.5] and thus, using

$$S_2(x) = 5x - 5.5$$

we get:

$$y(3) = 5\times 3 - 5.5 = 9.5$$

Verification
Let us verify the authenticity of our developed splines. For this, we do the simple calculations at all three knots and get confirmed based on,

$$S_0(1.5) = 3 = S_1(1.5)$$
$$S_1(2.5) = 7 = S_2(2.5)$$
$$S_2(3.5) = 12 = S_3(3.5)$$

8.6.2 QUADRATIC SPLINE

To construct a quadratic spline, we consider a polynomial of degree 2 (=k) in each subinterval $[x_i, x_{i+1}]$ as:

$$S_i(x) = y(x) = a_{0i} + a_{1i}(x - x_i) + a_{2i}(x - x_i)^2 \tag{8.85}$$

In which, a_0, a_1 and a_2 are constants to be determined, and $i = 0, 1, 2, ..., m-1$.

It is fairly true that each quadratic spline should go through two consecutive data points. Thus, by stated definition (see 'general definition of spline function'):

$$S_i(x_i) = y(x_i) \tag{8.86}$$

$$S_i(x_{i+1}) = y(x_{i+1}) \tag{8.87}$$

The general definition further reveals that $S^{(r)}$ is continuous everywhere in the interval $[x_0, x_m]$ for $0 \leq r \leq k-1$ (here $k=2$). It means, the quadratic splines are continuous and have continuous first derivatives, S'_i on $[x_0, x_m]$, particularly at the knots.

Now this $S'(x)$ is continuous on $[x_0, x_m]$ if:

$$S'_i(x_i) = d_i \tag{8.88}$$

$$S'_i(x_{i+1}) = d_{i+1} \tag{8.89}$$

Using all these conditions defined in Equations (8.86), (8.88) and (8.89), one obtains (derivation shown below):

$$S_i(x) = y(x_i) + d_i(x - x_i) + \frac{d_{i+1} - d_i}{2\,(x_{i+1} - x_i)}(x - x_i)^2 \tag{8.90}$$

where,

$$d_{i+1} = -d_i + 2\left[\frac{y(x_{i+1}) - y(x_i)}{x_{i+1} - x_i}\right] \qquad i = 0, 1, 2, ..., m-1 \qquad (8.91)^{14}$$

for which, d_0 is arbitrarily chosen.

Derivation of quadratic spline function given in Equation (8.90)

Let us start with Equation (8.85).

Finding a_{0i}

At $x = x_i$, Equation (8.85) yields:

$$a_{0i} = y(x_i) = y_i \tag{8.92}$$

Finding a_{1i}

Differentiating Equation (8.85),

$$S'_i(x) = a_{1i} + 2a_{2i}(x - x_i) \tag{8.93}$$

At $x = x_i$, the above equation yields:

$$S'_i(x_i) = a_{1i}$$

Using Equation (8.88), we have the second coefficient as:

$$a_{1i} = d_i \tag{8.94}$$

Finding a_{2i}

Similarly at $x = x_{i+1}$, Equation (8.93) yields:

$$S'_i(x_{i+1}) = a_{1i} + 2a_{2i}(x_{i+1} - x_i) \tag{8.95}$$

---

14 This recursive formula is obtained from Equation (8.90) at $x = x_{i+1}$ with using Equation (8.87).

Using Equation (8.89), we have the third coefficient as:

$$a_{2i} = \frac{d_{i+1} - d_i}{2\,(x_{i+1} - x_i)} \tag{8.96}$$

Substituting the expressions of these three coefficients derived above [i.e., a_{0i} in Equation (8.92), a_{1i} in (8.94) and a_{2i} in (8.96)] in Equation (8.85), one can directly obtain Equation (8.90).

Remarks

1. To start Equation (8.91), one needs d_0. Basically, there is degree of freedom equals 1. To make the system completely specified, we assume the first derivative as zero at the first point (i.e., at $x = x_0$). Accordingly, we have $d_0 = 0$ [see Equation (8.88)]. Note that there are a number of other choices you may have for d_0.
2. For quadratic spline, the curvature at each knot changes abruptly, which may lead to cause an undesired bend or distortion in the spline curve. Basically, at those knots, the second derivative[15] of the quadratic spline is discontinuous.

   To make the first and second derivatives continuous, we will use piecewise cubic polynomials in the subsequent part of our study.

Example 8.9 Quadratic spline

Find a set of quadratic splines that passes through the data given in Table 8.14.

TABLE 8.14 Data for quadratic spline

i	x_i	y_i
0	−1	5
1	0	3
2	3	1
3	6	4

Solution

Let us assume $d_0 = 0$. With this, from Equation (8.91),

$$d_1 = -d_0 + 2\left[\frac{y_1 - y_0}{x_1 - x_0}\right] = 0 + 2\left[\frac{3-5}{1}\right] = -4 \qquad \text{for } i = 0$$

$$d_2 = -d_1 + 2\left[\frac{y_2 - y_1}{x_2 - x_1}\right] = 4 + 2\left[\frac{1-3}{3}\right] = \frac{8}{3} \qquad \text{for } i = 1$$

$$d_3 = -d_2 + 2\left[\frac{y_3 - y_2}{x_3 - x_2}\right] = -\frac{8}{3} + 2\left[\frac{4-1}{6-3}\right] = -\frac{2}{3} \qquad \text{for } i = 2$$

15 The second derivative of a function f measures the concavity of the graph of f.

From Equation (8.90), we get the quadratic splines as:

$$S_0(x)=5+\frac{-4}{2}(x+1)^2=5-2(x+1)^2 \qquad \text{for } x\in[-1,\ 0]$$

$$S_1(x)=3-4x+\frac{\frac{8}{3}+4}{2(3-0)}x^2=3-4x+\frac{10}{9}x^2 \qquad \text{for } x\in[0,\ 3]$$

$$S_2(x)=1+\frac{8}{3}(x-3)+\frac{-\frac{2}{3}-\frac{8}{3}}{2(6-3)}(x-3)^2=1+\frac{8}{3}(x-3)-\frac{5}{9}(x-3)^2 \quad \text{for } x\in[3,\ 6]$$

Remarks

1. The above set of three quadratic splines is correct and for confirmation, we check with the followings: $S_0(0)=S_1(0)=3$ and $S_1(3)=S_2(3)=1$.
2. Notice that like a linear spline, a quadratic spline is formulated using two data points (instead of three). Same number of data points will also be used to construct a cubic spline discussed below.

8.6.3 Cubic Spline

Spline function

To construct a cubic spline, we adopt a polynomial of degree 3 (i.e., $k=3$) in each subinterval $[x_i, x_{i+1}]$ as:

$$S_i(x)=y(x)=a_{0i}+a_{1i}(x-x_i)+a_{2i}(x-x_i)^2+a_{3i}(x-x_i)^3 \tag{8.97}$$

for $i=0,1,2,\ldots,m-1$. Here, a_{0i}, a_{1i}, a_{2i} and a_{3i} are constants (total $4m$) to be determined. Accordingly,

$$\begin{aligned}
S_0(x)&=a_{00}+a_{10}(x-x_0)+a_{20}(x-x_0)^2+a_{30}(x-x_0)^3 && \text{for } i=0\\
S_1(x)&=a_{01}+a_{11}(x-x_1)+a_{21}(x-x_1)^2+a_{31}(x-x_1)^3 && \text{for } i=1\\
&\vdots\\
S_{m-1}(x)&=a_{0\ m-1}+a_{1\ m-1}(x-x_{m-1})+a_{2\ m-1}(x-x_{m-1})^2+a_{3\ m-1}(x-x_{m-1})^3 && \text{for } i=m-1
\end{aligned} \tag{8.98}$$

Obviously, there are total $4m$ unknown constants/coefficients and to find them, we will follow the 'general definition of spline function', along with defining some *boundary conditions.*

General conditions

Let us formulate the general conditions based on the 'general definition of spline function' given earlier.

(a) In order to fit the data points, $(x_0, y_0), (x_1, y_1), \ldots, (x_m, y_m)$, the cubic splines must satisfy:

$$S_i(x_i) = y(x_i) \qquad i = 0, 1, 2, \ldots, m-1 \tag{8.86}$$

$$S_{m-1}(x_m) = y(x_m) \tag{8.99}$$

Total $m + 1$ equations

(b) By definition, $S^{(r)}$ is continuous on $[x_0, x_m]$ for $0 \le r \le 2$. This reveals that:

$$S_{i+1}(x_{i+1}) = S_i(x_{i+1}) \qquad i = 0, 1, 2, \ldots, m-2 \text{ and } r = 0 \tag{8.100}$$

$$S'_{i+1}(x_{i+1}) = S'_i(x_{i+1}) \qquad i = 0, 1, 2, \ldots, m-2 \text{ and } r = 1 \tag{8.101}$$

$$S''_{i+1}(x_{i+1}) = S''_i(x_{i+1}) \qquad i = 0, 1, 2, \ldots, m-2 \text{ and } r = 2 \tag{8.102}$$

Total $3m - 3$ equations

Basically, the first condition represented in Equation (8.100) ensures that the cubic splines are to be continuous, and the next two conditions in Equations (8.101) and (8.102) lead to provide smoothness at the knots.

Boundary conditions

At this moment, we have total $4m-2$ $[=(m+1)+(3m-3)]$ equations for $4m$ unknowns (i.e., underdetermined system with degrees of freedom = 2). To make the resulting system completely specified, let us choose the following endpoint constraints:

$$S''_0(x_0) = S''_{m-1}(x_m) = 0 \tag{8.103}$$

which are often called the *free boundary conditions.*

Alternatively, one can explore with the *clamped boundary conditions* given as:

$$S'_0(x_0) = y'(x_0) \tag{8.104}$$

$$S'_{m-1}(x_m) = y'(x_m) \tag{8.105}$$

provided values of y' at x_0 and x_m are known.

Equations for finding coefficients

Using the general and boundary conditions presented above with Equation (8.97), we derive the following set of equations that will be employed subsequently to find the coefficients of cubic splines.

- From Equation (8.86) [i.e., $S_i(x_i) = y(x_i)$],

$$y_i = a_{0i} \qquad i = 0, 1, 2, \ldots, m-1 \tag{8.106}$$

- From Equation (8.100) [i.e., $S_{i+1}(x_{i+1}) = S_i(x_{i+1})$],

$$a_{0\,i+1} = a_{0i} + a_{1i}(\Delta x_i) + a_{2i}(\Delta x_i)^2 + a_{3i}(\Delta x_i)^3 \qquad i = 0, 1, 2, \ldots, m-2 \tag{8.107}$$

since,

$$S_{i+1}(x_{i+1}) = a_{0\ i+1} + a_{1\ i+1}(x_{i+1} - x_{i+1}) + a_{2\ i+1}(x_{i+1} - x_{i+1})^2 + a_{3\ i+1}(x_{i+1} - x_{i+1})^3 = a_{0\ i+1}$$

$$S_i(x_{i+1}) = a_{0i} + a_{1i}(x_{i+1} - x_i) + a_{2i}(x_{i+1} - x_i)^2 + a_{3i}(x_{i+1} - x_i)^3$$

where,

$$\Delta x_i = x_{i+1} - x_i$$

- From Equation (8.101) [i.e., $S'_{i+1}(x_{i+1}) = S'_i(x_{i+1})$],

$$a_{1\ i+1} = a_{1i} + 2a_{2i}(\Delta x_i) + 3a_{3i}(\Delta x_i)^2 \qquad i = 0, 1, 2, \ldots, m-2 \tag{8.108}$$

since,

$$S'_{i+1}(x_{i+1}) = a_{1\ i+1} + 2a_{2\ i+1}(x_{i+1} - x_{i+1}) + 3a_{3\ i+1}(x_{i+1} - x_{i+1})^2 = a_{1\ i+1}$$

$$S'_i(x_{i+1}) = a_{1i} + 2a_{2i}(x_{i+1} - x_i) + 3a_{3i}(x_{i+1} - x_i)^2 = a_{1i} + 2a_{2i}(\Delta x_i) + 3a_{3i}(\Delta x_i)^2$$

- From Equation (8.102) [i.e., $S''_{i+1}(x_{i+1}) = S''_i(x_{i+1})$],

$$a_{2\ i+1} = a_{2i} + 3a_{3i}(\Delta x_i) \qquad i = 0, 1, 2, \ldots, m-2 \tag{8.109}$$

since,

$$S''_{i+1}(x_{i+1}) = 2a_{2\ i+1} + 6a_{3\ i+1}(x_{i+1} - x_{i+1}) = 2a_{2\ i+1}$$

$$S''_i(x_{i+1}) = 2a_{2i} + 6a_{3i}(x_{i+1} - x_i)$$

- From Equation (8.103) [i.e., $S''_0(x_0) = S''_{m-1}(x_m) = 0$],

$$a_{20} = 0 \tag{8.110}$$

$$2a_{2\ m-1} + 6a_{3\ m-1}(\Delta x_{m-1}) = 0 \tag{8.111}$$

since,

$$S''_i(x_i) = 2a_{2i} + 6a_{3i}(x_i - x_i) \text{ and so, } S''_0(x_0) = 2a_{20} = 0$$

$$S''_i(x_{i+1}) = 2a_{2i} + 6a_{3i}(x_{i+1} - x_i) \text{ and so,}$$

$$S''_{m-1}(x_m) = 2a_{2\ m-1} + 6a_{3\ m-1}(x_m - x_{m-1}) = 0$$

Finding coefficients

Using the derived Equations (8.106)–(8.111), let us find all $4m$ coefficients involved in the cubic splines.

Finding a_{0i}

It is found out before as:

$$a_{0i} = y_i \qquad i = 0,1,2,\ldots,m-1 \tag{8.106}$$

Finding a_{2i}

Eliminating a_{1i} and a_{3i} from the set of equations derived in the last section, we get:

- For $i = 0$

$$a_{20} = 0 \quad \text{[previously we got]} \tag{8.110}$$

- For $i = 1, 2, ..., m-2$

$$(\Delta x_{i-1})a_{2\ i-1} + 2(\Delta x_i + \Delta x_{i-1})a_{2i} + (\Delta x_i)a_{2\ i+1} = \frac{3}{\Delta x_i}(a_{0\ i+1} - a_{0i}) - \frac{3}{\Delta x_{i-1}}(a_{0i} - a_{0\ i-1})$$

(8.112)[16]

- For $i = m-1$

$$(\Delta x_{m-2})a_{2\ m-2} + 2(\Delta x_{m-2} + \Delta x_{m-1})a_{2\ m-1} = \frac{3}{\Delta x_{m-1}}(y_m - a_{0\ m-1}) - \frac{3}{\Delta x_{m-2}}(a_{0\ m-1} - a_{0\ m-2})$$

(8.113)[12]

Consolidating these equations related to a_{2i} in matrix form:

$$A\,A_2 = B \tag{8.114}$$

where,

$$A = \begin{bmatrix} 1 & 0 & \cdots & \cdots & 0 & 0 \\ \Delta x_0 & 2(\Delta x_0 + \Delta x_1) & \Delta x_1 & \cdots & 0 & 0 \\ \cdot & \cdot & \cdot & \cdots & \cdot & \cdot \\ \cdot & \cdot & \cdot & \cdots & \cdot & \cdot \\ 0 & 0 & 0 & \cdots & \Delta x_{m-2} & 2(\Delta x_{m-2} + \Delta x_{m-1}) \end{bmatrix} \tag{8.115}$$

$$A_2 = \begin{bmatrix} a_{20} \\ a_{21} \\ \cdot \\ \cdot \\ a_{2\ m-1} \end{bmatrix} \tag{8.116}$$

$$B = \begin{bmatrix} 0 \\ \frac{3}{\Delta x_1}(a_{02} - a_{01}) - \frac{3}{\Delta x_0}(a_{01} - a_{00}) \\ \cdot \\ \cdot \\ \frac{3}{\Delta x_{m-1}}(y_m - a_{0\ m-1}) - \frac{3}{\Delta x_{m-2}}(a_{0\ m-1} - a_{0\ m-2}) \end{bmatrix} \tag{8.117}$$

Solving this linear tridiagonal (near) system [i.e., Equation (8.114)], one can get a_{2i}, $i = 0, 1, 2, ..., m-1$.

16 Formal derivation of this equation is shown in Appendix 8A.

Finding a_{3i}

- For $i = 0, 1, 2, ..., m-2$

$$a_{3i} = \frac{a_{2\ i+1} - a_{2i}}{3(\Delta x_i)} \qquad \text{[from Equation (8.109)]} \tag{8.118}$$

- For $i = m - 1$

$$a_{3\ m-1} = -\frac{a_{2\ m-1}}{3(\Delta x_{m-1})} \qquad \text{[same with Equation (A8.9)]} \tag{8.119}$$

Finding a_{1i}

- For $i = 0, 1, 2, ..., m-2$

$$a_{1i} = \frac{1}{\Delta x_i}(a_{0\ i+1} - a_{0i}) - \frac{\Delta x_i}{3}(a_{2\ i+1} + 2a_{2i}) \qquad \text{[same with Equation (A8.3)]} \tag{8.120}$$

- For $i = m - 1$

$$a_{1\ m-1} = a_{1\ m-2} + \Delta x_{m-2}(a_{2\ m-1} + a_{2\ m-2}) \qquad \text{[from Equation (A8.5)]} \tag{8.121}$$

Remarks

1. In a similar fashion, one can find a_0, a_1, a_2 and a_3 using the clamped boundary conditions [i.e., Equations (8.104) and (8.105)], instead of the free boundary conditions [i.e., Equation (8.103)].
2. Cubic splines are most widely used in practice. The shortcomings of both the linear and quadratic splines are elaborated before. Further, as stated, the quartic and higher-order splines may have instability problem inherent in higher-order polynomials.
3. As mentioned before, here we have formulated the splines based on the shifted power form of the polynomial function. Similarly, one can use the Lagrange polynomial, for which, the book by Mathews (2000) may be consulted.

Example 8.10 Cubic spline

Fit a set of 2 cubic splines, $S_0(x)$ and $S_1(x)$, to a semicircle described by:

$$y(x) = \sqrt{16-(x-6)^2} \tag{8.122}$$

with center at (6, 0) and radius = 4. For this, use the data points given in Table 8.15.

Table 8.15 Given data for cubic spline

i	x_i	y_i
0	2	0
1	6	4
2	10	0

Solution

From Equation (8.98),

$$S_0(x) = a_{00} + a_{10}(x-2) + a_{20}(x-2)^2 + a_{30}(x-2)^3 \qquad \text{for } x \in [2,\ 6]$$

$$S_1(x) = a_{01} + a_{11}(x-6) + a_{21}(x-6)^2 + a_{31}(x-6)^3 \qquad \text{for } x \in [6,\ 10]$$

Finding a_{0i}

From Equation (8.106),

$$a_{00} = y_0 = 0$$

$$a_{01} = y_1 = 4$$

Here, $m = 2$ and $\Delta x_0 = \Delta x_1 = 4$.

Finding a_{2i}

For the system Equation (8.114), we have:

$$A = \begin{bmatrix} 1 & 0 \\ 4 & 16 \end{bmatrix} \qquad \text{[from Equation (8.115)]}$$

$$A_2 = \begin{bmatrix} a_{20} \\ a_{21} \end{bmatrix} \qquad \text{[from Equation (8.116)]}$$

$$B = \begin{bmatrix} 0 \\ \frac{3}{4}(0-4) - \frac{3}{4}(4-0) \end{bmatrix} = \begin{bmatrix} 0 \\ -6 \end{bmatrix} \qquad \text{[from Equation (8.117)]}$$

From Equation (8.114),

$$\begin{bmatrix} 1 & 0 \\ 4 & 16 \end{bmatrix} \begin{bmatrix} a_{20} \\ a_{21} \end{bmatrix} = \begin{bmatrix} 0 \\ -6 \end{bmatrix}$$

Solving,

$$a_{20} = 0$$

$$a_{21} = -\frac{6}{16} = -0.375$$

Finding a_{3i} and a_{1i}

Then we calculate,

$$a_{30} = \frac{a_{21} - a_{20}}{3(\Delta x_0)} = \frac{-0.375}{12} = -0.03125 \qquad \text{[from Equation (8.118)]}$$

$$a_{31} = -\frac{a_{21}}{3(\Delta x_1)} = \frac{0.375}{12} = 0.03125 \qquad \text{[from Equation (8.119)]}$$

$$a_{10} = \frac{1}{\Delta x_0}(a_{01} - a_{00}) - \frac{\Delta x_0}{3}(a_{21} + 2a_{20})$$ [from Equation (8.120)]

$$= \frac{4}{4} - \frac{4}{3}(-0.375 + 0) = 1.5$$

$$a_{11} = a_{10} + \Delta x_0(a_{21} + a_{20})$$ [from Equation (8.121)]

$$= 1.5 + 4(-0.375) = 0$$

So, we get the cubic splines as:

$$S_0(x) = 1.5(x-2) - 0.03125(x-2)^3 \quad \text{for } x \in [2,\ 6]$$

$$S_1(x) = 4 - 0.375(x-6)^2 + 0.03125(x-6)^3 \quad \text{for } x \in [6,\ 10]$$

These two derived splines are compared with the semicircle described by Equation (8.122) in Figure 8.8.

Verification

To check the correctness of the developed splines, we would like to verify them with the following conditions with respect to an internal point, say $x = 6$:

i) $S_0(6) = 4 = S_1(6)$

ii) $S'_0(6) = 0 = S'_1(6)$

iii) $S''_0(6) = -0.75 = S''_1(6)$

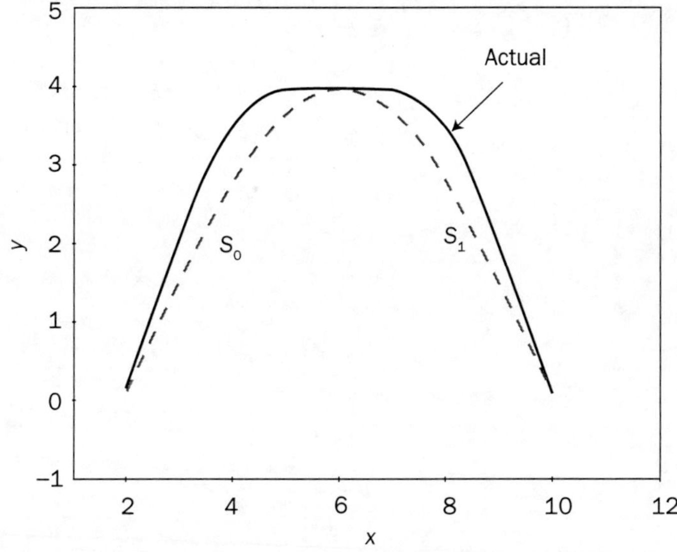

Figure 8.8 Comparing Equation (8.122) with two derived cubic splines, $S_0(x)$ and $S_1(x)$

Example 8.11 Comparing linear, quadratic and cubic splines

Develop the linear, quadratic and cubic splines based on the data given in Table 8.16, and compare their performance.

TABLE 8.16 Given data

i	x_i	y_i
0	3	4
1	5	2
2	8	4
3	10	1

Solution

Linear spline

Using Equation (8.83), we get:

$$S(x)=\begin{cases} S_0(x)=7-x & \text{for } x\in[3,\ 5] \\ S_1(x)=-\dfrac{4}{3}+\dfrac{2}{3}x & \text{for } x\in[5,\ 8] \\ S_2(x)=16-1.5x & \text{for } x\in[8,\ 10] \end{cases}$$

These linear spline functions are graphically illustrated in Figure 8.9a.

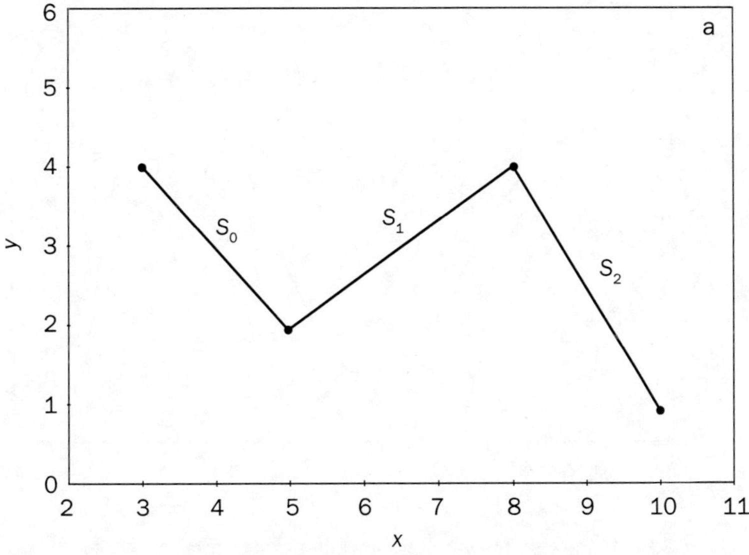

FIGURE 8.9a Piecewise polynomial interpolation (linear spline)

Quadratic spline

Let us assume $d_0 = 0$. With this, from Equation (8.91),

$$d_1 = -d_0 + 2\left[\frac{y_1 - y_0}{x_1 - x_0}\right] = -2 \qquad \text{for } i = 0$$

$$d_2 = -d_1 + 2\left[\frac{y_2 - y_1}{x_2 - x_1}\right] = \frac{10}{3} \qquad \text{for } i = 1$$

$$d_3 = -d_2 + 2\left[\frac{y_3 - y_2}{x_3 - x_2}\right] = -\frac{19}{3} \qquad \text{for } i = 2$$

From Equation (8.90), we get the quadratic splines as:

$$S_0(x) = 4 - \frac{1}{2}(x-3)^2 \qquad \text{for } x \in [3,\ 5]$$

$$S_1(x) = 12 - 2x + \frac{8}{9}(x-5)^2 \qquad \text{for } x \in [5,\ 8]$$

$$S_2(x) = -\frac{68}{3} + \frac{10}{3}x - \frac{29}{12}(x-8)^2 \qquad \text{for } x \in [8,\ 10]$$

With this, Figure 8.9b is produced to show how the developed quadratic splines fit the four data points.

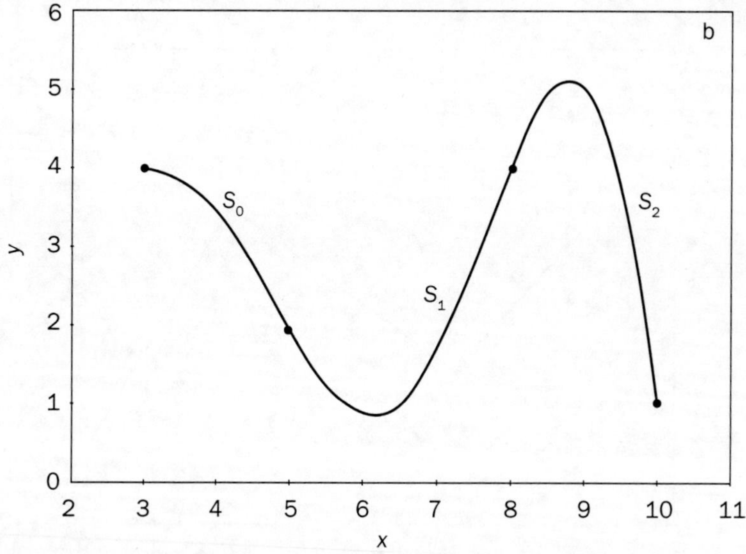

FIGURE 8.9b Piecewise polynomial interpolation (quadratic spline)

Cubic spline

From Equation (8.106),

$$a_{00} = y_0 = 4$$

$$a_{01} = y_1 = 2$$

$$a_{02} = y_2 = 4$$

Here, $m = 3$, $\Delta x_0 = \Delta x_2 = 2$ and $\Delta x_1 = 3$.

For the system Equation (8.114), we have:

$$\begin{bmatrix} 1 & 0 & 0 \\ 2 & 10 & 3 \\ 0 & 3 & 10 \end{bmatrix} \begin{bmatrix} a_{20} \\ a_{21} \\ a_{22} \end{bmatrix} = \begin{bmatrix} 0 \\ 5 \\ -6.5 \end{bmatrix}$$

Solving,

$$a_{20} = 0$$

$$a_{21} = 0.764$$

$$a_{22} = -0.879$$

Then we calculate,

$$a_{30} = \frac{a_{21} - a_{20}}{3(\Delta x_0)} = 0.1273$$

$$a_{31} = \frac{a_{22} - a_{21}}{3(\Delta x_1)} = -0.1825$$

$$a_{32} = -\frac{a_{22}}{3(\Delta x_2)} = 0.1465$$

$$a_{10} = \frac{1}{\Delta x_0}(a_{01} - a_{00}) - \frac{\Delta x_0}{3}(a_{21} + 2a_{20}) = -1.5093$$

$$a_{11} = \frac{1}{\Delta x_1}(a_{02} - a_{01}) - \frac{\Delta x_1}{3}(a_{22} + 2a_{21}) = 0.01767$$

$$a_{12} = a_{11} + \Delta x_1(a_{21} + a_{22}) = -0.3273$$

So, we get the cubic splines as:

$$S_0(x) = 4 - 1.5093(x-3) + 0.1273(x-3)^3 \qquad \text{for } x \in [3,\ 5]$$

$$S_1(x) = 2 + 0.01767(x-5) + 0.764(x-5)^2 - 0.1825(x-5)^3 \qquad \text{for } x \in [5,\ 8]$$

$$S_2(x) = 4 - 0.3273(x-8) - 0.879(x-8)^2 + 0.1465(x-8)^3 \qquad \text{for } x \in [8,\ 10]$$

With this, we produce Figure 8.9c for the cubic splines.

Remark

Note the following observations:

- As usual, for linear spline, there is no smoothness seen at the knots. The spline curve looks like a broken line.
- It is evident in Figure 8.9b that in the last two subintervals, the quadratic splines seem to swing too high. This is not there for cubic splines as shown in Figure 8.9c.

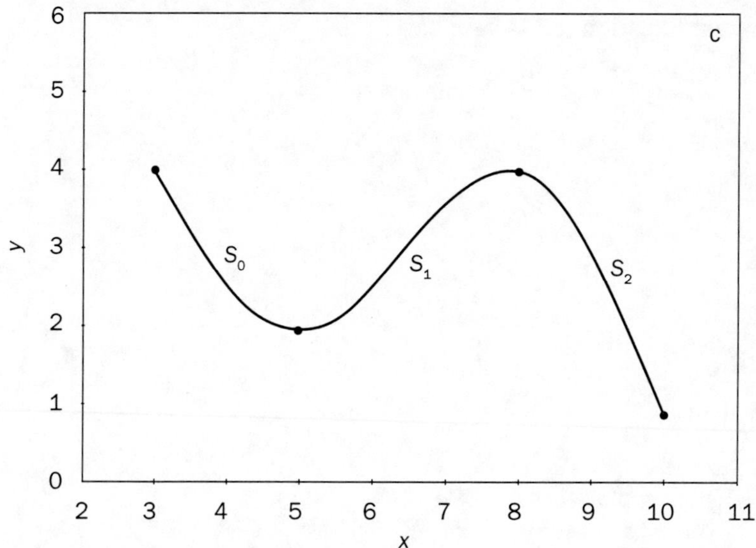

FIGURE 8.9c Piecewise polynomial interpolation (cubic spline)

8.7 Summary and Concluding Remarks

This chapter is devoted to fitting the interpolating polynomial based on the given set of discrete data points. Various interpolation schemes are formulated for both equally and unequally spaced data sets prior to their applications to several example systems. To avoid the oscillatory nature inherent in higher-order polynomials, piecewise interpolation schemes are subsequently developed with linear, quadratic and cubic splines. Before ending this chapter, these splines are illustrated with an example, showing their relative merits and demerits.

Appendix 8A

A8.1 Deriving Equation (8.29)

Let us use $y_m(x)$ to represent an interpolating polynomial of degree $\leq m$ and accordingly, Equation (8.21) is rewritten as:

$$y_m(x) = y[x_0] + y[x_1, x_0](x - x_0) + \ldots.. + y[x_m, x_{m-1}, \ldots., x_0](x - x_0)(x - x_1)\ldots(x - x_{m-1})$$

As stated before, for an interpolating polynomial:

$$y_m(x_i) = y(x_i) \qquad i = 0,\ 1,\ 2,\ \ldots,\ m$$

At any point ξ (other than $x_0, x_1, \ldots, x_m$), the error in interpolating polynomial is expressed as:

$$R_m(\xi) = y(\xi) - y_m(\xi)$$

Now consider a polynomial $y_{m+1}(x)$ of degree $\leq m+1$ that interpolates all points, $x_0, x_1, \ldots, x_m$, along with ξ. Accordingly,

$$y_{m+1}(x) = y[x_0] + y[x_1, x_0](x - x_0) + \ldots.. + y[x_m, x_{m-1}, \ldots.,\ x_0](x - x_0)(x - x_1)\ldots(x - x_{m-1}) +$$
$$y[\xi, x_m, x_{m-1}, \ldots.,\ x_0](x - x_0)(x - x_1)\ldots(x - x_m)$$

It yields:

$$y_{m+1}(x) = y_m(x) + y[\xi, x_m, x_{m-1}, \ldots.,\ x_0](x - x_0)(x - x_1)\ldots(x - x_m)$$

Since $y_{m+1}(x)$ interpolates ξ and so, $y_{m+1}(\xi) = y(\xi)$. With this, we have:

$$y(\xi) = y_{m+1}(\xi) = y_m(\xi) + y[\xi, x_m, x_{m-1}, \ldots.,\ x_0](\xi - x_0)(\xi - x_1)\ldots(\xi - x_m)$$

Rearranging,

$$R_m(\xi) = y(\xi) - y_m(\xi) = y\ [\xi,\ x_m,\ x_{m-1},\ \ldots.,\ x_0](\xi - x_0)(\xi - x_1)\ldots(\xi - x_m)$$

Derivation is complete here.

A8.2 Derivation of Equation (8.112)

Decreasing the index i by 1 in Equation (8.108),

$$a_{1i} = a_{1\ i-1} + 2a_{2\ i-1}(\Delta x_{i-1}) + 3a_{3\ i-1}(\Delta x_{i-1})^2 \qquad i = 1, 2, \ldots., m-1 \qquad \text{(A8.1)}$$

Similarly, Equation (8.109) yields:

$$a_{2i} = a_{2\ i-1} + 3a_{3\ i-1}(\Delta x_{i-1}) \qquad i = 1, 2, \ldots., m-1 \qquad \text{(A8.2)}$$

Solving Equation (8.109) for a_{3i} and substituting its value in Equation (8.107),

$$\begin{aligned} a_{1i} &= \frac{1}{\Delta x_i}(a_{0\ i+1} - a_{0i}) - \Delta x_i a_{2i} - \frac{\Delta x_i}{3}(u_{2\ i+1} - a_{2i}) \\ &= \frac{1}{\Delta x_i}(a_{0\ i+1} - a_{0i}) - \frac{\Delta x_i}{3}(a_{2\ i+1} + 2a_{2i}) \end{aligned} \qquad i = 0, 1, 2, \ldots., m-2 \qquad \text{(A8.3)}$$

Decreasing the index i by 1,

$$a_{1\ i-1} = \frac{1}{\Delta x_{i-1}}(a_{0i} - a_{0\ i-1}) - \frac{\Delta x_{i-1}}{3}(a_{2i} + 2a_{2\ i-1}) \qquad i = 1, 2, \ldots., m-1 \qquad \text{(A8.4)}$$

Similarly substitute Equation (A8.2) into (A8.1),

$$\begin{aligned} a_{1i} &= a_{1\ i-1} + 2a_{2\ i-1}(\Delta x_{i-1}) + \Delta x_{i-1}(a_{2i} - a_{2\ i-1}) \\ &= a_{1\ i-1} + \Delta x_{i-1}(a_{2i} + a_{2\ i-1}) \end{aligned} \qquad i = 1, 2, ..., m-1 \tag{A8.5}$$

Inserting a_{1i} [from Equation (A8.3)] and $a_{1\ i-1}$ [from Equation (A8.4)] into (A8.5):

$$\frac{a_{0\ i+1} - a_{0i}}{\Delta x_i} - \frac{\Delta x_i}{3}(a_{2\ i+1} + 2a_{2i}) = \frac{a_{0i} - a_{0\ i-1}}{\Delta x_{i-1}} - \frac{\Delta x_{i-1}}{3}(a_{2i} + 2a_{2\ i-1}) + \Delta x_{i-1}(a_{2i} + a_{2\ i-1}) \tag{A8.6}$$

for $i = 1, 2, ..., m-2$. It gives:

$$\frac{1}{3}\Delta x_{i-1}a_{2\ i-1} + a_{2i}\left(\frac{2}{3}\Delta x_i + \frac{2}{3}\Delta x_{i-1}\right) + \frac{1}{3}\Delta x_i a_{2\ i+1} = \frac{a_{0\ i+1} - a_{0i}}{\Delta x_i} - \frac{a_{0i} - a_{0\ i-1}}{\Delta x_{i-1}}$$

Further rearranging, we get Equation (8.112) for $i = 1, 2, ..., m-2$ and the derivation is complete.

A8.3 Derivation of Equation (8.113)

Let us start with Equation (8.99),

$$S_{m-1}(x_m) = y(x_m) \tag{8.99}$$

It gives:

$$a_{0\ m-1} + a_{1\ m-1}(x_m - x_{m-1}) + a_{2\ m-1}(x_m - x_{m-1})^2 + a_{3\ m-1}(x_m - x_{m-1})^3 = y_m$$

or,

$$a_{0\ m-1} + a_{1\ m-1}(\Delta x_{m-1}) + a_{2\ m-1}(\Delta x_{m-1})^2 + a_{3\ m-1}(\Delta x_{m-1})^3 = y_m \tag{A8.7}$$

Now we target to replace $a_{1\ m-1}$ and $a_{3\ m-1}$ by a_0 and a_2. Let us find the expressions for them.

Finding $a_{1\ m-1}$

Equation (A8.4) yields for $i = m-1$:

$$a_{1\ m-2} = \frac{1}{\Delta x_{m-2}}(a_{0\ m-1} - a_{0\ m-2}) - \frac{\Delta x_{m-2}}{3}(a_{2\ m-1} + 2a_{2\ m-2})$$

Substituting it into Equation (A8.5) for $i = m-1$:

$$\begin{aligned} a_{1\ m-1} &= a_{1\ m-2} + \Delta x_{m-2}(a_{2\ m-1} + a_{2\ m-2}) \\ &= \frac{1}{\Delta x_{m-2}}(a_{0\ m-1} - a_{0\ m-2}) - \frac{\Delta x_{m-2}}{3}(a_{2\ m-1} + 2a_{2\ m-2}) + \Delta x_{m-2}(a_{2\ m-1} + a_{2\ m-2}) \end{aligned}$$

Rearranging, we get:

$$a_{1\ m-1} = \frac{1}{\Delta x_{m-2}}(a_{0\ m-1} - a_{0\ m-2}) + \frac{\Delta x_{m-2}}{3}(2a_{2\ m-1} + a_{2\ m-2}) \quad \text{(A8.8)}$$

Finding $a_{3\ m-1}$

From Equation (8.111),

$$a_{3\ m-1} = -\frac{1}{3\Delta x_{m-1}} a_{2\ m-1} \quad \text{(A8.9)}$$

Plugging Equations (A8.8) and (A8.9) into (A8.7), and rearranging, one can easily get Equation (8.113).

Exercise

Review Questions

8.1 How to define an interpolating polynomial?

8.2 List the differences between:
- Least-squares method and polynomial interpolation
- Newton and Lagrange polynomial interpolation

8.3 When and why the interpolating polynomial is unique?

8.4 What is the main drawback of using the power form [Equation (7.25)] in polynomial interpolation?

8.5 Is there any difference between Newton forward divided difference and Newton forward difference scheme? Justify your answer.

8.6 Why cubic spline is better than quadratic spline?

8.7 Find the coefficients a_0, a_1 and a_2 for quadratic spline with $S''_0(x_0) = 0$ following the same methodology developed to find the coefficients for cubic spline (see Subsection 8.6.3).

8.8 For cubic spline [i.e., Equation (8.97)], find a_0, a_1, a_2 and a_3 using the clamped boundary conditions [i.e., Equations (8.104) and (8.105)].

8.9 Formulate the quadratic spline using:
- power form of polynomial
- Lagrange polynomial

8.10 List the equations involved in finding the coefficients of cubic spline for equispaced points.

Practice Problems

8.11 Are the following functions splines:

i) $$y(x) = \begin{cases} 4x & \text{for } x \in [0,\ 1] \\ 15x - 11 & \text{for } x \in [1,\ 2] \\ 7x + 5 & \text{for } x \in [2,\ 3] \end{cases}$$

ii) $$y(x) = \begin{cases} x + 2 & \text{for } x \in [-1,\ 1] \\ 2x + 1 & \text{for } x \in [1,\ 2] \\ 5x - 5 & \text{for } x \in [2,\ 3] \end{cases}$$

iii) $y(x)=\begin{cases}10x^2-3x+1 & \text{for } x\in[0,\ 1]\\ 7x^2+1 & \text{for } x\in[1,\ 2]\\ 5x^2+3x+3 & \text{for } x\in[2,\ 3]\end{cases}$

iv) $y(x)=\begin{cases}2-3.093(x-1)+0.573(x-1)^3 & \text{for } x\in[1,\ 5]\\ 5-0.389(x-5)+1.643(x-5)^2-0.325(x-5)^3 & \text{for } x\in[5,\ 7]\\ 1+0.271(x-7)-1.373(x-7)^2+0.3465(x-7)^3 & \text{for } x\in[7,\ 11]\end{cases}$

8.12 Using the Newton polynomial interpolation scheme (linear), find y at x = 2 based on the data given in Table 8.17.

Table 8.17 Given data

x	y
1	5
3	7

8.13 Using the Newton linear polynomial interpolation scheme, find y at x = 140. For this, the relevant data are given in Table 8.18.

Table 8.18 Given data

x	y
100	418.24
180	761.16

8.14 Using the Lagrange linear interpolation scheme, find y at x = 210 based on the given data in Table 8.19.

Table 8.19 Given data

x	y
200	1.1678
220	1.1892

8.15 The heat capacity (C_p) versus temperature (T) data are given for chloroform in Table 8.20.

Table 8.20 C_p versus T data

T (K)	C_p (J/mol.K)
300	66.01839
350	70.5569

Find temperature when C_p = 68.8128.

8.16 Construct the Newton quadratic polynomial that passes through the data points given in Table 8.21.

Table 8.21 Given data

x	y
−6	7
1	5
3	2

8.17 Develop the Newton quadratic polynomial that passes through the data points given in Table 8.22.

Table 8.22 Given data

x	y
1.15	2.13
2.37	3.68
5.19	2.62

8.18 Construct the Newton quadratic polynomial that passes through the data points given in Table 8.23.

Table 8.23 Given data

x	y
0.132	0.768
2.378	0.198
3.123	0.327

8.19 Develop the Newton quadratic polynomial that passes through the data points given in Table 8.24.

Table 8.24 Given data

x	y
6.32	9.18
10.38	12.74
19.37	2.38

Find y at $x = 10$ and 15.

8.20 Construct the Newton quadratic polynomial using the liquid molar volume versus temperature data given for methanol in Table 8.25.

Table 8.25 Given data

Temperature (K)	Liquid volume (cm^3/mol)
273.15	39.556
373.15	44.874
473.15	57.939

8.21 Construct the Newton forward divided difference interpolating polynomial (cubic) for the data given in Table 8.26.

Table 8.26 Pressure versus time data

Time (min)	Pressure (mm Hg)
2.5	10.5
5.0	12.5
10.0	15.8
15.0	18.0

8.22 For carbon dioxide, the heat capacity (C_p) versus temperature (T) data are recorded in Table 8.27.

Table 8.27 C_p versus T data for carbon dioxide

T (K)	C_p (J/mol.K)
300	37.25325
320	38.12632
350	39.37686
380	40.55897

Construct the Newton backward divided difference interpolating polynomial (cubic).

8.23 For chloroform, the enthalpy (H) versus temperature (T) data are recorded in Table 8.28.

Table 8.28 Enthalpy versus temperature data for chloroform

T (K)	H (J/mol)
300	14196.4042
325	15876.478
350	17613.2707
375	19403.1354
400	21242.5813

Construct a quartic polynomial by using the Newton divided difference interpolation.

8.24 The standard entropy (s^0) of air (ideal gas) is listed at various temperatures in Table 8.29 at 1 atm.

Table 8.29 Entropy data for ideal gas

T (°R)	s^0 (Btu/lbm.°R)
1000	1.78947
1040	1.79924
1080	1.80868
1120	1.81783
1160	1.82670

Find s^0 at $T = 1100$°R.

8.25 The specific volume (v) of superheated ammonia is listed at various temperatures in Table 8.30 at 400 psia.

Table 8.30 Specific volume of superheated ammonia

T (°F)	v (ft³/lbm)
200	0.8880
220	0.9338
240	0.9773
260	1.0192
280	1.0597
300	1.0992

Find v at $T = 250$°F.

8.26 The population of India has increased from 1961 to 2001 as shown in Table 8.31. Find the population in the year of 2000 by using the Newton divided difference formula.

Table 8.31 Population of India in every 10 years

Year	Population (in crore)
1961	43.92
1971	54.82
1981	68.33
1991	84.64
2001	102.87

8.27 Using the Newton divided difference formula, find a cubic interpolation to tan x based on the points, $x = 0, \frac{\pi}{2}, \frac{\pi}{3}$ and $\frac{\pi}{4}$.

8.28 Using the Newton–Gregory forward difference interpolation scheme, find $y(2.5)$ based on the data given in Table 8.32.

Table 8.32 Given data

i	x_i	y_i
0	1	1.23
1	2	2.18
2	3	3.27
3	4	4.23
4	5	5.87
5	6	6.72

8.29 Based on the given data set in Table 8.33, find *y*(2.5) using the Newton–Gregory forward difference interpolation scheme.

Table 8.33 Given data

i	x_i	y_i
0	2	−3.21
1	4	−1.73
2	6	−0.12
3	8	1.39
4	10	2.69

8.30 Using the enthalpy versus temperature data given for chloroform in Table 8.28, find the enthalpy at 360 K by employing the Newton–Gregory forward difference method.

8.31 Table 8.34 reports the concentration (*C*) versus time (*t*) in an adsorption process. Find concentration at t = 14.5 hr using the Newton–Gregory forward difference interpolation scheme.

Table 8.34 Concentration versus time data

i	t (hr)	C (ppm)
0	12	0
1	13	50
2	14	580
3	15	950
4	16	1420

8.32 Repeat the above problem to use the Newton–Gregory backward difference interpolating polynomial for finding the concentration at t = 14.5 hr.

8.33 Repeat Problem 8.26, in which, the population of India is given in every 10 years. Find the population in the year of 2000 by using the:

i) Newton–Gregory forward difference method
ii) Newton–Gregory backward difference method
iii) Lagrange method

8.34 Using the Newton–Gregory backward difference interpolating polynomial, find *y*(3.4) based on the data given in Table 8.35.

Table 8.35 Given data

i	x_i	y_i
0	0	1
1	1	2
2	2	4
3	3	6
4	4	8
5	5	10

8.35 Find the Lagrange interpolating polynomial of degree 2 when $y(0) = 1$, $y(2) = 3$ and $y(4) = 7$.

8.36 Repeat the above problem for Newton divided difference interpolating polynomial of degree 2 and compare the results.

8.37 Find the coefficients, a_0, a_1, a_2 and a_3, of the following formula:

$$Q(x) = a_0(x-1)^3 + a_1(x-1)^2 + a_2(x-1) + a_3$$

when,

$$Q(1) = y(1)$$
$$Q'(1) = y'(1)$$
$$Q(a) = y(a)$$
$$Q'(a) = y'(a)$$

8.38 Derive the Lagrange interpolating polynomial for the data points given at unequal intervals in Table 8.36.

TABLE 8.36 Given data

i	x_i	y_i
0	0	−1
1	1	1
2	2	6
3	3	26
4	5	126

Using the fitted polynomial, compute $y(4)$.

8.39 Construct the Lagrange interpolating polynomial for the data points given in Table 8.37.

TABLE 8.37 Given data

i	x_i	y_i
0	−3.8	3.18
1	−2.1	6.19
2	−1.0	10.21
3	2.0	13.64
4	3.6	15.39
5	4.9	17.31

Using the fitted polynomial, compute $y(4)$.

8.40 Construct a first-degree spline using the data given in Table 8.38.

Table 8.38 Given data

i	x_i	y_i
0	1.2	−3.173
1	3.7	−1.218
2	5.4	1.113
3	8.3	3.724
4	10	5.897

Use the resulting spline to approximate $y(7.5)$.

8.41 Construct a first-degree spline using the data given in Table 8.39.

Table 8.39 Given data

i	x_i	y_i
0	1	1
1	3	2.5
2	5	4
3	7	5.2
4	9	7.3
5	11	10

Use the resulting spline to approximate $y(10)$.

8.42 Table 8.40 lists the upward velocity (v) versus time (t) data for a rocket. Find the velocity at $t = 12$ sec developing the linear splines.

Table 8.40 Velocity versus time data

t (sec)	v (m/sec)
0	0
10	200
15	350
20	500
25	650

8.43 Fit the data given in Table 8.41 to a set of 3 quadratic splines.

TABLE 8.41 Data for 3 quadratic splines

x	y
0	4
2	8
4	16
6	38

8.44 Find the quadratic splines that pass through the data given in Table 8.42.

TABLE 8.42 Data for quadratic splines

x	y
−1	3
2	5
5	7
7	1

8.45 Find the quadratic splines that pass through the data given in Table 8.43.

TABLE 8.43 Data for quadratic splines

x	y
−2	3.24
3	5.19
7	8.32
9	2.60

8.46 Develop the quadratic splines by the use of the data given in Table 8.44.

TABLE 8.44 Data for quadratic splines

x	y
−3.21	5.16
−1.12	3.12
0	1.11
2.67	2.37
3.39	10.13

8.47 Construct a set of 2 cubic splines based on the data given in Table 8.45.

Table 8.45 Given data

i	x_i	y_i
0	–2	6
1	0	9
2	3	5

8.48 Construct a set of 3 cubic splines based on the data given in Table 8.46.

Table 8.46 Given data

i	x_i	y_i
0	0	0
1	1	0.5
2	2	3
3	3	2

8.49 Construct a set of 4 cubic splines based on the data given in Table 8.47.

Table 8.47 Given data

i	x_i	y_i
0	5	5
1	8	2
2	9	7
3	11	1
4	15	9

8.50 Develop a set of 4 cubic splines using the data given in Table 8.48.

Table 8.48 Data for cubic splines

x	y
–5	2.139
0	3.784
4	9.810
7	15.632
9	28.651

Find $y(5)$ and $y(8)$.

8.51 The population of the United States starting from 1900 is given in Table 8.49. Fit a set of cubic splines and find the population in 1965.

TABLE 8.49 Population of the United States

Year	Population
1900	75,994,570
1920	105,710,623
1940	131,669,270
1960	179,323,170
1980	226,504,820

8.52 Develop the linear, quadratic and cubic splines based on the data given in Table 8.50, and compare them graphically.

TABLE 8.50 Given data

i	x_i	y_i
0	1	6
1	3	3
2	6	6
3	8	2

8.53 Use the data set given in Table 8.40 to compare the results (in producing a plot) between the linear, quadratic and cubic splines.

8.54 Use the data set given in Table 8.49 to compare the results (in a comparative plot) of the linear, quadratic and cubic splines.

REFERENCE

Mathews, J. H. (2000). *Numerical Methods for Mathematics, Science, and Engineering*, 2nd ed., New Delhi: Prentice-Hall.

9

Numerical Integration

"The experimental verification of a theory concerning any natural phenomenon generally rests on the result of an integration." – Joseph William Mellor

Key Learning Objectives

- Reviewing the basics of numerical integration
- Knowing commonly used Newton–Cotes integration technique
- Studying Richardson extrapolation for Romberg integration
- Knowing Gauss quadrature for more efficient integration
- Learning how to choose a suitable numerical integration method

9.1 Introduction

Here, our primary objective is to compute the integral I of the following form:

$$I = \int_a^b f(x)dx \tag{9.1}$$

It basically stands for the integral of the function, $f(x)$[1] with respect to x (independent variable) over a given interval $[a, b]$. Prior to doing integration, we need the fitted function, $f(x)$. In the last chapter, we discussed various methods to determine this function based on the specified data points. Here we fix our subsequent goal to do the integration of $f(x)$ having its polynomial form.

The integral I in Equation (9.1) geometrically represents the area under the curve $y = f(x)$ for a certain range of x. As shown in Figure 9.1, it is basically the area of the region bounded by the curve $y = f(x)$, the lines $x = a = x_0$ and $x = b = x_n$, and the x-axis. On the other hand, the definite integral I in Equation (9.1) can be evaluated algebraically having the function $f(x)$ as a polynomial. This scheme is often called the *numerical quadrature.*

1 This is called *integrand.*

The *numerical integration* or *quadrature*[2] formula is used to obtain approximate answers for definite integrals, which are either impossible or difficult to evaluate analytically. This is true particularly when the function is complicated, which is not so uncommon in realistic examples. Moreover, we employ the numerical integration approach to integrate an unknown function, for which, only a few sample points are available.

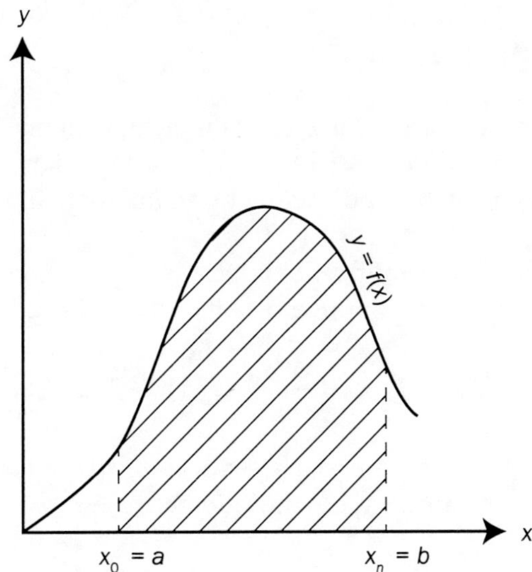

FIGURE 9.1 Area under the curve $y = f(x)$ over $[a, b]$

Numerical integration has various applications in several disciplines. A popular example includes the determination of enthalpy (H) at any temperature (T) from:

$$H = H^0 + \int_{T_0}^{T} C_p \, dT$$

In which, C_p is the heat capacity that is usually available as a polynomial function of T. This apart, there are many other applications in mass transfer, reaction engineering, and so on.

9.2 Methods of Numerical Integration

We will explore various ways of approximating the integral of a function over a given domain. With this goal, here the following integration approaches we would like to cover:

- Newton–Cotes method
  - Trapezoidal rule
  - Simpson 1/3 rule
  - Simpson 3/8 rule
  - Boole rule

2 Quadrature, although frequently used today as a synonym of numerical integration, is actually an archaic term, which means the formation of a square with the same area as some curvilinear figure.

- Romberg integration
- Gauss–Legendre quadrature
  - 1-point formula
  - 2-points formula
  - 3-points formula

9.3 Newton–Cotes Integration

As stated, quadrature formulation is done based on the polynomial interpolation. *Newton–Cotes*[3] scheme is the most widely used numerical integration strategy. It works with replacing a function *f*(x) or a tabulated data set by an interpolating polynomial to estimate the integral.

With this, the Newton forward difference interpolating polynomial [Equation (8.52)] of equally spaced data is to be used to approximate *f*(x). It has the following mth-order form:

$$y=f(x)=y_0+\Delta y_0 s+\frac{\Delta^2 y_0}{2!}s(s-1)+.....+\frac{\Delta^m y_0}{m!}s(s-1)(s-2)....(s-m+1)+R_m(\xi) \tag{9.2}$$

Recall that it is indeed the *Newton–Gregory forward difference* interpolation formula. As usual, $y_0 = y(x_0)$ and the remainder (i.e., truncation error) is:

$$R_m(\xi)=\frac{y^{(m+1)}(\xi)}{(m+1)!}(\Delta x)^{m+1}s(s-1)......(s-m) \tag{8.64}$$

With this background, we will discuss the following Newton–Cotes integration methods: trapezoidal rule, Simpson rule and Boole rule. Note that these are actually called the *closed* Newton–Cotes formula[4] since we use the two end points of the integration domain, namely $x_0 = a$ and $x_n = b$.

9.4 Trapezoidal Rule (m = 1)

The Newton–Cotes integration scheme starts with the trapezoidal rule, which is one of the simplest methods employed to find the area under a curve. The trapezoidal rule is developed on the basis of the linear interpolation of the function (i.e., first-degree polynomial). Accordingly, from Equation (9.2) with m = 1:

$$y=f(x)=y_0+\Delta y_0 s+\frac{y''(\xi)}{2!}(\Delta x)^2 s(s-1) \tag{9.3}$$

---

3 Named after Sir Isaac Newton, PRS (1643–1727) and English mathematician Roger Cotes, FRS (1682–1716).

4 Open Newton–Cotes formulas do not include the end points (a and b) as nodes/abscissas.

Notice that the last right-hand term refers to the remainder (truncation error). Plugging the above equation in Equation (9.1) over the interval $[x_0, x_1]$:

$$I = \int_{x_0}^{x_1} f(x)dx = \int_{x_0}^{x_1}\left[y_0 + \Delta y_0 s + \frac{y''(\xi)}{2}(\Delta x)^2 s(s-1) \right] dx \tag{9.4}$$

Here the two endpoints are $a = x_0$ and $b = x_1$. Let us represent the right hand terms of the above equation in terms of s (and its derivative), instead of x (and its derivative). Since [from Equation (8.49)]

$$s = \frac{x - x_0}{\Delta x}$$

thus,

$$dx = \Delta x.ds$$

Here, both x_0 and Δx are the fixed quantities.

Equation (9.4) yields:

$$I = \Delta x \int_{s=0}^{s=1}\left[y_0 + \Delta y_0 s + \frac{y''(\xi)}{2}(\Delta x)^2 s(s-1) \right] ds$$

Integrating and then applying Equation (8.38) [i.e., $\Delta y_i = y_{i+1} - y_i$],

$$I = \frac{\Delta x}{2}(y_0 + y_1) - \frac{(\Delta x)^3}{12} f''(\xi) \tag{9.5}$$

Neglecting the last term (i.e., the error),

$$I = \frac{\Delta x}{2}(y_0 + y_1) \tag{9.6}$$

This is the formula for the *trapezoidal rule.*

Deriving Trapezoidal Rule Geometrically

The trapezoidal rule is derived above by algebraically integrating the equation of a straight line that connects the two points. Alternatively, one can derive the same formula by finding the area of the trapezoid (Figure 9.2) bounded by $x = x_0$, $x = x_1$, straight line T and x-axis.

Referring to Figure 9.2, the area of trapezium is:

$$\frac{\Delta x}{2}(y_0 + y_1)$$

This is basically an approximation of the integral I, which means:

$$I \approx \frac{\Delta x}{2}(y_0 + y_1)$$

It is nothing but the trapezoidal rule. This scheme is also called as the *trapezoid* or *trapezium rule* since this formula is equivalent to the area of a trapezium.

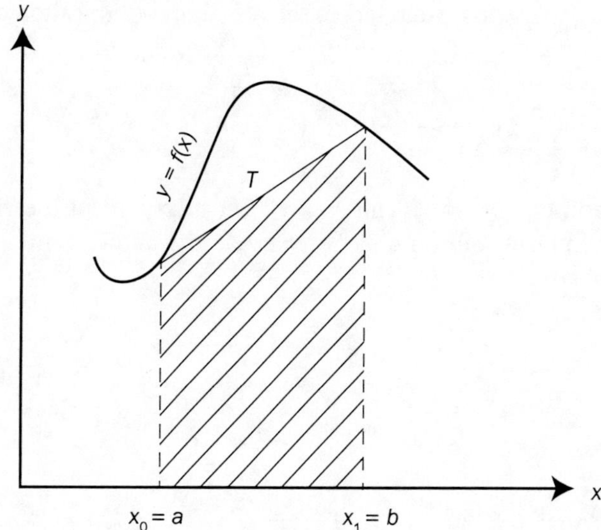

FIGURE 9.2 Illustrating the trapezoidal rule

Rewriting Equation (9.6) with reference to Figure 9.2:

$$I=\int_{x_0}^{x_1} f(x)dx=\frac{\Delta x}{2}(y_0+y_1)$$

$$=(b-a)\times\frac{(y_0+y_1)}{2}$$

$$=\text{width}\times\text{average height}$$

It is worth emphasizing that all the Newton–Cotes formulas follow this general format, in which, *width* corresponds to the difference of two end limits. These formulas only differ in terms of the *average height*.

Truncation error

In the trapezoidal rule depicted in Figure 9.2, the area under the curve is approximated by that under a straight line. Consequently, this leads to generate an error (i.e., truncation error) and it is pictorially illustrated in Figure 9.3.

Let us represent this truncation error by a mathematical expression. Recall that the last term in Equation (9.5) corresponds to this error of a single application of the trapezoidal rule:

$$R(\xi)=-\frac{(\Delta x)^3}{12}f''(\xi) \tag{9.7}$$

Indeed, it gives an idea of how the error depends on the step size Δx (discussed later with Table 9.2) and the smoothness of the function.

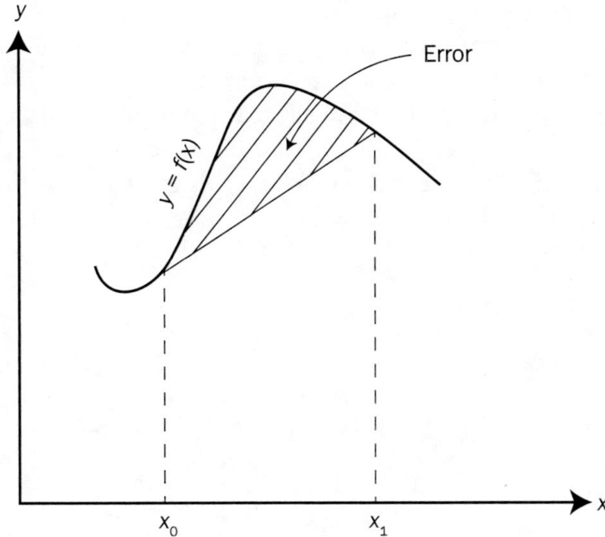

FIGURE 9.3 Truncation error in trapezoidal rule

Remarks

1. Instead of $f(x)$, if the data points (x_0, y_0) and (x_1, y_1) are given, one can also use the trapezoidal rule [Equation (9.6)] to evaluate the integral.
2. It is obvious from Equation (9.7) that the truncation error scales as a cube of the interval size, Δx. Accordingly, if this size is halved (i.e., $\Delta x/2$) as shown in Figure 9.4, the truncation error gets reduced by a factor of 8.
3. The trapezoidal rule provides cent percent accuracy (i.e., exact result) when $f''(\xi)=0$, $\xi \in [x_0, x_1]$ in Equation (9.5). It is possible only if $y = f(x)$ is a straight line (i.e., $m = 1$) and thus, the degree of precision of the trapezoidal formula is said to be 1.

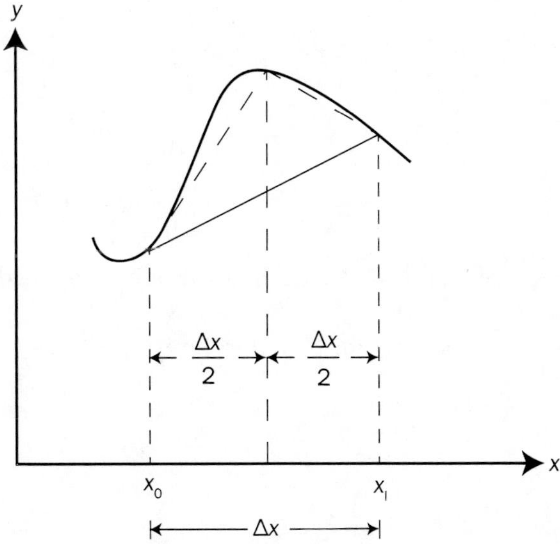

FIGURE 9.4 Halving interval size in trapezoidal rule

Multiple or Composite Trapezoidal Rule

The trapezoidal rule derived above is with using a single interpolant over the entire interval $x \in [a, b]$. In order to improve the accuracy, one can divide this integration interval from a to b into a set of segments/subintervals (Figure 9.5) and then apply the formula to each of those subintervals in a piecewise fashion. Adding the areas of individual segments, the integral can be obtained for the entire interval. This is referred as the *multiple* or *composite trapezoidal rule.*

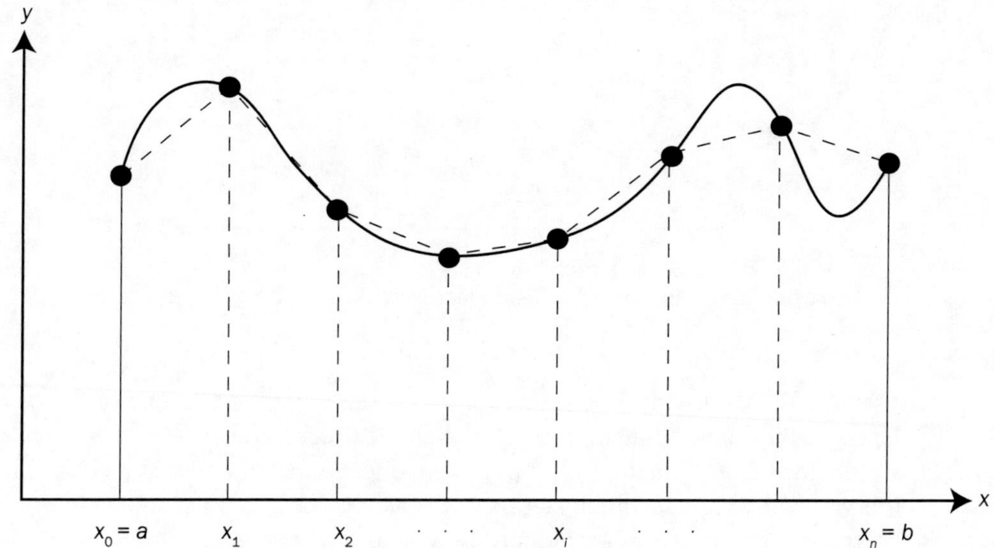

FIGURE 9.5 Illustrating multiple trapezoidal rule

Following this concept of composite rule, we have:

$$I = \int_{x_0}^{x_1} f(x)dx + \int_{x_1}^{x_2} f(x)dx + ... + \int_{x_{n-1}}^{x_n} f(x)dx \tag{9.8}$$

Applying Equation (9.5) to all n subintervals,

$$I = \frac{\Delta x}{2}(y_0 + y_1) + \frac{\Delta x}{2}(y_1 + y_2) + ... + \frac{\Delta x}{2}(y_{n-1} + y_n) - \frac{(\Delta x)^3}{12}\sum_{i=0}^{n-1} f''(\xi_i) \tag{9.9}$$

where, $x_i \leq \xi_i \leq x_{i+1}$. Notice that the last term in right side is obtained by summing up the individual errors for each subinterval.

For the equispaced *nodes* (or *abscissas* or *quadrature points*),

$$x_i = x_0 + i\Delta x \qquad \text{for } i = 0, 1, 2, ..., n \tag{9.10a}$$

that gives:

$$\Delta x = \frac{x_n - x_0}{n} = \frac{b - a}{n} \tag{9.10b}$$

Rearranging Equation (9.9) with the use of mean value theorem,

$$I=\int_a^b f(x)dx=\frac{\Delta x}{2}[y_0+2y_1+2y_2+.....+2y_{n-1}+y_n]-\frac{(\Delta x)^3}{12}n f''(\bar{\xi}) \tag{9.11}$$

where, $\bar{\xi}\in[a,\ b]$. This yields the *multiple* or *composite trapezoidal formula* as:

$$I=\int_a^b f(x)dx=\frac{\Delta x}{2}[y_0+2y_1+2y_2+.....+2y_{n-1}+y_n] \tag{9.12}$$

Using Equation (9.10b), the composite formula can further be written as:

$$I=\frac{\Delta x}{2}\left[y_0+2\sum_{i=1}^{n-1}y_i+y_n\right]$$

$$=\underbrace{(b-a)}_{\text{width}}\ \underbrace{\frac{\left[y_0+2\sum_{i=1}^{n-1}y_i+y_n\right]}{2n}}_{\text{average height}}$$

Truncation error

The truncation error involved in the composite trapezoidal rule is quite obvious in Equation (9.11) as:

$$R(\xi)=-\frac{(\Delta x)^3}{12}n f''(\bar{\xi})$$

Combining with Equation (9.10b), we get:

$$R(\xi)=-\frac{(b-a)}{12}(\Delta x)^2 f''(\bar{\xi})\equiv o[(\Delta x)^2] \tag{9.13}$$

It is fairly true that we do not know ξ at which the error needs to be evaluated. Thus, it is often recommended to take the mean or average value of $f''(\bar{\xi})$ for the entire interval.

Remarks

1. The composite trapezoidal rule [Equation (9.12)] is obtained by typically fitting the straight lines to the points $(x_0,\ y_0)$ and $(x_1,\ y_1)$ first, then to $(x_1,\ y_1)$ and $(x_2,\ y_2)$, and so on. It indicates that this composite version is applicable to both the odd and even number of subintervals.
2. It is quite obvious that the composite version of the trapezoidal rule corresponds to $n\geq 2$. For a single application of the trapezoidal rule, $n=1$.
3. Notice from Equation (9.13) that the composite trapezoidal rule is second-order accurate. Recall that this trapezoidal rule in single application [Equation (9.7)] is third-order accurate. This issue concerning the local truncation error versus global truncation error is detailed in Chapter 1.

Example 9.1 Trapezoidal rule

(a) Compute the integral:

$$I = \int_0^3 (x^2+1)dx$$

using the trapezoidal rule ($n = 1$) and composite trapezoidal rule with $n = 3$
(b) Compare the numerical results with the analytical value
(c) Estimate the truncation error in both cases (i.e., $n = 1$ and 3)

Solution
In this problem, the integrand is:

$$f(x) = x^2 + 1$$

For this, we do the following calculations.

(a) Trapezoidal rule with $n = 1$
From Equation (9.10b),

$$\Delta x = \frac{b-a}{n} = \frac{3-0}{1} = 3$$

Using Equation (9.6), we get the integral as:

$$I = \frac{\Delta x}{2}(y_0 + y_1) = \frac{3}{2}(1+10) = 16.5$$

Composite trapezoidal rule with $n = 3$
In this case,

$$\Delta x = \frac{b-a}{n} = \frac{3-0}{3} = 1$$

Using Equation (9.12), we get the integral as:

$$I = \frac{\Delta x}{2}(y_0 + 2y_1 + 2y_2 + y_3) = \frac{1}{2}(1+4+10+10) = 12.5$$

(b) Analytical solution

$$I = \int_0^3 (x^2+1)dx = \left[\frac{x^3}{3} + x\right]_0^3 = 9+3 = 12$$

One can now easily compare the numerical and analytical values.

(c) Truncation error
For the trapezoidal rule with $n = 1$, the truncation error is:
$R = 16.5 - 12 = 4.5$ (i.e., 37.5%)

One can obtain the same value (absolute) from Equation (9.7) since $f'' = 2$.

Similarly, for the multiple trapezoidal rule with $n = 3$,

$R = 12.5 - 12 = 0.5$ (i.e., 4.17%)

The same value (absolute) one can obtain from Equation (9.13) with $f'' = 2$.

Example 9.2 Composite trapezoidal rule

Use the composite trapezoidal rule to compute the integral:

$$I = \int_{0}^{0.3} f(x)dx$$

based on the data given in Table 9.1.

TABLE 9.1 Given data for multiple trapezoidal rule

x	f(x)
0	1
0.1	3
0.2	2
0.3	5

Solution

In this problem,

$$\Delta x = \frac{b-a}{n} = \frac{0.3-0}{3} = 0.1$$

Using Equation (9.12),

$$I = \frac{\Delta x}{2}(y_0 + 2y_1 + 2y_2 + y_3) = \frac{0.1}{2}(1+6+4+5) = 0.8$$

Remark

It is shown that instead of $f(x)$, if the data points are given, one can also use the trapezoidal rule to evaluate the integral.

Example 9.3 Composite trapezoidal rule (error analysis)

For the following integral,

$$I = \int_{2}^{5} \frac{6}{x} dx$$

(a) find the number of subintervals (n) and the step size (Δx), for which, the truncation error remains less than or equal to 10^{-6}.

(b) determine the number of evaluations of $f(x)$.

Solution

(a) Here the integrand is:

$$f(x)=\frac{6}{x}$$

The derivatives of $f(x)$ are:

$$f'(x)=-\frac{6}{x^2} \qquad f''(x)=\frac{12}{x^3}$$

The maximum value of $|f''(x)|$ in the interval [2,5] is obtained at $x_0=a=2$. It reveals that:

$$|f''(\xi)|\leq|f''(2)|=\frac{3}{2}$$

for $2\leq\xi\leq5$.

Based on Equation (9.13), the error bound is obtained as:

$$|R(\xi)|=\left|-\frac{(b-a)}{12}(\Delta x)^2 f''(\bar{\xi})\right|\leq\frac{3}{12}\frac{3}{2}(\Delta x)^2=\frac{9}{24}(\Delta x)^2$$

Using Equation (9.10b) [i.e., $\Delta x = 3/n$],

$$\frac{9}{24}\left(\frac{3}{n}\right)^2\leq10^{-6}$$

It gives:

$$n\geq1837.12$$

It is true that n must be an integer, and thus, we adopt $n=1838$, which corresponds to $\Delta x=3/n=0.001632208923$.

(b) The composite trapezoidal rule with $n=1838$ involves total 1839 evaluations of $f(x)$ that is obvious from Equation (9.12).

A Short Note

The results of composite trapezoidal rule with varying the number of subintervals (n) are summarized in Table 9.2 for Example 9.1. It is evident that the error decreases with the increase of the number of subintervals. In fact, when n is doubled, the truncation error gets quartered. The reason behind this one can find from the following expression [obtained by rewriting Equation (9.13)]:

$$R(\xi)=-\frac{(b-a)^3}{12n^2}f''(\bar{\xi})$$

that R is inversely related with n^2.

At the same time, we notice that the rate of convergence is gradually slowing down with n. It is also reflected in Figure 9.6. In order to improve the convergence rate and accuracy level, the higher-order Newton–Cotes formulas are developed in the subsequent sections.

TABLE 9.2 Results of composite trapezoidal rule

n	Δx	I	%error
2	1.5	13.125	9.38
3	1	12.5	4.17
4	0.75	12.28	2.33 (~ ¼ of 9.38%)
5	0.6	12.18	1.5
6	0.5	12.125	1.04 (~ ¼ of 4.17%)
.	.	.	.
.	.	.	.
10	0.3	12.045	0.375 (= ¼ of 1.5%)

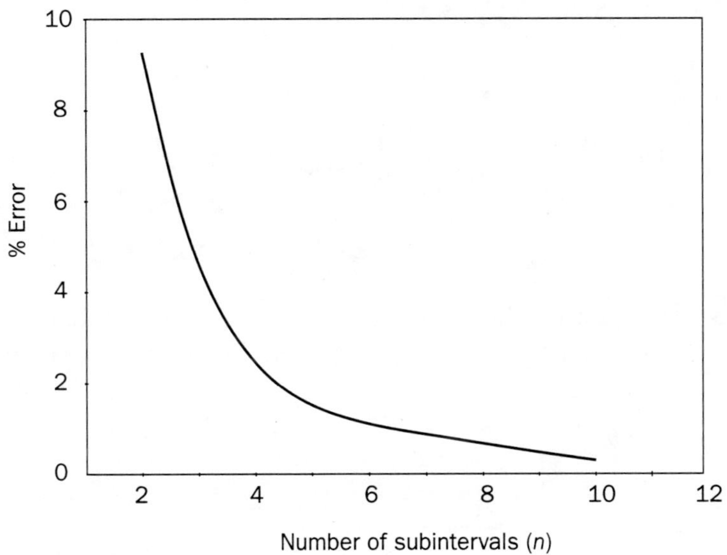

FIGURE 9.6 Rate of convergence

9.5 Simpson 1/3 Rule ($m = 2$)

In the case of trapezoidal rule, two end points of the region of integration are joined with a straight line. To improve the accuracy as well as the rate of convergence, an intermediate point (i.e., midpoint) is included in between the two end points as shown in Figure 9.7. Then, a quadratic curve or parabola (instead of a straight line used in trapezoidal rule) is employed to fit to these three points. This scheme is popularly called the *Simpson 1/3 rule.*[5]

5 Named after the British mathematician Thomas Simpson, FRS (1710–1761).

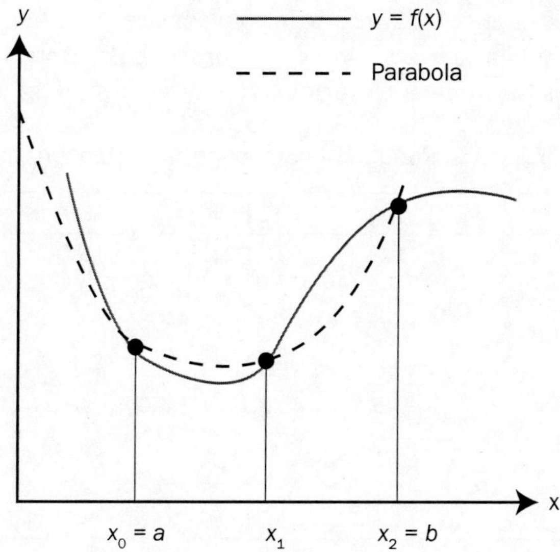

FIGURE 9.7 Illustrating Simpson 1/3 rule

From Equation (9.2) with $m = 3$:

$$y=f(x)=y_0+\Delta y_0 s+\frac{\Delta^2 y_0}{2!}s(s-1)+\frac{\Delta^3 y_0}{3!}s(s-1)(s-2)$$
$$+\frac{y^{(4)}(\xi)}{4!}(\Delta x)^4 s(s-1)(s-2)(s-3) \tag{9.14}$$

Perhaps you are wondering why $m = 3$ (instead of $m = 2$) is considered for the approximation by a quadratic curve; this will be clear in a few moments [one can visit Equation (9.17) in advance].

Referring to Figure 9.7,

$$\Delta x = \frac{x_2 - x_0}{2} = \frac{b-a}{2} \tag{9.15}$$

Obviously, the step size is half of the original domain length. So the integral over the interval $[x_0, x_2]$,

$$I = \int_a^b f(x)dx$$
$$= \int_{x_0}^{x_2}\left[y_0+\Delta y_0 s+\frac{\Delta^2 y_0}{2}s(s-1)+\frac{\Delta^3 y_0}{6}s(s-1)(s-2)+\frac{f^{(4)}(\xi)}{24}(\Delta x)^4 s(s-1)(s-2)(s-3)\right]dx$$
$$= \Delta x\int_{s=0}^{s=2}\left[y_0+\Delta y_0 s+\frac{\Delta^2 y_0}{2}s(s-1)+\frac{\Delta^3 y_0}{6}s(s-1)(s-2)+\frac{f^{(4)}(\xi)}{24}(\Delta x)^4 s(s-1)(s-2)(s-3)\right]ds \tag{9.16}$$

Since,

$$\int_0^2 s(s-1)(s-2)ds = \left[\frac{s^4}{4} - s^3 + s^2\right]_0^2 = 0 \tag{9.17}$$

we have from Equation (9.16),

$$I = \Delta x\left[2y_0 + 2\Delta y_0 + \frac{\Delta^2 y_0}{3} - \frac{1}{90} f^{(4)}(\xi)(\Delta x)^4\right] \tag{9.18}$$

Again we know from Equations (8.38) and (8.39), respectively,

$$\Delta y_i = y_{i+1} - y_i$$
$$\Delta^2 y_i = \Delta(\Delta y_i) = \Delta(y_{i+1} - y_i) = y_{i+2} - 2y_{i+1} + y_i$$

Plugging in Equation (9.18),

$$I = \frac{\Delta x}{3}[y_0 + 4y_1 + y_2] - \frac{1}{90} f^{(4)}(\xi)(\Delta x)^5 \tag{9.19}$$

This yields the *Simpson 1/3 rule* as:

$$I = \int_{x_0}^{x_2} f(x)dx = \frac{\Delta x}{3}\left[y_0 + 4y_1 + y_2\right] \tag{9.20}$$

Since Δx is multiplied by '1/3', thus it is added in the name of Simpson rule.

Truncation error

From Equation (9.19),

$$R(\xi) = -\frac{1}{90} f^{(4)}(\xi)(\Delta x)^5 \tag{9.21}$$

which indicates that the error in the Simpson 1/3 rule scales as $(\Delta x)^5$.

Remarks

1. The Simpson 1/3 rule provides a higher degree of accuracy [i.e., $(\Delta x)^5$] than the trapezoidal rule [i.e., $(\Delta x)^3$]. This improvement in accuracy is achieved at the expense of only one additional function evaluation (i.e., y_2).
2. It is quite true that the Simpson 1/3 rule can secure cent percent accuracy (i.e., $R = 0$) when $y = f(x)$ is a parabola. One can also confirm that for a parabola, $f^{(4)}(\xi) = 0$ and thus, Equation (9.21) yields $R = 0$.
3. The beauty involved in Simpson 1/3 rule is that one can additionally have an exact result for cubic equations even though it is derived from a parabola. Thus, the degree of precision of the Simpson 1/3 formula is said to be 3 (i.e., $m = 3$).

The reason behind this appealing advantage is that the truncation error (R) gets vanished in Equation (9.21) since $f^{(4)}(\xi)=0$ for cubic equations. Actually this error in Simpson rule is proportional to the fourth derivative of f, rather than being proportional to the third derivative, and this is attributed to Equation (9.17).

Note that one can have such additional level of precision for all even values of m (i.e., m = 2 for Simpson 1/3 rule, m = 4 for Boole rule).

Composite Simpson 1/3 Rule

It is quite straightforward to extend Equation (9.20) for the composite Simpson 1/3 rule as:

$$\begin{aligned} I &= \int_a^b f(x)dx = \int_{x_0}^{x_2} f(x)dx + \int_{x_2}^{x_4} f(x)dx + ... + \int_{x_{n-2}}^{x_n} f(x)dx \\ &= \frac{\Delta x}{3}\left[y_0+4y_1+2y_2+4y_3+2y_4+.....+2y_{n-2}+4y_{n-1}+y_n\right] \end{aligned} \tag{9.22}$$

To obtain this equation, one can fit parabolas to the points (x_0, y_0), (x_1, y_1) and (x_2, y_2) first, then to (x_2, y_2), (x_3, y_3) and (x_4, y_4), and so on. It indicates that the above equation requires an even number of subintervals (n).

Rearranging Equation (9.22),

$$I = \frac{\Delta x}{3}\left[y_0+4(y_1+y_3+...+y_{n-1})+2(y_2+y_4+...+y_{n-2})+y_n\right] \tag{9.23}$$

with

$$\Delta x = \frac{x_n - x_0}{n} = \frac{b-a}{n}$$

Equation (9.23) represents the *composite* or *multiple Simpson 1/3 rule.*

This composite formula can further be written as:

$$\begin{aligned} I &= \frac{\Delta x}{3}\left[y_0+4\sum_{i=1,3,5,..}^{n-1} y_i+2\sum_{i=2,4,6,...}^{n-2} y_i+y_n\right] \\ &= (b-a)\,\frac{\left[y_0+4\sum_{i=1,3,5,..}^{n-1} y_i+2\sum_{i=2,4,6,...}^{n-2} y_i+y_n\right]}{3n} \end{aligned}$$

Truncation error

For the composite Simpson 1/3 rule with even n,

$$R(\xi) = -\frac{1}{90}(\Delta x)^5\sum_{i=0}^{\frac{n}{2}-1} f^{(4)}(\xi_i) \qquad x_{2i} \le \xi_i \le x_{2i+2}$$

$$= -\frac{1}{90}(\Delta x)^5\frac{n}{2}f^{(4)}(\bar{\xi}) \qquad \bar{\xi} \in [x_0, x_n]$$

It gives:

$$R(\xi) = -\frac{(b-a)}{180}(\Delta x)^4 f^{(4)}(\bar{\xi}) \equiv o[(\Delta x)^4] \tag{9.24}$$

As usual, $a = x_0$ and $b = x_n$.

Remarks

1. As stated, in the multiple Simpson 1/3 rule, the number of subintervals (n) must be even (i.e., odd number of equispaced points) and it is valid for $n \geq 4$.
2. This composite rule is fourth-order accurate that is evident in Equation (9.24).
3. This composite 1/3 rule provides a higher degree of accuracy [i.e., $(\Delta x)^4$] than the composite trapezoidal rule [i.e., $(\Delta x)^2$]. Equivalently, to secure the same level of accuracy, the trapezoidal rule requires larger number of function evaluations (i.e., more round-off error) than the Simpson 1/3 rule (see Example 9.7).

Example 9.4 Simpson 1/3 rule (revisit Example 9.1)

Compute the integral:

$$I = \int_0^3 (x^2+1)dx$$

using the Simpson 1/3 rule.

Solution

From Equation (9.15),

$$\Delta x = \frac{b-a}{2} = 1.5$$

Using Equation (9.20),

$$I = \frac{\Delta x}{3}(y_0 + 4y_1 + y_2) = \frac{1.5}{3}(1+13+10) = 12$$

So we get the exact solution.

Remark

This same problem was solved before with the use of composite trapezoidal rule. As shown in Table 9.2, the trapezoidal rule fails to provide exact solution even with $n = 10$ (i.e., total 11 evaluations of the function involved), whereas the Simpson 1/3 rule secures cent percent accuracy with only 3 evaluations of that function.

Example 9.5 Simpson 1/3 rule (cubic polynomial)

Compute the integral:

$$I = \int_1^7 (7.5x^3 + 3x^2 + 5)dx$$

using the Simpson 1/3 rule.

Solution

Here,

$$\Delta x = \frac{b-a}{2} = 3$$

Using Equation (9.20),

$$I = \frac{\Delta x}{3}(y_0 + 4y_1 + y_2) = \frac{3}{3}(15.5 + 2132 + 2724.5) = 4872$$

The analytical solution is:

$$I = \int_1^7 (7.5x^3 + 3x^2 + 5)dx = \left[\frac{7.5x^4}{4} + x^3 + 5x\right]_1^7 = 4872$$

Obviously, the Simpson 1/3 rule secures the exact solution. This example confirms that the Simpson rule is capable of providing exact result for cubic equations as well even though it is derived from a parabola.

Example 9.6 Composite Simpson 1/3 rule (quartic polynomial)

Compute the integral:

$$I = \int_0^3 (x^4 + 1)dx$$

using the composite Simpson 1/3 rule with $n = 4$.

Solution

Here,

$$\Delta x = \frac{b-a}{n} = \frac{3}{4}$$

Using Equation (9.23),

$$I = \frac{\Delta x}{3}(y_0 + 4y_1 + 2y_2 + 4y_3 + y_4)$$
$$= 0.25(1 + 5.26563 + 12.125 + 106.51563 + 82) = 51.726565$$

The analytical solution is:

$$I = \int_0^3 (x^4 + 1)dx = \left[\frac{x^5}{5} + x\right]_0^3 = 51.6$$

It gives an error of 0.126565 (= 51.726565 – 51.6). Further, let us use Equation (9.24) and get,

$$|R(\xi)| = \frac{(x_n - x_0)}{180}(\Delta x)^4 f^{(4)}(\xi) = \frac{3}{180}\left(\frac{3}{4}\right)^4 24 = 0.126563$$

Obviously, we get close value for truncation error with that obtained before. This slight difference is there due to rounding off the numbers.

Example 9.7 Composite Simpson 1/3 rule (error analysis)

Revisit Example 9.3 with the following integral:

$$I = \int_2^5 \frac{6}{x} dx$$

(a) Find the number of subintervals (n) and the step size (Δx), for which, the truncation error should be less than or equal to 10^{-6}.

(b) Determine the number of evaluations of $f(x)$.

Solution

(a) Here the integrand is:

$$f(x) = \frac{6}{x}$$

and thus,

$$f^{(4)}(x) = \frac{144}{x^5}$$

So, one can have the maximum value of $\left|f^{(4)}(x)\right|$ at $a = 2$ over the interval [2,5]. Accordingly, for $2 \le \xi \le 5$, we fix the bound as:

$$\left|f^{(4)}(\xi)\right| \le \left|f^{(4)}(2)\right| = \frac{9}{2}$$

Based on Equation (9.24), one can write:

$$\left|R(\xi)\right| = \left|-\frac{(b-a)}{180}(\Delta x)^4 f^{(4)}(\bar{\xi})\right| \le \frac{3}{180}\frac{9}{2}(\Delta x)^4 = \frac{3}{40}(\Delta x)^4$$

Inserting $\Delta x = 3/n$,

$$\frac{3}{40}\left(\frac{3}{n}\right)^4 \le 10^{-6}$$

It gives:

$n \ge 49.65$

It is true that n must be a positive integer (and even)[6], and thus, we adopt n = 50, which corresponds to $\Delta x = 3/n = 0.06$.

(b) The multiple Simpson 1/3 rule with $n = 50$ involves total 51 evaluations of $f(x)$. Recall that for this same integral, the composite trapezoidal rule required total 1839 evaluations (see Example 9.3). It confirms the superiority of the Simpson 1/3 rule over the trapezoidal rule.

6 To ensure it as an even number, one can consider $n = 2M$ and thus, $\Delta x = 3/n = 3/2M$. With this, we find $M \ge 24.82$ (≈ 25) and the corresponding $\Delta x = 0.06$.

9.6 Higher-Order Newton–Cotes Formulas ($m > 2$)

Our next assignment is to find the Newton–Cotes integration formulas based on the interpolating polynomial with $m > 2$. In the following, the higher-order formulas are only highlighted since their derivation is easy and similar to that of Simpson 1/3 or trapezoidal rule shown before.

9.6.1 Simpson 3/8 Rule

For $m = 3$, we can have:

$$I=\int_{x_0}^{x_3} f(x)dx=\frac{x_3-x_0}{8}\left[y_0+3y_1+3y_2+y_3\right]-\frac{3}{80}f^{(4)}(\xi)(\Delta x)^5$$

for $x_0 \leq \xi \leq x_3$. It gives the *Simpson 3/8 rule*:

$$I=\int_{x_0}^{x_3} f(x)dx=\frac{3\Delta x}{8}\left[y_0+3y_1+3y_2+y_3\right] \tag{9.25}$$

with

$$\Delta x=\frac{x_3-x_0}{3}$$

$$R(\xi)=-\frac{3}{80}f^{(4)}(\xi)(\Delta x)^5$$

This Simpson 3/8 rule is called so because Δx is multiplied with '3/8'. Extension to its composite form is quite straightforward and thus, avoided here.

Remarks

1. Notice that it has the same order of accuracy [i.e., $(\Delta x)^5$] as the Simpson 1/3 rule.
2. However, comparing Equation (9.25) with (9.20), one can see that the Simpson 1/3 rule requires one less function evaluation compared to the Simpson 3/8 rule. This is an advantage of Simpson 1/3 over 3/8 rule.
3. Recall that the composite Simpson 1/3 rule is applicable to even number of subintervals. Thus the necessity of Simpson 3/8 rule sometimes arises when the number of subintervals is odd (elaborated later).

Example 9.8 Precision of Simpson 3/8 rule

Find the degree of precision of Simpson 3/8 rule.

Solution

Let us adopt the five test functions, namely $f(x) = 1, x, x^2, x^3$ and x^4 over the interval [0,1]. Applying the Simpson 3/8 rule, we get the results as documented in Table 9.3.

Table 9.3 Results of Simpson 3/8 rule

Integral	Exact value	Numerical value using Simpson 3/8 rule
$\int_0^1 1\,dx$	1	$\frac{3}{8}.\frac{1}{3}[1+3\times1+3\times1+1]=1$
$\int_0^1 x\,dx$	$\frac{1}{2}$	$\frac{3}{8}.\frac{1}{3}\left[0+3\times\frac{1}{3}+3\times\frac{2}{3}+1\right]=\frac{1}{2}$
$\int_0^1 x^2\,dx$	$\frac{1}{3}$	$\frac{3}{8}.\frac{1}{3}\left[0+3\times\frac{1}{9}+3\times\frac{4}{9}+1\right]=\frac{1}{3}$
$\int_0^1 x^3\,dx$	$\frac{1}{4}$	$\frac{3}{8}.\frac{1}{3}\left[0+3\times\frac{1}{27}+3\times\frac{8}{27}+1\right]=\frac{1}{4}$
$\int_0^1 x^4\,dx$	$\frac{1}{5}$	$\frac{3}{8}.\frac{1}{3}\left[0+3\times\frac{1}{81}+3\times\frac{16}{81}+1\right]=\frac{11}{54}$

It is obvious that the Simpson 3/8 rule, like the Simpson 1/3 formula, provides the exact value for the function having the power of x up to 3. Thus the degree of precision of the Simpson 3/8 rule is $m = 3$.

9.6.2 Boole Rule

Similarly, for $m = 4$, one can obtain:

$$I=\int_{x_0}^{x_4} f(x)dx=\frac{2\Delta x}{45}\left[7y_0+32y_1+12y_2+32y_3+7y_4\right]-\frac{8}{945}f^{(6)}(\xi)(\Delta x)^7$$

for $x_0 \leq \xi \leq x_4$. It gives the *Boole rule:*[7]

$$I=\int_{x_0}^{x_4} f(x)dx=\frac{2\Delta x}{45}\left[7y_0+32y_1+12y_2+32y_3+7y_4\right] \tag{9.26}$$

with

$$\Delta x=\frac{x_4-x_0}{4}$$

$$R(\xi)=-\frac{8}{945}f^{(6)}(\xi)(\Delta x)^7$$

7 Named after English mathematician George Boole (1815–1864).

Remarks

1. This Boole rule has an accuracy of $o[(\Delta x)^7]$.
2. As indicated before, the Newton–Cotes method with $m = 4$ has the same order of accuracy with that with $m = 5$. To confirm it, one can derive the following form for $m = 5$ as:

$$I = \int_{x_0}^{x_5} f(x)dx = \frac{5\Delta x}{288}\left[19y_0 + 75y_1 + 50y_2 + 50y_3 + 75y_4 + 19y_5\right] - \frac{275}{12096} f^{(6)}(\xi)(\Delta x)^7 \tag{9.27}$$

with

$$\Delta x = \frac{x_5 - x_0}{5}$$

3. Although we have the same order of accuracy [i.e., $o[(\Delta x)^7]$] for the Newton–Cotes formulas with $m = 4$ and $m = 5$, the former scheme involves one less function evaluation over the later one in each application.

Extension of these higher-order Newton–Cotes formulas to their composite forms is left for your exercise. Anyway, let us list these numerical integration schemes in Table 9.4 in a comparative form.

9.7 Numerical Integration with Unequally Spaced Points

So far we have discussed the Newton–Cotes scheme for equally spaced points. But there are many practical situations when we have experimental data available with unequal segments. For this, we would like to explore the numerical integration with the use of Newton–Cotes scheme developed previously. Let us pick up the trapezoidal rule. Use this scheme in each segment, add the results and get:

$$I = \frac{\Delta x_1}{2}(y_0 + y_1) + \frac{\Delta x_2}{2}(y_1 + y_2) + ... + \frac{\Delta x_n}{2}(y_{n-1} + y_n) \tag{9.28}$$

In which, Δx_i is the width of ith segment. Notice that it is same with the conventional trapezoidal rule (composite) and only difference is that here Δx_i is not constant. This simple numerical strategy is illustrated in the following example.

Table 9.4 Newton–Cotes integration formulas

Method	Integration formula	Order of interpolating polynomial (*m*)	Truncation error (*R*)
Trapezoidal	$\int_{x_0}^{x_1} f(x)dx = \frac{\Delta x}{2}(y_0 + y_1)$	1	$-\frac{(\Delta x)^3}{12} f''(\xi)$
Composite trapezoidal	$\int_{x_0}^{x_n} f(x)dx = \frac{\Delta x}{2}[y_0 + 2y_1 + 2y_2 + ... + 2y_{n-1} + y_n]$	1	$-\frac{(x_n - x_0)}{12}(\Delta x)^2 f''(\bar{\xi})$
Simpson 1/3	$\int_{x_0}^{x_2} f(x)dx = \frac{\Delta x}{3}\left[y_0 + 4y_1 + y_2\right]$	3	$-\frac{1}{90} f^{(4)}(\xi)(\Delta x)^5$
Composite Simpson 1/3 (*n* must be multiple of 2)	$\int_{x_0}^{x_n} f(x)dx = \frac{\Delta x}{3}\left[\begin{matrix} y_0 + 4(y_1 + y_3 + ... + y_{n-1}) + \\ 2(y_2 + y_4 + ... + y_{n-2}) + y_n \end{matrix}\right]$	3	$-\frac{(x_n - x_0)}{180}(\Delta x)^4 f^{(4)}(\bar{\xi})$
Simpson 3/8	$\int_{x_0}^{x_3} f(x)dx = \frac{3\Delta x}{8}\left[y_0 + 3y_1 + 3y_2 + y_3\right]$	3	$-\frac{3}{80} f^{(4)}(\xi)(\Delta x)^5$
Composite Simpson 3/8 (*n* must be multiple of 3)	$\int_{x_0}^{x_n} f(x)dx = \frac{3\Delta x}{8}\left[\begin{matrix} y_0 + 3(y_1 + y_2 + y_4 + y_5 \cdots + y_{n-2} + y_{n-1}) + \\ 2(y_3 + y_6 + ... + y_{n-3}) + y_n \end{matrix}\right]$	3	$-\frac{(x_n - x_0)}{80}(\Delta x)^4 f^{(4)}(\bar{\xi})$
Boole rule	$\int_{x_0}^{x_4} f(x)dx = \frac{2\Delta x}{45}\left[7y_0 + 32y_1 + 12y_2 + 32y_3 + 7y_4\right]$	5	$-\frac{8}{945} f^{(6)}(\xi)(\Delta x)^7$

Example 9.9 Numerical integration with unequally spaced data points

Compute the integral for the unequally spaced data points given in Table 9.5.

TABLE 9.5 Given data

x	y
0	2
1	1
4	3
5	5
7	2

Solution

Applying Equation (9.28), we get the integral as:

$$I = \frac{1}{2}(2+1) + \frac{3}{2}(1+3) + \frac{1}{2}(3+5) + \frac{2}{2}(5+2) = 18.5$$

9.8 RECOMMENDATIONS

The Newton–Cotes scheme is discussed as a useful method for numerical integration when either a function or the tabulated values are specified. When the function is directly given, one can attempt to achieve the desired accuracy by generating as many values of $f(x)$ as are required for the integration formula. In contrast, there is no such flexibility one can have for the given tabulated data since they are fixed at limited number points.

Higher-order integration formula (e.g., Boole or Weddle rule) offers lower truncation error but with larger rounding error. Moreover, they often suffer from a problem called the *polynomial wiggle* (large oscillation between the samples). It is experienced that the Simpson rules are simply enough to provide the desired accuracy in most practical applications. For further improvement of accuracy level, one can use their composite versions. At this point, let us note that for even number of subintervals, one would prefer to use the Simpson 1/3 rule. On the other hand, for odd number of segments (say for total 9 segments), instead of trapezoidal rule (that provides large truncation error), it is recommended to adopt either Simpson 3/8 rule (for all 9 segments) or the combination of Simpson 1/3 (for first 6 segments) and 3/8 rule (for last 3 segments).

If one requires high accuracy, there are integration methods, namely the Romberg integration and the Gauss quadrature formula, which are usually preferred when the function is known. In the following, we will discuss these two efficient schemes.

9.9 Romberg Integration

The Romberg integration[8] approach is a popular and elegant iterative method proposed for numerical integration of functions. It is developed by combining the composite trapezoidal (or Simpson) rule and the Richardson extrapolation. The trapezoidal (or Simpson) rule is preferred here because of its low order that leads to reduce the error. These composite rules are discussed previously and let us now briefly know the Richardson extrapolation before performing Romberg integration.

9.9.1 Richardson Extrapolation

Let's I be the value of the integral,

$$\int_a^b f(x)dx$$

There are various approximation procedures, in which, a step size Δx is first picked up and then an approximation $I(\Delta x)$ is generated to some desired quantity I as:

$$I = I(\Delta x) + a_1(\Delta x)^m + a_2(\Delta x)^{m+1} + \ldots\ldots \tag{9.29}$$

Here, a_1 and a_2 are constants (independent of Δx), and m is the *order of error*. As indicated, $I(\Delta x)$ is the estimate of I with Δx (not $I \times \Delta x$), whereas $a_1(\Delta x)^m$ means $a_1 \times (\Delta x)^m$. To understand this form of equation at this stage, one can compare it with the composite trapezoidal rule [say, Equation (9.11)].

Using the same approach with spacing $\frac{\Delta x}{2}$, one can have a slightly more accurate estimate, $I\left(\frac{\Delta x}{2}\right)$ for the same quantity I. Accordingly,

$$I = I\left(\frac{\Delta x}{2}\right) + a_1\left(\frac{\Delta x}{2}\right)^m + a_2\left(\frac{\Delta x}{2}\right)^{m+1} + \ldots\ldots \tag{9.30}$$

Multiply Equation (9.30) by 2^m and then subtract Equation (9.29),

$$I = \left[\frac{2^m I\left(\frac{\Delta x}{2}\right) - I(\Delta x)}{2^m - 1}\right] - \frac{u_2(\Delta x)^{m+1}}{2^{m+1} - 2} \tag{9.31}$$

Neglecting the last term [i.e., error ~ $o[(\Delta x)^{m+1}]$],

8 Named after German mathematician and physicist Werner Romberg (1909–2003).

$$I = \left[\frac{2^m I\left(\frac{\Delta x}{2}\right) - I(\Delta x)}{2^m - 1} \right] \tag{9.32}$$

Remarks

1. Equation (9.32) provides a better accuracy for I than the other two earlier estimates, $I(\Delta x)$ and $I\left(\frac{\Delta x}{2}\right)$; clearly, the order of error has been increased by 1 (from m to $m+1$). Generation of such improved approximation for I from its two values with different Δx is referred to as the *Richardson extrapolation.*[9]
2. In a similar fashion, one can use Δx, $\frac{\Delta x}{2}$ and $\frac{\Delta x}{4}$ to get further improvement in accuracy level [i.e., error ~ $o[(\Delta x)^{m+2}]$].
3. Equation (9.32) is rewritten as:

$$I = \int_a^b f(x)dx = \left[\frac{2^m I_2 - I_1}{2^m - 1} \right] \tag{9.33}$$

Obviously, I_1 denotes the numerical integration with spacing Δx and I_2 is for spacing $\frac{\Delta x}{2}$. Let us further modify the representation of I based on the spacing $\Delta x, \frac{\Delta x}{2}, \frac{\Delta x}{4}, \frac{\Delta x}{8}, \ldots$, respectively, as $I_1^0, I_2^0, I_3^0, I_4^0, \ldots$. As before, subscript denotes the integration with different spacing (i.e., Δx corresponds to subscript 1, $\frac{\Delta x}{2}$ to 2 and so on) and superscript denotes the iteration index (i.e., superscript 0 indicates the initial approximation, 1 the first iteration and so on).

9 Named after the English mathematician and physicist Lewis Fry Richardson, FRS (1881–1953).

Arranging all these I's in a tabulated form, we have the Romberg array as:

$$
\begin{array}{ccccccc}
I_1^0 & & & & & & \\
I_2^0 & I_2^1 & & & & & \\
I_3^0 & I_3^1 & I_3^2 & & & & \\
I_4^0 & I_4^1 & I_4^2 & I_4^3 & & & \\
. & . & . & . & . & & \\
. & . & . & . & . & . & \\
I_{j+1}^0 & I_{j+1}^1 & I_{j+1}^2 & I_{j+1}^3 & . & . & I_{j+1}^j
\end{array}
$$

The last estimate, I_{j+1}^j is noted as our final value of the integral, I provided it meets the stopping criterion.

9.9.2 Romberg Integration Algorithm

As stated, when this Richardson extrapolation scheme is applied for integration with the use of the composite trapezoidal (or Simpson) rule, it is called the *Romberg integration*. To develop this integration algorithm, let us start with the following integral,

$$I = \int_a^b f(x)dx \tag{9.1}$$

Considering n_1 number of subintervals in the composite trapezoidal rule, that is:

$$\Delta x_1 = \frac{b-a}{n_1}$$

we have,

$$I_1 = \frac{\Delta x_1}{2}\left[f(a) + 2\sum_{i=1}^{n_1-1} f(x_i) + f(b) \right] \tag{9.34}$$

with the truncation error [from Equation (9.13)],

$$R_1 = -\frac{(b-a)}{12}(\Delta x_1)^2 \overline{f}'' \tag{9.35}$$

Combining the above two equations, we have:

$$I \approx I_1 + R_1 \equiv I_1 + a_1(\Delta x_1)^2 \tag{9.36}$$

which is an equivalent form of Equation (9.29). This is true if $\bar{f}''$ is independent of the step size.

In the next phase, we consider a smaller step size as:

$$\Delta x_2 = \frac{b-a}{n_2}$$

Here, n_2 is the number of subintervals. With this,

$$I_2 = \frac{\Delta x_2}{2}\left[f(a) + 2\sum_{i=1}^{n_2-1} f(x_i) + f(b) \right] \tag{9.37}$$

and

$$R_2 = -\frac{(b-a)}{12}(\Delta x_2)^2 \bar{f}'' \tag{9.38}$$

Combining the above two equations,

$$I \approx I_2 + R_2 \tag{9.39}$$

Notice that both the R_1 and R_2 scale as $o\,[(\Delta x)^2]$.

Equating the actual I's,

$$I \approx I_1 + R_1 \approx I_2 + R_2 \tag{9.40}$$

Using Equations (9.35) and (9.38),

$$\frac{R_1}{R_2} = \left(\frac{\Delta x_1}{\Delta x_2} \right)^2 \tag{9.41}$$

provided $\bar{f}''$ is independent of the step size. With this, one can proceed without prior knowledge of second derivative of the function.

Now Equation (9.40) yields:

$$I_1 + R_2 \left(\frac{\Delta x_1}{\Delta x_2} \right)^2 = I_2 + R_2$$

It gives:

$$R_2 = \frac{I_1 - I_2}{1 - \left(\frac{\Delta x_1}{\Delta x_2} \right)^2} \tag{9.42}$$

Accordingly,

$$I = I_2 + R_2 = I_2 + \frac{I_1 - I_2}{1 - \left(\frac{\Delta x_1}{\Delta x_2}\right)^2}$$

Simplifying,

$$I = \frac{I_1 - \left(\frac{\Delta x_1}{\Delta x_2}\right)^2 I_2}{1 - \left(\frac{\Delta x_1}{\Delta x_2}\right)^2} \tag{9.43}$$

As stated, integrals I_1 and I_2 have the error of $o\,[(\Delta x)^2]$ [see Equations (9.35) and (9.38)]. However, it can be shown that the error of I in Equation (9.43) is $o\,[(\Delta x)^4]$. For this, one needs to consider an alternate expression (Ralston and Rabinowitz, 1978):

$$I = \frac{\Delta x}{2}\left[f(a) + 2\sum_{i=1}^{n-1} f(a + i\Delta x) + f(b)\right] - \frac{(\Delta x)^2}{12}[f'(b) - f'(a)] + \frac{(b-a)}{720}(\Delta x)^4 \bar{f}^{(4)}$$

instead of the conventional trapezoidal rule [Equation (9.36)]. Anyway, it reveals that by combining the two applications of the trapezoidal rule having error $o\,[(\Delta x)^2]$, we can compute a third estimate with error $o\,[(\Delta x)^4]$, thereby improving the order of accuracy.

Equation (9.43) can be rewritten as:

$$I = \frac{I_1 - \left(\frac{n_2}{n_1}\right)^2 I_2}{1 - \left(\frac{n_2}{n_1}\right)^2} \tag{9.44}$$

Let us consider a special case where the interval is halved. It means, $\Delta x_1 = 2\Delta x_2$ (i.e., $n_2 = 2n_1$), and thus, the Romberg integration yields:

$$I = \frac{4I_2 - I_1}{4 - 1}$$

which can be written as:

$$I_2^1 = \frac{4I_2^0 - I_1^0}{4 - 1} \tag{9.45}$$

For kth iteration, similarly we have:

$$I_{j+1}^k = \frac{4^k I_{j+1}^{k-1} - I_j^{k-1}}{4^k - 1} \qquad j = 1, 2, 3, \ldots \tag{9.46}$$

This is the *general form of the Romberg integration* for composite trapezoidal rule. Here, I_{j+1}^{k-1} and I_j^{k-1} refer to the more and less accurate integrals, respectively, and I_{j+1}^{k} denotes the improved integral. With this and as indicated before, the suffix $j+1$ corresponds to more accurate and j to less accurate estimates. Further, the iteration index (superscript k) corresponds to the original trapezoidal rule estimates [i.e., $o\,[(\Delta x)^2]$] when $k = 0$. Similarly, $k = 1$ refers to the estimates with $o\,[(\Delta x)^4]$, $k = 2$ refers to the estimates with $o\,[(\Delta x)^6]$ and so on.

Remarks

1. In the same fashion as in composite trapezoidal rule, the kth iteration of the Romberg integration yields the following form with the use of composite Simpson 1/3 rule:

$$I_{j+1}^{k} = \frac{4^{k+1} I_{j+1}^{k-1} - I_j^{k-1}}{4^{k+1} - 1} \qquad j = 1, 2, 3, \cdots \tag{9.47}$$

2. Romberg integration is more efficient than the trapezoidal or Simpson rule. For attaining a desired accuracy, this approach requires less number of function evaluations than those of the trapezoidal (or Simpson) rule.
3. Like any other iterative method, one needs to define a stopping or termination criterion for the Romberg integration.

Example 9.10 Romberg integration with trapezoidal rule

Compute the integral:

$$I = \int_0^1 \frac{1}{5+x}\,dx$$

using the Romberg integration with composite trapezoidal rule for spacing Δx, $\frac{\Delta x}{2}$, $\frac{\Delta x}{4}$ and $\frac{\Delta x}{8}$.

Solution

The initial approximations (≡ 0th iteration) for the integral I with various number of subintervals, n are computed first by using the trapezoidal rule and they are documented in Table 9.6.

Table 9.6 Initial approximations

$n=1$ $\Delta x=1$	$I_1^0=\frac{\Delta x}{2}[f(0)+f(1)]=\frac{1}{2}\left[\frac{1}{5}+\frac{1}{6}\right]=0.183333$
$n=2$ $\Delta x=\frac{1}{2}=0.5$	$I_2^0=\frac{\Delta x}{2}[f(0)+2f(0.5)+f(1)]=\frac{0.5}{2}\left[\frac{1}{5}+\frac{2}{5.5}+\frac{1}{6}\right]=0.182576$
$n=4$ $\Delta x=\frac{1}{4}=0.25$	$I_3^0=\frac{\Delta x}{2}[f(0)+2\{f(0.25)+f(0.5)+f(0.75)\}+f(1)]=0.182385$
$n=8$ $\Delta x=\frac{1}{8}=0.125$	$I_4^0=\frac{\Delta x}{2}\left[f(0)+2\begin{Bmatrix}f(0.125)+f(0.25)+f(0.375)+f(0.5)\\+f(0.625)+f(0.75)+f(0.875)\end{Bmatrix}+f(1)\right]=0.182337$

Let us next use the Romberg integration formula developed previously:

$$I_{j+1}^k=\frac{4^k I_{j+1}^{k-1}-I_j^{k-1}}{4^k-1} \qquad j=1,2,3 \tag{9.46}$$

- For first iteration (k = 1):

$$j=1 \qquad I_2^1=\frac{4I_2^0-I_1^0}{4-1}=0.182324$$

$$j=2 \qquad I_3^1=\frac{4I_3^0-I_2^0}{4-1}=0.182321$$

$$j=3 \qquad I_4^1=\frac{4I_4^0-I_3^0}{4-1}=0.182321$$

- For second iteration (k = 2):

$$j=2 \qquad I_3^2=\frac{4^2I_3^1-I_2^1}{4^2-1}=0.182321$$

$$j=3 \qquad I_4^2=\frac{4^2I_4^1-I_3^1}{4^2-1}=0.182321$$

- For third (last) iteration (k = 3):

$$j=3 \qquad I_4^3=\frac{4^3I_4^2-I_3^2}{4^3-1}=0.182321$$

These produced results are further briefed in Table 9.7.

TABLE 9.7 Results of Romberg iterations

Δx	0th iteration (trapezoidal)	1st iteration (Romberg)	2nd iteration (Romberg)	3rd iteration (Romberg)
1	0.183333			
0.5	0.182576	0.182324		
0.25	0.182385	0.182321	0.182321	
0.125	0.182337	0.182321	0.182321	0.182321

Notice that the last two estimates (= 0.182321) match upto apparently six decimal points, and if so, the final solution 0.182321 is at least correct upto six decimal places. In fact, our numerical value is same with the analytical value.

Remarks

1. To achieve the desired accuracy of 10^{-6}, as shown above, the Romberg integration with trapezoidal rule requires only 9 function evaluations. On the other hand, the composite trapezoidal rule involves total 38 function evaluations for achieving the same accuracy level (found out by following the same procedure discussed previously in Example 9.3). It clearly indicates that the Romberg integration outperforms the conventional trapezoidal rule.
2. For this example problem, further note that the Romberg integration can provide the same value of 0.182321 with using Δx, $\frac{\Delta x}{2}$ and $\frac{\Delta x}{4}$; there is no need to use the extra spacing $\frac{\Delta x}{8}$. With this, the Romberg integration involves only 5 function evaluations (instead of 9).

9.10 GAUSS–LEGENDRE QUADRATURE

All the quadrature formulas discussed so far in the preceding sections are represented in the following general form:

$$I = \int_a^b f(x)dx \approx \sum_{i=0}^{n} w_i f(x_i) \tag{9.48}$$

in which, w_i are the *weights* and x_i the *nodes* (or Gauss points) distributed within the limits of integration $[a, b]$.

At this point, we should note that for the Newton–Cotes formula, there are equally spaced nodes x_i (i.e., $x_i = x_0 + i\Delta x$, $i = 0, 1, 2, ..., n$) with $w = 1$. In contrast, the Gauss quadrature formula employs unequally spaced nodes aiming to minimize the error involved in the approximation. Figure 9.8 depicts the difference between the trapezoidal rule and Gauss quadrature with two points. It is evident that in case of trapezoidal rule, the integral is approximated by the area under the straight line joining the function values at the ends of the integration interval $[a,b]$. Thus, this rule results in a reasonably large error. On the other hand, if we are free to evaluate the area under a straight line connecting any two points (cleverly chosen) on the curve, an improved estimate of the integral can be obtained. This is the basic idea of the *Gauss quadrature*. Here we would like to discuss a particular Gauss quadrature formula that is often called the *Gauss–Legendre* formula.[10]

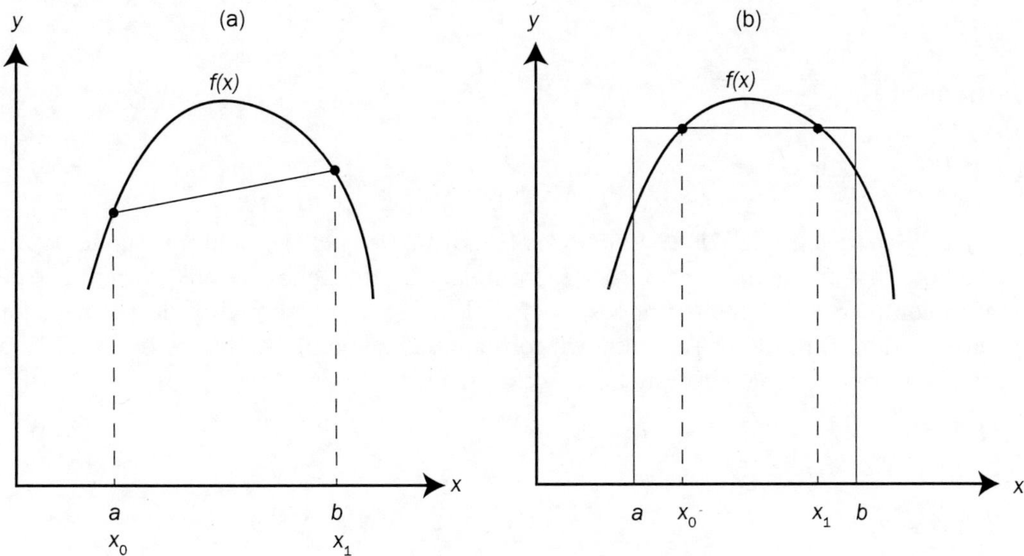

FIGURE 9.8 (a) Trapezoidal rule, and (b) Gauss quadrature

9.10.1 Deriving Gauss–Legendre Quadrature Formula

Let us start by changing the interval of integration from $[a,b]$ to $[-1,\ 1]$. For this, the variable x gets changed to z through the following linear transformation:

$$x=\frac{b-a}{2}z+\frac{b+a}{2} \tag{9.49}$$

Differentiating,

$$dx=\frac{b-a}{2}dz \tag{9.50}$$

Accordingly, Equation (9.48) gets the following form:

$$I=\int_a^b f(x)\,dx=\frac{b-a}{2}\int_{-1}^{1} g(z)\,dz \tag{9.51}$$

10 Named after German mathematician and physicist Johann Carl Friedrich Gauss (1777–1855), and French mathematician Adrien-Marie Legendre (1752–1833).

where,

$$g(z) = f\left(\frac{(b-a)z+(b+a)}{2}\right)$$

With this, we proceed to derive the Gauss–Legendre quadrature formula for the following integral:

$$\int_{-1}^{1} g(z)\, dz$$

It is written as:

$$\int_{-1}^{1} g(z)\, dz \approx \sum_{i=0}^{n} w_i g(z_i) \tag{9.52}$$

Obviously, there are total $2n+2$ unknowns [i.e., $n+1$ weights (w_i) and $n+1$ node points (z_i)]. Note that unlike the Newton–Cotes formula, here the nodes are unknown. It is true that to compute n weights (or nodes), we need to use a polynomial of degree $n-1$ for approximation. Thus for $2n+2$ unknown constants, a polynomial of degree $2n+1$ needs to be utilized to approximate the function. Accordingly,

$$g(z) = a_0 p_0(z) + a_1 p_1(z) + ... + a_{2n+1} p_{2n+1}(z) \tag{9.53}$$

where,

$$p_j(z) = z^j$$

The above two equations are extended from Equation (8.1).

The left hand side (LHS) of Equation (9.52):

$$\text{LHS} = \int_{-1}^{1} g(z)\, dz = \int_{-1}^{1} (a_0 + a_1 z + a_2 z^2 + ... + a_{2n+1} z^{2n+1})\, dz$$

$$= 2a_0 + \frac{2}{3}a_2 + \frac{2}{5}a_4 + ...$$

Again, the right hand side (RHS) is:

$$\begin{aligned}\text{RHS} = \sum_{i=0}^{n} w_i g(z_i) = {} & w_0(a_0 + a_1 z_0 + a_2 z_0^2 + ... + a_{2n+1} z_0^{2n+1}) \\ & + w_1(a_0 + a_1 z_1 + a_2 z_1^2 + ... + a_{2n+1} z_1^{2n+1}) \\ & + w_2(a_0 + a_1 z_2 + a_2 z_2^2 + ... + a_{2n+1} z_2^{2n+1}) \\ & \quad . \\ & \quad . \\ & \quad . \\ & + w_n(a_0 + a_1 z_n + a_2 z_n^2 + ... + a_{2n+1} z_n^{2n+1})\end{aligned}$$

Comparing LHS with RHS:

$$w_0+w_1+w_2+.....+w_n=2$$

$$w_0z_0+w_1z_1+w_2z_2+.....+w_nz_n=0$$

$$w_0z_0^2+w_1z_1^2+w_2z_2^2+.....+w_nz_n^2=\frac{2}{3}$$

$$w_0z_0^3+w_1z_1^3+w_2z_2^3+.....+w_nz_n^3=0 \qquad (9.54)$$

.

.

.

$$w_0z_0^{2n+1}+w_1z_1^{2n+1}+w_2z_2^{2n+1}+.....+w_nz_n^{2n+1}=0$$

This is the *general form* for n-points Gauss–Legendre scheme that includes $2n + 2$ equations for finding $n + 1$ weights and $n + 1$ nodes.

Remarks

1. Gauss quadrature is more efficient method than the Newton–Cotes formula. However, this Gauss quadrature requires information about the function at all Gauss points/nodes that we choose freely. When the function is readily available, this is not an issue. But if tabulated data are there, instead of function, it is fine as long as the table includes the data at all Gauss points.
2. Solving a large set of non-linear algebraic Equations (9.54), particularly for higher-order Gauss quadrature, is a herculean task.

Based on this general formulation, we will next discuss a few sample cases below.

1-point Gauss–Legendre quadrature (n = 0)

For $n = 0$, we have total 2 unknowns [$=2n + 2$, which include 1 weight (w_0) and 1 node (z_0)]. To determine them exactly, we use Equation (9.54) to get:

$$w_0 = 2$$

$$w_0z_0 = 0$$

Solving,

$$w_0 = 2$$

$$z_0 = 0$$

As a result, Equation (9.52) gives the following *Gauss–Legendre 1-point* formula as:

$$\int_{-1}^{1} g(z)\,dz = w_0 g(z_0) = 2g(0) \qquad (9.55)$$

Remark

This 1-point method involves only one function evaluation.

2-points Gauss–Legendre quadrature (n = 1)

For $n = 1$, there are total 4 unknowns (i.e., 2 weights and 2 nodes). To determine them exactly, we use Equation (9.54) to get:

$$w_0 + w_1 = 2 \tag{9.56a}$$

$$w_0 z_0 + w_1 z_1 = 0 \tag{9.56b}$$

$$w_0 z_0^2 + w_1 z_1^2 = \frac{2}{3} \tag{9.56c}$$

$$w_0 z_0^3 + w_1 z_1^3 = 0 \tag{9.56d}$$

Multiplying Equation (9.56b) by z_1^2 and then subtracting from (9.56d),

$$w_0 z_0 (z_0^2 - z_1^2) = 0$$

It gives:

$$w_0 z_0 (z_0 + z_1)(z_0 - z_1) = 0$$

Based on this form of equation, there are four possibilities: $w_0 = 0$ or $z_0 = 0$ or $z_0 = z_1$ or $z_0 = -z_1$. For the first three, the remaining equations [in Equation set (9.56)] do not hold. Thus, we adopt the last option, $z_0 = -z_1$.

Accordingly, from Equation (9.56b):

$$w_0 z_0 - w_1 z_0 = 0$$

Using the above equation (i.e., $w_0 = w_1$) in Equation (9.56a):

$$w_0 = w_1 = 1$$

With this, Equation (9.56c) yields:

$$z_0^2 + z_1^2 = \frac{2}{3}$$

Since $z_0 = -z_1$, we get:

$$z_0 = \frac{1}{\sqrt{3}}$$

$$z_1 = -\frac{1}{\sqrt{3}}$$

Finding the weights (w_0 and w_1) and nodes (z_0 and z_1), we get the following *Gauss–Legendre 2-points* formula as:

$$\int_{-1}^{1} g(z)\, dz = w_0 g(z_0) + w_1 g(z_1) = g\left(\frac{1}{\sqrt{3}}\right) + g\left(-\frac{1}{\sqrt{3}}\right) \tag{9.57}$$

Remarks

1. This 2-points method involves two function evaluations.
2. One can show that this 2-points formula is exact for all polynomials of degree ≤ 3 $(=2n+1)$ since it is exact for the polynomial in Equation (9.56). Thus, it has degree of precision = 3.

Example 9.11 2-points Gauss quadrature (cubic polynomial)

Show that 2-points Gauss–Legendre rule is exact for:

$$\int_{-1}^{1}\left(x^3+x^2+x+1\right)dx$$

Solution

Here, the interval of integration is directly given as $[-1, 1]$ and thus, no linear transformation of x to z is needed. Applying 2-points Gauss quadrature formula [Equation (9.57)]:

$$\int_{-1}^{1} g(z)\,dz = g\left(\frac{1}{\sqrt{3}}\right)+g\left(-\frac{1}{\sqrt{3}}\right)$$

we get:

$$\int_{-1}^{1}\left(x^3+x^2+x+1\right)dx=\left(\frac{1}{3\sqrt{3}}+\frac{1}{3}+\frac{1}{\sqrt{3}}+1\right)+\left(-\frac{1}{3\sqrt{3}}+\frac{1}{3}-\frac{1}{\sqrt{3}}+1\right)=\frac{8}{3}$$

The exact solution is:

$$\int_{-1}^{1}\left(x^3+x^2+x+1\right)dx=\left[\frac{x^4}{4}+\frac{x^3}{3}+\frac{x^2}{2}+x\right]_{-1}^{1}=\frac{8}{3}$$

It shows that 2-points Gauss quadrature formula is exact for the polynomials of degree 3. In the similar way, you can verify for the polynomials of degree < 3.

Remark

For this cubic polynomial, 2-points formula secures an exact solution on the basis of 2 function evaluations. To attain the same performance (exact solution), the Simpson 1/3 rule requires 3 function evaluations. It confirms that 2-points formula is more efficient than the Newton–Cotes method.

3-points Gauss–Legendre quadrature ($n = 2$)

For $n = 2$, there are total 6 unknowns (i.e., 3 weights and 3 nodes). To determine them exactly, we use Equation (9.54) to get:

$$w_0 + w_1 + w_2 = 2$$

$$w_0 z_0 + w_1 z_1 + w_2 z_2 = 0$$

$$w_0 z_0^2 + w_1 z_1^2 + w_2 z_2^2 = \frac{2}{3}$$

$$w_0 z_0^3 + w_1 z_1^3 + w_2 z_2^3 = 0$$

$$w_0 z_0^4 + w_1 z_1^4 + w_2 z_2^4 = \frac{2}{5}$$

$$w_0 z_0^5 + w_1 z_1^5 + w_2 z_2^5 = 0$$

We are not going to solve this set of non-linear equations. The unique solution given by Gauss is provided below.

$$z_0 = -\sqrt{\frac{3}{5}} \qquad z_1 = 0 \qquad z_2 = \sqrt{\frac{3}{5}}$$

$$w_0 = \frac{5}{9} \qquad w_1 = \frac{8}{9} \qquad w_2 = \frac{5}{9}$$

The *Gauss–Legendre 3-points* formula is accordingly obtained as:

$$\begin{aligned}\int_{-1}^{1} g(z)\,dz &= w_0 g(z_0) + w_1 g(z_1) + w_2 g(z_2) \\ &= \frac{5}{9} g\left(-\sqrt{\frac{3}{5}}\right) + \frac{8}{9} g(0) + \frac{5}{9} g\left(\sqrt{\frac{3}{5}}\right)\end{aligned} \tag{9.58}$$

Remarks

1. This 3-points Gauss–Legendre method involves three function evaluations.
2. This 3-points formula is exact for all polynomials of degree ≤5 (=$2n + 1$).
3. As indicated, the Gauss quadrature requires the function to evaluate it at unevenly spaced points within the integration interval. Thus, without knowing this function, one cannot apply this method. Further, this integration method is not suited for the problems, for which, the tabulated data set is available.

Example 9.12 Gauss–Legendre integration

i) Determine the integral:

$$\int_0^1 \frac{1}{5+x}\,dx$$

using 1-point, 2-points and 3-points formula

ii) Compare the Gauss quadrature results with exact value

iii) Compare the performance of the composite trapezoidal rule, Romberg integration and 3-points Gauss quadrature for the desired tolerance of 10^{-6}.

Solution

i) To convert the interval [0, 1] to [−1, 1], we insert $a = 0$ and $b = 1$ in Equation (9.49):

$$x = \frac{1}{2}z + \frac{1}{2}$$

So,

$$dx = \frac{dz}{2}$$

1-point formula

Based on Equation (9.55):

$$\int_0^1 \frac{1}{5+x}dx = \int_{-1}^{1} \frac{1}{5+\left(\frac{z}{2}+\frac{1}{2}\right)}\left(\frac{dz}{2}\right)$$

$$= \frac{1}{2}\int_{-1}^{1} \frac{1}{5+\left(\frac{z}{2}+\frac{1}{2}\right)}dz$$

$$= \frac{1}{2}2g(0)$$

$$= 0.18181818$$

2-points formula

From Equation (9.57):

$$\int_0^1 \frac{1}{5+x}dx = \frac{1}{2}\int_{-1}^{1} \frac{1}{5+\left(\frac{z}{2}+\frac{1}{2}\right)}dz$$

$$= \frac{1}{2}\left[g\left(\frac{1}{\sqrt{3}}\right)+g\left(-\frac{1}{\sqrt{3}}\right)\right]$$

$$= \frac{1}{2}\left[\frac{1}{5.5+\frac{1}{2\sqrt{3}}}+\frac{1}{5.5-\frac{1}{2\sqrt{3}}}\right]$$

$$= 0.18232044$$

3-points formula

From Equation (9.58):

$$\int_0^1 \frac{1}{5+x}dx = \frac{1}{2}\left[\frac{5}{9}g\left(-\sqrt{\frac{3}{5}}\right)+\frac{8}{9}g(0)+\frac{5}{9}g\left(\sqrt{\frac{3}{5}}\right)\right]$$

$$=\frac{1}{2}\left[\frac{5}{9}\frac{1}{5.5-0.5\sqrt{\frac{3}{5}}}+\frac{8}{9}\frac{1}{5.5}+\frac{5}{9}\frac{1}{5.5+0.5\sqrt{\frac{3}{5}}}\right]$$

$$=0.18232155$$

ii) **Comparing with exact value**

The exact value of the integral is 0.18232155. Obviously, 2-points formula secures better performance than 1-point formula. Again, among the three Gauss quadratures, 3-points formula shows the best performance with providing exact value of the integral.

iii) For the desired tolerance of 10^{-6}, the comparative performances are highlighted below:

- the composite trapezoidal rule involves total 38 function evaluations (shown before in Example 9.10)
- the Romberg integration with trapezoidal rule requires 5 function evaluations (shown in Example 9.10)
- 3-points Gauss quadrature method requires 3 function evaluations.

Obviously, the Gauss–Legendre quadrature provides the best performance followed by the Romberg integration and then the composite trapezoidal rule.

9.10.2 Truncation Error

Equation (9.52) can be written in its actual form as:

$$\int_{-1}^{1} g(z)\,dz = \sum_{i=0}^{n} w_i\, g(z_i) + R_{2n+1} \tag{9.59}$$

As mentioned before, the Gauss quadrature is exact for the polynomials up to degree $(2n+1)$. It indicates:

$$R_{2n+1}=0 \qquad \text{when } g(z)=z^j,\ j=0,1,2,\ldots,2n+1$$

On the other hand,

$$R_{2n+1}\neq 0 \qquad \text{when } g(z)=z^{2n+2}$$

With this, we have:

$$R_{2n+1}=\frac{C}{(2n+2)!}g^{(2n+2)}(\xi) \tag{9.60}$$

for $\xi \in [-1, 1]$. Here the *error constant* is [based on Equation (9.52)]:

$$\begin{aligned} C &= \int_{-1}^{1} g(z)\, dz - \sum_{i=0}^{n} w_i g(z_i) \\ &= \int_{-1}^{1} z^{2n+2}\, dz - \sum_{i=0}^{n} w_i z_i^{2n+2} \end{aligned} \tag{9.61}$$

Example 9.13 Deriving error of 2-points Gauss quadrature

Derive the expression for truncation error of 2-points Gauss–Legendre quadrature.

Solution

For 2-points Gauss quadrature ($n = 1$), the error is [from Equation (9.60)]:

$$R_3 = \frac{C}{4!} g^{(4)}(\xi)$$

for which, the error constant is [from Equation (9.61)]:

$$\begin{aligned} C &= \int_{-1}^{1} z^4\, dz - (w_0 z_0^4 + w_1 z_1^4) \\ &= \frac{2}{5} - \left(\frac{1}{9} + \frac{1}{9}\right) = \frac{8}{45} \end{aligned}$$

So the error of 2-points Gauss quadrature is:

$$\begin{aligned} R_3 &= \frac{C}{4!} g^{(4)}(\xi) \\ &= \frac{1}{135} g^{(4)}(\xi) \end{aligned}$$

In a similar fashion, one can find the error of 3-points Gauss quadrature as:

$$R_5 = \frac{1}{15750} g^{(6)}(\xi)$$

in which,

$$\begin{aligned} C &= \int_{-1}^{1} z^6\, dz - (w_0 z_0^6 + w_1 z_1^6 + w_2 z_2^6) \\ &= \frac{8}{175} \end{aligned}$$

Table 9.8 List of Gauss–Legendre quadrature formulas

No. of points $(n+1)$	Gauss point (z_i)	Weight (w_i)	Formula: $\int_{-1}^{1} g(z)\,dz = \sum_{i=0}^{n} w_i g(z_i)$	Truncation error
1	0	2	$2g(0)$	$\frac{1}{3}g^{(2)}(\xi)$
2	$\pm\frac{1}{\sqrt{3}}$	1	$g\left(-\frac{1}{\sqrt{3}}\right)+g\left(\frac{1}{\sqrt{3}}\right)$	$\frac{1}{135}g^{(4)}(\xi)$
3	0 $\pm\sqrt{\frac{3}{5}}$	$\frac{8}{9}$ $\frac{5}{9}$	$\frac{5}{9}g\left(-\sqrt{\frac{3}{5}}\right)+\frac{8}{9}g(0)+\frac{5}{9}g\left(\sqrt{\frac{3}{5}}\right)$	$\frac{1}{15750}g^{(6)}(\xi)$
4	$\pm\sqrt{\frac{3-2\sqrt{\frac{6}{5}}}{7}}$ $\pm\sqrt{\frac{3+2\sqrt{\frac{6}{5}}}{7}}$	$\frac{18+\sqrt{30}}{36}$ $\frac{18-\sqrt{30}}{36}$	$\frac{18+\sqrt{30}}{36}g\left(\sqrt{\frac{3-2\sqrt{\frac{6}{5}}}{7}}\right)+\frac{18+\sqrt{30}}{36}g\left(-\sqrt{\frac{3-2\sqrt{\frac{6}{5}}}{7}}\right)+$ $\frac{18-\sqrt{30}}{36}g\left(\sqrt{\frac{3+2\sqrt{\frac{6}{5}}}{7}}\right)+\frac{18-\sqrt{30}}{36}g\left(-\sqrt{\frac{3+2\sqrt{\frac{6}{5}}}{7}}\right)$	$\frac{1}{3472875}g^{(8)}(\xi)$

Contd

Contd

No. of points $(n+1)$	Gauss point (z_i)	Weight (w_i)	Formula: $\int_{-1}^{1} g(z)\,dz = \sum_{i=0}^{n} w_i g(z_i)$	Truncation error
5	0 $-\frac{1}{3}\sqrt{5-2\sqrt{\frac{10}{7}}}$ $\pm\frac{1}{3}\sqrt{5+2\sqrt{\frac{10}{7}}}$	$\frac{128}{225}$ $\frac{322+13\sqrt{70}}{900}$ $\frac{322-13\sqrt{70}}{900}$	$\frac{322+13\sqrt{70}}{900} g\left(-\frac{1}{3}\sqrt{5-2\sqrt{\frac{10}{7}}}\right)+\frac{322+13\sqrt{70}}{900}\times$ $g\left(\frac{1}{3}\sqrt{5-2\sqrt{\frac{10}{7}}}\right)+\frac{128}{225}g(0)+\frac{322-13\sqrt{70}}{900}\times$ $g\left(-\frac{1}{3}\sqrt{5+2\sqrt{\frac{10}{7}}}\right)+\frac{322-13\sqrt{70}}{900} g\left(\frac{1}{3}\sqrt{5+2\sqrt{\frac{10}{7}}}\right)$	$\frac{1}{1237732650} g^{(10)}(\xi)$

A Short Note

For 1-point and 2-points Gauss quadrature formulas, it is quite simple to find the location of the Gauss points (z_i) and their weights (w_i). However, formulating higher-points formulas is not so simple because of the involvement of a large set of non-linear algebraic equations. Table 9.8 lists the Gauss quadrature formulas with a few higher-order schemes for your ready reference.

Like the trapezoidal or Simpson rule, the Gauss–Legendre quadrature also has the composite formula, which is not discussed here to avoid repetitions. It is similarly derived by applying the quadrature formula to each subinterval.

9.11 Summary and Concluding Remarks

In this chapter, we have studied the numerical integration performed to approximate the definite integrals. For this, the three different classes of integration methods, namely Newton–Cotes, Romberg integration and Gauss quadrature, are discussed in details. The truncation error involved in each of these integration strategies is also covered. Illustrating these numerical techniques by various examples, we have compared them to identify the relative merits and demerits, which are helpful in selecting a suitable method for a particular integral.

Exercise

Review Questions

9.1 Why and when one might prefer to do integration numerically rather than analytically?

9.2 As the order of the Newton–Cotes formula is increased, what happens in terms of error?

9.3 When do the trapezoidal and Simpson 1/3 rule provide exact value and why?

9.4 Why does the Simpson 1/3 rule provide exact result when $y = f(x)$ is a cubic equation?

9.5 The trapezoidal rule and its composite version are third-order and second-order accurate, respectively. Does it mean that the former approach is more accurate than the later one?

9.6 What are the basic differences between the three numerical integration strategies, Newton–Cotes, Romberg integration and Gauss quadrature? List their relative merits and demerits.

9.7 Develop the guidelines for selecting a suitable numerical integration method against various cases commonly encountered in practice.

9.8 The rectangle or midpoint rule over the interval $[x_0, x_1]$ is:

$$I = \int_{x_0}^{x_1} f(x)dx = \Delta x\, f\left(x_0 + \frac{\Delta x}{2}\right) + \frac{(\Delta x)^3}{24} f''(\xi)$$

where,

$$\Delta x = \frac{(x_1 - x_0)}{2}$$

i) Derive the above formula through the Taylor series expansion of $F(x)$ (the anti-derivative of $f(x)$) about $\left(x_0+\frac{\Delta x}{2}\right)$.

ii) Develop the composite rectangle rule over $[a,b]$.

9.9 Find the degree of precision of:

i) the trapezoidal rule as $m = 1$

ii) the Simpson 1/3 rule as $m = 3$

iii) the Simpson 3/8 rule as $m = 3$

9.10 Derive the Newton–Cotes integration formula for

i) $m = 5$ (fifth-order)

ii) $m = 6$ (sixth-order)

9.11 Find the expression for truncation error of the Newton–Cotes integration formula with $m = 5$.

9.12 Develop the composite formula for the Boole rule.

9.13 Develop the Romberg integration formula by combining the Richardson extrapolation with:

i) the composite Simpson 1/3 rule

ii) the composite Simpson 3/8 rule

9.14 Derive the expression for truncation error of 1-point Gauss–Legendre quadrature.

9.15 Is there any type of polynomial, for which, 1-point Gauss quadrature can provide exact solution?

9.16 For Gauss quadrature, show that the truncation error is:

$$R_{2n+1} = 0 \qquad \text{when } g(z) = z^{2n+1}$$

9.17 Let us consider an initial value problem (IVP) with:

$$\frac{dx}{dt} = f(x,t) \qquad x(t_0) = x_0$$

Integrating,

$$\int_{x_0}^{x} dx = \int_{t_0}^{t} f(x,t)dt$$

It yields:

$$x = x_0 + \int_{t_0}^{t} f(x,t)dt$$

Approximate the shaded area shown in Figure 9.9 as a rectangle and come out with the formula for forward difference (FD) and backward difference (BD) method.

Practice Problems

9.18 i) Compute the integral:

$$I = \int_{1}^{3} (3x+5)dx$$

using the trapezoidal rule ($n = 1$)

ii) Compare the numerical and analytical value. Is there any difference? Why?

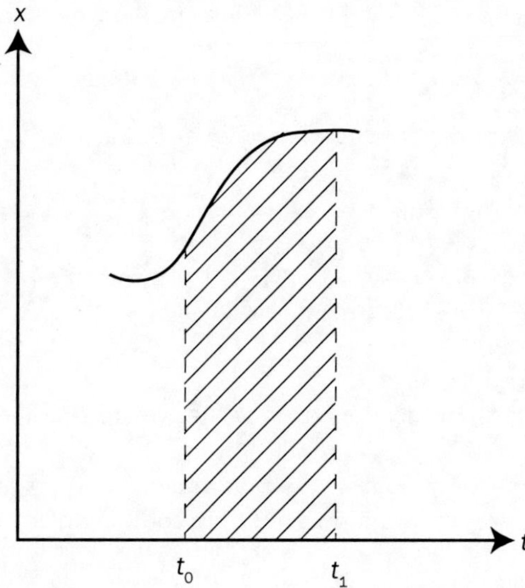

FIGURE 9.9 Area under the curve

9.19 i) Compute the integral:

$$I = \int_2^5 (x^2 + 5x)dx$$

using the trapezoidal rule ($n = 1$) and composite trapezoidal rule with $n = 4$

ii) Compare the numerical results with the exact value

iii) Estimate the truncation error in both cases (i.e., $n = 1$ and 4)

9.20 i) Compute the integral:

$$I = \int_0^2 \left(x^3 + 4x^2 + 5\right)dx$$

using the trapezoidal rule ($n = 1$) and composite trapezoidal rule with $n = 5$

ii) Compare the numerical results with the analytical value

iii) Estimate the truncation error in both cases (i.e., $n = 1$ and 5)

9.21 Compute the integral:

$$I = \int_1^2 (1 - x + \ln x)dx$$

using the composite trapezoidal rule with $n = 5$.

9.22 Using the composite trapezoidal rule, find

$$\int_0^2 x^{1/3}\, dx$$

with $\Delta x = 0.5$.

9.23 Approximate the integral:

$$\int_0^5 \frac{e^{-x}+5}{\sin x}dx$$

using the composite trapezoidal rule with $\Delta x = 1$.

9.24 Using the trapezoidal rule, estimate the integral:

$$I = \int_0^2 f(x)dx$$

based on the data given in Table 9.9.

TABLE **9.9** Given data

x	f(x)
0	2.5
1	1.5
2	0.5

9.25 Use the composite trapezoidal rule to approximate the integral:

$$\int_{-2}^{2} f(x)dx$$

based on the data given in Table 9.10.

TABLE **9.10** Given data

x	f(x)
−2	0.136
−1	0.425
0	1.387
1	1.012
2	3.578

9.26 Using the composite Simpson 1/3 rule, find

$$\int_1^5 (3+\sin\sqrt{x})dx$$

with 9 sample points. Find the exact value and the truncation error.

9.27 Use the composite Simpson 1/3 rule to approximate the integral

$$\int_0^5 f(x)dx$$

based on the data given in Table 9.11.

Table 9.11 Given data

x	$f(x)$
0	7.132
1	6.543
2	4.319
3	3.162
4	5.389
5	10.314

9.28 Approximate the integral

$$\int_{60}^{300} C_p dT$$

based on the heat capacity (C_p) versus temperature (T) data given in Table 9.12 for copper.

Table 9.12 Given data for copper

T (K)	C_p (J/mol.K)
60	8.595
120	18.25
180	21.94
240	23.60
300	24.45

Use the composite Simpson 1/3 rule.

9.29 Compute the integral:

$$I = \int_{-2}^{2} \frac{1}{(1+x)^3} dx$$

using the composite trapezoidal rule ($n = 9$) and the composite Simpson 1/3 rule with $n = 6$.

9.30 Compute the integral:

$$I = \int_{0}^{4} \frac{2}{\sqrt[4]{x^3}} dx$$

using the composite trapezoidal rule ($n = 20$) and the composite Simpson 1/3 rule with $n = 10$.

9.31 Estimate the integral:

$$I = \int_{0}^{2} \exp(5x^2) dx$$

using the Simpson 3/8 rule with $n = 9$.

9.32 Compute the integral:

$$\int_0^2 x\ln(x)\,dx$$

using the composite trapezoidal rule ($n = 5$) and the Simpson 3/8 rule ($n = 6$), and compare the results.

9.33 Find:

$$I = 6\int_0^1 x^5\,dx$$

to compare the results of the trapezoidal, the Simpson 1/3, the Simpson 3/8 and the Boole rule.

9.34 Compute the integral:

$$I = \int_2^5 (5x^2 + x)dx$$

using the

i) Simpson 3/8 rule
ii) Boole rule

9.35 Find the number of subintervals (n) and step size (Δx) required to approximate the following integral:

$$I = \int_4^5 \frac{1}{9-x}dx$$

with an error less than or equal to 10^{-5} by using the

i) multiple trapezoidal rule
ii) multiple Simpson 1/3 rule
iii) multiple Simpson 3/8 rule

9.36 Find the number of subintervals (n) and Δx to approximate the following integral:

$$I = \int_0^2 (e^x \sin 2x)dx$$

with an error less than or equal to 10^{-5} by using the

i) multiple trapezoidal rule
ii) multiple Simpson 1/3 rule
iii) multiple Simpson 3/8 rule

9.37 Consider the following expression (an equivalent form of composite trapezoidal rule given by Ralston and Rabinowitz, 1978):

$$I = \frac{\Delta x}{2}\left[f(a) + 2\sum_{i=1}^{n-1} f(a + i\Delta x) + f(b)\right] - \frac{(\Delta x)^2}{12}[f'(b) - f'(a)] + \frac{(b-a)}{720}(\Delta x)^4 \bar{f}^{(4)}$$

Using the Richardson extrapolation, find the third estimate with error $o\,[(\Delta x)^4]$.

9.38 Determine the integral:

$$I = \int_0^1 \frac{9}{1+x^2} dx$$

using the Romberg integration with composite trapezoidal rule for spacing $\Delta x, \frac{\Delta x}{2}, \frac{\Delta x}{4}, \frac{\Delta x}{8}$ and $\frac{\Delta x}{16}$.

9.39 Solve the above problem by using the Romberg integration with composite Simpson 1/3 rule for spacing $\Delta x, \frac{\Delta x}{2}, \frac{\Delta x}{4}$ and $\frac{\Delta x}{8}$.

9.40 Estimate the integral:

$$I = \int_0^2 \exp(-x^2) dx$$

using the trapezoidal rule with n = 1, 2 and 4. Show the improvement in results by further using the Romberg integration.

9.41 Show that the 3-points Gauss–Legendre rule is exact for:

$$\int_{-1}^1 6x^5 \, dx$$

9.42 Estimate the integral:

$$I = \int_0^2 \exp(5x^2) dx$$

using the 3-points Gauss–Legendre quadrature formula.

9.43 i) Determine the integral:

$$\int_1^2 \frac{1}{1+x} dx$$

using the 1-point, 2-points and 3-points formula

ii) Compare the Gauss quadrature results with exact value

iii) Compare the performance of the composite Simpson 1/3 rule, Romberg integration and 3-points Gauss quadrature in terms of the number of function evaluations (tol = 10^{-6}).

9.44 Estimate the integral:

$$\int_0^2 [1-\exp(-2x)] dx$$

using the 3-points Gauss–Legendre quadrature formula.

9.45 Compute the integral:

$$\int_0^1 \frac{1}{4+x^2} dx$$

using the 4-points Gauss–Legendre quadrature formula (use information given in Table 9.8).

9.46 Estimate the integral:

$$\int_0^2 \exp(-x^2)dx$$

using the 4-points Gauss–Legendre quadrature formula (use information given in Table 9.8).

Reference

Ralston, A. and Rabinowitz, P. (1978). *A First Course in Numerical Analysis*, 2nd ed., New York: Dover Publications.

Index